This book surveys the state of the art in the theory of combinatorial games, that is, games not involving chance or hidden information. Enthusiasts will find a wide variety of exciting topics, from a trailblazing presentation of scoring to solutions of three piece ending positions of bidding chess. Theories and techniques in many subfields are covered, such as universality, Wythoff Nim variations, misère play, partizan bidding (a.k.a. Richman games), loopy games, and the algebra of placement games. Also included are an updated list of unsolved problems, extremely efficient algorithms for Taking and Breaking games, a historical exposition of binary numbers and games by David Singmaster, chromatic Nim variations, renormalization for combinatorial games, and a survey of temperature theory by Elwyn Berlekamp, one of the founders of the field.

Mathematical Sciences Research Institute
Publications

70

Games of No Chance 5

Mathematical Sciences Research Institute Publications

1 Freed/Uhlenbeck: *Instantons and Four-Manifolds*, second edition
2 Chern (ed.): *Seminar on Nonlinear Partial Differential Equations*
3 Lepowsky/Mandelstam/Singer (eds.): *Vertex Operators in Mathematics and Physics*
4 Kac (ed.): *Infinite Dimensional Groups with Applications*
5 Blackadar: *K-Theory for Operator Algebras*, second edition
6 Moore (ed.): *Group Representations, Ergodic Theory, Operator Algebras, and Mathematical Physics*
7 Chorin/Majda (eds.): *Wave Motion: Theory, Modelling, and Computation*
8 Gersten (ed.): *Essays in Group Theory*
9 Moore/Schochet: *Global Analysis on Foliated Spaces*, second edition
10–11 Drasin/Earle/Gehring/Kra/Marden (eds.): *Holomorphic Functions and Moduli*
12–13 Ni/Peletier/Serrin (eds.): *Nonlinear Diffusion Equations and Their Equilibrium States*
14 Goodman/de la Harpe/Jones: *Coxeter Graphs and Towers of Algebras*
15 Hochster/Huneke/Sally (eds.): *Commutative Algebra*
16 Ihara/Ribet/Serre (eds.): *Galois Groups over* $\mathbb{Q}$
17 Concus/Finn/Hoffman (eds.): *Geometric Analysis and Computer Graphics*
18 Bryant/Chern/Gardner/Goldschmidt/Griffiths: *Exterior Differential Systems*
19 Alperin (ed.): *Arboreal Group Theory*
20 Dazord/Weinstein (eds.): *Symplectic Geometry, Groupoids, and Integrable Systems*
21 Moschovakis (ed.): *Logic from Computer Science*
22 Ratiu (ed.): *The Geometry of Hamiltonian Systems*
23 Baumslag/Miller (eds.): *Algorithms and Classification in Combinatorial Group Theory*
24 Montgomery/Small (eds.): *Noncommutative Rings*
25 Akbulut/King: *Topology of Real Algebraic Sets*
26 Judah/Just/Woodin (eds.): *Set Theory of the Continuum*
27 Carlsson/Cohen/Hsiang/Jones (eds.): *Algebraic Topology and Its Applications*
28 Clemens/Kollár (eds.): *Current Topics in Complex Algebraic Geometry*
29 Nowakowski (ed.): *Games of No Chance*
30 Grove/Petersen (eds.): *Comparison Geometry*
31 Levy (ed.): *Flavors of Geometry*
32 Cecil/Chern (eds.): *Tight and Taut Submanifolds*
33 Axler/McCarthy/Sarason (eds.): *Holomorphic Spaces*
34 Ball/Milman (eds.): *Convex Geometric Analysis*
35 Levy (ed.): *The Eightfold Way*
36 Gavosto/Krantz/McCallum (eds.): *Contemporary Issues in Mathematics Education*
37 Schneider/Siu (eds.): *Several Complex Variables*
38 Billera/Björner/Green/Simion/Stanley (eds.): *New Perspectives in Geometric Combinatorics*
39 Haskell/Pillay/Steinhorn (eds.): *Model Theory, Algebra, and Geometry*
40 Bleher/Its (eds.): *Random Matrix Models and Their Applications*
41 Schneps (ed.): *Galois Groups and Fundamental Groups*
42 Nowakowski (ed.): *More Games of No Chance*
43 Montgomery/Schneider (eds.): *New Directions in Hopf Algebras*
44 Buhler/Stevenhagen (eds.): *Algorithmic Number Theory: Lattices, Number Fields, Curves and Cryptography*
45 Jensen/Ledet/Yui: *Generic Polynomials: Constructive Aspects of the Inverse Galois Problem*
46 Rockmore/Healy (eds.): *Modern Signal Processing*
47 Uhlmann (ed.): *Inside Out: Inverse Problems and Applications*
48 Gross/Kotiuga: *Electromagnetic Theory and Computation: A Topological Approach*
49 Darmon/Zhang (eds.): *Heegner Points and Rankin L-Series*
50 Bao/Bryant/Chern/Shen (eds.): *A Sampler of Riemann–Finsler Geometry*
51 Avramov/Green/Huneke/Smith/Sturmfels (eds.): *Trends in Commutative Algebra*
52 Goodman/Pach/Welzl (eds.): *Combinatorial and Computational Geometry*
53 Schoenfeld (ed.): *Assessing Mathematical Proficiency*
54 Hasselblatt (ed.): *Dynamics, Ergodic Theory, and Geometry*
55 Pinsky/Birnir (eds.): *Probability, Geometry and Integrable Systems*
56 Albert/Nowakowski (eds.): *Games of No Chance 3*
57 Kirsten/Williams (eds.): *A Window into Zeta and Modular Physics*
58 Friedman/Hunsicker/Libgober/Maxim (eds.): *Topology of Stratified Spaces*
59 Caporaso/M^C^Kernan/Mustaţă/Popa (eds.): *Current Developments in Algebraic Geometry*
60 Uhlmann (ed.): *Inverse Problems and Applications: Inside Out II*
61 Breuillard/Oh (eds.): *Thin Groups and Superstrong Approximation*
62 Eguchi/Eliashberg/Maeda (eds.): *Symplectic, Poisson, and Noncommutative Geometry*
63 Nowakowski (ed.): *Games of No Chance 4*
64 Bellamy/Rogalski/Schedler/Stafford/Wemyss: *Noncommutative Algebraic Geometry*
65 Deift/Forrester (eds.): *Random Matrix Theory, Interacting Particle Systems, and Integrable Systems*
66 Corvino (ed.): *Aspects of Mathematical Relativity*
67–68 Eisenbud/Iyengar/Singh/Stafford/Van den Bergh (eds.): *Commutative Algebra and Noncommutative Algebraic Geometry*
70 Larsson (ed.): *Games of No Chance 5*

Games of No Chance 5

Urban Larsson
Technion – Israel Institute of Technology

CAMBRIDGE
UNIVERSITY PRESS

Urban Larsson
urban031@gmail.com

Silvio Levy (*Series Editor*)
Mathematical Sciences Research Institute
levy@msp.org

The Mathematical Sciences Research Institute wishes to acknowledge support by the National Science Foundation and the *Pacific Journal of Mathematics* for the publication of this series.

CAMBRIDGE
UNIVERSITY PRESS

University Printing House, Cambridge CB2 8BS, United Kingdom
One Liberty Plaza, 20th Floor, New York, NY 10006, USA
477 Williamstown Road, Port Melbourne, VIC 3207, Australia
314-321 3rd Floor, Plot 3, Splendor Forum, Jasola District Centre, New Delhi - 110025, India
79 Anson Road, #06-04/06, Singapore, 079906

Cambridge University Press is part of the University of Cambridge.

It furthers the University's mission by disseminating knowledge in the pursuit of education, learning and research at the highest international levels of excellence.

www.cambridge.org
Information on this title: www.cambridge.org/9781107129542

First published 2019

Printed in the United Kingdom by TJ International Ltd. Padstow Cornwall

A catalogue record for this publication is available from the British Library.

Library of Congress Cataloging-in-Publication data

ISBN 9781108485807 Hardback 9781108713658 Paperback

Games of No Chance 5
MSRI Publications
Volume **70**, 2017

Contents

Games of No Chance 5
MSRI Publications
Volume **70**, 2017

About this book

URBAN LARSSON

This book consists of 23 invited, original peer-reviewed papers in Combinatorial Game Theory (CGT) [5; 11; 46][1] — seven surveys and sixteen research papers. This is the fifth volume in the subseries Games Of No Chance (GONC) of the Mathematical Sciences Research Institute Publications. The name emphasizes these volumes' focus on play with no dice and no hidden cards, situating them in the landscape of game theory at large, where incomplete and/or imperfect information is common. Considering our class of games, *perfect play* can in theory be computed, and thus we include games such as CHESS, GO[2] and CHECKERS, but not YAHTZEE, BACKGAMMON and POKER.

Another characterizing feature is that combinatorial games are usually zero-sum, typically win-loss situations, although in some games draws are also possible. Players alternate in moving, so for any game description, we include a move flag of who starts. Game positions can be very sensitive to this move flag, and a common question is, given a combinatorial game, if I offer you to start, should you accept?

Often it is better to start, but not always. In the popular game of GO, the second player is rewarded a "komi" advantage of about 6.5 points before the game begins. In CHESS it is also regarded that White has a slight advantage. In neither of these games there is a mathematical proof, of this believed advantage, but since the games have been played for thousands of years, the belief seems well founded through overwhelming play-evidence.

[1]This book was initiated at the Combinatorial Game Theory Workshop, January 2011, at the Banff International Research Station (BIRS). As usual, this workshop attracted many researchers from Asia, Europe and North America, and it was organized by Richard Nowakowski (Dalhousie University), Elwyn Berlekamp (University of California, Berkeley), Aviezri Fraenkel (Weizmann Institute of Science), Martin Mueller (University of Alberta), and Tristan Cazenave (Paris-Dauphine University).

[2]On page 8, Carlos Santos reviews briefly DeepMind's AI advances of AlphaZero, a generalization of AlphaGo Zero, which recently beat the previously highest ranked CHESS program Stockfish, after just a few hours of training, alas using a massive computing power.

There are play-games which are also math-games.[3] The first player loses the game TWENTYONE. The rules are as follows: start with the number 21. The players alternate in subtracting 1 or 2 from the current number. If you start, then (in perfect play) the other player will "complement your move modulo 3", and win after 7 such rounds. Here, the game is specially *designed* to be a second player win.

We include three papers (13 on p. 313, 14 on p. 333 and 19 on p. 403) in the spirit of mechanism design in game theory; here, given a candidate set of P-positions[4], related to Beatty's classical theorem [2; 3], these contributions construct three classes of game rules with this set as the set of P-positions. The problem originates in the traditional combinatorial game of WYTHOFF NIM [53], which has a Beatty type solution on the modulus the Golden section. Generalizations of this game have begun to accumulate a lot of work, and we present the first comprehensive survey related to the heritage of Wilhelm Abraham Wythoff (paper 2, p. 35).

Singmaster famously proved [48] that almost no combinatorial game is a second player win. In this volume, we have a contribution by Singmaster (paper 7, p. 207), where he surveys the history of binary arithmetic in connection with puzzles, that is "one-player games" such as the CHINESE RINGS and the TOWER OF HANOI [27].

The game of HEX [25] is a classical game related to Brouwer's fixed-point theorem: two players compete in being the first to connect opposite sides of a hexagonal grid. A convention is often added, to compensate for the first player's advantage, namely, immediately after the first move the second player is given the opportunity to switch players. We have an amazing contribution of the theory of HEX in this book, by its current master, Ryan Hayward (paper 17, p. 387).

The game of CHOMP [9] has become famous for various reasons. Two players alternate to "chomp" pieces from a chocolate bar, by pointing at one piece and

[3]This is an informal distinction, but can be helpful to some extent: typical play-games are GO, CHESS, HEX, CHECKERS, KONANE, BLOKUS, FOUR IN A ROW, MANCALA, TIC TAC TOE, DOTS&BOXES, FOX&GEESE etc, while typical Math-games are DOMINEERING, HACKENBUSH, NIM, HEX, FOX&GEESE, WYTHOFF NIM, EUCLID, FIBONACCI NIM, etc. So, for example HEX and FOX&GEESE belong to both classes. One way of classification is to use the literature to claim membership of the latter class, while membership in the former class is due to being a popular social game. Other, more formal ways of classification may be suitable depending on purpose, but at least play-games should exclude games of values such as OMEGA; on the other hand many loopy/cyclic games are perfectly playable for human players. The distinction can be important when we aspire to build game rules, knowing beforehand "the solution"; see discussion on page 12 related to papers 13, 14 and 19 in this book.

[4]The class of second player win positions is usually called P-positions (*P*revious player wins), and they can be recursively computed for a game with a finite number of positions, starting with the terminal(s).

eating everything to the right and above, trying to avoid eating the lower left poisonous piece. The first player has a winning move, and the argument is as follows. The first player chomps the single right uppermost piece.[5] Now, any move the second player makes, could instead have been played also by the first player. Therefore the second player cannot have a winning strategy in this game. Of course, this argument does not give a clue of how to play. No general rigorous method is known, even for three row CHOMP, but to great surprise a method in physics, called renormalization, gives an estimate of where the first winning move must be [20]. This result, among others, were also published in a previous volume in this series (3.349 in the Index). Here, we are happy to report yet another finding where the renormalization approach gives precise estimates of solutions of a novel generalization of WYTHOFF NIM, called LINEAR NIMHOFF (paper 15, p. 343), indicating that in an infinite class of "linear" Wythoff extensions, all second player winning positions are distributed along thin lines.

The theory of combinatorial games was initiated by Charles Leonard Bouton in 1901 [8], when he discovered that elementary binary arithmetics solves the game of NIM, in a way that any finite number of heaps can be replaced by one single heap. For the rules of this game, see page 171 in Siegel's article (paper 6, p. 169). The next major development was in the 1930s, when Roland Percival Sprague [50] and Patrick Michael Grundy [23] independently discovered that any impartial 2-player game with the normal ending convention (last move wins) is equivalent to a one heap NIM position (and this development is also surveyed in Siegel's chapter). Again, observe how brilliant and surprising this is. Played on its own, of course, the one heap NIM game is a ridiculous thing — if the heap is nonempty, you win by removing all pebbles, and otherwise you already lost — but, by playing in a "disjunctive sum" with other NIM heaps, then this simple game encodes *any* other game in its class, and the class is huge (!), and moreover, the simple arithmetics of solving NIM then suffices to solve any such game. So, "there is something going on here" (attempting to read the minds of Sprague and Grundy). Although, this class of games is solved in theory, computationally, the games are often hard, and we include a famous yet unsolved problem in this book, presented by Grossman (paper 16, p. 373).

The next big discovery occurred in the 1950s when John Milnor [42] and Olof Hanner [24] developed a similar theory for a wide class of scoring-play games (highest score wins) without zugzwangs, that is games where it is never (!) bad to move first. Although "scoring games" are standard in game theory

---

[5] It has been conjectured that, in a natural generalization of CHOMP (SUBSET TAKEAWAY), "taking the largest element" is a winning move [22], but more recently counterexamples were found [10]. Note also that, given a quadratic chocolate bar of size larger than 1, you will win if you chomp off all except the lower row and left column.

at large, via concepts such as "utility", "revenue" etc., in CGT, the main idea usually concerns the "move ability"; when is it beneficial to have move options? In fact Milnor's games belong to a class of scoring games where either no player can move, or both players can move, so they remain closer to "economic games" for these two reasons.

In this volume, we proudly present Stewart's eye-opener on the full class of scoring games (paper 22, p. 447), where he pinpoints the difficulty of analyzing the full class; the problem boils down to a subset of games, where you want to start, but you do not have any move option (!). Those readers who study the misère play convention (last move loses) would acknowledge with a nod, that this situation usually induces severe complications. Many standard CGT (normal-play) tools fail.

These type of problems are discussed in three survey papers in this volume: papers 6 (p. 169) and 4 (p. 113) on impartial and partisan misere play developments respectively, and paper 3 (p. 89) on recent progress in scoring-play. In fact, a theory has recently been developed for those scoring games which exclude exactly Stewart's problematic games, the class of guaranteed scoring games [35; 34], and it is shown that the normal-play games are order embedded into this class. In this landscape, intersecting scoring- with normal-play, we find also the master pieces on DOTS&BOXES [6] and Mathematical GO [7].

The huge leap forward was in the 1970s-80s when Elwyn Berlekamp, John Horton Conway and Richard Kenneth Guy developed the normal-play theory to encompass so called partizan games, where players do not necessarily have the same move options [11; 5]. They adapted Milnor's definitions of disjunctive sum and game comparison [42], which was inspired by the apparent decomposition of GO positions into independent components towards the end of play.

Let G, H be normal-play games (without draws). Then $G \geq H$ if, for any normal-play game X, Left wins the game $H+X$ then Left wins the game $G+X$.

The intuition is as follows: let us imagine that you (playing Left) are in the middle of a complex game, a game which is decomposed in several (a finite number) of components — you are allowed to play only in one component at the time — and get an offer by a passerby to exchange one of the game components for another one. Let us say, your game is $G+X$, where X denotes a complex part of the game that you do not quite understand, and the passerby offers the game H in exchange for G, both much simpler games. Should you accept this offer?

One of the main theorems of normal-play CGT without draws [11] is that you can ignore the complicated X component, and simply play out the game $G-H = G+(-H)$, where the negative denotes that the players have swapped positions; then check whether you win this game when the opponent starts, which is the same as checking whether $G - H \geq 0$.[6]

In this book we have a unique contribution (paper 10, p. 271), by Carvalho and Santos, where the authors describe a ruleset, a modification of the traditional Hawaian ruleset KONANE to "PORTUGUESE KONANE", which has a position in each equivalence class of *short* (acyclic games with finite ranks and out-degrees) normal-play games. One could think of this as a CGT analogy of a universal Turing machine: one ruleset encodes it all. This relates to computational complexity. Again, we can imagine CGT relatives to simple universal machines, such as Emil Post's classical Tag productions; many extremely simple rulesets, such as OFFICERS (paper 16, p. 373) has so far defied all solution attempts by human and computer. In fact, recent development in the field contributes three classes of Turing complete classes of combinatorial games [16; 38; 39]. In this volume (paper 9, p. 299), Burke and George show that a generalization of NIM on a graph is PSPACE-complete, a more common hardness measure for combinatorial games [26] (see also 3.3 in the Index).

Some games are cyclic (or loopy), i.e., they have infinite game trees, and such games can be hard to analyze, although FOX&GEESE is an example of a ruleset, where analysis has been fruitful [5].[7] Moreover, one full class of games is fully understood; for the class of loopy normal-play impartial games (on finitely many positions), a complete theory is known. The first solution was given by Smith [49], and then using a more constructive algorithmic approach, Fraenkel and Yesha [19] generalize the classical Sprague–Grundy theory by letting "infinities", enumerate the loopy game values.[8] In this volume (paper 21,

[6]Let us illustrate with an example: the game component is $G = *$, and the game offer is "up" is defined by $H = \uparrow = \{0 \mid *\}$, where the game (class) "$*$" denotes a NIM heap of size one, and where "0" denotes the equivalence class containing as simplest element the empty game, that is the game where no player can move. Suppose that you are playing Left. In this case, you should not exchange G for H. The reason for this is the following: play the game $G - H$, and ask the other player, Right, to start. The negative of "up" is "down", which is the game $-H = \downarrow = \{* \mid 0\}$. That is, the test is to let Right start the game $* + \{* \mid 0\} = \{\{* \mid 0\}, 0 \mid \{* \mid 0\}, *\}$. Right has a good move. Which one?

[7]The original 1982 version of Winning Ways included a lot of examples of loopy games, including subclasses such as stoppers and enders, and more complicated examples, such as BACH'S CAROUSEL. The usual rules of canonical forms still apply to stoppers. In the second edition of Winning Ways, the much-enlarged chapter on FOX&GEESE pretty much solved all initial positions of that game, and many others, thanks also to Siegel's popular program CGSuite [47], much of which he developed in the course of those studies. So loopy games have long played a prominent role in the core content of CGT.

[8]As a personal note by the editor, optimal play may be infinite, but a slightest mistake by your opponent may lead to your easy victory; throughout childhood I played hundreds of games of the traditional game of PICARIA (which was proved drawn in [37]) an extension of THREE MEN MORRIS, both cyclic generalizations of TIC-TAC-TOE, and those plays always concluded with a winner.

p. 439) Sarkar establishes an infinite class of drawn positions of the classical CGT ruleset PHUTBALL, so here, players are indifferent to an offer of choosing side, and playing first or second. See also 3.91 in the Index (and 3.125) for a brilliant introduction to this topic.

We have more pioneers in this book. Some rulesets encourage questions of the form "how many game positions are there?"; this holds true for the class of *placement games* (papers 9 on p. 259, 11 on p. 285 and 12 on p. 297), where relations with simplicial complexes and generating functions are skillfully exposed by Brown et al., Faridi, Huntemann and Nowakowski. This type of games are particularly appealing to "conjoin", which is demonstrated by Huggan and Nowakowski in paper 18 (p. 395). In paper 23 (p. 469), Weimerskirch presents a CGT framework which generalizes the normal and misère conventions and ingeniously includes the notion of disjunctive sum, which brings us back to the heat of the matter; Berlekamp gives a splendid performance in surveying the temperature of the field (paper 1, p. 21). How urgent is it to move in the game component X? He also shows how urgency, and temperature, can be precisely captured by playing the original game in conjunction with an idealized stack of coupons.

The measure of "importance to move" is also captured in the setting of bidding games (paper 20, p. 421), where each play consists in two phases, first make your bid, and if you win the bid you get to move, hence mixing in "auction play" a popular subject in algorithmic game theory to the setting of combinatorial games.

Since its start in the early 1990s, this series of books has captured much of the core of the CGT-development. To celebrate its 20th anniversary, and as suggested by Elwyn Berlekamp, we include an index of all published GONC papers, compiled by Silvio Levy.[9]

An elementary introduction to combinatorial games is contained in the first part of paper 6 (p. 169), by Aaron Siegel, before he plunges into the complexity of the misère quotients and more; see also his current state of the art reference [46], a monumental contribution to the field of combinatorial games.

The book is divided into two sections: Survey articles and Research articles. In the Survey section, we find:

(1) Temperatures of games and coupons (Berlekamp)

(2) Wythoff visions (Duchêne, Fraenkel, Gurvich, Ho, Kimberling, Larsson)

[9]The previous books are *Games of no chance*, volume **29** in the MSRI Publications series (1998), *More games of no chance*, **42** (2002), *Games of no chance 3*, **56** (2009), and *Games of no chance 4*, **66** (2015); all were edited by Richard J. Nowakowski, GONC 3 jointly with Michael H. Albert.

(3) Scoring games: the state of play (Larsson, Nowakowski, Santos)

(4) Restricted developments in partizan misère game theory (Milley, Renault)

(5) Unsolved problems in combinatorial games (Nowakowski)

(6) Misère games and misère quotients (Siegel)

(7) An historical tour of binary and tours (Singmaster)

The Research papers are:

(8) A note on polynomial profiles of placement games (Brown et al.)

(9) A PSPACE-complete Graph Nim (Burke, George)

(10) A nontrivial surjective map onto the short Conway group (Carvalho, Santos)

(11) Games and complexes I: Transformation via ideals (Faridi, Huntemann, Nowakowski)

(12) Games and complexes II: Weight games and Kruskal–Katona type bounds (Faridi, Huntemann, Nowakowski)

(13) Chromatic Nim finds a game for your solution (Fischer, Larsson)

(14) Take-away games on Beatty's theorem and the notion of k-invariance (Fraenkel, Larsson)

(15) Geometric analysis of a generalized Wythoff game (Friedman, Garrabrant, Landsberg, Larsson, Phipps-Morgan). Related to work in these volumes: Friedman, Landsberg (3.349)

(16) Searching for periodicity in Officers (Grossman)

(17) Good pass moves in no-draw HyperHex: two proverbs (Hayward). Related work in these volumes: Anshelevich (2.151); Hayward (3.151); Payne, Robeva (4.207); Henderson, Hayward (4.129)

(18) Conjoined games: Go-cut and Sno-Go (Huggan, Nowakowski)

(19) Impartial games whose rulesets produce given continued fractions, (Larsson, Weimerskirch)

(20) Endgames in Bidding Chess (Larsson, Wästlund). Related work (a.k.a. Richman games) in these volumes: Lazarus, Loeb, Propp, Ullman (1.427, 1.439) (the field initiator); Payne, Robeva (4.207)

(21) Scoring play combinatorial games (Stewart)

(22) Phutball draws (Sarkar). Related work in these volumes: Demaine, Demaine, Eppstein (2.351); Grossman Nowakowski (2.361); Siegel (3.91)

(23) Generalized misère play (Weimerskirch)

Before we move on, we would like to say a few words about the recent blooming development of AI and deep neural networks in playing combinatorial games. Thanks to Carlos Santos for contributing this discussion:

> DeepMind team published an arXiv-preprint (December 5th, 2017) about AlphaZero, a computer program developed to play GO, and generalized to play CHESS (and SHOGI). Within 24 hours, it achieved an outstanding level of play.
>
> AlphaZero was trained with no opening theory or endgame tables. Comparing with the previous Monte Carlo algorithms, AlphaZero used just 80 000 positions per second, whereas Stockfish used 70 million. Even so, it won against Stockfish: in 100 games AlphaZero scored 25 wins and 25 draws with White, while with Black it scored 3 wins and 47 draws. It didn't lose a game, with the final score 64:36. The 9th game of the match showed an amazing attacking player with profound positional play. The 10th game was a masterpiece with identical characteristics.
>
> Therefore, as humans know, sometimes less is more! It seems a historical moment, AlphaZero Chess presents a very good "human" CHESS style. But with the incredible power of precise calculations.
>
> Garry Kasparov said "It is a remarkable achievement, even if we should have expected it after AlphaGo."

Acknowledgements: Many thanks to Richard Nowakowski, Svenja Huntemann and Melissa Huggan for assisting in editorial tasks on this volume. Thanks also to Elwyn Berlekamp, Argyrios Deligkas, Reshef Meir and Carlos Santos for several helpful comments and suggestions on this preface.

Overview of contents

Let us overview the contributions of this volume in their various contexts.

Surveys. We highlight some popular CGT topics.

Paper 1 (p. 21). *Temperatures of games and coupons* (Berlekamp)

For each game position, the normal-play theory defines a temperature, which is a measure of the importance of the next move [5; 46]. Here the author discusses many classical rulesets and divides them into "temperature classes" visualized by thermographic spectral lines. He comments on these games, starting with the simplest games at the lowest temperatures and continues on upwards into the hotter ones.

This is a survey, which includes some new results, using coupons to indicate "how hot a game is". In this way, normal-play games become economic-style games, whose outcome is determined by scores. Moreover, with the normal termination rule, it is hard to define a good strategy for the losing player, as all of his possible strategies will lose against an optimal opponent. In this new approach both players have well-defined optimal strategies.

Paper 2 (p. 35). *Wythoff visions* (Duchêne, Fraenkel, Gurvich, Ho, Kimberling, Larsson)

This is the first comprehensive survey of WYTHOFF NIM type games and related sequences. The game of WYTHOFF NIM (Wythoff 1907) was originally played on two heaps of stones, with rules: remove the same number of stones from both heaps or any number from exactly one of the heaps. It became famous because of its elegant solution via the Golden ratio and the floor function [53], and its popularity increased further via its more geometric interpretation CORNER THE LADY (played with a single queen of CHESS, Isaacs 1963), which inspired numerous variations such as "disjunctive sum play" (using several Queens of CHESS, "blocking maneuvers" and "linear extensions". Its richness never ceases to surprise; here we emphasize connections with various fields such as number theory, combinatorics on words and recently also cellular automata and physics.

Paper 3 (p. 89). *Scoring games: the state of play* (Larsson, Nowakowski, Santos)

Scoring-play combinatorial games started with Milnor and Hanner in the 1950s [42; 24]. Recent progress by Ettinger [15], Johnson [29], Larsson, Nowakowski, Santos [34; 35] and Stewart [51] (paper 21, p. 439) motivates a survey on the subject. There are similarities with classical settings in normal-play and misère-play, but the subject is richer than both these combined. This survey has a threefold purpose, first to survey the existing Combinatorial Game Theory in the area, such as disjunctive sum, game comparison, game reduction and game values. Secondly, and this is a novelty, important ideas, in relation with normal- and misère-play rulesets are introduced: e.g., when does a scoring-play ruleset have more interesting behavior than a normal-play ditto? The last topic in this survey is a discussion of many scoring combinatorial rulesets from Graph Theory, which have not yet been studied in the broader CGT context.

Paper 4 (p. 113). *Restricted developments in partizan misère game theory* (Milley, Renault)

Although initially designed only for impartial games, the restricted misère analysis works equally well for partizan games. The study of restricted partizan misère games began with the doctoral theses of Allen [1] and Ottaway [43].

This brilliant survey highlights the most significant results from recent research, including canonical forms of partizan misère games, the invertibility of games under restricted misère-play, and applications to specific partizan misère versions of Nim, Kayles, and Hackenbush [5].

Paper 5 (p. 125). *Unsolved problems in combinatorial games* (Nowakowski)

This survey is an updated version of Nowakowski's popular list of unsolved combinatorial games' problems.

Paper 6 (p. 169). *Misère games and misère quotients* (Siegel)

These comprehensive lecture notes concern impartial games and contain a complete course in the theory of misère quotients, illustrated through many examples. They are based on a short course offered at the Weizmann Institute of Science in 2006. In normal-play there are just six games with birthday smaller than six, whereas in misère-play, there are 4171780, and on day six there are more than $2^{4171779}$. Hence, the full theory of misère games modulo "=" is not so useful; we cannot hope for much structure if we require that games behave identically in any context. Through the theory of misère quotients, it suffices to understand how a ruleset position interacts with other positions in the same ruleset.

The notes begin with the basic definitions of combinatorial game theory and a proof of the Sprague–Grundy Theorem [50; 23], then proceeds via misère quotients to the full proof of the Guy–Smith–Plambeck Periodicity Theorem [44]. It concludes with a discussion of major open problems and directions for future research.

Paper 7 (p. 207). *An historical tour of binary and tours* (Singmaster)

This original survey by Singmaster, and edited by Larsson, dwells on topics in recreational mathematics in connection with binary representations and paths on graphs.

Highlights include the history of the CHINESE RINGS and its Gray code solution, which are elaborated in various directions, one of which leads to an appealing result on the classical TOWER OF HANOI puzzle [27]. Singmaster determines the "distance" of an arbitrary TOWER OF HANOI position to a terminal tower (see also [28]).

Workshop topics. Some contributions were presented at the workshop or were authored in direct response to workshop questions.

Placement games and simplicial complexes. New theory for *placement games* were discussed at the workshop, and resulted in four papers. Early work counted the number of positions in the game of GO [16], although this game is not strictly a placement game, because, by playing, pieces can be removed. In a placement game pieces cannot be moved or removed, only "placed".

Paper 8 (p. 243). *A note on polynomial profiles of placement games* (Brown et al.)

In this paper, placement games are recognized as a class of games. A game-specific polynomial encodes the number of distinct positions of games such as COL, SNORT [5] and NOGO (A variation of GO), provided they are played on a single path. The enumeration uses a bijection between the positions and the independent sets of an associated graph.

Paper 11 (p. 285). *Games and Complexes I: Transformation via Ideals* (Faridi, Huntemann, Nowakowski)

The authors show that for a given board and placement game, played on a graph, there are two associated simplicial complexes, defined via ideals on square-free monomials. The first simplicial complex is in terms of the legal positions, the second in terms of the illegal positions.

Paper 12 (p. 297). *Games and Complexes II: Weight games and Kruskal–Katona type bounds* (Faridi, Huntemann, Nowakowski)

This paper continues the work on simplicial complexes in relation with placement games. In the subclass of *strong* placement games, every move sequence between legal positions is legal. Moreover, for the class of *weight games*, game pieces may cover several neighboring nodes. By combining these classes, they find upper bounds on the number of positions with i pieces, or equivalently the number of faces with i vertices.

Paper 18 (p. 395). *Conjoined games: Go-cut and Sno-Go* (Huggan, Nowakowski)

This is a study of a recently introduced operation on rulesets, which generalizes math-games such as BUILDING NIM [13] and traditional play-games such as

NINE MEN MORRIS (which is a draw on a standard game board; see 1.101 in Index) and PICARIA [37]. In a pair of *conjoined games*, the first ruleset is played to a terminal position, then play continues under the second ruleset. They find the outcome classes for two such games, GO-CUT and SNO-GO, played on a strip.

Games and number theory. Fraenkel posed an intruiging problem at the workshop: "describe nice/short rulesets for games with so-called complementary Beatty sequences [2; 3] as sets of P-positions". The problem is the opposite of the main field of research in this area, which is, given some ruleset, to search for its set of P-positions.

This question resulted in three contributions. All but one requires a partial disclosure of the P-positions in the game rules. The one that does not depend on this studies a sub-class of Beatty sequences, described by periodic continued fractions.

A relaxation of Fraenkel's problem, to find so-called *invariant* rulesets [14], was recently resolved, by using a novel $\star$-operator [36], but without satisfying the proviso of "short".

We make a yet very informal distinction between rule sets for "math games" versus "play games"; rules (and/or solutions) in the former class appeals in particular to mathematicians, whereas rules in the latter class are more generally playable. These type of problems have also been discussed in (the introduction of) a recent PhD thesis [31]. Note the similarity with the field of mechanism design (or reverse game theory) in economics and game theory.

Paper 13 (p. 313). *Chromatic Nim finds a game for your solution* (Fischer, Larsson)

This contribution produces truly playable rules: "the rules can be understood by a five-year-old". This is done by a bi-coloring of the tokens in the stacks, thus mimicking one of the Beatty-type patterns on each heap. The rules are just classic NIM, with a proviso on the color of the top tokens in the current heaps.

Inspired by the resolution of the original problem, the authors continue by developing a general theory for this class of 2-heap games, and they study a "minimal sequence" for this theory, namely the famous Prouhet–Thue–Morse sequence.

Paper 14 (p. 333). *Take-away games on Beatty's theorem and the notion of k-invariance* (Fraenkel, Larsson)

This paper studies various short math-game rules, with the proviso that for the full game rules one must keep track of the number of tokens in the heaps, and moreover one of the Beatty sequences is revealed in giving the game rules.

The notion of k-invariance is introduced, and the games in this paper are 2-invariant. The succinctness of the game rules, together with their *invariance number*, seems to be related with the complexity of the proposed set of P-positions, and the paper ends with an intriguing problem interconnecting number-theory and computer science through game theory.

Paper 19 (p. 403). *Impartial games whose rulesets produce given continued fractions* (Larsson, Weimerskirch)

In this contribution, the moduli of the complementary Beatty sequences are given by the continued fractions $(1; k, 1, k, 1, \ldots)$ and $(k+1; k, 1, k, 1, \ldots)$, respectively.

The authors describe (short math-game) rules that satisfy two criteria: they are given by a closed formula and/or a simple recurrence, and they are invariant, that is, the available moves do not depend on the position played from (for all options with nonnegative coordinates).

The solution involves Sturmian word and morphism constructions of the Beatty Sequences [41]. Relating back to the problem in the previous paper, in this case, the smallest invariance number is possible with low complexity of game rules.

Classical GONC subjects. These four contributions (and more) are follow-ups to earlier GONC papers; indexed at the end of this issue.

Paper 17 (p. 387). *Good pass moves in no-draw HyperHex: two proverbs* (Hayward)

Hayward defines the concept of a "good move" and notes that in HEX, as in GO, it is not always true that your opponent's good move is your good move. The author studies a generalization of classical games: HYPERHEX is the hypergraph generalization of variants of HEX, where each player has a list of win-sets, and wins by coloring all cells of any of her win-sets. He finds a condition for this game where each move is good for both players, and he reminds us that "it is never too late to play a good move".

Previous HEX papers in the GONC series: Anshelevich (2.151); Hayward (3.151); Payne, Robeva (4.207); Henderson, Hayward (4.129).

Paper 15 (p. 343). *Geometric analysis of a generalized Wythoff game* (Friedman, Garrabrant, Landsberg, Larsson, Phipps-Morgan)

Using methods from physics, namely *renormalization* [20], the authors study a class of combinatorial games, LINEAR NIMHOFF, for which a probabilistic geometry explains the global behavior of observed outcome patterns. Let us highlight two novelties: (1) By axiomatizing the "renormalization properties", they present a rigorous method, where the question of existence is left for a future study. (2) Via a related *reorganization* property, they demonstrate precisely when game rules can be omitted, because they do not contribute to the global geometry.

Moreover, "symmetric rules" sometimes cause an underlying geometry to transform into quasi log-periodic fluctuations, centered in the proposed probabilistic geometry solution. As a special case, the method provides a solution to a class of games which has defied all previous analysis, namely the class GENERALIZED DIAGONAL WYTHOFF NIM [32; 33].

The GONC-series contributed a field pioneer on the subject of renormalization in combinatorial games: Friedman, Landsberg (3.349).

Paper 20 (p. 421). *Endgames in Bidding Chess* (Larsson, Wästlund)

A bidding combinatorial game is a combinatorial game, where instead of alternating play, the players bid for the opportunity to move. Richman showed in the 1980s that, in a finite setting for impartial games, the bidding games are equivalent to "random turn games" (1.427 and 1.439 in Index) and [40].

Develin and Payne claim on page 3 in their comprehensive paper on discrete bidding games [12] that "...the basic results and arguments of Richman game theory go through unchanged for partisan games...". Here, the authors develop the theory for partisan Bidding games, and show that this claim is not true in general. It is demonstrated that in the special case of 3-piece endgames in BIDDING CHESS [4], the claim holds. This is an area which recently has attracted research from the larger games' community, and where zero-sum has been generalized to general-sum [30].

The earlier GONC-papers about Bidding games (a.k.a. Richman games) are: Lazarus, Loeb, Propp, Ullman (1.427, 1.439) (the field initiater); Payne, Robeva (4.207).

Paper 22 (p. 447). *Phutball draws* (Sarkar)

The game PHILOSOPHER'S PHUTBALL was introduced by Conway et al. and appeared in Winning Ways [5]. It is usually played on a GO-board with one black stone (the ball) and the remaining stones white (the chaps). A move consists in either jumping a line of chaps with the ball and removing them, or placing a new chap on the board. The goals are the top and bottom edge of the board, respectively.

This game appeared twice in *More Games of no chance*; first: it is NP-complete to decide if one has a "win in one" (Demaine et al.), and secondly, the one-dimensional restriction was analyzed for a restricted class (Grossman et al.). In *Games of no chance 3* there is a survey paper on cyclic games by Siegel. He claims that, in one-dimensional PHUTBALL, it is not even known if both players have a drawing strategy. In this paper, the author finds an infinite class of drawn 2-dimensional positions.

Previous GONC-papers on PHUTBALL: Demaine, Demaine, Eppstein (2.351); Grossman Nowakowski (2.361); Siegel (3.91)

Conceptualizers. We have some aspiring trailblazers.

Paper 10 (p. 271). *A nontrivial surjective map onto the short Conway group* (Santos, Carvalho)

This is a unique contribution to classical combinatorial game theory. Every impartial game is equivalent to a unique NIM-heap, but no partizan game with an analogous property was known before this volume.

The authors demonstrate universality of a specific ruleset, namely a modification of the classical Hawaiian game of KONANE (see also 3.287 in the Index), and thereby generalizing [45]. Each short loop-free game value in normal-play theory can be described by a position in this game.

Paper 22 (p. 447). *Scoring play combinatorial games* (Stewart)

Stewart introduces a general class of scoring play games (generalizing Milnor's nonnegative incentive games [42] and Ettinger's dicot games [15]), and he points towards the difficulty of having a full class of scoring games. Namely, the troublesome games are those games where you benefit by not being able to move. Indeed his observation already inspired development of the theory of guaranteed scoring games [34; 35], where exactly those type of games are excluded; see also the related scoring play survey in this proceedings.

Stewart studies also the special case of "impartial" (or symmetric) Scoring games. He defines a variation of the 'Sprague–Grundy theory' for take-and-break scoring play, and studies analogies of subtraction- and octal games in this setting. He finishes off with several accompanying conjectures.

Paper 23 (p. 469). *Generalized misère-play* (Weimerskirch)

Weimerskirch develops a framework by which to view classical impartial games as an infinite array of game boards, or lattice points. Instead of thinking of normal-play and misère-play as differing in their winning condition, they are here viewed as differing in which set of positions are "in the field of play". This approach leads to a novel class of generalizations, and moreover, a separate notion of "disjunctive sum" becomes obsolete, because it is inherent in the lattice point definition of a position.

Computational aspects. Hardness problems lie at the heart of combinatorial game theory.

Paper 16 (p. 373). *Searching for periodicity in Officers* (Grossman)

OFFICERS [5] is a take-and-break game in which a move consists of removing a bean from a heap and leaving the remaining beans from that heap in exactly one or two nonempty heaps. It is an open question as to whether or not the Sprague–Grundy values of this game are eventually periodic; answering this question in the positive for take-and-break games generally requires computing enough values to explicitly find the period, but appearances of large "rare values" has continued to perplex the games community.

The presented method, which involves two novel parallelization strategies, is general and can be applied to other take-and-break games. The trick is to exploit the "rare values". If an oracle told us they are finitely many, then since the remaining values are bounded in size, one could conclude that OFFICERS were periodic. Without an oracle, Grossman shows how to exploit the rare value phenomenon to accelerate the computation, and after computation of more than 140 trillion values, no periodicity has been found.

Paper 9 (p. 259). *A PSPACE-complete Graph Nim* (Burke, George)

The game of NEIGHBORING NIM is played on graphs; each node contains a heap of pebbles and a move consists of removing some pebbles from some node then moving along a neighboring edge. This game is a generalization of a wide class of games including GEOGRAPHY [18] and NIM. The authors use methods from GEOGRAPHY to prove that the general game is PSPACE-hard and that a restricted variant is PSPACE-complete.

References

[1] M. R. Allen, *An investigation of partizan misère games*, PhD thesis, Dalhousie University (2009).

[2] S. Beatty, Problem 3173, *Amer. Math. Monthly* 33 (1926) 159.

[3] S. Beatty, A. Ostrowski, J. Hyslop, A. C. Aitken, "Solutions to Problem 3173", *Amer. Math. Monthly* 34 (1927) 159–160.

[4] J. Beasley, Bidding Chess, *Variant Chess* 7 (2008), no. 57, 42–44.

[5] E. R. Berlekamp, J. H. Conway, R. K. Guy, *Winning Ways for your Mathematical Plays,* 2nd ed., A K Peters, Ltd. (2001–2004) Volumes 1–4.

[6] E. R. Berlekamp, *The Dots-and-Boxes Game: Sophisticated Child's Play*, A K Peters, Natick, MA (2000).

[7] E. R. Berlekamp, D. Wolfe, *Mathematical Go: Chilling Gets the Last Point*, A K Peters, Natick, MA (1994).

[8] C. L. Bouton, Nim, a game with a complete mathematical theory, *Annals of Mathematics* 3 (1901-1902) 35–39.

[9] A. E. Brouwer, Chomp, http://www.win.tue.nl/~aeb/games/chomp.html.

[10] A. E. Brouwer, J. D. Christensen, Counterexamples to Conjectures About Subset Takeaway and Counting Linear Extensions of a Boolean Lattice, Order (2017) https://doi.org/10.1007/s11083-017-9431-6

[11] J. H. Conway, *On Numbers and Games*, 2nd ed., A K Peters, Ltd. (2001).

[12] M. Develin, S. Payne, Discrete bidding games, *Electron. J. Combin.* 17 (2010), #R85.

[13] E. Duchene, M. Dufour, S. Heubach, U. Larsson, Building nim, *Int. J. Game Theory* (2016).

[14] E. Duchêne, M. Rigo, Invariant games, *Theoret. Comput. Sci.*, 411 (34–36) (2010) 3169–3180.

[15] M. Ettinger, *Topics in Combinatorial Games*, Phd Thesis, University of Wisconsin (1996).

[16] G. E. Farr. The go polynomials of a graph. *Theoret. Comp. Sci.* (2003) 306:1–18.

[17] A. Fink, Lattice games without rational strategies *J. Combin. Theory, Series A* 119 (2012) 450–459.

[18] A. S. Fraenkel and S. Simonson, Geography, *Theoret. Comput. Sci.* (Math Games) 110 (1993) 197–214.

[19] A. S. Fraenkel, Y. Yesha, The generalized Sprague–Grundy function and its invariance under certain mappings,*J. Combin. Theory, Ser. A* 43 (1986) 165–177.

[20] E. J. Friedman, A. S. Landsberg, Nonlinear dynamics in combinatorial games: Renormalizing Chomp, *CHAOS* 17 (2007).

[21] D. Gale, A curious Nim-type game, Amer. Math. Monthly 81 (1974) 876–879.

[22] D. Gale, A. Neyman, Nim-type games, *Int. J. Game Theory* (1982) 11:17.

[23] P. M. Grundy, Mathematics and Games, *Eureka* **2** (1939) 6–8.

[24] Olof Hanner, Mean play of sums of positional games. *Pacific J. Math.*, 9 (1959) 81–99.

[25] P. Hein. Vil de laere Polygon? Politiken Newspaper (December 26 1942).

[26] R. A. Hearn and E. D. Demaine, *Games, Puzzles, and Computation*, A. K. Peters (2009).

[27] A. M. Hinz, S. Klavžar, U. Milutinović, and C. Petr, *The Tower of Hanoi—Myths and Maths*, Birkhäuser, Basel (2013).

[28] A. M. Hinz, Shortest paths between regular states of the Tower of Hanoi, *Inform. Sci.* 63, (1992) 173–181.

[29] W. Johnson, The combinatorial game theory of well-tempered scoring games *Int. J. Game Theory* (2014) 415–438.

[30] G. Kalai, R. Meir, M. Tennenholz, Bidding games and efficient allocations, in Proceedings of the Sixteenth ACM Conference on Economics and Computation (2017) 113–130.

[31] U. Larsson, *Impartial games and recursive functions*, PhD thesis, Chalmers and University of Goteborg (2013).

[32] U. Larsson, A generalized diagonal Wythoff Nim, *INTEGERS* 12 (2012) #G02.

[33] U. Larsson, Splitting sequences and Wythoff Nim extensions, *J. Integer Seq.*, 17 (2014) Article 14.5.7.

[34] U. Larsson, R. J. Nowakowski, C. P. Santos, Games with guaranteed scores and waiting moves, *Int. J. Game Theory* (2017).

[35] U. Larsson, J. P. Neto, R. J. Nowakowski, C. P. Santos, Guaranteed scoring games, *Electron. J. Combin.*, 23 (2016) #P3.27.

[36] U. Larsson, P. Hegarty, A. S. Fraenkel, Invariant and dual subtraction games resolving the Duchêne–Rigo conjecture, *Theoret. Comput. Sci.*, Vol. 412 (2011) 729–735.

[37] U. Larsson, I. Rocha, Eternal Picaria, *Recr. Math. Magazine*, 4(7) (2017).

[38] U. Larsson, Impartial games emulating one-dimensional cellular automata and undecidability, *J. Combin. Theory, Ser. A*, 120 (2013) 1116–1130.

[39] U. Larsson, J. Wästlund, From heaps of matches to the limits of computability, *Electron. J. Combin.*, 20 (2013) P41.

[40] A. Lazarus, D. Loeb, J. Propp, W. Stromquist, D. Ullman, Combinatorial games under auction play, *Games Econom. Behav.* 27 (1999) 229–264.

[41] M. Lothaire, Sturmian Words. *Algebraic Combinatorics on Words*. Cambridge: Cambridge University Press (2002).

[42] J. Milnor, Sums of positional games, *Ann. of Math. Stud.* (Contributions to the Theory of Games, H. W. Kuhn and A. W. Tucker, eds.), Princeton **2**(28) (1953) 291–301.

[43] P. Ottaway, Combinatorial games with restricted options under normal and misère play, PhD thesis, Dalhousie University (2009).

[44] T. E. Plambeck, Taming the wild in impartial combinatorial games, *INTEGERS*, 5 (2005) #G5.

[45] C. P. Santos, J. N. Silva, Konane has infinite nim dimension, *INTEGERS* (2008), #G02.

[46] A. N. Siegel, *Combinatorial Game Theory*, American Math. Society (2013).

[47] A. N. Siegel, *Combinatorial Game Suite*, http://cgsuit.sourceforge.net.

[48] D. Singmaster, Almost all games are first person games, *Eureka* **41**, 33–37 (1981).

[49] C. A. B. Smith, Graphs and composite games, *J. Combin. Theory, Ser. A* 1 (1966) 51–81.

[50] R. P. Sprague, Über mathematische kampfspiele, *Tôhoku Mathematical Journal* **41** (1935) 438–444.

[51] F. Stewart, *Scoring Play Combinatorial Games*, Phd Thesis, University of Dundee (2011). Also paper 22 in this volume, p. 447.

[52] D. Zeilberger, Three-Rowed CHOMP, Adv. Applied Math. 26 (2001) 168–179.

[53] W. A. Wythoff, A modification of the game of Nim, *Nieuw Arch. Wisk.* **7** (1907) 199–202.

urban031@gmail.com

Department of Industrial Engineering and Management, Technion - Israel Institute of Technology, Haifa, Israel

Survey Articles

Games of No Chance 5
MSRI Publications
Volume **70**, 2017

Temperatures of games and coupons

ELWYN BERLEKAMP

This paper gives an overview of many popular combinatorial games and the temperatures which occur in them. It also includes an explanation of Coupons, a temperature-related construction which has proved very useful in the study of relatively complicated combinatorial games such as Go and Amazons.

Overview

In its broadest sense, *combinatorial game theory* (CGT) is the study of two-person, perfect information games of no chance. For each position in such a game, the theory defines a temperature, which is a measure of the importance of the next move. CGT differs from economic game theory, which emphasizes multiplayer games including elements of chance and imperfect information. Most economic games focus on maximizing some payoff or score; CGT was originally more concerned with getting the last move, but it now also applies to games whose outcomes are determined by scores.

CGT is a branch of mathematics. It seeks to find and understand strategies which can provably succeed against any opposition. This differs from the primary goal of human or computer competitors, who are more focused on making fewer serious mistakes than their opponents. CGT seeks to understand *every* position, including composed problems. It assigns no special importance to any official "opening" position, nor to who gets the first move. Each position is treated as its own game, and both possibilities for who moves next are given appropriate consideration. Most CGT results employ the "divide and conquer" methodology:

(1) partition the board into disjoint regions;

(2) analyze each region, condensing it into an appropriate data structure;

(3) analyze the entire board position as the (disjunctive) sum of these disjoint regions.

The results are so interesting that many combinatorial game theorists now also play and analyze hybrid games, which are sums of positions in different

MSC2010: 91A46.
Keywords: combinatorial, games, temperature, Coupons, Go, Amazons.

games. Such a hybrid sum is called a gallimaufry.

The most successful application of CGT to Anglo-American checkers has been to composed problems (e.g., [Berlekamp 2002]). Elkies [1996] has successfully applied CGT to composed chess problems, and even to at least one position which occurred in a world championship chess match. But in most historical games of chess and checkers, every position that occurs is already as well understood by players who know no CGT as by those who do. But in every one of the other games considered in this paper, most, and sometimes even all, well-played games pass through a sequence of endgame positions about which CGT provides significant extra insights to those who have learned it. In some cases, it provides a complete solution. Details of the histories and rule variants of checkers and many other ancient and modern combinatorial games may be readily found from numerous sources on the web.

Introductions to CGT may be found in *Winning Ways* (WW) [Berlekamp, Conway and Guy 2001–2004] and in [Albert, Nowakowski and Wolfe 2007]. Siegel [2013] provides the major mathematical results and their proofs. His Appendix C is an excellent historical summary of how this subject has evolved from its early roots in ancient board games and in recreational mathematics.

Temperatures

In Winning Ways, the temperature of a game was viewed as a specific number, determined as the base of its thermograph's mast. But in cases where the lower portion of the mast coincides with one or both of its walls, it is now considered more convenient to allow the Left-temperature and the Right-temperature to be viewed as intervals of numbers, whose lower endpoints coincide at the value originally called "the" temperature. It is a measure of the importance of the next move. It can be computed by thermography, a graphical method described in WW and extended in [Berlekamp 1996].

Table 1 lists several combinatorial games and the known temperatures of their positions, listed in approximately descending order of their hottest known positions with finite temperature. More information about these games is summarized in Table 2. The reader is challenged to find improvements and/or corrections to these tables!

I'll now comment on these games, starting with the simplest games at the lowest temperatures and continuing on upwards into the hotter ones.

Temperature -1

Only integers have temperature -1. The "cutcake" family of games at this temperature has solutions which provide challenging examples for beginners.

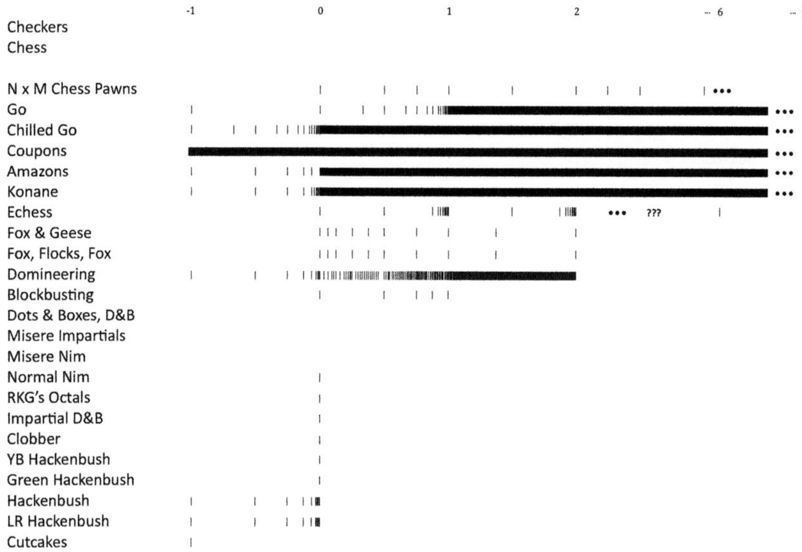

Table 1. Known temperatures of several combinatorial games.

Temperatures −1 to 0

The values of all such normal finite games are numbers. Blue-Red Hackenbush, (alias LR Hackenbush) is the best outstanding example. A basic theorem states that if G is a game for which there are one or more numbers greater than all of G's Left followers, G^L, and less than all of its Right followers, G^R, then G is the *simplest* such number. Among integers, the simplest is the one of least magnitude, and among other numbers, the simplest is the dyadic rational, $J/2^K$, with smallest nonnegative denominator, K. Several structural theorems in WW provide polynomial-time algorithms for sums of strings, trees, spiders, and several other classes of LR Hackenbush positions. But when applied to "Redwood beds", they yield a proof of NP-hardness, as indicated in Table 2.

Temperature 0

After subtracting out its numerical mean, every game of temperature 0 becomes a nonzero infinitesimal. So every game of temperature 0 is number-ish, where "ish" can be viewed as an abbreviation of "infinitesimally shifted". A game of temperature 0 is confused with at most one number, which is its mean.

The most common nonzero infinitesimal, by far, is the game STAR $= \{0 \mid 0\}$, denoted by $*$. It is confused with 0. The most common positive infinitesimal is UP $= \{0 \mid *\}$, denoted by $\uparrow$. Up turns out to be the first of many orders of infinitesimals, several of which appear in the complete, explicit solution of

Combinatorial Games	Guestimated # of Players	Origin	Winner	Loopy ?	★	NP hard ?	Most relevant theory
Checkers	10^8	~3000 BCE	N	L		n/a	c
Chess	10^8	~1810		L		n/a	c
N x M Chess Pawns	10^2	1996	N	-	*8191	?	a c
Go (Total)	10^8				*		
Chinese Weiqi		~2000 BCE	S	?	*	NP	c d e+ f′ g
Taiwan [Ing] Goe		~2000 BCE	S	?	*	NP	c d e+ f′ g
Korean Baduk		~ 800 CE	S	L	*	NP	c d e+ f′ g
Japanese Go		~ 800 CE	S	L	*	NP	c d e+ f′ g
American Go		~1930s	S	L	*	NP	c d e+ f′ g
Mathematical Go	10^4	1989	N	L	*	NP	w c d e+ f′ g
Chilled Go	10^3	1989	N	L	*3	NP	d e f g
Coupons	10^2	1997	S	-	0	P	g
Amazons	10^4	1988	N	-	*3	NP	c d e g
Konane	10^5	Medieval	N	-	★	NP	c d e h
Echess	10^2	~2000 CE			0		w c e+
Fox & Geese, F&G[1]	10^5	Medieval	N	-	?	P?	
Fox, Flocks, Fox	10^2	2002	N	L	?	?	c
Domineering	10^4	~1980	N	-	*3	?	c d e+
Blockbusting	10^3	1984	N	-	*	P	w e+
Dots & Boxes, D&B	10^6	1889	S	-	-	NP	c
Misere Impartials	10^6	<1600	M	-	-	?	c′ i
Misere Nim	10^6	<1600	M	-	-	P	a
Normal Nim	10^6	<1600	N	-	★	P	a
RKG's Octals	10^5	~1950	N	-	?	?	ab c
Impartial D&B	10^4	1967	N	-	?	?	ab c
Clobber	10^3	2001	N	-	?	?	c f
YB Hackenbush	10^2	1993	N	-	*	P	c f
Green Hackenbush	10^4	1971	N	-	★	P	a c
Hackenbush	10^4	1971	N	-	★	NP	c d f
LR Hackenbush	10^4	1970	N	-	0	NP	c
Cutcakes	10^3	1970	N	-	0	P	c

[1] alias Fox & Hounds

N = Normal rule, player unable to move loses
M = Misere rule, player unable to move wins
S = Scored

Entries in the column headed "★" show *K for the largest K known to exist within that game. ★ = Remote star

w	Warming inverts chilling
a	Bouton
b	Grundy-Guy
c	Canonical
c′	Misere canonical
d	Thermography
e	Heating
f	Atomic weights
f′	Liberties as combinatorial games
g	Coupons and orthodoxy
h	Universality
i	Finite Semi-groups

Table 2. Additional information of games listed in Table 1.

sums of strings of Yellow-Brown Hackenbush. A class of much-smaller positive infinitesimals are called *tinies*. Their negatives are called *minies*, although sometimes the word "tinies" may refer to both.

All finite impartial games are infinitesimal; each of them is equivalent to a green Hackenbush string. The value of the string of length N is called $*$ if $N = 1$; otherwise it is called $*N$. There is an important impartial infinitesimal called *remote star*, ★, which is an easy first-player win when played with any sum of finite stars. All members of a common class of infinitesimals, including all that occur in Hackenbush, can be approximated (to within at most 1 or 2), by an integer multiple of $\uparrow$s. This integer is called its atomic weight. Using a definition which includes ★ in a basic way, the atomic weight becomes a homomorphism from infinitesimals to all games. The importance of atomic weights is exemplified in a popular game called *Clobber*, all of whose positions are infinitesimals but whose atomic weights span a very wide variety of games.

Although the outcome of conventional (i.e., partizan) Dots and Boxes is determined by scores, many of its positions depend on the outcome of impartial Dots and Boxes, in which the loser is the player who is unable to complete his turn. Both variants of Dots and Boxes have an unconventional rule: you must continue making moves until you complete your turn with a move which completes no box(es). This gives rise to a peculiar value, called *loony*, denoted by ☽. It has the property that the next player to move can either play the loony to zero and complete his turn, or play the loony to zero and continue to move elsewhere. ☽ is an idempotent, ☽ + ☽ = ☽. It is a win for the first player: OUTCOME(☽) = first player, unlike OUTCOME(0) = second player. But when added to any nonzero game, either impartial or partizan, ☽ is negligible. It has even less effect than any tiny, in the sense that OUTCOME(G+☽) = OUTCOME(G) unless $G =$ 0. So ☽ is an infinitesimal, and like all infinitesimals, its temperature must be 0.

Temperatures up thru 1

Cooling is a homomorphism that reduces temperature while preserving the mean. Cooling by any positive amount transforms all infinitesimals to 0. In general, it is therefore a many-to-one homomorphism.

Intrigued by $2 \times N$ and $3 \times N$ Domineering, Berlekamp [1988] invented a simplified related game called Blockbusting, which is played on $1 \times N$ strips, each of whose ends is either L or R. I discovered that all but one of Blockbusting's positions become a number when *chilled* (i.e., cooled by one degree), whence the recursions to solve chilled Blockbusting are much simpler than the computations to solve Blockbusting. Almost all values of chilled Blockbusting are all dyadic rationals, and the sequences of them are arithmeticoperiodic.

Whereas in typical games cooling and chilling are many-to-one homomorphisms, Blockbusting satisfies two atypical conditions: All scores in Blockbusting are integers and $*$ is the only infinitesimal. Hence, chilling is only a two-to-one homomorphism. The parity turns out to be resolvable, and so chilling is reversible by an appropriately defined operator called *warming*. This yielded explicit tractable expressions for all Blockbusting positions.

Temperatures up thru 2

Domineering. Further refinements of heating operators and their application to Blockbusting values led to precise solutions of infinitely many $2 \times N$ and $3 \times N$

Domineering games, and to many others within "ish". These analyses remain far from complete. Kim [1996] included constructions and analyses of several new sequences and a summary of all classes of positions whose values were then known explicitly. More temperatures were found by Shankar and Sridharan [2005]. Drummond-Cole [2004] discovered a Domineering position which has temperature 2, the hottest value yet found.

Drummond-Cole [2005] composed a position of value $*2$, which could not be realized by legal moves starting from any empty rectangular board. Another position of value $*2$, which was realizable, was found by Uiterwijk and Barton [2015]. This explains Domineering's $*3$ in Table 2.

Fox and Geese. Although there are loopy positions in Fox and Geese, they are all stoppers with tractable canonical forms. The simplest two of these are $\text{OVER} = \{0 \mid \text{OVER}\}$ and $\text{OFF} = \{\ \mid \text{OFF}\}$. Every position in which the fox has escaped is easily seen to have value OFF, so the outcome of any Fox and Geese position on any $N \times 8$ board is reduced to deciding whether the value of the position is finite (a win for the Geese), or OFF (a win for the Fox), or infinitely HOT as in $\{x \mid \text{OFF}\}$, (a win for the next player to move). OVER and OFF are idempotents. The temperature of $\{x \mid \text{OFF}\}$ is ON.

With reasonable play, Uiterwijk and Barton [2015] showed that by retreating when appropriate, the Fox can easily avoid getting trapped anywhere except the bottom edge of the board, or possibly in the side square of the double-corner adjacent to the bottom of the board. So modifying the rules to prohibit the Fox from moving into a simple trap along any other side or top edge has no significant effect on the game. It is this modified variation of Fox and Geese whose known temperatures are shown in Table 1. If one also allows composed positions in which the Fox is about to be trapped along the side edge on an $N \times 8$ board, then there is such a position which has $G^L = 2N$; $G^R < 2$; and temperature exceeding $N - 1$.

Fox, Flocks, Fox. This game is the sum of two games, in one of which a traditional flock of four black geese seeks to trap a white fox, and another in which four white geese seek to trap a black fox on the other side of the board. It can be conveniently played on a single 8×8 board, viewed as two 4×8 boards. When the two starting positions are symmetric, their values are putative negatives of each other. Second player can survive indefinitely by responding to each move with its opposite, so the question becomes whether first player can force a draw. He can iff a loopy position (i.e., OVER, or OFF) appears anywhere in the canonical form. The game of Fox, Flocks, Fox provides interest and relevance in the canonical values of Fox and Geese, many of which are given in Chapter 20 of 2nd edition of WW for four geese versus a fox on an $N \times 8$ board. The temperatures of these positions are shown in Table 1, but that list might be incomplete.

Temperatures slightly higher

Entrepreneurial chess (Echess) by Berlekamp and Low [2017] is naturally played on an infinite board, comprising one quarter of the entire plane. Its hottest known finite temperature is $5\,\frac{5}{8}$. Many Echess positions have relatively tractable canonical values. When cooled or quenched (i.e., cooled by 2), many become a number plus an infinitesimal of integral atomic weight. When played alone, the outcome depends only on the stopping positions. Each of the finite stopping positions is an integer plus an additional OVER. As in Blockbusting and Go, there is another warming operator which inverts chilling.

Echess becomes more interesting when played as a component of a gallimaufry. The set of Echess' thermographic spectral lines shown in Table 1 is conjectured to be incomplete.

Temperatures yet higher

Coupons. As we look at earlier positions of a well-played endgame of Amazons or Go, we often find positions which are both hotter and more complicated.

Coupons were initially developed to facilitate more quantitative discussions with expert human Go players. It turned out that they are also helpful in studying other games, including Amazons.

When played in isolation, the game of Coupons is degenerate because there is never more than one legal move, which is to take the top coupon. The game becomes much more interesting when played as a summand added to another game, such as Amazons or Go. However, the analyses of Coupon Amazons and Coupon Go both depend on the following analysis of an ideal stack of Coupons played in isolation.

This game is played with a stack of coupons. Each coupon has a value printed on its face. The coupons in the stack are in monotonic nonascending order. The *ideal stack* consists of two large substacks. The bottom substack is a large number of coupons of the same terminal temperature, T_0. The top substack, in ascending order, contains coupons of these values: $T_0 + \frac{1}{2}\delta$, $T_0 + \frac{3}{2}\delta$, $T_0 + \frac{5}{2}\delta, \ldots, T_{\text{top}}$. Every consecutive pair of coupons above T_0 has values which differ by the same amount, δ. At any intermediate stage of play, when the top coupon has value C, the stack is said to have an *ambient temperature* of $(C + \frac{1}{2}\delta)$, unless the top coupon has value T_0, in which case the ambient temperature is T_0. In all other cases, the ambient temperature is the average of the previous coupon and the next coupon. We also define a *current komi*, whose value is half the current ambient temperature.

Consider a game of Coupons played on a consecutive subset of the ideal stack described above, with the ambient temperature running from T_2 down to T_1. Without komis, a simple calculation shows that the net sum of all those consecutive coupons will be $\pm\frac{1}{2}T_2 \pm \frac{1}{2}T_1$, where the signs depend on who moves first and who moves last, respectively. To make the game fair, we need the komis. The interval's initial komi of magnitude $\frac{1}{2}T_2$ is assigned to the opponent of the player who makes the first move, at an ambient temperature of T_2; its terminal komi, of magnitude $\frac{1}{2}T_1$, is assigned to the opponent of the player who makes the final move in this interval at an ambient temperature of T_1. When the game is over, your score is the sum of all coupons you have taken, including komis. When this ideal interval coupon game is played in isolation, the net final score will be precisely zero.

When Coupons are played with any chosen (possibly composed) starting position, the ideal initial temperature should be large enough that both players will take a few coupons before either chooses to play on the board. The terminal temperature should be $T_0 = -1$, and the number of coupons of this temperature should exceed the number of empty squares on the Amazons position. When the terminal temperature is reached, at every turn the player will prefer to fill a point of his territory on the board rather than take the -1 point coupon. So when the game eventually ends, all scores on the board will have been converted into the coupons. The winner is the player with the higher score. The terminal komi obviates any advantage or disadvantage of getting the last move. So a tie is a possible outcome. This has effectively converted a combinatorial game with normal termination rule into an economic-style game whose outcome is determined by scores. One advantage of this viewpoint is that *both* players now have well-defined optimal strategies. With the normal termination rule, it is hard to define a good strategy for the losing player, as all of his possible strategies will lose against an optimal opponent.

Expediting play with thicker stacks. To simplify the analysis when Coupons are added to another game, it is convenient to let δ to be a small nonnegative number. If δ is small but positive, the number of coupon moves will be so large that it is convenient to expedite the game by the following procedure, which has no effect on the eventual score.

Whenever the players take three coupons on consecutive turns, the game is interrupted. (The reason that we do not interrupt after only two coupons is that in some ko positions in Go, one player may use coupons as ko threats). The opponent of the player who took the third of these three coupons is awarded the current komi. If the ambient temperature is T_0, the game is terminated. If not, each player is required to submit a sealed bid $\geq T_0$. The bid must be the temperature of a coupon which, if play continued, it would be his turn to take. The winning bidder and his bid are announced. Larger coupons are removed from the stack. The opponent of the winning bidder is awarded the new komi. Unless the new temperature is T_0, the winning bidder is required to make a move on the board, and play resumes. If instead the winning bid is T_0, then the winning bidder may either take a coupon or play on the board, and play resumes until three consecutive coupons of value T_0 are taken, at which point the terminal komi is awarded and the game ends.

Infinitely thick stacks. Using these conventions, it is feasible to play with $\delta = 0$. There is no physical stack of coupons. Instead, only the current ambient temperature is relevant. Initially it is so large that the first three moves take coupons, followed by a komi, an auction, another komi, and the first move on the board. The players may then either play on the board or take a coupon at this temperature. After three consecutive coupons are taken at the current temperature, a komi is awarded; a bidding auction yields a new lower temperature, at which another komi is awarded and play resumes. If the auction ends in a tie, you may let your opponent break it arbitrarily.

Encores. The first play at a negative ambient temperature begins a phase of the game called the *encore*. As explained in Berlekamp and Wolfe [1994], the encore is often lengthy, tedious, and dull. The easiest way for players to avoid it is to agree on a forecast of how it would turn out if played. This is usually straightforward when the temperature is 0. However, there are rare examples in which even very good players may not agree. Some Amazonian territories are defective. Other Amazonian territories may have a value that might not be obvious even to very good players, such as the value $\frac{1}{16}$ in [Snatzke 2002]. In Coupon Go, even though the score on the board at the beginning of an encore is an integer, the appendix of the 1989 Japanese rules contain several interesting examples in which the value of that score has been debated. The mathematically simple procedures stated in this paper resolve all such disputes by continued play of the encore. In the very rare cases in which different dialects of Go yield different scores, the results of the encore with infinitely thick coupons tend to be more consistent with Chinese or American scoring than with Japanese scoring.

Orthodoxy. A theorem states that if $\delta = 0$, the optimum final score is the mean value of the board's starting position. Each player has a strategy which ensures an outcome at least that good for him. Moves which are consistent with any such strategy are called *orthodox* moves. The orthodox viewpoint yields much simpler game graphs than the canonical viewpoint. It also sometimes enables refinement of the decomposition of the board into "independent" regions. Even when playing a Go endgame in a traditional way, without coupons, orthodox accounting in Berlekamp [1996] facilitates a prediction of the final net score and an itemization of how much of this score is due to each region of the board and how much is dependent on who gets the next move. Refinements facilitate locally computable quantitative estimates of the values of kos. For further discussion of orthodoxy see this link: https://math.berkeley.edu/~berlek/pubs/videos.html.

In Amazons, Berlekamp [2000] composed a hard problem featuring four opposing pairs of Amazons, each pair in a region of size 11×2 or smaller; Snatzke [2002] built a database big enough to analyze each of them. Several had canonical forms with many thousands of positions, yet their thermographs were very simple, with temperatures up to about 5. In Go, a team of three expert combinatorial gamesmen and two Go players of the highest rank (9p) spent several months analyzing a position they had encountered 66 moves before the end of a full game. It was published by Spight [2002]. When they had played it, they had estimated the temperature as slightly under 4, but analysis showed it was actually about 5. The determination of the temperature of another region entailed the compilation and study of a computer database of over 20000 positions. The answer was 3.42.

Amazons

Amazons is a conventional loop-free combinatorial game with the normal ending condition. Hence, it has no positions of infinite temperature, although on large boards it contains positions with arbitrarily large temperatures. It has numbers and interesting infinitesimals. It includes positions which have very complicated canonical values but simple orthodox values. Its largest known negative temperature is $-\frac{1}{16}$, found independently by Snatzke [2002] and Tegos [2002].

Tegos [2002] also found positions on a 4×4 board with positive temperatures whose denominators were 256. Song and Mueller [2015] provide more results and references, with a primary focus on who can win from certain conventional starting positions on rectangular boards of various sizes. The definitive list of positive Amazonian temperatures will evidently require more resolution than the printing constraints of this journal can provide in Table 1.

Go

As explained in the Appendices of Berlekamp and Wolfe [1994], there are many dialects of the rules of Go. The mathematical study of this game is complicated by the common occurrence of positions called “kos”. More information about that can be found in [Berlekamp and Kim 1996] and in [Spight 2003].

Unlike chess, where White gets the first move, in Go it is Black who gets the first move. In chess, it is now widely believed that despite White’s advantage, Black has a reasonable chance of getting a draw. But in Go, modern experts believe that if uncompensated, Black’s first-move advantage would be decisive.

Hence, in most modern professional tournaments White is given a special compensation of 6.5 points, which is added to his score. This is called the "komi". If we presume that the temperature of the empty 19×19 Go board is about 13, then this "komi" plays approximately the same role in conventional Go as what we call the "initial komi" in Coupon Go. Berlekamp [1996] explains how the absence of other komis and coupons in conventional Go corresponds to comparable terms in an "orthodox accounting" which, in principle, itemizes the score of a well-played Go endgame in terms of each move and each region of the board.

Chess pawns

This is a degenerate form of chess, in which all pieces are pawns, and the "normal" objective is to get the last move.

$N \times 1$ chess pawns is a degenerate form of chess pawns, in which there is only one file. Its positions provide all of the temperatures shown in the relevant row of Table 1. In Table 2, the construction of $*8191$ is described by Elkies [2002].

Chess

Noam Elkies [1996; 2002] has shown that some real chess positions, including a nontrivial one that occurred in a world championship match, simplify to sums of positions in the simpler game I'm now calling "chess pawns". Carlos Santos [2015] has composed more real chess positions which can be solved by CGT.

Due to the very unusual termination rules of conventional chess, including such notions as "stalemate", it isn't clear how temperature could be meaningfully defined.

Anglo-American checkers

Although relatively few native English speakers realize it, there are many variations of checkers now popular in different countries of the world. In most countries in continental Europe, checkers has "flying kings", who can jump opposing checkers at some distance away on the same diagonal. Board sizes of 10×10 rather than 8×8 are also common. Some people even regard Konane as the Hawaiian variation of the checkers family.

Since Anglo-American checkers positions only rarely, if ever, decompose into sums of disjoint regions, there has been little, if any, study of temperatures of positions in this game, so I've left this row of Table 1 blank.

Nevertheless, the temperature theory of CGT has been very successfully applied to at least one composed gallimaufry problem including a checkers position (i.e., "four games for Gardner" in [Berlekamp 2002]). Surprisingly, the results of CGT are so robust that the solution of that gallimaufry is independent of what initially might appear to be important details of the rules:

(1) What is the goal of a game whose components include such diverse components as Go and chess?

(2) What is the scope of the compulsory capture rule in checkers? Does it compel the opponent to take the capture immediately, or does it only prevent him from making any other move on the checker board.

(3) Can a move elsewhere, perhaps in chess, be used as a ko threat in Go?

Open problems

(1) Debug and extend the entries in the existing rows of Table 1.

(2) Insert more rows into Table 1. Obvious candidates include more restricted versions of some games already listed there. In particular, most of the Konane positions constructed by Santos and Nuno-Silva [2008] have checkers of both colors on both colors of squares, although in all positions that can arise from the ancients' official starting position, black checkers can only occupy black squares and white checkers can only occupy white squares. So composed Konane problems can be partitioned into two sets: unrestricted Portugese Konane and restricted ancient Hawaiian Konane.

(3) Several families of Domineering positions are known, each containing an infinite number of different temperatures. Compose a sum of them which maximizes the difference between the orthodox result and the result when played optimally.

(4) Make a similar study of which atomic weights occur in which infinitesimal games. Hopefully, this would include corridors in Go, many of which are infinitesimal but not "all-small".

References

[Albert, Nowakowski and Wolfe 2007] M. Albert, R. Nowakowski, and D. Wolfe, *Lessons in play: an introduction to combinatorial game theory*, A K Peters, Wellesley, MA, 2007.

[Berlekamp 1988] E. R. Berlekamp, "Blockbusting and domineering", *J. Combin. Theory Ser. A* **49**:1 (1988), 67–116. MR

[Berlekamp 1996] E. Berlekamp, "The economist's view of combinatorial games", pp. 365–405 in *Games of no chance*, edited by R. Nowakowski, Math. Sci. Res. Inst. Publ. **29**, Cambridge University Press, 1996.

[Berlekamp 2000] E. Berlekamp, "Sums of $N \times 2$ amazons", pp. 1–34 in *Game theory, optimal stopping, probability and statistics*, edited by F. T. Bruss and L. L. Cam, IMS Lecture Notes – Monograph Series **35**, Institute of Mathematical Statistics, Beachwood, OH, 2000.

[Berlekamp 2002] E. Berlekamp, "The 4G4G4G4G4 problems and solution", pp. 231–241 in *More games of no chance*, edited by R. Nowakowski, Math. Sci. Res. Inst. Publ. **42**, 2002.

[Berlekamp and Kim 1996] E. Berlekamp and Y. Kim, "Where is the 'Thousand-Dollar Ko'?", pp. 203–226 in *Games of no chance*, edited by R. Nowakowski, Math. Sci. Res. Inst. Publ. **29**, Cambridge Univ. Press, 1996.

[Berlekamp and Low 2017] E. R. Berlekamp and R. Low, "Entrepreneurial Chess", *Internat. J. Game Theory* (online publication May 2017).

[Berlekamp and Wolfe 1994] E. Berlekamp and D. Wolfe, *Mathematical Go*, A K Peters, Wellesley, MA, 1994.

[Berlekamp, Conway and Guy 2001–2004] E. Berlekamp, J. H. Conway, and R. K. Guy, *Winning ways for your mathematical plays*, A. K. Peters, Wellesley, MA, 2001–2004. Four volumes: 2001, 2002, 2003, 2004.

[Drummond-Cole 2004] G. C. Drummond-Cole, "Temperature two in Domineering", unpublished note, 2004.

[Drummond-Cole 2005] G. C. Drummond-Cole, "Positions of value *2 in generalized domineering and chess", *Integers* **5**:1 (2005), G6, 13.

[Elkies 1996] N. D. Elkies, "On numbers and endgames", pp. 135–150 in *Games of no chance*, edited by R. Nowakowski, Math. Sci. Res. Inst. Publ. **29**, Cambridge University Press, 1996.

[Elkies 2002] N. D. Elkies, "Higher nimbers in pawn endgames on large chessboards", pp. 61–105 in *More games of no chance*, edited by R. Nowakowski, Math. Sci. Res. Inst. Publ. **42**, Cambridge University Press, 2002.

[Kim 1996] Y. Kim, "New values in Domineering", *Theoret. Comput. Sci.* **156** (1996), 263–280.

[Santos 2015] C. P. Santos, "Presentation at Momath MOVES", presentation, August 3 2015.

[Santos and Nuno-Silva 2008] C. P. Santos and J. Nuno-Silva, "Konane has infinite nim-dimension", *Integers (Comb. Games Sect.)* **8**:1 (2008), G02.

[Shankar and Sridharan 2005] A. Shankar and M. Sridharan, "New temperatures in Domineering", *Integers (Comb. Games Sect.)* **5** (2005), G04.

[Siegel 2013] A. N. Siegel, *Combinatorial game theory*, American Mathematical Society, Providence, RI, 2013.

[Snatzke 2002] R. G. Snatzke, "Exhaustive search in the game Amazons", pp. 261–278 in *More games of no chance*, edited by R. Nowakowski, Math. Sci. Res. Inst. Publ. **42**, Cambridge University Press, 2002.

[Song and Mueller 2015] J. Song and M. Mueller, "An enchanced solver for the game of Amazons", *Trans. Comput. Intelligence AI Games* **7**:1 (2015), 16–27.

[Spight 2002] B. Spight, "Go thermography: the 4/12/98 Jiang–Rui endgame", pp. 89–105 in *More games of no chance*, edited by R. Nowakowski, Math. Sci. Res. Inst. Publ. **42**, Cambridge University Press, 2002.

[Spight 2003] W. L. Spight, "Evaluating kos in a neutral threat environment: preliminary results", pp. 413–428 in *Computers and games* (CG 2002) (Edmonton, AB), edited by J. Schaeffer et al., Lect. Notes Comput. Sci. **2883**, Springer, Berlin, 2003. [Problem E4].

[Tegos 2002] T. Tegos, *Shooting the last arrow*, Master's thesis, University of Alberta, Edmonton, AB, 2002. See especially Chapter 7.

[Uiterwijk and Barton 2015] J. W. H. M. Uiterwijk and M. Barton, "New results for domineering from combinatorial game theory endgame databases", *Theoret. Comput. Sci.* **592** (2015), 72–86.

berlek@gmail.com *Elwyn and Jennifer Berlekamp Foundation, Oakland, CA, United States*

Games of No Chance 5
MSRI Publications
Volume **70**, 2017

Wythoff visions

ERIC DUCHÊNE, AVIEZRI S. FRAENKEL, VLADIMIR GURVICH,
NHAN BAO HO, CLARK KIMBERLING AND URBAN LARSSON

Six authors tell their stories from their encounters with the famous combinatorial game WYTHOFF NIM and its sequences, including a short survey on exactly covering systems. The volume of the mathematical study of this game is 59% of that of the most ubiquitous game CHESS (MathSciNet). The former originated in 1907, the latter in antiquity. Thus the mathematical study of WYTHOFF NIM may surpass that of CHESS!

1. A modification of the game of NIM

The game of NIM only preceded Wythoff's modification by a few years. By the famous theory of Sprague and Grundy some decades later, NIM drew a lot of attention. WYTHOFF NIM (here also called Wythoff's game), on the other hand, only became regularly revisited towards the end of the 20th century, but its winning strategy is related to Fibonacci's old discovery of the evolution of a rabbit population. The subject has been receiving more attention in recent decades thanks in part to new studies of WYTHOFF NIM and its variants by Fraenkel, and investigations into related sequences and arrays by Kimberling. In this paper we provide surveys of six different aspects of this subject by six of its current masters. Let us recall Wythoff's original definition of the game, given in item 1 in his paper [158]:

> *The game is played by two persons. Two piles of counters are placed on the table, the number of each pile being arbitrary. The players play alternately and either take from one of the piles an arbitrary number of counters or from both piles an equal number. The player who takes up the last counter or counters, wins.*

Wythoff proceeds by designating the *safe* positions (P-positions in current jargon) of his game, first by noting that the heaps are unordered, which implies

Duchêne was supported by the ANR-14-CE25-0006 project of the French National Research Agency. Larsson was supported by the Killam Trust.
MSC2010: primary 91-02; secondary 11-02.
Keywords: combinatorial game, golden ratio, impartial game, Wythoff Nim.

n	0	1	2	3	4	5	6	7	8	9	10	11	12	13	14
A_n	0	1	3	4	6	8	9	11	12	14	16	17	19	21	22
B_n	0	2	5	7	10	13	15	18	20	23	26	28	31	34	36

Table 1. The first few terms of the A and B sequences.

that (x, y) is safe if and only if (y, x) is, where x and y denote the respective number of counters in each pile. Then he proceeds to the nowadays celebrated *minimal exclusive algorithm* (mex), but without giving it a name. Let U be a finite subset of the nonnegative integers. Then the minimal excludant of U, mex U, is the smallest nonnegative integer not in U.

Theorem 1 (recursive characterization of WYTHOFF NIM's P-positions). *Let* $\{(A_n, B_n), (B_n, A_n) : n \geq 0\}$ *be the set of P-positions of* WYTHOFF NIM. *Then, for all* $n \geq 0$,

$$A_n = \text{mex}\{A_i, B_i : 0 \leq i < n\}, \tag{1}$$

$$B_n = A_n + n. \tag{2}$$

We display the first few terms of the sequences in Table 1.

This result implies an exponential-time winning strategy in the input size $\log(xy)$ of an arbitrary input position (x, y). Wythoff's game has become famous because of the algebraic characterization of the P-positions (item 6 of [158]), the solution via the floor function and the golden ratio, which implies a polynomial-time winning strategy.

Theorem 2 (algebraic characterization of WYTHOFF NIM's P-positions). *A combination of pile sizes of* WYTHOFF NIM *is a P-position if and only if it is of the form* $\lfloor \varphi n \rfloor, \lfloor \varphi^2 n \rfloor$, *for some nonnegative integer* n, *and where* $\varphi = \frac{1}{2}(1+\sqrt{5})$ *is the golden ratio.*

It also became famous because of the difficulty to compute the nonzero Sprague–Grundy values [69; 147] (see Section 6 for a definition and some recent progress), and the problem to find a generalization to several heaps that preserves properties of WYTHOFF NIM (see Section 4.3).

We will often encounter the more general concept of a *Beatty sequence* [7]. Let α denote a positive irrational. Then $(\lfloor \alpha n \rfloor)$ is a Beatty sequence, where n ranges over the positive integers. Two or more sets of positive integers are *complementary* if each positive integer occurs in precisely one of them. Such systems are also known under the name of *exactly covering systems*, or *splitting systems*. It is a well-known result that the sets $\{\lfloor \alpha n \rfloor\}$ and $\{\lfloor \beta n \rfloor\}$, where n ranges over the positive integers, are complementary if and only if α, β are positive irrationals satisfying $\alpha^{-1} + \beta^{-1} = 1$. Such a pair of sequences $(\lfloor \alpha n \rfloor)$, $(\lfloor \beta n \rfloor)$ is often called a pair of *complementary Beatty sequences*, although the result was discovered by Rayleigh in the book *The theory of sound* [138] (without giving a proof) and independently proved by Hyslop, Ostrowski and Aitken [8]. WYTHOFF NIM's P-positions give a special case of this, where α is the golden ratio.

In a survey paper [24] concerning the golden ratio, phyllotaxis and WYTHOFF NIM in 1953, Coxeter sketches a simple proof of Theorem 2, recalling the proof of Hyslop and Ostrowski [8], using also Theorem 1. He omits elaborating on the formula for B_n (a similar shortcoming appears in Wythoff's original proof). Namely, the mex-property for A_n holds by the (graph theory) kernel property of the P-positions of an impartial game. For the relation $B_n = A_n + n$, however, an inductive argument of a fill-rule property of diagonal parents of P-positions is also required. It is a geometric argument, and we display the idea in Figure 1, noting that if (x, y) is a P-position, then for example $(x+t, y+t)$ is an N-position for all $t > 0$. This *fill-rule property* is further generalized in a renormalization approach of GENERALIZED DIAGONAL WYTHOFF NIM and LINEAR NIMHOFF; see Section 8.

WYTHOFF NIM has been considered in the context of phyllotaxis more recently in Chapter 17 in the book "Symmetry in Plants" [82] and here the *maximal Fibonacci representation* is used to represent Wythoff's sequences (Adamson's Wythoff Wheel [82, Chapter 17]), while we more often encounter the *minimal Fibonacci representation* (often called the Zeckendorff [159] representation, although it was discovered by Ostrowski [134] and Lekkerkerker [121]) together with left and right shifts, e.g., [140; 38].

In the book *Theory of graphs*, by C. Berge [22], R. P. Isaac mentions the game of WYTHOFF NIM in Example 1 in Section 6, but played with a queen of chess; this variation of the game is often called "Corner the Lady" [60]. Without reference, the winning strategy is mentioned, and for the first time, a nice illustration of the fill rule of the diagonal moves is presented. The picture also gives the initial Grundy values of Wythoff's game. The section concerns "Nim-type games" and illustrates "the kernel of a graph" idea for positions of Grundy value 0.

In 1959, Connell [20] restricted Wythoff's game by requiring to take a multiple of $b \geq 1$ from a single pile. For the P-positions, he obtained b pairs of sequences, each pair consisting of two complementary nonhomogenous Beatty sequences [36].

In 1968, Holladay [79] extended Wythoff's game to a k-WYTHOFF NIM, by the extension of the diagonal rule:

> *take from both piles, but do not take more than k more counters from one pile than from the other.*

This game gives a simple but elegant generalization of both the minimal exclusive description in Theorem 1, and also the algebraic description for the

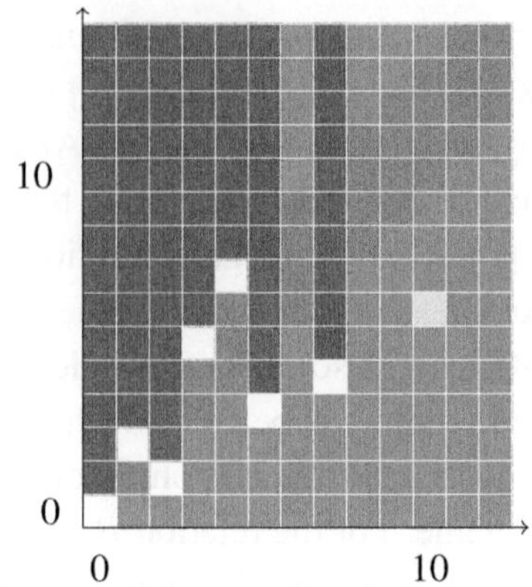

Figure 1. The P-position (6, 10) (green) of WYTHOFF NIM has been computed via a fill rule algorithm, using the P-positions closer to the origin (light). We also indicated the symmetric P-position (10, 6) (yellow). We colored vertical (blue) and diagonal (red) N-positions that have these smaller P-positions as options. View the old P-positions as sources of light, and the N-positions as colored light-beams: the green and yellow cells will next become light sources and invoke new colored beams, giving birth to new green and yellow cells, and so on. Note that the diagonal red beam is tightly packed, and this is part of the induction hypothesis, whereas the vertical beams leave a-periodic gaps, related to the famous rabbit-birth algorithm of Fibonacci: one baby rabbit if and only if we find two *upper* green cells (P-positions) within the same gap (so the picture illustrates an immature rabbit). We omitted the horizontal beams in the picture because they are not required in the algorithm. Neither are the diagonal beams below the line $y = x$ needed. The fill-rule idea gives a nice conjecture for a generalization of WYTHOFF NIM (see also Figure 7).

P-positions in Theorem 2. In fact, he defines four variations to these game rules with exactly the same set of P-positions. The k-Wythoff game was revisited by Fraenkel [38], where computational aspects and connections to continued fractions are emphasized. Holladay also studies related games where at most k counters may be removed from both piles, or any number from just one pile, but these variations do not invoke the fill-rule-property, and therefore neither do they generalize Wythoff's characterizations of the P-positions.

In 1973, Fraenkel and Borosh [49] generalized Wythoff's game in a way that includes both Connell's and Holladay's games, preserving the complementary (nonhomogeneous) Beatty sequences property, and in 2009 Larsson [106; 111]

13	8	5	3	2	1	n	8	5	3	2	1
					1	1					1
				1	0	**2**				1	0
			1	0	0	3				1	1
			1	0	1	4			1	0	1
		1	0	0	0	**5**			1	1	0
		1	0	0	1	6			1	1	1
		1	0	1	0	**7**		1	0	1	0
	1	0	0	0	0	8		1	0	1	1
	1	0	0	0	1	9		1	1	0	1
	1	0	0	1	0	**10**		1	1	1	0
	1	0	1	0	0	11		1	1	1	1
	1	0	1	0	1	12	1	0	1	0	1
1	0	0	0	0	0	**13**	1	0	1	1	0

Table 2. The minimal (no two consecutive 1s) and maximal (no two consecutive 0s) Fibonacci representation of the first few integers, respectively. The bold numbers are the ones in the B sequence. These representations satisfy a number of interesting number theoretical properties. Note that the numbers in the A sequence have even number of rightmost "0"s in the minimal representation. In the maximal representation they are the ones ending in a "1". The so-called left-shift property holds for both representations: the number $B(n)$ is obtained by shifting the digits of $A(n)$ one step to the left and putting a "0" as the least significant bit.

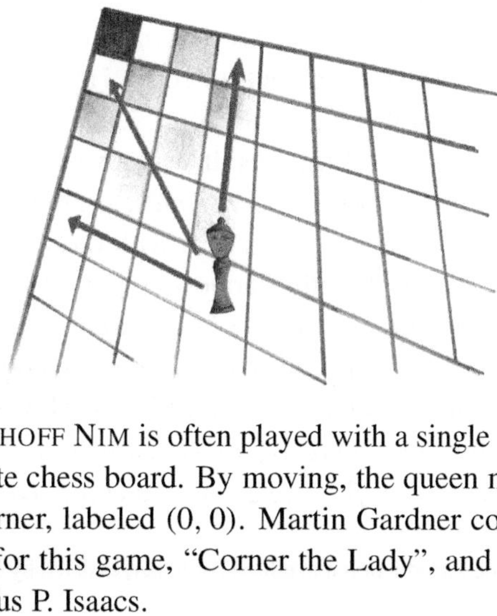

Figure 2. WYTHOFF NIM is often played with a single queen of chess on a semi-infinite chess board. By moving, the queen must get closer to the single corner, labeled (0, 0). Martin Gardner coined the other classical name for this game, "Corner the Lady", and attributed this variation to Rufus P. Isaacs.

found yet three such games, with distinct complementary Beatty sequences. See Sections 4.1 and 8.1, respectively.

As we already noted, Wythoff's game is closely connected to complementary and to disjoint integer, rational and irrational Beatty sequences, a concept which generalizes arithmetic progressions. Such sequences are considered in many papers, with or without references to Wythoff's game, and often concerning exactly covering systems. In Section 2, we review some of this work, which has partly been inspired by Wythoff's sequences but also stem from diverse origins. Starting with E. Duchêne's vision in Section 3, the author's contributions are presented in alphabetical order[1].

2. Exactly covering sequences with some game applications

Obvious exactly covering systems, i.e., partitions of the set of positive integers, are arithmetic sequences, such as $2n-1$, $2n$, $n \geq 1$; or $4n-3$, $4n-1$, $4n$, $n \geq 1$. In these two examples, the two largest *moduli* (2 and 4 respectively) are the same. This is a general property of exactly covering arithmetic sequences: If all the moduli α_i are integers with $m \geq 2$ and $\alpha_1 \leq \alpha_2 \leq \cdots \leq \alpha_m$, then $\alpha_{m-1} = \alpha_m$. A proof using complex numbers and roots of unity was given by Mirsky, Newman, Davenport and Rado — see Erdös [33]. Since every math paper — even a survey paper! — should contain at least one proof, we present here a "proof from the Book" of their result.

Let $\{na_i + b_i : 1 \leq i \leq m;\ n = 1, 2 \ldots\}$ be an exactly covering system of $m \geq 2$ arithmetic sequences, where $a_1 \leq \cdots \leq a_m$. Consider the generating function $z^{b_i}/(1-z^{a_i}) = \sum_{n\geq 1} z^{na_i+b_i}$. The fact that the system is exactly covering is expressed by the identity $\sum_{i=1}^{m} z^{b_i}/(1-z^{a_i}) = z/(1-z)$. Let ξ be a primitive a_m-th root of unity and let $z \to \xi$. If $a_{m-1} < a_m$, then the only unbounded term in the identity is $z^{b_m}/(1-z^{a_m})$, a contradiction. Hence $a_{m-1} = a_m$. (See [102] for another application of this proof method.).

A first elementary proof of this result was given independently by Berger et al. [14] and by Simpson [141]. *Beatty sequences* are normally associated with irrational *moduli* α, β. Recent studies deal with rational moduli α, β. Clearly if $a/b \neq g/h$ are rational, then the sequences $\{\lfloor na/b \rfloor\}$ and $\{\lfloor ng/h \rfloor\}$ cannot be complementary, since $kbg \times a/b = kha \times g/h = kag$ for all $k \geq 1$. Also the former sequence is missing the integers $ka-1$ and the latter $kg-1$, so both are missing the integers $kag-1$ for all $k \geq 1$. However, complementarity can be maintained for the *nonhomogeneous* case: In [36; 133], necessary and sufficient conditions on α, γ, β, δ are given so that the sequences $\{\lfloor n\alpha + \gamma \rfloor\}$ and

[1]U. Larsson wrote the introduction and edited the paper, assisted by E. Duchêne, A. S. Fraenkel and N. B. Ho.

$\{\lfloor n\beta + \delta \rfloor\}$ are complementary — for both irrational moduli and rational moduli. We are not aware of any previous work in this direction, except that in Bang [4] necessary and sufficient conditions are given for $\{\lfloor n\alpha \rfloor\} \supseteq \{\lfloor n\beta \rfloor\}$ to hold, both for the case α, β irrational and the case α, β rational. Results of this sort also appear in Niven [131], for the homogeneous case only. In Skolem [143; 144], the homogeneous and nonhomogeneous cases are studied, but only for α and β irrational. Incidentally, Skolem set out from the point of view of Steiner systems [148] and discovered Wythoff's sequences, but without making the connection to Wythoff Nim. For related work, see also [66; 12; 13].

Uspensky [155] demonstrated, by using a well-known theorem of Kronecker, that if we have $k > 1$ homogeneous Beatty sequences with real moduli $a, b, \ldots$ that partition the positive integers, then $k = 2$, $1/a + 1/b = 1$ and a, b are irrational. Graham [64] later demonstrated that $k = 2$ by elementary means in a one page proof.

These investigations spawned the following interesting conjecture [37] (see also Erdös and Graham [34]): If the system $\bigcup_{i=1}^{m}\{\lfloor n\alpha_i + \gamma_i \rfloor\}$, $n = 1, 2, \ldots$ splits the positive integers with $m \geq 3$ and $\alpha_1 < \alpha_2 < \cdots < \alpha_m$, then

$$\alpha_i = (2^m - 1)/2^{m-i}, \quad i = 1, \ldots, m. \tag{3}$$

Graham [65] showed that if all the m moduli are irrational and $m \geq 3$, then two moduli are equal. Thus distinct integer moduli or irrational moduli cannot exist for $m \geq 2$ or $m \geq 3$ respectively in a splitting system.

The conjecture was proved for $m = 3$ by Morikawa [126], $m = 4$ by Altman et al. [3], for all $3 \leq m \leq 6$ by Tijdeman [154] and for $m = 7$ by Barát and Varjú [5] and was generalized by Graham and O'Bryant [67]. Other partial results were given by Morikawa [127] and Simpson [142]. Many others have contributed partial results; see Tijdeman [153] for a detailed history. The conjecture has some applications in job scheduling and related industrial engineering areas (just-in-time manufacturing); see e.g., Altman et al. [3], Brauner and Jost [17]; also in [123; 156]. However, the conjecture itself has not been settled.

So, this is a problem that has been solved for the integers, has been solved for the irrationals, and is wide open for the rationals!

The conjecture provides a challenge to find game rules using the sequences as candidate sets of P-positions [48]. Thus, for example, for the rat (rat — rational) game, the P-positions are $\{(\lfloor 7n/4 \rfloor, \lfloor 7n/2 - 1 \rfloor, 7n - 3), n = 1, 2, \ldots\} \cup \{0\}$. For the related mouse game on two piles, the use for an invariant analogue, a mouse trap [112], became apparent, which brings us to a modern trend in CGT. A typical interest in CGT is, given a finite rule set describing a game, find its P-positions, or also, when possible, its Sprague–Grundy function. A modern trend is to reverse this process: given a subsequence R of nonnegative vectors, is there a game whose set of P-positions is precisely R? Suppose we gave a family of games for which the moves and the outcomes have the same description, for example t-tuples of nonnegative integers (for WYTHOFF NIM $t = 2$). Any such game for which some move cannot be made from all game-positions (sometimes because it would be a move connecting two P-positions) is a *variant* game (e.g., the rat and mouse games). Duchêne and Rigo [31] conjectured that if R is the set of numbers produced by a pair of complementary homogeneous Beatty sequences (with irrational moduli), then there is an *invariant* game whose set of P-positions is R, together with (0, 0). Larsson et al. proved a generalization thereof [116]. Informally, a game is invariant if every move can be done from every position, provided only that the result is a game position. Much earlier Golumb [63] defined the notion of a *vector subtraction game*, which is an instance of the family of invariant games, including WYTHOFF NIM and many other impartial heap games. The *move-size dynamic* games FIBONACCI NIM [157; 117; 118] and IMITATION NIM [105], also have winning strategies related to the P-sequences of WYTHOFF NIM, although they are noninvariant in this sense. From the games' perspective it leads us to a general territory, with many open problems: when do exactly covering sequences or variations hereof provide efficient procedures for the outcomes of nice/short combinatorial games? This type of question is addressed also in three research papers in this book [114; 115; 119], in response to a question of Fraenkel at the BIRS workshop in CGT 2011.

Before we move on, one should note that Stolarsky has contributed some interesting papers related to Beatty sequences; for example, one with Porta and Fraenkel [56], where many curious identities involving φ (= golden section) are proved. For example, the reals $\{n\varphi\}$ are closed under ordinary multiplication, where $\{x\}$ is the fractional part of x. In fact, $\{m\varphi\}\{n\varphi\} = \{k\varphi\}$, where $k = mn - m\lfloor\varphi\rfloor - n\lfloor m\varphi\rfloor$. Another more recent contribution of Stolarsky and Kimberling [100] concerns interesting kinds of convergence. A sequence (x_n) *converges deviously* to L if, in addition to converging to L, it is true that for every real B, there exists $\ell \neq L$ such that $x_n = \ell$ for more than B numbers n. For example, let

$$g(n) = \frac{n}{\varphi\lfloor n/\varphi\rfloor},$$

where $\varphi = \frac{1}{2}(1+\sqrt{5})$. The sequence $(\lfloor n/\varphi\rfloor)$ is an example of a *slow Beatty sequence*, and $(g(n))$ converges deviously to 1. Next, suppose that sequences w_1 and w_2 partition the positive integers. Suppose further that (a_n) is a sequence such that $a_{w_1(n)} \to L_1$ and $a_{w_2(n)} \to L_2$, so that (a_n) converges if and only if $L_1 = L_2$. Otherwise, (a_n) *diverges partitionally* on w. Define

$$h(n) = n(g(n+1) - g(n)),$$

and for irrational $t > 1$, let $w_1(n) = \lfloor nt \rfloor$ and $w_2(n) = \lfloor nt/(t-1) \rfloor$, these being complementary Beatty sequences. Then $(h(n))$ is partitionally divergent, with $h(w_1(n)) \to 1 - t$ and $h(w_2(n)) \to 1$.[2]

3. WYTHOFF NIM seen as a vector subtraction game, and solved with infinite words

Written by Eric Duchêne

In this section, I have chosen to consider WYTHOFF NIM under two "visions":

- WYTHOFF NIM can be considered as an instance of the more general *vector subtraction games* introduced by Golomb in [63]. A large set of variants of WYTHOFF NIM found in the literature can be seen as instances of Golomb's game. In Section 3.1, I will give a couple of personal results obtained for some particular cases of this game.
- My second vision is about the link between WYTHOFF NIM and Fibonacci, and more precisely the Fibonacci word. Variants of WYTHOFF NIM based on other words, such as the so-called *Tribonacci word* will be described.

3.1. WYTHOFF NIM *as a vector subtraction game.* In [63], Golomb introduced the notion of *t-vector subtraction games*. Given t piles of counters, a *position* of such a game is a t-tuple of nonnegative integers, corresponding to the number of counters in each pile. A *move* is also a t-tuple of nonnegative integers corresponding to the number of counters that are removed from each pile. Let $p = (p_1, \dots, p_t)$ be a position and $m = (m_1, \dots, m_t)$ be a nonzero move. The move m can be applied to the position p provided that $m \leq p$, i.e., for all i, $m_i \leq p_i$. The position resulting from the application of m is the t-tuple $p - m$. Given a set $\mathcal{M}$ of allowed moves and a starting position p, two players alternately apply a move from $\mathcal{M}$. The first player unable to move loses the game. Clearly, WYTHOFF NIM is an instance of the vector subtraction game with

$$\mathcal{M}_{WYT} = \{(0, i), (i, 0), (i, i) : i > 0\}.$$

In the literature, several games can be seen as instances of the vector subtraction game. In particular for $t = 2$, some of them have a set $\mathcal{M}$ corresponding to a proper subset (restriction) or superset (extension) of WYTHOFF NIM. Some of these games will be detailed in the following section.

Remark 3. Note that in [31], such games are also called take-away *invariant* games, since the allowed moves do not depend on the initial position (i.e., if $m \in \mathcal{M}$, playing $p - m$ is allowed for any p provided $m \leq p$). If invariant

[2]A.S. Fraenkel and U. Larsson wrote this section, and the last paragraph was composed by A.S. Fraenkel and C. Kimberling.

take-away games and vector subtraction games are equivalent, the notion of invariance is devoted to be expanded in a more general context than the one of take-away games.

2-vector subtraction games. Some instances of the 2-vector subtraction game will be considered further in the current paper. The first one is Connell's game [20] (see Section 4.1), defined as

$$\mathcal{M}_{Con}(b) = \{(0, i), (i, 0) : i = kb, k > 0\} \cup \{(i, i) : i > 0\}.$$

In other words, Connell's game (of parameter b) is a restriction of WYTHOFF NIM where the Nim moves must be multiples of b. Connell proved that the P-positions of his game can be seen as a set of b pairs of homogeneous Beatty sequences $(A_{i,n}, B_{i,n})$ for $i = 0, \ldots, b-1$, such that for all $n \geq 0$,

$$A_{i,n} = \left\lfloor \left(n + \frac{i}{b}\right)\left(1 + \frac{1}{\alpha}\right) \right\rfloor, \quad B_{i,n} = \left\lfloor \left(n + \frac{i}{b}\right)\left(1 + \alpha\right) \right\rfloor,$$

where $\alpha = \frac{1}{2}(b + \sqrt{b^2 + 4})$.

Another example is the case of t-WYTHOFF NIM defined by Fraenkel [38], where

$$\mathcal{M}_{GW}(t) = \{(0, i), (i, 0) : i > 0\} \cup \{(i, j) : |i - j| < t,\ i, j > 0\}.$$

For this game, the P-positions can be characterized with an algebraic formula close to the one of WYTHOFF NIM (this formula will be given in Section 4.1).

The games NIM(a, b) [73] (see Section 5), MAHARAJA NIM [120] (see Section 8.4) are other invariant WYTHOFF NIM variations. One can also mention the recent game WYT(F) [58], where f is a given function $\mathbb{N} \to \mathbb{N}$, and defined as follows:

$$\mathcal{M}_{WYT}(f) = \{(0, i), (i, 0) : i > 0\} \cup \{(i, j) : 1 \leq i \leq j < f(k)\}.$$

When the number of moves is finite, one can mention the game where the allowed vectors correspond to the moves of a knight in chess. In that case, the Grundy function was proved to be periodic [6]. The same kind of result is also true for the king and its powers [28] (in other words, this is WYTHOFF NIM where one can remove at most k counters per heap, for a given k).

In [28], Duchêne and Gravier have considered a restriction of WYTHOFF NIM, namely the $[a, b]$-vector game (a, b being two positive integers), which can be expressed as a particular family of 2-vector subtraction games:

$$\mathcal{M}(a, b) = \{(0, i), (i, 0), (ia, ib) : i > 0\}.$$

If $a \neq b$, it is proved that the P-positions of this game are exactly the set $\{(i, i) : i \geq 0\}$. The situation is more tricky when $a = b$. For $a = b = 1$, the game is equivalent to WYTHOFF NIM. In [28], an acceptable exponential algorithm is given to compute the P-positions of the [2, 2]-game. Yet, it seems to us that a closed formula should be available, since it is conjectured that the P-positions follow the progression of $\lfloor \frac{1}{4}(n(3+\sqrt{17})) \rfloor$. Note that the $[a, b]$-game can be naturally extended to n heaps, including the most natural extension of WYTHOFF NIM:

> *Given n piles of counters, both players alternately take either from one of the piles an arbitrary number of counters or from all piles an equal number. The player who takes up the last counter or counters, wins.*

When n is odd, it was shown in [28] that this game has the same set of P-positions as NIM. When n is even, the resolution remains open (except for $n = 2$, i.e., WYTHOFF NIM).

Given a subset K of $\mathbb{N}$, another natural restriction of WYTHOFF NIM, namely WYT_K, is the following instance of the 2-vector subtraction game:

$$\mathcal{M}_{WYT}(K) = \{(0, i), (i, 0) : i > 0\} \cup \{(k, k) : k \in K\}.$$

The game WYT_K with $K = \mathbb{N}$ is WYTHOFF NIM. In [26; 53], this game has been investigated for $|K| = 1$. In such a case, the P-positions are known. A full characterization of the $\mathcal{G}$-function is even proved for $K = \{2^k\}$ for some $k \geq 0$, and also for every subset of the powers of 2 including 1. In a certain manner, this characterization shows that these WYTHOFF NIM restrictions are "closer" to NIM than WYTHOFF NIM, since their Grundy functions behave like the one of NIM (i.e., a latin square with a strong regularity).

In [31], another set of 2-vector subtraction games is considered:

$$\mathcal{M}_{DR}(k) = \mathcal{M}_{WYT} \setminus \{(2i, 2i) \mid 0 < i < k\} \cup \{(2k+1, 2k+2), (2k+2, 2k+1)\}.$$

In other terms, these games can be described as follows:

> *Given a positive integer k, either take a positive number from a single pile, or (i, i) from both as in* WYTHOFF NIM, *or $2k+1$, $k > 0$ from one and $2k+2$ from the other, except that the* WYTHOFF NIM *moves of taking $(2i, 2i)$, $i < k$ from both are disallowed.*

For this set of games, it was proved [31] that the P-positions can be expressed as a pair of complementary Beatty sequences, as it is the case for WYTHOFF NIM. More precisely:

Theorem 4. *The P-positions of the game $\mathcal{M}_{DR}(k)$ are of the form $(\lfloor n\alpha_k \rfloor, \lfloor n\beta_k \rfloor)_{n \geq 0}$, where α_k is the quadratic irrational number having $(1; \overline{1, k})$ as continued fraction expansion, and $1/\alpha_k + 1/\beta_k = 1$.*

Remark 5. For $t > 2$, there are few instances of the t-vector subtraction that are considered in the literature. Some of them were mentioned in [28].

3.2. WYTHOFF NIM *as the "Fibonacci game"*. In [29], Duchêne and Rigo introduced a new characterization of the P-positions of WYTHOFF NIM, which deals with the Fibonacci word (see e.g. [9]). Given a two-letter alphabet $\{a, b\}$, take the morphism $\phi : \{a, b\} \to \{a, b\}^*$ defined as follows:

$$\phi(a) = ab, \quad \phi(b) = a.$$

By iterating this morphism from a, we get the famous *Fibonacci word* $w_F = (w_n)_{n\geq 1} = \lim_{n\to+\infty} \phi^n(a)$,

$$w_F = abaababaabaababaababaabaababa\ldots$$

In this word, we will use the convention that the first letter has index 1. For $X = A, B$ (resp. $x = a, b$), we define the sets

$$X = \{X_1 < X_2 < \cdots\} = \{n \in \mathbb{N} \mid w_n = x\}.$$

Roughly speaking, the indices of the letters a (resp. b) in w_F correspond to the sequence (A_n) (resp. (B_n)). In addition, we set $A_0 = B_0 = 0$. According to this definition, the P-positions of WYTHOFF NIM exactly correspond to the sequence (A_n, B_n).

Remark 6. Note that a similar characterization has been obtained [30] for the P-positions of k-WYTHOFF NIM using the morphism

$$\phi'(a) = a^k b, \quad \phi'(b) = a.$$

Since the P-positions of WYTHOFF NIM are correlated to the Fibonacci word, a natural question arose: does there exist a 3-heap game whose P-positions can be coded by the so-called *Tribonacci word* w_T, defined as the unique fixed-point of the morphism $\psi : \{a, b, c\} \to \{a, b, c\}^*$, starting from a:

$$\psi(a) = ab, \quad \psi(b) = ac, \quad \psi(c) = a.$$

Hence w_T starts with

$$w_T = abacabaabacababacabaabacabacabaabacababaca\ldots$$

The first values of the sequence (A_n, B_n, C_n) derived from the Tribonacci word are given in Table 3. In [29], a 3-heap game is built, whose P-positions exactly correspond to the sequence (A_n, B_n, C_n) (with all their permutations adjoined). This game has been called the TRIBONACCI GAME:

Given 3 piles of counters, the rules are the following:

- *Any positive number of tokens from up to two piles can be removed.*

n	0	1	2	3	4	5	6	7	8	9	10	11	12	13	14	15
A_n	0	1	3	5	7	8	10	12	14	16	18	20	21	23	25	27
B_n	0	2	6	9	13	15	19	22	26	30	33	37	39	43	46	50
C_n	0	4	11	17	24	28	35	41	48	55	61	68	72	79	85	92

Table 3. First values of the sequences $(A_n)_{n\geq 0}$, $(B_n)_{n\geq 0}$ and $(C_n)_{n\geq 0}$.

- *Let* α, β, γ *be three positive integers such that*

$$2\max\{\alpha, \beta, \gamma\} \leq \alpha + \beta + \gamma.$$

Then one can remove α (*resp.* β, γ) *from the first* (*resp. second, third*) *pile.*

- *Let* $\beta > 2\alpha > 0$. *From position* (a, b, c) *one can remove the same number* α *of counters from any two piles and* β *counters from the unchosen one with the following condition. If* a' (*resp.* b', c') *denotes the number of counters in the pile which contained* a (*resp.* b, c) *tokens before the move, then the configuration*

$$a' < c' < b'$$

is not allowed.

In [29; 62; 30; 31], several take-away games were deeply investigated with the use of such words. In many cases, deciding whether a given position is N or P is proved to be polynomial thanks to a numeration system derived from the underlying morphism.

Remark 7. Note that the TRIBONACCI GAME is not invariant (i.e., it is not an instance of the 3-vector subtraction game). In [32], the authors provide an algorithm which decides whether invariant rules could have been proposed to fit this set of P-positions.

4. Some of the ramified depths and wisdoms of WYTHOFF NIM

Written by Aviezri S. Fraenkel

This section is a partial survey — *partial* in two senses: It is tailored to our own partial taste (not impartial), and it contains only a small part of the appetizing WYTHOFF NIM curiosities (not comprehensive). In contrast — and in a third sense of partiality — it contains only one study, in Section 4.2, of partial games, more often termed partizan games. Otherwise only impartial games are surveyed: occasionally we refer to properties of all games, not just WYTHOFF NIM. Then all *impartial* games are meant.

Notation 8. The set of all P-positions of a game is denoted $\mathcal{P}$; the set of all its N-positions is $\mathcal{N}$.

There are extensions and restrictions of WYTHOFF NIM.

4.1. *Three 2-pile extensions.*

(i) A Nim move restriction. Connell [20] restricted Wythoff's game by requiring to take a multiple of $b \geq 1$ from a single pile. For the P-positions, he obtained b pairs of sequences, each pair consisting of two complementary nonhomogenous Beatty sequences [36].

(ii) Nim-move restriction and diagonal move extension/restriction. In [49] we analyzed the following generalization of Connell's game, dubbed bt-WYTHOFF NIM NIM: for fixed positive integer parameters, b and t, remove a positive multiple of b tokens from a *single* pile, or $k > 0$ from one and $\ell > 0$ from the other, provided that $|k - \ell| < bt$, $k - \ell \equiv 0 \pmod b$.

(iii) A diagonal move extension, i.e., t-WYTHOFF NIM:

> *The moves are of two types: remove any positive number from a* single *pile* (NIM *move*), *or* $k > 0$ *from one and* $\ell > 0$ *from the other, provided that* $|k - \ell| < t$, *where* t *is a fixed positive integer parameter* (diagonal *move*).

Notice that in (ii), taking $b = 1$ gives t-WYTHOFF NIM and taking $t = 1$ gives Connell's game; in (iii), taking $t = 1$ is classical WYTHOFF NIM, where the *same* amount has to be taken from both piles.

In [38] three strategies are presented for computing the P-positions of t-WYTHOFF NIM:

- Recursive. $\mathcal{P} = \bigcup_{i=0}^{\infty}\{(A_i, B_i)\}$, where $A_n = \operatorname{mex}\{A_i, B_i : 0 \leq i < n\}$, $B_n = A_n + tn$, $n \geq 0$.
- Algebraic. $\mathcal{P} = \bigcup_{n=0}^{\infty}\{(\lfloor n\alpha \rfloor, \lfloor n\beta \rfloor)\}$, where $\alpha^{-1} + (\alpha + t)^{-1} = 1$, $\beta = \alpha + t$, so $\alpha = \frac{1}{2}(2 - t + \sqrt{t^2 + 4})$, $\beta = \frac{1}{2}(2 + t + \sqrt{t^2 + 4})$.
- Arithmetic. In the exotic numeration system whose basis elements are the numerators of the convergents of the simple continued fraction expansion $\alpha = [1, t, t, t, \ldots]$, the numbers A_i end in an even number of 0s; B_i is the "left shift" of A_i, that is, it is the representation A_i with a 0 adjoined at the end of the representation. For further details see [38].

The input size of any t-WYTHOFF NIM-position (x, y) is $\log(xy)$. The computation for deciding whether $(x, y) \in \mathcal{P}$ is exponential for the first of the three strategies, but polynomial for the last two.

4.2. ***Misère t*-WYTHOFF NIM.** In [39], t-WYTHOFF NIM in misère play was studied. As for normal play, recursive, algebraic and arithmetic strategies were given. Let S_1 and S_2 denote the P-positions for normal and misère play respectively. Curiously, for $t = 1$, $S_1 = S_2$ except for the first two P-positions, where $(A_0, B_0) = (2, 2)$, $(A_1, B_1) = (0, 1)$ and $S_1 \cap S_2 = \varnothing$ for all $t > 1$. Specifically,

- Recursive. For $t = 1$, $(A_0, B_0) = (2, 2)$, $A_n = \operatorname{mex}\{A_i, B_i : 0 \le i < n\}$, $B_n = A_n + n$, $n \ge 1$. For $t > 1$, $A_n = \operatorname{mex}\{A_i, B_i : 0 \le i < n\}$, $B_n = A_n + n + 1$, $n \ge 0$.
- Algebraic. For $t = 1$, $(A_0, B_0) = (2, 2)$, $(A_1, B_1) = (0, 1)$, $A_n = \left\lfloor \frac{1}{2}n(1+\sqrt{5})\right\rfloor$, $B_n = \left\lfloor \frac{1}{2}n(3+\sqrt{5})\right\rfloor$, $n \ge 2$. For $t > 1$, $A_n = \lfloor n\alpha + \gamma \rfloor$, $B_n = \lfloor n\beta + \delta \rfloor$, $n \ge 2$, where $\alpha = \frac{1}{2}(2 - t + \sqrt{t^2+4})$, $\beta = \alpha + t$, $\gamma = \alpha^{-1}$, $\delta = \gamma + 1$.
- Arithmetic. For $t = 1$, $(A_0, B_0) = (2, 2)$, $(A_1, B_1) = (0, 1)$; for $n \ge 2$, A_n, B_n are the same as for normal play. For $t > 1$, The characterization is too long to state here — see [39].

Also a subset of binary trees, dubbed *cedar trees*, was constructed and used for conducting generalized searches and consolidating the three strategies of WYTHOFF NIM in both normal and misère play.

4.3. ***Multipile* WYTHOFF NIM.** When our interest in combinatorial games first arose, we noticed that WYTHOFF NIM seems to be rather more difficult than NIM in at least two aspects, though both are acyclic two-player games with perfect information and no chance moves:

(i) Computation of the Sprague–Grundy function g.

(ii) Generalization to more than two piles. No generalization seemed to preserve the properties $B_n = A_n + n$ for the P-positions of some two piles, and the role of $\varphi := (1+\sqrt{5})/2$ (golden ratio) in the strategy.

Study of the 1-values of g and related aspects was done in [15]. Some light was shed on the approximate distance from the nonzero g-values to the 0s by Nivasch [130]. In Dress, Flammenkamp, Pink's work [25], the additive periodicity of the Sprague–Grundy function of WYTHOFF NIM was established: the g-function of each row minus its saltus is periodic. A much simpler proof was given independently by Landman [103] (see also Section 6.1). All of these studies attest to the difficulty of computing the g-function of WYTHOFF NIM.

We asked the experts for an explanation of this discrepancy. We were told that it is due to the nondisjunctive move of taking from both piles simultaneously. We tested this claim by replacing the diagonal move by taking k from one, ℓ from the other for any $k \ne \ell$. To our surprise we saw that the experts were wrong: the P-position strategy remained precisely as that of NIM (though not necessarily the nonzero g-values). This was true for both $k \neq \ell$ fixed, say $(k, \ell) = (4, 7)$, or selecting any $k \neq \ell$ at each move.

Many authors attempted to generalize WYTHOFF NIM to multipile WYTHOFF NIM by taking the same number from couples or triples or all the piles, and many similar variations. None of those preserved the properties (ii) above.

It turns out that taking the same number of tokens from both piles in WYTHOFF NIM is a red herring! Rather, taking k from one and ℓ from the other pile such that $k \oplus \ell = 0$ (so $k = \ell$), where $\oplus$ denotes Nim-addition, is the key for understanding the nature of WYTHOFF NIM: The couples (k, k) are the P-positions of 2-pile NIM. Adjoining them as moves necessarily destroys the P-positions of NIM, since $\mathcal{P}$ of any game is an independent set. The independence follows from the fundamental properties of any acyclic game:

$$u \in \mathcal{P} \Longleftrightarrow F(u) \subseteq \mathcal{N}, \quad u \in \mathcal{N} \Longleftrightarrow F(u) \cap \mathcal{P} \neq \varnothing,$$

where $F(u)$ denotes the set of direct followers of position u in the game. More precise information is given in [16] and [52, §5.1].

These observations led us to the conclusion that the proper set of diagonal moves for N-pile WYTHOFF NIM is the set of P-positions of N-pile NIM (only moves that leave nonnegative pile sizes). This, in turn, led to our conjectures stated below. Its budding is in [40, §6]. See Nowakowski and Guy [75]; see also [41].

Define an N-pile WYTHOFF NIM game as follows:

Given $N \geq 2$ piles of finitely many tokens, whose sizes are $p_1, \dots, p_N$. The moves are to take any positive number of tokens from a single pile or to take $(a_1, \dots, a_N) \in \mathbb{Z}^N_{\geq 0}$ from all the piles — a_i from the i-th pile — subject to the conditions:

(1) $a_i > 0$ *for some* i,

(2) $a_i \leq p_i$ *for all* i,

(3) $a_1 \oplus \dots \oplus a_N = 0$.

The player making the last move wins and the opponent loses.

Notice that WYTHOFF NIM is the case $N = 2$.

For $N \geq 3$, denote the P-positions for N-pile WYTHOFF NIM by

$$(A^1, \dots, A^{N-2}, A_n^{N-1}, A_n^N), \quad A^{N-2} \leq A_n^{N-1} \leq A_n^N$$

and $A_n^{N-1} < A_{n+1}^{N-1}$ for all $n \geq 0$. The notation is intended to imply that $A^1, \dots, A^{N-2}$ are fixed.

Conjecture 9. *There exists an integer m_1, depending only on $A^1, \dots, A^{N-2}$, such that $A_n^{N-1} = \operatorname{mex}(\{A_i^{N-1}, A_i^N : 0 \leq i < n\} \cup T)$, $A_n^N = A_n^{N-1} + n$ for all $n \geq m_1$ where T is a (small) set of integers depending only on $A^1, \dots, A^{N-2}$.*

Conjecture 10. *There exist integers m_2, a such that $A_n^{N-1} = \lfloor n\varphi \rfloor + a + \epsilon_n$ and $A_n^N = A_n^{N-1} + n$, $-1 \le \epsilon_n \le 1$ for all $n \ge m_2$.*

Both conjectures were proved by Sun and Zeilberger [152] for the special case $N = 3$ and $1 \le A^1 \le 10$. In [52] it was shown, inter alia, that Conjecture 9 implies Conjecture 10. This was also proved, inter alia, by Sun [151, Corollary 4.6]. (In his abstract — not in the paper itself — it is stated erroneously that the two conjectures are proven to be equivalent.) See also Coxeter's work [24], and [54; 59; 40].

4.4. ***Bridges between* NIM *and* WYTHOFF NIM.** Motivated by the difficulty to compute the Sprague–Grundy function g for WYTHOFF NIM and the ease to do the same for NIM, we attempted to bridge these two games with in-between games. In NIMHOFF (hybrid of NIM and WYTHOFF NIM) [53], the diagonal move is restricted in various ways. A closed formula for the Sprague–Grundy function g is given for most of these games. A generalized Nim-sum is given to guarantee the polynomiality of g for these games. A second bridge between the two games is established in [26], which continues the above first bridge. The diagonal moves are restricted by taking k from both piles only if k belongs to a predetermined given set K. The P-positions are computed; it is determined under what conditions on K is $(a_j, a_j + j) \in \mathcal{P}$; the g-function is computed, and the regularity properties of g are studied.

4.5. ***The game of* END-WYTHOFF.** Motivated by the game END-NIM of Albert and Nowakowski [1], we studied END-WYTHOFF in normal play [57]:

> *A position in* END-WYTHOFF *is a vector of finitely many piles of finitely many tokens. Two players alternate in taking a positive number of tokens from either end-pile ("burning-the-candle-at-both-ends"), or taking the same positive number of tokens from both ends.*

A recursive characterization of the P-positions (a_i, K, b_i) is presented. For special cases of the vector K of middle-piles, the recursive characterization can be improved. It is also shown that $b_i - a_i = i$ for sufficiently large i (which holds for all i in WYTHOFF NIM). Further, it is shown that if K is a P-position, then (a, K, b) is a P-position if and only if (a, b) is a P-position of WYTHOFF NIM. Finally, a polynomial algorithm is given for computing the P-positions (a_i, K, b_i).

4.6. ***Extensions, restrictions of* WYTHOFF NIM *preserving its P-positions.*** In the paper [27], we show that no strict subset of rules of WYTHOFF NIM is the ruleset of a game having the same set of P-positions as WYTHOFF NIM [27]. On the other hand, we characterize all moves that can be adjoined while preserving the set of P-positions of WYTHOFF NIM. Testing if a move belongs to such an extended set of rules is shown to be doable in polynomial time.

Many arguments rely on the infinite Fibonacci word, automatic sequences and the corresponding numeration system. With these tools, we provide new two-dimensional morphisms generating an infinite picture encoding P-positions of WYTHOFF NIM and moves that can be adjoined.

4.7. *Rat games.* The general rat game considered here is played on $m \geq 2$ piles. The k-th component of its P-positions has the form

$$\left\lfloor \frac{2^m - 1}{2^{m-k}} n \right\rfloor - 2^{k-1} + 1, \quad k = 1, \ldots, m; \quad n = 1, 2, \ldots \tag{4}$$

The m terms $\lfloor (2^m - 1)/2^{m-k} \rfloor$, $k = 1, \ldots, m$ are called the *moduli* of the system (4).

Rat games (rat—rational) are studied in [48]. For $m = 3$, the rat game is played on 3 piles of tokens. Positions are denoted (x, y, z) with $0 \leq x \leq y \leq z$, and moves $(x, y, z) \to (u, v, w)$, where also $0 \leq u \leq v \leq w$. The following is the essence of the game rules:

(I) Take any positive number of tokens from up to 2 piles.

(II) Take $\ell > 0$ from the x pile, $k > 0$ from the y pile, and an arbitrary positive number from the z pile, subject to the constraint $|k - \ell| < a$, where

$$a = \begin{cases} 1 & \text{if } y - x \not\equiv 0 \pmod 7, \\ 2 & \text{if } y - x \equiv 0 \pmod 7. \end{cases}$$

(III) Take $\ell > 0$ from the x pile, $k > 0$ from the z pile, and an arbitrary positive number from the y pile, subject to the constraint $|k - \ell| < b$, where $b = 3$ if $w = u$; otherwise,

$$b = \begin{cases} 5 & \text{if } w - u \not\equiv 4 \pmod 7, \\ 6 & \text{if } w - u \equiv 4 \pmod 7. \end{cases}$$

Also for $m = 2$ (the "mouse" game), game rules were given there. But for $m \geq 4$, we didn't find "nice" game rules.

I have shown [37] that for every $m \geq 2$, the m sequences of the form (4) split the positive integers into m nonintersecting complementary sets. I further conjectured that this splitting is unique: it is the only system that splits the positive integers with *distinct* moduli [41]; see also Erdös and Lin [66]; and Erdös and Graham [34, p. 19]. The motivation for this study is thus 4-fold:

(i) to try a games approach, which might help to settle the conjecture;

(ii) demonstrate existence of a take-away game whose P-positions depend on *rational* numbers;

(iii) find another analyzable non-NIM take-away game played on more than 2 piles; and

(iv) present another challenge of finding “nice” game rules, given the game’s P-positions. (Such a challenge is implicit in Duchêne and Rigo [31] and Larsson et al. [116])

Since “nice” game rules based on a given set of P-positions were mentioned, I feel that a tentative definition thereof should be given here, though this is a survey: The game rules are *nice* if they depend on at most a finite number of the P-positions or range values of functions thereof. Notice that according to this definition, the game rules given above for $m = 3$ are nice. (Nice game rules may, perhaps, be called *invariant* game rules.)

4.8. RATWYT. This is another game played with rational numbers (rat — rational, Wyt — Wythoff). Given a rational number p/q in lowest terms, a *step* is defined by

$$\frac{p}{q} \to \frac{p-q}{q},$$

if $p/q \geq 1$, otherwise

$$\frac{p}{q} \to \frac{p}{q-p}.$$

> RATWYT [47] *is played on a pair of reduced rational numbers* $(p_1/q_1, p_2/q_2)$. *A move consists of either doing any positive number of steps to precisely one of the rationals, or doing the same number of steps to both. The first player unable to play* (*because both numerators are* 0) *loses.*

A winning strategy using the Calkin Wilf tree [18] is given.

4.9. *Games played by Boole and Galois.* In [42] we proved the following:

Theorem 11. *Let* $S = \bigcup_{i\geq 0}(a_i, b_i)$, *where for all* $n \geq 0$,

$$a_n = \text{mex}\{a_i, b_i : 0 \leq i < n\},$$

$b_0 = 0$, *and for all* $n > 0$,

$$b_n = f(a_{n-1}, b_{n-1}, a_n) + b_{n-1} + a_n - a_{n-1}.$$

If f is positive, monotone and semiadditive (*defined in* [42, §4]), *then S is the set of* P-*positions of a general* 2-*pile subtraction game with* **constraint function** f, *and the sequences* $A = \{a_i\}_{i\geq 0}$, $B = \{b_i\}_{i\geq 0}$ *have the following properties*:

(i) *they partition the positive integers*;

(ii) $b_{n+1} - b_n \geq 2$ *for all* $n \geq 0$;

(iii) $a_{n+1} - a_n \in \{1, 2\}$ *for all* $n \geq 0$.

The case $f = t$ is t-WYTHOFF NIM considered above.

In [42] we illustrated Theorem 11 with a collection of games based on sequences A and B, including known ones such as Prouhet–Thue–Morse, Hofstadter sequence, and mainly on new sequences. In [44] we gave an assortment of games based on constraint functions over Boolean variables or GF(2) (Galois).

4.10. WYTHOFF-*like games*. We introduced a class of variants of WYTHOFF NIM whose diagonal move is constrained by a function f [58]:

> *Three types of functions f are considered: f a constant, f strictly increasing and superadditive, $f(k) = \sum_{i=0}^{s} a_i k^i$ a polynomial of degree $s > 1$ with nonnegative integer coefficients and $a_0 > 0$.*

A function from the nonnegative integers to the nonnegative integers is *superadditive* if it satisfies $f(k) \geq k$ and $f(k+\ell) \geq f(k) + f(\ell)$ for all $k, \ell \geq 0$. The P-positions are pairs (A_n, B_n), $n \geq 0$, where A_n is computed by the mex function, and B_n is a function of A_n.

4.11. *Harnessing the unwieldy* mex *function*. A pair of integer sequences that splits the positive integers is often—especially in the context of combinatorial game theory such as WYTHOFF NIM-like games—defined recursively by $a_n = \text{mex}\{a_i, b_i : 0 \leq i < n\}$, $b_n = a_n + c_n$ $(n \geq 0)$. A typical problem is this: given integers $0 \leq x \leq y$, decide whether $x = a_n$, $y = b_n$. For general functions c_n, the best known algorithm for this decision problem is exponential in the input size $|\Omega(\log x + \log y)|$.

In [55] we produced a polynomial-time algorithm for solving this problem for the case of approximately linear functions c_n. We call the sequence $\bigcup_{i \geq 0} c_i$ *approximately linear* if there exist real constants α, u_1, u_2 such that $u_1 \leq c_n - n\alpha \leq u_2$ for all $n \geq 0$.

This result solves constructively and efficiently the complexity question of a number of previously analyzed take-away games of various authors.

4.12. *Translations of* WYTHOFF NIM*'s P-positions*. In [51], the translation phenomenon of the P-positions of WYTHOFF NIM was studied. The question was whether there exists a variant of WYTHOFF NIM whose P-positions, except for a finite number, are translations of those of WYTHOFF NIM, forming the set

$$S \cup \{(\lfloor \varphi n \rfloor + k, \lfloor \varphi^2 n \rfloor + k) : n \geq n_0\},$$

where $k \neq 0$, $n_0 \geq 0$ and S is a finite set of pairs of integers. Two variants of WYTHOFF NIM that answer the question for all positive integers k were established:

> *Given $k \geq 1$, in the variant called $\mathcal{W}_k$, each move is either removing a number of tokens from a single pile or removing an equal number of*

tokens from both piles, provided that none of the resulting piles has size less than k: the move from (a, b) *to* $(a-i, b-i)$ *with* $\min(a-i, b-i) < k$ *is not allowed.*

The P-positions of $\mathcal{W}_k$ form the set

$$\{(i, i) : 0 \leq i < k\} \cup \{(\lfloor \varphi n \rfloor + k, \lfloor \varphi^2 n \rfloor + k) : n \geq 0\}.$$

A variant of $\mathcal{W}_k$ that also exhibits the translation phenomenon is then introduced. Let $0 \leq j \leq k$.

In the variant $\mathcal{W}_{j,k}$*, each move from a position* (a, b) *with* $a \leq b$ *is either removing a number of tokens from a single pile or removing an equal number* $i > 0$ *of tokens from both piles provided that the resulting position* $(a-i, b-i)$ *satisfies both* $a - i \geq j$ *and* $b - i \geq k$*.*

Notice that the two games $\mathcal{W}_{k,k}$ and $\mathcal{W}_k$ are identical. The set $\mathcal{P}$ of $\mathcal{W}_{j,k}$ is identical to that of $\mathcal{W}_k$ for all $j \leq k$. It is important to note that the set $\mathcal{P}$ of $\mathcal{W}_{j,k}$ depends only on k.

The other variant of WYTHOFF NIM, called $\mathcal{T}_k$, is as follows:

From a position (a, b) *with* $a \leq b$*, one can either*

(i) *remove a positive number of tokens from a single pile, or*

(ii) *remove an equal positive number, say* s*, of tokens from both piles provided that* $a - s > 0$ *and*

$$\left| \left\lfloor \frac{b-s}{a-s} \right\rfloor - \left\lfloor \frac{b}{a} \right\rfloor \right| \leq k.$$

Note that the diagonal move (ii) is a restriction of the diagonal move of WYTHOFF NIM. In this move, the condition $a - s > 0$ guarantees that the ratio $(b-s)/(a-s)$ is defined. Thus, when making a diagonal move in $\mathcal{T}_k$, one must ensure that the difference between the ratios of the bigger entry over the smaller entry before and after the move must not exceed k.

Consider the special case $k = \infty$. The game $\mathcal{T}_\infty$ is the variant of WYTHOFF NIM in which the only restriction is that the diagonal move cannot make any pile empty.

The following general question has been proposed: Does there exist a variant of WYTHOFF NIM whose P-positions, except possibly a finite number, are $(A_n + k, B_n + l)$ for some fixed integers $k \neq l$ [51]?

4.13. WYTHOFF NIM *and* EUCLID. The game EUCLID, like WYTHOFF NIM, is played on two piles of tokens, though we usually play it, equivalently, on a pair of positive integers. Unlike WYTHOFF NIM, the integers remain positive throughout; a move consists of decreasing the larger number by any positive multiple of the smaller, as long as the result remains positive. The player first unable to move loses; see Lengyel [122].

Two exotic characterizations of the Sprague–Grundy function (g-function) values of EUCLID's game, in terms of the winning strategy of t-WYTHOFF NIM, are given in [43]. A novel polynomial-time algorithm for computing the g-function for EUCLID is given in Nivasch [129].

5. The game NIM(a, b); recursive solution, asymptotic, and polynomial algorithm based on the Perron–Frobenius theory

Written by Vladimir Gurvich

For any positive integer a and b, a game NIM(a, b) was introduced in [73] as follows:

Two piles contain x and y matches. Two players alternate turns. By one move, it is allowed to take x' and y' matches from these two piles such that

$$\begin{gathered} 0 \le x' \le x, \quad 0 \le y' \le y, \quad 0 < x' + y', \\ \textit{and either } |x' - y'| < a \quad \textit{or} \quad \min(x', y') < b. \end{gathered} \tag{5}$$

In other words, a player can take "approximately equal" (differing by at most $a - 1$) numbers of matches from both piles or any number of matches from one pile but at most $b - 1$ from the other. This game, NIM(a, b), extends further the game NIM(a) = NIM$(a, 1)$ considered by Fraenkel [38; 39], which, in its turn, is a generalization of the classic game NIM$(1, 1)$ introduced by Wythoff [158]; see also [24].

A position of NIM(a, b) is a nonnegative integer pair (x, y). Due to obvious symmetry, positions (x, y) and (y, x) are equivalent. By default, we will assume that $x \le y$.

Obviously, $(0, 0)$ is a unique terminal position. By definition, the player entering this position is the winner in the *normal* version of the game and (s)he is the loser in its *misére* version.

The normal version of NIM(a, b) was solved in [73]. It was shown that the P-positions (x_n, y_n) are characterized by the recursion

$$x_n = \text{mex}_b(\{x_i, y_i \mid 0 \le i < n\}), \qquad y_n = x_n + an; \qquad n \ge 0, \tag{6}$$

where $x_n \le y_n$ and the function mex$_b$ is defined as follows:

Given a finite nonempty subset $S \subset \mathbb{Z}_+$ of m nonnegative integers, let us order S and extend it by $s_{m+1} = \infty$ and by $s_0 = -b$ to get the sequence $s_0 < s_1 < \cdots < s_m < s_{m+1}$. Obviously there is a unique minimum i such that $s_{i+1} - s_i > b$. By definition, let us set mex$_b(S) = s_i + b$; in particular, mex$_b(\varnothing) = 0$.

n	0	1	2	3	4	5	6	7	8	9
x_n	0	1	3	4	6	8	9	11	12	14
y_n	0	2	5	7	10	13	15	18	20	23

n	0	1	2	3	4	5	6	7	8	9
x_n	0	1	2	4	5	7	8	9	11	12
y_n	0	3	6	10	13	17	20	23	27	30

Table 4. P-positions of NIM(a, b) for $(a = b = 1)$, and $(a = 2, b = 1)$, respectively.

It is easily seen that mex_b is well-defined and for $b = 1$ it is exactly the classic minimum excludant mex, which assigns to S the (unique) minimum nonnegative integer missing in S. Thus, $\text{mex}_1 = \text{mex}$ and (6) turns into the recursive solution of NIM$(a, 1)$ given by Fraenkel [38; 39].

The first ten P-positions (with $x \le y$) of the games NIM$(1, 1)$ and NIM$(2, 1)$ are given in Table 4.

Furthermore, Fraenkel solved the recursion for NIM$(a, 1)$ and got the following explicit formula for (x_n, y_n): Let $\alpha_a = \frac{1}{2}(2 - a + \sqrt{a^2 + 4})$ be the (unique) positive root of the quadratic equation $\hat{z}^2 + (a - 2)\hat{z} - a = 0$, or equivalently, $1/\hat{z} + 1/(\hat{z} + a) = 1$. In particular, $\alpha_1 = \frac{1}{2}(1 + \sqrt{5})$ is the *golden section* and $\alpha_2 = \sqrt{2}$. Then, it follows that for all $n \in \mathbb{Z}_+$ we have

$$x_n = \lfloor \alpha_a n \rfloor \quad \text{and} \quad y_n = x_n + an \equiv \lfloor n(\alpha_a + a) \rfloor. \tag{7}$$

This recursion implies the asymptotic

$$\lim_{n\to\infty} x_n(a)/n = \alpha_a \quad \text{and} \quad \lim_{n\to\infty} y_n(a)/n = \alpha_a + a.$$

As it was mentioned in [38], the explicit formula (7) solves the game in linear time, in contrast to recursion (6), which provides only an exponential algorithm. Yet, it looks too difficult to solve (6) explicitly when $b > 1$, because of the following bounds from [73]:

$$b \le x_{n+1} - x_n \le 2b \quad \text{and} \quad b + a \le y_{n+1} - y_n \le 2b + a. \tag{8}$$

For $b = 1$ the difference $x_{n+1} - x_n$ is either 1 or 2, and thus $\alpha_a n$ is a good approximation of x_n. When $b > 1$, it seems harder to find a similar estimate, since the bound of (8) for $x_{n+1} - x_n$ is looser. Although no closed form expressions for x_n and y_n are known in case $b > 1$ (yet), in [11], these values were computed (and thus NIM(a, b) solved) by a polynomial time algorithm based on the Perron–Frobenius theory.

n	0	1	2	3	4	5	6	7	8	9
x_n	0	2	5	9	11	14	17	21	25	27
y_n	0	3	7	12	15	19	23	28	33	36

n	0	1	2	3	4	5	6	7	8	9
x_n	0	3	8	11	15	20	26	29	33	36
y_n	0	5	12	17	23	30	38	43	49	54

Table 5. P-positions of NIM(a, b) for $(a=1, b=2)$ and $(a=2, b=3)$, respectively.

The first ten P-positions (with $x \le y$) are given in Table 5 for $(a = 1, b = 2)$ and $(a = 2, b = 3)$.

The linear asymptotic still holds, not only for $b = 1$ but for $b > 1$ as well. In [73] it was conjectured that the limits $\ell(a, b) = \lim_{n\to\infty} x_n(a, b)/n$ exist for all positive integers a, b and are irrational algebraic numbers. This conjecture was proven in [11]; moreover, the following explicit formula for the limiting values was obtained: The limit $\ell(a, b)$ exists for all positive integers a, b and, when they are coprime ($\gcd(a, b) = 1$), it is given by the fraction $\ell(a, b) = a/(r - 1)$, where $r > 1$ is a unique positive real root of the polynomial

$$P(z) = z^{b+1} - z - 1 - \sum_{i=1}^{a-1} z^{\lceil ib/a \rceil}, \tag{9}$$

which is the characteristic polynomial of a nonnegative integer $(b+1) \times (b+1)$ matrix M that depends only on a and b. For any (coprime) a and b there exist unique integers α and β such that $\alpha \ge 0$, $0 < \beta \le b$ and $a = \alpha b + \beta$. For example, if $b = 1$ then $\alpha = a - 1$ and $\beta = 1$. The entries $M_{i,j}$ of M for $i, j \in \{0, 1, \ldots, b\}$ are defined by

$$M_{i,j} = \begin{cases} \alpha & \text{if } i = 0 \text{ and } 0 \le j \le b - \beta, \\ \alpha + 1 & \text{if } i = 0 \text{ and } b - \beta < j \le b, \\ 1 & \text{if } i > 0 \text{ and } (j + a - i \bmod b) = 0, \\ 0 & \text{if } i > 0 \text{ and } (j + a - i \bmod b) \ne 0. \end{cases} \tag{10}$$

Note that, for all j, $M_{0,j} = \lfloor (a+j-1)/b \rfloor$. By the Perron–Frobenius theorem, we have $|r'| < r$ for any other root r' of $P(z)$.

The case $\gcd(a, b) > 1$ is easily reduced to the case $\gcd(a, b) = 1$ considered above, since, as it was shown in [73], $x_n(a, b)$ (and, hence $y_n(a, b)$ and $\ell(a, b)$

$i \downarrow j \rightarrow$	0	1	$\cdots$	$b-\beta-1$	$b-\beta$	$b-\beta+1$	$\cdots$	$b-1$	b
0	α	α	$\cdots$	α	α	$\alpha+1$	$\cdots$	$\alpha+1$	$\alpha+1$
1	0	0	$\cdots$	0	0	1	$\cdots$	0	0
$\vdots$	$\vdots$	$\vdots$	$\ddots$	$\vdots$	$\vdots$	$\vdots$	$\ddots$	$\vdots$	$\vdots$
$\beta-1$	0	0	$\cdots$	0	0	0	$\cdots$	1	0
β	1	0	$\cdots$	0	0	0	$\cdots$	0	1
$\beta+1$	0	1	$\cdots$	0	0	0	$\cdots$	0	0
$\vdots$	$\vdots$	$\vdots$	$\ddots$	$\vdots$	$\vdots$	$\vdots$	$\ddots$	$\vdots$	$\vdots$
$b-1$	0	0	$\cdots$	1	0	0	$\cdots$	0	0
b	0	0	$\cdots$	0	1	0	$\cdots$	0	0

Table 6. The matrix M.

as well) are uniform functions of a and b; that is,

$$\begin{aligned} x_n(ka, kb) = kx_n(a, b), \quad y_n(ka, kb) = ky_n(a, b), \\ \text{and} \quad \ell(ka, kb) = k\ell(a, b). \end{aligned} \tag{11}$$

The main results of [11] were derived with help of the Perron–Frobenius theorem and the Collatz–Wielandt formula for the nonnegative matrices; see [124, Chapter 8]. Alternatively, these results can be derived from the Cauchy–Ostrovsky theorem; see [136, Theorems 1.1.3 and 1.1.4] and verify that our polynomial $P(z)$ satisfies all conditions of the latter.

For the joint consideration of the normal and misère versions of an impartial game we refer the reader to the books [6] and [19, Chapter 12]. This approach was applied to $\text{NIM}(a, 1)$ in [39] and to $\text{NIM}(a, b)$ in [72; 73; 74].

However, the results differ in the cases $a = 1$ and $a > 1$. In the case $a = 1$ (for any $b \geq 1$) the set of P-positions P^N and P^M (of the normal and misère versions, respectively) "almost coincide". More precisely, their symmetric difference consists of only six positions:

$$P^N \setminus P^M = \{(0, 0), (b, b+1), (b+1, b)\}, \quad \text{while} \quad P^M \setminus P^N = \{(0, 1), (1, 0), (b+1, b+1)\}.$$

This result was obtained in [39] for $b = 1$ and extended for any positive integer b in [73; 74].

Also in [72; 74] the following three properties were shown for $a = 1$:

(i) from any position of $P^M \setminus P^N$ there is a move to $P^N \setminus P^M$;

(ii) from any nonterminal position of $P^N \setminus P^M$, that is, from $(b, b+1)$ or $(b+1, b)$, there is a move to $P^M \setminus P^N$;

(iii) from any position $(x, y) \notin P^N \cup P^M$, either each set P^N and P^M can be reached in one move, or none of them can.

In the case $a > 1$ (for any $b \geq 1$), the kernel of the misère version is defined by the recursion

$$\begin{aligned} \tilde{x}_n &= \text{mex}_b(\{\tilde{x}_i, \tilde{y}_i \,|\, 0 \leq i < n\}), \\ \tilde{y}_n &= \tilde{x}_n + an + 1; \quad n \in \mathbb{Z}_+. \end{aligned} \tag{12}$$

This formula was proven in [39] for $b = 1$ and extended to any positive integer b in [73]. Let us notice that formulas (6) and (12) differ just slightly. Comparing them we immediately conclude that for any integer $a > 1$ and $b \geq 1$ the sets of P-positions of the normal and misère versions are disjoint, in contrast to the case $a = 1$; see [72] for more details and, in particular, for the cases $a = 0$ or $b = 0$. According to terminology of [70; 72; 74], NIM(a, b) is a *strongly miserable* game when $a > 1$ and it is *miserable* (but not strongly) when $a = 1$.

Some conclusions and open problems. Two main recursions (6) and (12) are deterministic, yet their solutions (the kernels, or equivalently, the P-positions of the normal and misère versions of NIM(a, b)) behave in a "pseudochaotic way" when $b > 1$. For which other combinatorial games do their kernels demonstrate such behavior? It seems that the four-parametric game NIM$(a, b;\, p, q)$, introduced in [71], is a good candidate. This game is a generalization of NIM(a, b) and Larsson's NIM(a, p) from [76; 111]; see also [105; 106]. Yet, the class in question might be much larger.

Both recursions (6) and (12) can be solved by a polynomial algorithm based on the Perron–Frobenius theorem. Which other recursions can be solved in such a way?

For $b = 1$ the solutions of both recursions are given by closed formulas, while for $b > 1$ this is hardly possible.

Cases $a = 1$ and $a > 1$ also differ substantially. In the first case, the symmetric difference $P^N \,\Delta\, P^M$ consists of only six positions, while in the second case these two sets are disjoint, $P^N \cap P^M = \varnothing$. In [72; 74] such two types of games are named miserable and strongly miserable, respectively, and simple characterizations for both classes are obtained.

Do recursions (6) and (12) or similar ones have other applications, perhaps beyond game theory?

Acknowledgements. This study has been partially funded by the Russian Academic Excellence Project '5-100'. The author is also thankful to Vladimir Oudalov, Endre Boros, and Urban Larsson for many helpful remarks.

6. Sprague–Grundy values and preserving P-positions

Written by Nhan Bao Ho

Recall the definition of Sprague–Grundy values [69; 147] for impartial normal-play games: the value of a position x is the smallest nonnegative integer not in the set of values of the positions that can be reached from x by one move. Thus, the recurrence starts with the value zero of each terminal (final) position. Moreover, a position is a P-position if and only if it has value zero.

Section 6.1 was composed by N. B. Ho and U. Larsson, and Section 6.2 by N. B. Ho.

6.1. ***Additive periodicity of* WYTHOFF NIM*'s Sprague–Grundy function.*** We give an overview of Landman's FSM-based proof of arithmetic periodicity of Sprague–Grundy values of WYTHOFF NIM [103]. An FSM (or finite state machine) is a restricted Turing machine which has a finite memory, and moreover it can only move in one direction, and it cannot print. Its simple features implies that its states must eventually enter a cycle, at least if there is no structured input (i.e., the new state depends only on the previous state).

We are concerned with the Sprague–Grundy function G of WYTHOFF NIM along a fixed y-coordinate, indicated with the horizontal dashed red line in Figure 3, and we want to show arithmetic periodicity. It suffices to show that an FSM, with no input, can compute a function $H(x, y) = G(x, y) - x + 2y$. As we remarked in the previous paragraph, then H must be periodic and hence G must be arithmetic periodic. One can prove that G satisfies the inequalities $x - 2y \le G(x, y) \le x + y$, and hence H is bounded, as a function of x, $0 \le H(x, y) \le 3y$. As an FSM has finite memory, this bound is necessary. Moreover, Landman shows that it suffices for establishing periodicity. For each $0 \le y_i \le y$, it suffices to register the H-values in the sets L'_i (bounded left), D_i (down) and S_i (slant), as in Figure 3.

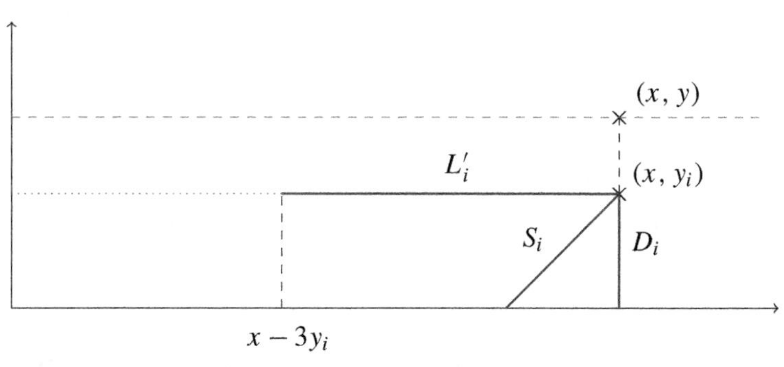

Figure 3. An FSM-analysis of Grundy values of WYTHOFF NIM.

These can be stored in three bit strings of length $3y_i + 1$, where a "1" at index j indicates membership of H-value j. Then the NOR operator finds the smallest index for which there is no "1" in either string, and $G(x, y_i)$ can be computed via $H(x, y_i)$ for each i. Of course, the machine is never concerned with the actual Grundy values, and even if modified with a printing ability, it could not print out G-values because they will become arbitrarily large and hence no finite memory could represent them.

To update the bit strings to binary code $H(x+1, y_i)$, the machine shifts the entries and drops the left most entry. Hence an FSM can simulate the H-function along any given y-coordinate, and so H must be ultimately periodic. The number of bit strings is $O(y)$ and each bit string is of length $O(y)$. Hence the machine requires at most $O(2^{y^2})$ states. Thus the Sprague–Grundy function is ultimately arithmetic periodic along any given y-coordinate (or x-coordinate, by symmetry of WYTHOFF NIM).

Note that a key ingredient for Landman's method is the establishment of the bounds for $G(x, y)$. Using similar bounds, one can apply this technique for other 2-pile NIM-like games. One such example was analyzed in [81].

6.2. ***Two variants that preserve P-positions of* WYTHOFF NIM.** Duchêne et al. [27] characterize modifications of WYTHOFF NIM that preserves its P-positions (also see Section 4.6). In the context of that work, modifications of WYTHOFF NIM are invariant in the sense that if the move $(a, b) \to (a-i, b-j)$ is allowed then the move $(a, b) \to (a-j, b-i)$ is also allowed, provided that $a \geq j$ and $b \geq i$.

In [77], the author studies two noninvariant modifications, one extension and one restriction, of WYTHOFF NIM preserving its P-positions.

> *In the restriction called* $\mathcal{R}$-WYTHOFF, *the constraint is that removing tokens from the smaller pile is not allowed.*

In other words, from an $\mathcal{R}$-WYTHOFF position (a, b) with $a \leq b$, one can move either $(a, b) \to (a, b-i)$ or $(a, b) \to (a-i, b-i)$. The author also proves that there is no restriction of $\mathcal{R}$-WYTHOFF that preserves $\mathcal{R}$-WYTHOFF's P-positions.

> *In the extension called* $\mathcal{E}$-WYTHOFF, *along with original moves of* WYTHOFF NIM, *there exists an extra move of the form* $(a, b) \to (a-k, b-l)$ *in which* $a \leq b$ *and* $l < k$.

The author also establishes positions whose Sprague–Grundy values are 1 of both games as follows:

$$\{(2,2), (4,6), (\lfloor \phi n \rfloor - 1, \lfloor \phi n \rfloor + n - 1) \mid n \geq 1, n \neq 2\} \quad \text{for } \mathcal{R}\text{-WYTHOFF};$$

$$\{(\lfloor \phi n \rfloor - 1, \lfloor \phi n \rfloor + n - 1) \mid n \geq 1\} \quad \text{for } \mathcal{E}\text{-WYTHOFF}.$$

These positions are remarkably close to the P-positions of WYTHOFF NIM, being determined by a translation except for three initial positions in the first case.

The author also proves the following property of Sprague–Grundy functions for both games: for any nonnegative integer a and Sprague–Grundy value g, there exists b such that $\mathcal{G}(a, b) = g$. Actually, this property is equivalent to the following feature of the sequence of positions whose Sprague–Grundy values are g. Let $((a_n, b_n))_{n \geq 0}$ be the sequence of positions whose Sprague–Grundy values are g, in which $a_i < a_j$ if $i < j$. Then the set $\{a_n, b_n \mid n \geq 0\}$ contains every nonnegative integer.

Another feature of Sprague–Grundy values of both $\mathcal{R}$-WYTHOFF and $\mathcal{E}$-WYTHOFF NIM is the additive periodicity of the sequence $(\mathcal{G}(a, n))_{n \geq 0}$ in the sense that there exist n_0 and $p > 0$ such that $\mathcal{G}(a, n + p) = \mathcal{G}(a, n) + p$ for all $n \geq n_0$. This type of periodicity of WYTHOFF NIM is discussed in Section 6.1.

7. The Wythoff array and associated arrays and sequences

Written by Clark Kimberling

In response to an invitation, the author surveys the Wythoff array and its many associates, concentrating on his own contributions and those of others.

7.1. *The Wythoff array.* During the first year of existence of the Fibonacci Association, one of the founders published a short article [2] in the first volume of *The Fibonacci Quarterly*. There, Brother Alfred Brousseau discusses an ordering of the set of *all* Fibonacci sequences of positive integers. He concludes with these words: "The above approach in representing Fibonacci sequences and ordering them is all by way of suggestion. There are doubtless other ways of achieving the same objective. It would be very helpful if additional proposals were aired before a final standard is adopted."

A second ordering of the positive Fibonacci sequences appears in Kenneth Stolarsky's one-page article [149], in which he introduces an ordering in a form now known as a *Stolarsky array*. Three years later, David Morrison, then a student at Harvard University, published another ordering [128]. If — to borrow Brother Alfred's words — there is a "final standard", this must be it. In Morrison's ordering, the Fibonacci sequences appear as rows of the array shown in Table 7.

Morrison named this array after Wythoff because the rows consist of Wythoff pairs — these being the winning pairs for Wythoff's game. For example, in row 1, the Wythoff pairs are $(1, 2)$, $(3, 5)$, $(8, 13)$, ...; in row 2, they are $(4, 7)$, $(11, 18)$, $(29, 47)$, ...; and so on. The Wythoff pairs are given by $(\lfloor n\alpha \rfloor, \lfloor n\alpha^2 \rfloor)$, where $\alpha = \frac{1}{2}(1 + \sqrt{5})$, the golden ratio. Properties of the Wythoff array W include the following:

(1) Every row is a Fibonacci sequence; i.e., the recurrence $x_n = x_{n-1} + x_{n-2}$ holds.

(2) The rows extend indefinitely to the left by "precursion" ($x_{n-2} = x_n - x_{n-1}$), resulting in an array that contains every Fibonacci sequence of integers [83].

The array W is an interspersion and a dispersion. The first of these means, briefly, that every row is interspersed by every other row, and the second means that the first column can be used to disperse its complement using certain iterated compositions. More precise definitions [84] follow:

An array $A = A(i, j)$ of positive integers is an *interspersion* if

(I1) every positive integer occurs exactly once in A;

(I2) every row of A is an increasing sequence;

(I3) every column of A is an increasing sequence;

(I4) (u_i) and (v_j) are distinct rows of A, and if i and h are indices for which $u_i < v_h < u_{i+1}$, then $u_{i+1} < v_{h+1} < u_{i+2}$.

To define *dispersion,* suppose that s is an increasing sequence of positive integers, that the complement t of s is infinite, and that $t(1) = 1$. The dispersion of s is the array whose n-th row is

$$t(n), s(t(n)), s(s(t(n))), s(s(s(t(n)))), \dots$$

The Wythoff array is the dispersion of the complement of its first column; indeed, the odd-numbered columns form the lower Wythoff sequence, and the rest of W forms the upper Wythoff sequence. The main theorem on interspersions and dispersions is that they are equivalent [84].

Another property of W is observed when, for each n, we write the number of the row of W that contains n, resulting in this sequence:

$$f = (1, 1, 1, 2, 1, 3, 2, 1, 4, 3, 2, 5, 1, 6, 4, 3, 7, 2, 8, 5, 1, \dots).$$

1	2	3	5	8	13	21	34	55	89	144	⋯
4	7	11	18	29	47	76	123	199	322	521	⋯
6	10	16	26	42	68	110	178	288	466	754	⋯
9	15	24	39	63	102	165	267	432	699	1131	⋯
12	20	32	52	84	136	220	356	576	932	1508	⋯
14	23	37	60	97	157	254	411	665	1076	1741	⋯
17	28	45	73	118	191	309	500	809	1309	2118	⋯
19	31	50	81	131	212	343	555	898	1453	2351	⋯
⋮	⋮	⋮	⋮	⋮	⋮	⋮	⋮	⋮	⋮	⋮	

Table 7. The Wythoff array.

1	2	5	13	34	89	233	610	1597	⋯
3	7	18	47	123	322	843	2207	5778	⋯
4	10	26	68	178	466	1220	3194	8362	⋯
6	15	39	102	267	699	1830	4791	12543	⋯
8	20	52	136	356	932	2440	6388	16724	⋯
9	23	60	157	411	1076	2817	7375	19308	⋯
⋮	⋮	⋮	⋮	⋮	⋮	⋮	⋮	⋮	

Table 8. The Wythoff difference array.

Deleting the first occurrence of each positive integer leaves the same sequence f. Because this upper-trim operation can be repeated indefinitely and always returns f, this sequence, and any other that arises similarly from an interspersion, is called a *fractal sequence* [89]. The lower-trim of f, obtained by deleting all the 0s from $f - 1$, is also a fractal sequence.

There are several modifications of the Wythoff array W that appear in the Online Encyclopedia of Integer Sequences (OEIS) [132]. Aside from W itself, which is indexed as $A033513$ in the OEIS, the left-justified Wythoff array, $A165357$, formed from W by left extending each row to a pair (a, b) such that $a > b$, has the following property: every (a, b) satisfying $a > b \geq 0$ occurs exactly once, and every (c, d) satisfying $0 \leq c \leq d$ occurs exactly once.

A second array obtained from W is the Wythoff difference array D ([93] and $A080164$), formed by differences between Wythoff pairs in W; see Table 8.

Properties of D include the following:

(1) The difference between adjacent terms in every column is a Fibonacci number.

(2) Every term of column 1 of W is in column 1 of D.

(3) Every term in a row of D, except the first, is in the corresponding row of W.

(4) D is an interspersion.

(5) D is the dispersion of the upper Wythoff sequence $A001950$, whereas W is the dispersion of the lower Wythoff sequence $A000201$.

7.2. ***Zeckendorf arrays and the Wythoff array.*** Every positive integer n is a sum of Fibonacci numbers, no two of which are consecutive. This unique sum is known as the Zeckendorf representation of n. The *Zeckendorf array* $Z = Z(i, j)$ is defined [87] as follows: column j of Z is the increasing sequence of all n whose Zeckendorf representation the least term is F_{j+1}. For example, row 1 of Z is given by

$$z(1,1) = 1 = F_2, \quad z(1,2) = 2 = F_3, \ldots, z(1,j) = F_{j+1}, \ldots$$

When working with Zeckendorf representations, it is often helpful to refer to a shift function s defined from a Zeckendorf representation as follows:

$$n = \sum_{i=1}^{\infty} c_i F_{i+1} \Longrightarrow s(n) = \sum_{i=1}^{\infty} c_i F_{i+2}.$$

Using s, it is proved [87] that the Zeckendorf array is identical to the Wythoff array. Related developments, including higher-order Zeckendorf arrays and Zeckendorf/Wythoff trees, are found in Lang's work [104]; Bicknell-Johnson's paper [10]; a paper on Fibonacci Phyllotaxis [146] by Spears, Bicknell-Johnson, and Yan; Cooper's work [23]; and Ericksen and Anderson's paper [35]. Other related arrays are discussed in Hegarty and Larsson [76] and Kimberling [85; 86; 88; 97].

7.3. *Lower and upper Wythoff sequences.* The winning solutions of Wythoff's game are the previously mentioned pairs $(\lfloor n\alpha \rfloor, \lfloor n\alpha^2 \rfloor)$. Separating the components gives the lower Wythoff sequence, $L = (\lfloor n\alpha \rfloor) = A000201$ and the upper Wythoff sequence $U = (\lfloor n\alpha^2 \rfloor) = A001950$. Clearly, the terms of L fill the odd numbered columns of the Wythoff array, and those of U, the even numbered columns.

It is easy to write out terms of L and U without reference to an irrational number or Wythoff's game. Consider the rows in Table 9.

The first step is to write row 1 and then to place 1 below the 1 in row 1. Add the two numbers to get $U(1) = 2$. Thereafter, take each $L(n)$ to be the least positive integer not yet in rows 2 and 3, and take $U(n) = n + L(n)$.

A related procedure, from a comment by Roland Schroeder at *A*000201, produces L from a Mancala-type game as follows: n stacks of chips are aligned and numbered from left to right as #1, #2, #3, etc., with stack #n consisting initially of n chips. One step in the game consists of transferring from the leftmost stack all of its chips so that the stacks to the right each gain 1 chip until one of two things happens: either there are no more chips, or otherwise, the leftover chips are used to create new stacks, one chip per stack, lined up to the right of the stacks already present. The game continues until there are n stacks, no two of which have the same number of chips. The number of steps for the whole game is $L(n)$.

A variant of the Schroeder–Mancala game is described in a proposal by Ron Knott, with a solution by Sam Northshield [101]: "As an infinite Mancala game,

n	1	2	3	4	5	6	7	8	9	10	11	12	⋯
$L(n)$	1	3	4	6	8	9	11	12	14	16	17	19	⋯
$U(n)$	2	5	7	10	13	15	18	20	23	26	28	31	⋯

Table 9. Values of n, L, and U.

suppose a line of pots contains pebbles, 1 in the first, 2 in the second, and n in the n-th, without end. The pebbles are taken from the leftmost nonempty pot and added, one per pot, to the pots to the right. Prove that the number of pebbles in pot n as it is emptied is $\lfloor n\varphi \rfloor$, where φ is the golden ratio $\frac{1}{2}(1+\sqrt{5})$."

Another method for generating both L and U is by sending light rays through a certain 2-dimensional array of half-silvered mirrors (the type used in the Michelson–Morley experiment) and seeing which integer points on the coordinate axes become illuminated. See Porta and Stolarsky [135].

Yet another method for generating both L and U is to decree that $L(1) = 1$, that $a(n+1) = a(n)+2$ if $a(n)$ is already determined, and that $a(n+1) = a(n)+1$ otherwise. This procedure can be generalized (e.g., [94], $A184117$) to produce Beatty sequences other than L and U. See also [96] and [98].

Quite a different approach to L and U is to arrange in increasing order all the numbers j/α and k/α^2 (or equivalently, $j\alpha$ and k), so that the list begins with

$$\frac{1}{\alpha^2}, \frac{1}{\alpha}, \frac{2}{\alpha^2}, \frac{3}{\alpha^2}, \frac{2}{\alpha}, \frac{4}{\alpha^2}, \frac{3}{\alpha}, \frac{5}{\alpha^2}, \frac{6}{\alpha^2}, \frac{4}{\alpha}, \frac{7}{\alpha^2}. \frac{8}{\alpha^2}, \frac{5}{\alpha}, \frac{9}{\alpha^2}, \frac{5}{\alpha}.$$

Here, for every n, the position of n/α is $\lfloor n\alpha^2 \rfloor$, and that of n/α^2 is $\lfloor n\alpha \rfloor$. The same method works for many pairs of irrational numbers: if $r > 1$ and $1/r + 1/s = 1$, then the positions of n/r and n/s in the joint ranking of all j/r and k/s are $\lfloor ns \rfloor$ and $\lfloor nr \rfloor$, respectively. This may be the shortest route for introducing pairs of Beatty sequences and proving that they partition the positive integers. The method extends to more than two sequences; e.g., Paul Hanna's three-way splitting of $\mathbb{N}$ using a zero γ of $\gamma^3 = \gamma^2 + \gamma + 1$, at $A184820$–$A184822$. See also $A187950$.

It would be of interest to see more of algebraic irrationalities of degree greater than 2 playing a role in this subject. See also Section 3.2, [27], and [137]. But it is the case that the original Wythoff pairs have in fact a (somewhat hidden) quartic algebraic structure. See [150] and [139].

The lower and upper Wythoff sequences both occur in connection with both the greedy and lazy Fibonacci representations of positive integers. We have already discussed the "greedy" case, since the greedy algorithm simply finds the Zeckendorf representation, for which the numbers in L are those whose representation ends with an even number of 0s, and, of course, those in U that end with an odd number of 0s.

Another way to find Zeckendorf representations starts with the sequence $(n)_{\text{base 2}}$: simply delete every term that contains "11", so that the remaining terms comprise $(n)_{\text{Zeckendorf}}$. Analogously, deleting every term that contains "00" leaves $(n)_{\text{lazy}}$. It can be shown that

$$(\#\text{ terms in } (n)_{\text{Zeckendorf}}) \leq (\#\text{ terms in } (n)_{\text{lazy}}),$$

as in $A095792$; indeed, the Zeckendorf representation is often called the minimal Fibonacci representation, and the lazy, the maximal Fibonacci representation.

An interesting way to present lazy Fibonacci representations is as a graph consisting of two components, L^* and U^*, each being a binary tree. The tree L^* is rooted in 1, of which the children are $1+2$ and $1+3$. The children of $1+2$ are $1+2+3$ and $1+2+5$, and the children of $1+3$ are $1+3+5$ and $1+3+8$; in general, the children of each

$$m = F_{i_1} + F_{i_2} + \cdots + F_{i_k}, \quad \text{where} \quad i_1 < i_2 < \cdots < i_k,$$

are $m + F_{i_k+1}$ and $m + F_{i_k+2}$. The same rule of generation applies starting with the root 2 of U^*. The numbers in L^* and U^*, taken in order as generated, form the sequences $A255773$ and $A255774$, which are permutations of L and U, respectively; see also $A095903$.

Among the zero-one sequences known as the infinite Fibonacci word is

$$A003849 = (0, 1, 0, 0, 1, 0, 1, \ldots),$$

definable as the fixed point of the morphism $0 \to 01$, $1 \to 0$, starting with 0. The lower Wythoff sequence L tells the positions of 0 in $A003849$, and U the positions of 1. Let S denote the infinite Fibonacci word $A003849$, and let $S(n) = (s(1), s(2), \ldots, s(n))$ be the initial segment of S that has length n. Then $S(n)$ occurs infinitely many times in S. We ask where each appearance starts and answer as follows: for $n = 1$, the segment consists solely of 0, and starts at positions given by L; for $n = 2$ and $n = 3$, the segment starts at positions given by the Wythoff AA numbers $A003622$; for $n = 4, 5, 6$, the segment starts at the Wythoff AAA numbers $A134859$; for the next F_5 numbers ($n = 7, 8, 9, 10, 11$), the segments start at positions given by the Wythoff AAAA numbers $A151915$, and so on. See $A246354$.

Other appearances of U and L are as solutions to complementary equations, as introduced in [91]; that is, equations that can be put into the form $f(a, b) = 0$, where a and b are complementary sequences of positive integers. Four examples, for which the initial condition is $a(1) = 1$ and the unique solution is $a = L$ (or equilvalently, $b = U$) are shown here:

(1) $a(a(n)) = b(n) - 1$,

(2) $a(b(n)) = a(n) + b(n)$,

(3) $b(a(n)) = a(n) + b(n) - 1$,

(4) $b(b(n)) = a(n) + 2b(n)$.

These four equations are used as lemmas for developing more elaborate complementary equations in which the columns of the Wythoff array play a central role

[50; 91; 92; 29; 46; 45; 95]. In particular, Fraenkel introduces the game of *flora* in [45].

7.4. *Wythoff-related trees.* Let T_1 be the tree generated by these rules: the root is 1, and for each node x, the children are $\lfloor nx \rfloor$ and $\lfloor nx^2 \rfloor$. The first four generations of T_1 are given by

$$\{1\}, \quad \{2\}, \quad \{3, 5\}, \quad \{4, 7, 8, 13\}, \quad \{6, 10, 11, 18, 12, 20, 21, 34\}.$$

In T_1, every Wythoff pair except $(1, 2)$ occurs as a pair of children. Taken in order of appearance in T_1, the numbers comprise $A074049$.

Next, let T_2 be the tree, essentially $A052499$, generated [90] as follows: $1 \in T_2$, and if $x \in T_2$, then $2x \in T_2$ and $4x - 1 \in T_2$. When all the terms of T_2 are arranged in increasing order and the initial 1 is removed, the remaining even numbers are in positions $1, 3, 4, 6, 8, \ldots$; i.e., the lower Wythoff sequence L and the odd numbers are in positions given by U.

The next tree T_3 contains every integer exactly once. Here, $0 \in T_3$, and if $x \in T_3$, then $2x \in T_3$ and $1 - x \in T_3$, and duplicates are deleted as they occur. As in $A232723$, the numbers in order of generation are

$$0, 1, 2, 4, -1, 8, -3, -2, 16, -7, -6, -4, 3, 32, \ldots$$

The even integers occupy the positions given by L, and the odds by U.

Another tree T_4 gives an ordering of the positive rational numbers. To generate T_4, start with 1, and if $x \in T_4$, then $x + 1 \in T_4$ and $1/x \in T_4$, and duplicates are deleted as they occur. The first few fractions in T_4 are

$$\tfrac{1}{1}, \tfrac{2}{1}, \tfrac{3}{1}, \tfrac{1}{2}, \tfrac{4}{1}, \tfrac{1}{3}, \tfrac{3}{2}, \tfrac{5}{1}, \tfrac{1}{4}, \tfrac{4}{3}, \tfrac{5}{2}, \tfrac{2}{3}, \tfrac{6}{1}.$$

Here, the positions of the positive integers comprise row 1 of the Wythoff array W and the positions of the numbers $n + \frac{1}{2}$ comprise row 2. In general, the positions of denominators ($A226080$) congruent to r mod n, where $0 < r < n$ and $\gcd(n, r) = 1$, comprise a row of W.

In T_4, taken as a sequence, the fractions ≤ 1 occupy positions given by $U - 1$, and those > 1 by $L - 1$. Other trees, including trees consisting of all the rational numbers, all the Gaussian integers, and all the Gaussian rational numbers, are introduced in [99]. See $A226080$ for an overview of such trees.

8. Goldilocks principle in combinatorial games

Written by Urban Larsson

WYTHOFF NIM interacts well with number theory, computer science, physics, biology and more. Its generosity lets us ask questions about rearranging game rules, or we may begin by looking at Wythoff-type sequences or recurrences, and search for games with those sequences as P-positions. We show that new theories get revealed by altering old patterns, keeping some properties and shifting others — usually not too many at a time, just enough to be able to recognize some new features, and not losing sight of known ones. The game truly inspires a multitude of experiments, founded in its connection with the golden mean.[3]

8.1. IMITATION NIM *and* BLOCKING WYTHOFF NIM. In this section we describe various dynamic restrictions of 2-heap NIM and WYTHOFF NIM respectively. In the first variation, the previous move(s) gives the restriction whereas in the subsequent examples the restriction is a temporary blocking of options imposed by the other player (but independently of the previous moves).

The game of IMITATION NIM [105] is a move-size dynamic restriction of the classical game of NIM on two piles:

> *Suppose that the previous player removed x tokens from the smaller heap (any heap if they have equal size). Then the next player may not remove x tokens from the larger heap. In case of a starting position (with no move dynamic restriction), any nim-type move is allowed.*

Notice that by this move restriction, the winning strategy of 2-pile NIM is altered. For example, the player who moves from the position $(1, 1)$ will lose in NIM, but win in IMITATION NIM (independent of previous move). It turns out that, regarded as starting positions, the P-positions correspond to those of WYTHOFF NIM. To begin to see this, note that the position $(1, 2)$ does not have a winning move (independent of move-size dynamics). Namely the possible options are the N-positions $(0, 2; 1)$, $(1, 0)$, $(1, 1)$. Here, the move-size dynamic notation $(x, y; z)$, with $x \le y - z$, means that the nim move $(x, y - z)$ is not allowed. Note for example that $(0, 1; 1)$ is a terminal P-position, but $(0, 1)$ is an N-position. Hence, the P-equivalence with WYTHOFF NIM only regards the starting positions of IMITATION NIM.

The game generalizes nicely. Suppose that $k - 1$ consecutive imitations from one and the same player are allowed, but not the k-th one. For example, with $k = 2$ and $0 < x \le y$, suppose that the three most recent moves were $(x, y) \to (x - z, y) \to (x - z, y - z) \to (x - z - w, y - z)$, alternating between the two players. Then precisely the move to $(x - z - w, y - z - w)$ is prohibited.

[3] According to Wikipedia there are many interpretations of Goldilocks principle, of which I pick just two: "In ancient Greek philosophy, especially that of Aristotle, the golden mean or golden middle way or Goldilocks Theory is the desirable middle between two extremes, one of excess and the other of deficiency" and "In cognitive science and developmental psychology, the Goldilocks principle refers to an infant's [my remark: or scientist's?] preference to attend to events which are neither too simple nor too complex according to their current representation of the world." My tentative interpretation: if we depart too far from the rules of Wythoff Nim, we see either unordered chaos or else trivial regularity, but in its neighborhood there is a rich and thriving environment.

The P-positions of this k-IMITATION NIM correspond to those of a variation of WYTHOFF NIM with a certain *k-blocking* on the diagonal options [76].

The author has studied three *blocking* variations of WYTHOFF NIM. The first game has the same set of P-positions as those of k-IMITATION NIM.

> *Let k be a positive integer. The first blocking variation is as* WYTHOFF NIM, *with one exception: the previous player may, before the current player moves, block off* $k-1$ *of the diagonal type options and declare them forbidden. When a player has moved, any blocking maneuver is forgotten.*

Thus the parameter $k = 1$ gives WYTHOFF NIM. The P-positions of an m-Wythoff type generalization of this game approximate closely pairs of complementary homogeneous Betty sequences of the form

$$\left(\left\lfloor n\frac{\sqrt{m^2+4k^2}+2k-m}{2k}\right\rfloor, \left\lfloor n\frac{\sqrt{m^2+4k^2}+2k+m}{2k}\right\rfloor\right),$$

for positive integers n. However, there is no Beatty-type solution to this game for $k > 1$ [105; 55, Appendix].

Combinatorial games with a blocking maneuver, or so-called Muller Twist, were proposed via the game Quarto in "Mensa Best Mind Games Award" in 1993. Later the idea appeared in the literature [80; 145; 61].

Since a blocking maneuver on the diagonal type options gives rise to interesting sequences of integers [76], we set out to find other natural blocking maneuvers of WYTHOFF NIM. In another study [111], blocking is instead exclusive to the NIM-type options (and three P-equivalent games are defined). The P-positions of these games can be described *exactly* via k complementary pairs of (nonhomogenous) Beatty sequences for all blocking parameters, generalizing simultaneously the P-positions of Connell's and Holladay's games (in the following formulas, put $m = 1$, and $k = 1$ respectively). For positive integers x, let

$$\phi(x) = \frac{2-x+\sqrt{x^2+4}}{2}.$$

The Beatty sequences are $a = (a_n)$ and $b = (b_n)$, for nonnegative integers n, where

$$a_n = a_n^{m,k} = \left\lfloor \frac{n\phi(mk)}{k} \right\rfloor \quad \text{and} \quad b_n = b_n^{m,k} = \left\lfloor \frac{n(\phi(mk)+mk)}{k} \right\rfloor.$$

For a third variation [107], blocking is allowed on any option of WYTHOFF NIM:

> *Let k be a positive integer. The game of* k-BLOCKING WYTHOFF NIM *is played as* WYTHOFF NIM, *except that, before the current player*

moves, the previous player may block off at most $k-1$ options. After a move, any blocking maneuver is forgotten.

In the case of an unrestricted blocking maneuver, an exact formula for the P-positions is known for the case where at most one option may be blocked. For this game, the *upper* P-positions have *split* into two sequences of P-positions, one with slope ϕ, similar to the Beatty type formula for WYTHOFF NIM, and the other with slope 2. A position (x, y) is *upper* if $y \geq x$. Somewhat surprisingly, there is a closed formula expression for the P-positions of the game with at most two blocked options, but then the game becomes harder.

Consider a game with $k = 5$, where the queen is now at $(3, 3)$; yellow in Figure 4. It is player A's turn, and player B is blocking the four positions

$$\{(0, 0), (1, 1), (0, 3), (3, 0)\}$$

(dark brown and light olive). This leaves A with the options

$$\{(3, 1), (3, 2), (2, 2), (2, 3), (1, 3)\}$$

(each is black or blue). Regardless of which of these A chooses, B will then have at least five winning moves to choose from (ones marked yellow, or light, medium, or dark olive). These are winning moves because it is possible when moving there to block all possible moves of the other player and thereby immediately win. Therefore player B will win.

In general this game is hard to analyze, but nevertheless quite interesting due to an elegant self-organization for large blocking parameters, and due to the fact that a generalization of the outcomes can be simulated by a cellular automaton [21]; see Figure 5. We cite, from the abstract:

> "As k becomes large, parts of the pattern of winning positions converge to recurring chaotic patterns that are independent of k. The patterns for large k display a surprising amount of self-organization at many scales."

This self-organization is illustrated in Figure 6. The cellular automaton (CA for short) computes the "palace numbers"$-k$ (the palace number of a given position counts the number of P-positions among the options of Wythoff's queen). It updates diamond shaped cells in parallel, and time is running southeast; see Figure 6.

The main result connecting the CA to the combinatorial game [21] is this: the k-BLOCKING WYTHOFF NIM position (x, y) is a P-position if and only if the CA gives a negative value at that position when the CA is started from an initial condition defined by

$$CA(x, y) = \begin{cases} k & x < 0 \text{ and } y < 0, \\ 0 & x < 0 \text{ and } y \geq 0, \\ 0 & x \geq 0 \text{ and } y < 0. \end{cases}$$

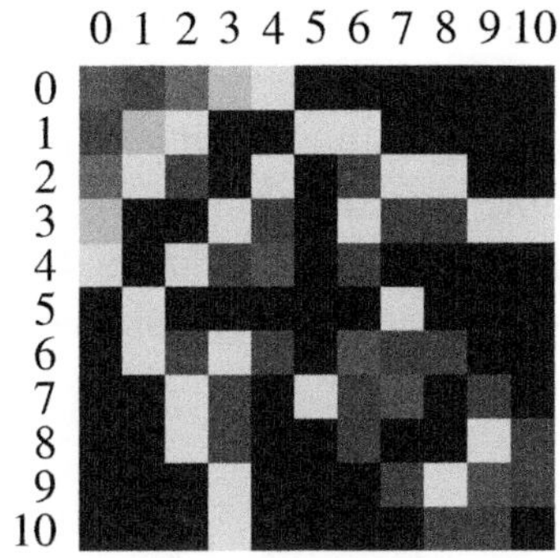

Figure 4. 5-BLOCKING WYTHOFF NIM: (0, 0) is the upper left position on this 11×11 chessboard, and the queen is allowed to move north, west, or northwest. The *palace number* is the number of *palaces* (P-positions) visible, as shown in this picture, and the *surplus number* is the palace number minus k (in this case $k = 5$): -5 = brown, -4 = dark olive, -3 = olive, -2 = light olive, -1 = yellow, 0 = black, 1 = blue, 2 = indigo, and in general, yellowish-olive colors are winning moves (the queen wants to move to her palaces and eat olives) and bluish colors are losing moves.

The behavior of our game/CA generalizes Wythoff Nim in various ways, but mostly proofs are hard to catch. For example: in Figure 6 (the leftmost picture), the uppermost white colored "threads" of P-positions have horizontal thickness "a small Fibonacci number", and, for each such thread, the number of P-positions to the left remains constant.

8.2. ***A* GENERALIZED DIAGONAL WYTHOFF NIM *and splitting beams of P-positions.*** The P-positions of NIM lie on the single *beam* of slope 1, whereas those of WYTHOFF NIM lie on the beams of slopes ϕ and ϕ^{-1}. Therefore, going from NIM to WYTHOFF NIM has *split* the single *P-beam* in NIM into two new P-beams for WYTHOFF NIM of distinct slopes. Let p, q be positive integers. If we adjoin to the game of WYTHOFF NIM new moves of the form (pt, qt) and (qt, pt), for all positive integers t, will the upper P-positions of the new game, denoted (p, q)-GDWN, split once again into two new distinct slopes?

The first paper on a GENERALIZED DIAGONAL WYTHOFF NIM (GDWN) [108] proves that the ratio of the coordinates of the upper P-positions of this game do not have a unique accumulation point if $(p, q) = (1, 2)$ (see also the rightmost picture in Figure 10). Via experimental results it was also conjectured that the upper P-positions of (p, q)-GDWN "split" if and only if (p, q) is either a Wythoff pair or a dual Wythoff pair, that is of the form $(p, q) = (\lfloor \phi n \rfloor, \lfloor n\phi^2 \rfloor)$ or $(\lceil \phi n \rceil, \lceil n\phi^2 \rceil)$, for n a positive integer.

The conjecture was subsequently proved [110] for $(1, 2)$-GDWN. Two discoveries made this possible. We sketch the idea: Suppose that the sequences (a_i) and (b_i) satisfy a certain property W (roughly $\{(a_i, b_i)\}$ represents the upper P-positions of some WYTHOFF NIM *extension*, where "extension" means that

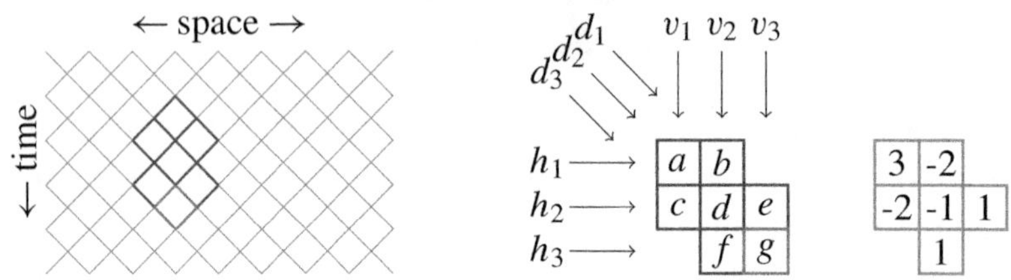

Figure 5. The updates of BLOCKING THE QUEEN's cellular automaton (for references to colors, see an online version of the survey). The red g-cell's value is computed according to the formula $g = a - b - c + e + f + p$, where a, b, c, e, f are the values of previous states, and p represents the total contribution of the *palace compensation terms*. The green squares (to the right) correspond to the blue cells (in the middle) and show the palace compensation terms. If a blue cell contains a palace (a negative value), then the corresponding palace compensation term is "added"; for example, if $a = -1, b = -2, c = 0, d = 2, e = -1, f = 0$, then $p = 3 - 2 + 1 = 2$ which gives $g = -1 + 2 - 1 + p = 2$. The initial condition of our cellular automaton is given in text.

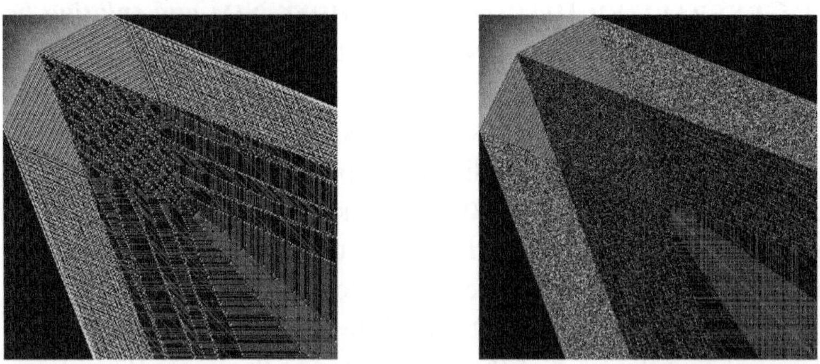

Figure 6. Self-organized regions in 100-BLOCKING WYTHOFF NIM (left, 300×300 region) and 1000-BLOCKING WYTHOFF NIM (right, 3000×3000 region). Here, the upper left corner is the $(0, 0)$-position.

Figure 7. The fill-rule properties of the rules $(0, t)$, (t, t) and $(3t, 5t)$ respectively, computed for the game (3,5)-GDWN. Apart from the striking dynamics of the fill rules, initial fluctuations give rise to "quasi-log-periodicity" visible in the two left most pictures, with a scale factor ≈ 1.48. For a further discussion and some conjectures, see "Geometric Analysis of a Generalized Wythoff Game" in these proceedings.

new "moves" may be adjoined but no moves have been removed). Then

$$\frac{\#\{i > 0 \mid a_i < n\}}{n} \geq \phi^{-1} - o(1) \tag{13}$$

and

$$\frac{\#\{i > 0 \mid b_i < n\}}{n} \leq \phi^{-2} + o(1),$$

where n tends to infinity. The bound (13) was used to prove that there is a positive lower asymptotic density of x-coordinates of P-positions above the line $y = 2x$, and it was demonstrated that this implies that the upper P-positions $\{(a_n, b_n)\}$ of $(1, 2)$-GDWN split.

Moreover, the conjecture is that there are precisely two accumulation points for the upper P-beams, namely to the ratio of coordinates $1.477\ldots$ and $2.247\ldots$ respectively; see Figure 10, the rightmost picture. The conjecture has been further strengthened in the *Linear Nimhoff* project (geometric analysis of a generalized Wythoff game) in this book. This leads us to dwell a bit upon the idea of a "fill-rule property" of a class of (Wythoff) Nim extensions. Central to the hypothesis built in that paper is a renormalization idea from physics. Figure 7 shows some of $(3, 5)$-GDWN's behavior: N-positions obtained by given fill-rules in the respective regions are colored black. The P-positions are not visible, but we can see their impact on the horizontal, $(1, 1)$-diagonal and $(3, 5)$-diagonal N-positions respectively. Each game rule independently and completely fills one designated region with N-positions (the regions are obtained by renormalization equations).

8.3. **Maharaja Nim** ***and a dictionary process.*** Maharaja Nim [120] is an *extension* of Wythoff Nim, and a restriction of $(1, 2)$-GDWN, where

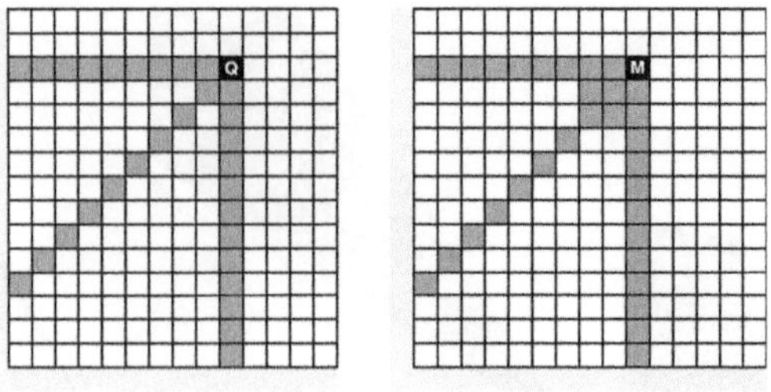

Figure 8. The rules of WYTHOFF NIM, left, and MAHARAJA NIM, right.

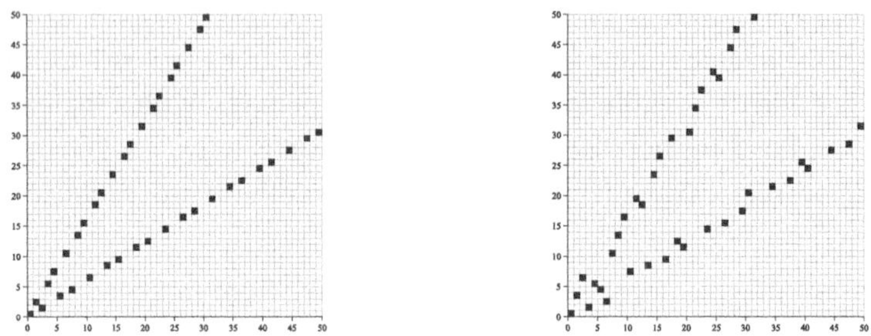

Figure 9. The initial P-positions of WYTHOFF NIM, left, and MAHARAJA NIM, right.

the queen and knight of chess are combined in one and the same piece, the Maharaja *(no coordinate increases by moving). See also Figure 8.*

It is clear that the P-positions of WYTHOFF NIM will be altered for this game. Namely, the "smallest" nonzero P-positions of WYTHOFF NIM are $(1, 2)$ and $(2, 1)$, corresponding precisely to the new move options introduced for MAHARAJA NIM.

Figure 9 illustrates that there is indeed a lot of reordering of P-positions in going from WYTHOFF NIM to MAHARAJA NIM. Nevertheless, the P-positions remain within a bounded distance to the half-lines of slopes ϕ^{-1} and ϕ respectively. This is established by relating the upper P-positions to a certain dictionary process on binary words (a process that is proved Turing complete in [120]). The dictionary is constructed by generalizing the "fill-rule property" of WYTHOFF NIM. A binary sequence indicates whether the x-th P-position (x, y_x) is below or above the main diagonal. After a few initial bits, whenever each difference in an interval $0 < y_x - x < n$ is obtained for some $x > 0$, then any new word is written to the dictionary, and the read-head resets. Iteratively, each new word is translated by computing the upper P-positions and using the fill rule property (and the new upper P-positions are copied symmetrically to the lower P-positions).

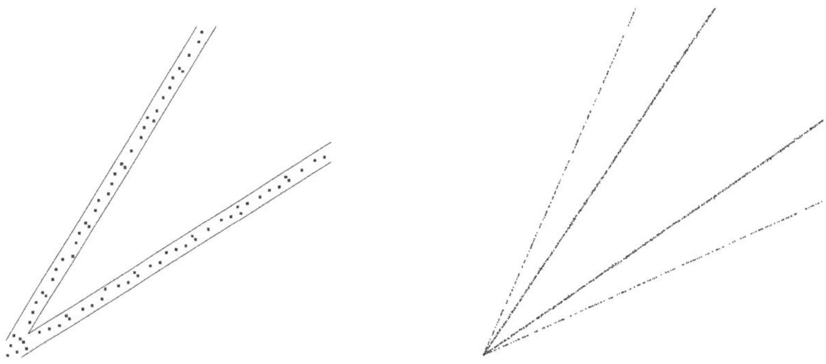

Figure 10. Initial P-positions for MAHARAJA NIM and (1, 2)-GDWN respectively. The leftmost picture shows that MAHARAJA NIM's upper P-positions fits within a narrow stripe of slope ϕ. On the other hand, the P-positions of GDWN to the right will eventually depart from any such stripe, no matter how wide we make it; a renormalization approach (from this book) suggests that the slopes of the upper pair of P-beams of (1, 2)-GDWN satisfy a pair of fourth degree equations, converging to 1.47779977... and 2.24772558... respectively.

In this way, it is proved that a short Dictionary exists (less than 15 words), and by some lemmata an $O(1)$ bound is obtained as shown in Figure 10, left.

The proof depends on a relaxation of an already very nice result [55] (they require (y_n) increasing), as follows. Suppose (x_n) and (y_n) are complementary sequences of positive integers with (x_n) increasing. Suppose further that there is a positive real constant δ such that, for all n, $y_n - x_n = \delta n + O(1)$. Then there are constants $1 < \alpha < 2 < \beta$ such that, for all n, $x_n - \alpha n = O(1)$ and $y_n - \beta n = O(1)$. See [77] for a variation of MAHARAJA NIM with a surprising connection to Lucas representations of positive integers.

8.4. ***A game creating operator and Wythoff Nim.*** The game of WYTHOFF NIM motivated the definition of a certain *game creating* operator[4], also dubbed the $\star$-operator [116] of impartial vector subtraction games (see Section 3 for a discussion of this general class and its relation to a bigger class of "invariant games"). Let G be a vector subtraction game in any finite dimension. Then $G^\star = \mathcal{P}(G) \setminus \{\mathbf{0}\}$ is another vector subtraction game. If $G = G^{\star\star}$, then we say that G is reflexive. It is easy to see that WYTHOFF NIM is not reflexive. However, WYTHOFF NIM's set of nonzero P-positions constitutes a reflexive vector subtraction game; in Figure 12, the game (WYTHOFF NIM*)*, which is P-equivalent to WYTHOFF NIM [116], is displayed.

[4]Thanks to Silvia Heubach for proposing the new name.

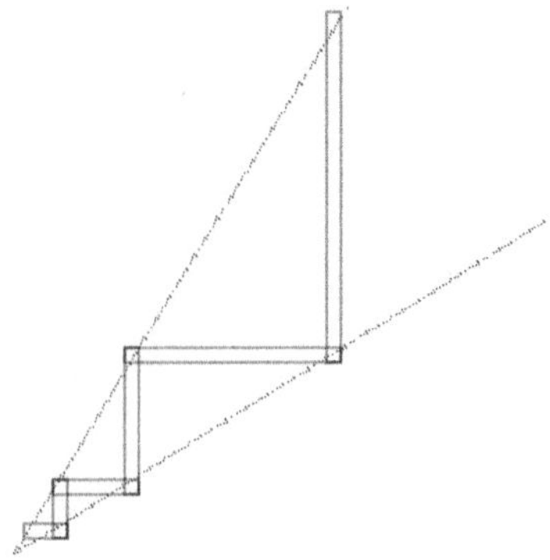

Figure 11. An open question: is it possible to decide in polynomial time whether a given position is in P for MAHARAJA NIM? A "telescope" with focus $O(1)$ and reflectors along the lines ϕn and n/ϕ attempts to determine the outcome (P or N) of some position (x, y) at the top of the picture. The method is successful for a similar game called (2, 3)-MAHARAJA NIM [113]. (It gives the correct value for all extensions of WYTHOFF NIM with a finite nonterminating converging dictionary). The focus is kept sufficiently wide (a constant) to provide correct translations in each step. The number of steps is linear in $\log(xy)$.

More generally, in [31] it was conjectured that, given a pair of complementary Beatty sequences (a_i) and (b_i) (as described in the Section 1), there is an invariant subtraction game for which the P-positions constitute precisely all the pairs (a_i, b_i) and (b_i, a_i), together with the terminal position $(0, 0)$. This was subsequently proved in more generality [116].

8.5. *Wythoff partizan subtraction.* One can play a one heap partizan subtraction game using the Wythoff sequences as infinite subtraction sets [125], and this generalizes to the class of *complementary subtraction games* (take your favorite sequence of positive integers as Left's subtraction set and let Right subtract any other positive integer). Here Left subtracts numbers from the lower Wythoff sequence and Right removes numbers from the upper sequence.

This is an example where game outcomes are almost trivial; it is easy to see that each heap is either a Left- or Next-player win (Left wins from heaps in the lower sequence and the Next player wins from heaps in the upper sequence). Playing this game as a disjunctive sum of several heaps leads to a yet rare encounter of Wythoff's sequences with Conway's famous theory for partizan normal-play games. Moreover, one benefits greatly by using a novel theory for game approximation: namely each heap size is either a number or a *reduced canonical form* [68] (equivalence classes ignoring infinitesimals) switch [125].

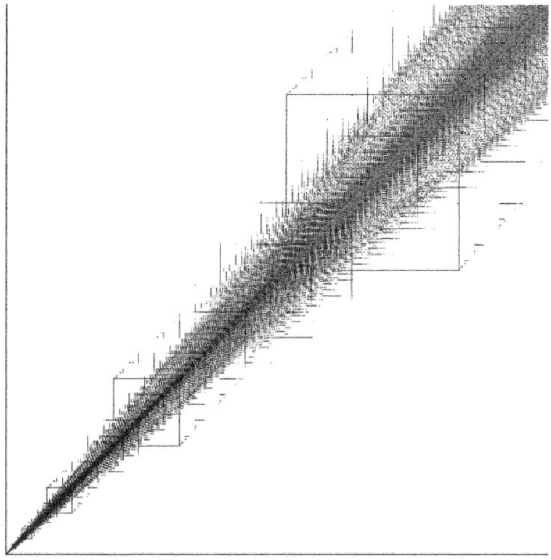

Figure 12. The initial P-positions of the game (WYTHOFF NIM)* (coordinates less than 5000), or equivalently $(0, 0)$ (the lower left corner) together with the moves of the game (WYTHOFF NIM)$^{\star\star} \neq$ WYTHOFF NIM (!). The overall pattern remains a mystery, although one can prove that its chaotic/self-organized part is contained between half lines from the origin of slopes ϕ^{-1} and ϕ. Moreover, a characterization of infinitely many log-periodic positions has been obtained [109]. (WYTHOFF NIM)** is not easily viewed as a "play game", although it is P-equivalent to WYTHOFF NIM. However, the former game has a very nice property, which is absent in WYTHOFF NIM, namely that it is reflexive; that is, (WYTHOFF NIM)**=(WYTHOFF NIM)^{2k*} for all $k \geq 1$. Thus, "simplest" rules do not always give the "nicest" game properties.

The result is accomplished by numerous intertwinings of the classical Fibonacci words and sequences.

Acknowledgements

Thanks to Silvio Levy and Blake Knoll for their great patience with our final adjustments and corrections of these Wythoff visions.

References

[1] M. Albert, R. J. Nowakowski, The game of End-Nim. In honor of Aviezri Fraenkel on the occasion of his 70th birthday. *Electron. J. Combin.* 8 (2001), no. 2, Research Paper 1, 12 pp.

[2] B. U. Alfred, On the ordering of Fibonacci sequences, *Fibonacci Quart.* 1 (1963) 43–46.

[3] E. Altman, B. Gaujal, A. Hordijk, Balanced sequences and optimal routing, *J. ACM* 47 (2000) no. 4, 752–775.

[4] Th. Bang, On the sequence $\lfloor n\alpha \rfloor$, $n = 1, 2, \ldots$, Supplementary note to the preceding paper by Th. Skolem, *Math. Scand.* 5 (1957) 69–76.

[5] J. Barát, P. P. Varjú, Partitioning the positive integers to seven Beatty sequences, *Indag. Math.* (N. S.) 14 (2003) no. 2, 149–161.

[6] E. R. Berlekamp, J. H. Conway, R. K. Guy, *Winning ways for your mathematical plays*, Vol.1–4, 2nd ed., A K Peters, Natick, MA, 2001–2004.

[7] S. Beatty, Problem 3173, *Amer. Math. Monthly*, 33 (1926) 159.

[8] S. Beatty, A. Ostrowski, J. Hyslop, A. C. Aitken, "Solutions to problem 3173", *Amer. Math. Monthly* 34 (1927) no. 3, 159–160.

[9] J. Berstel, P. Séébold, A characterization of Sturmian morphisms, in: A. M. Borzyszkowski, S. Sokolowski (Eds.), *Mathematical Foundations of Computer Science*, LNCS 711, Springer, Berlin, (1993) 281–290.

[10] M. Bicknell-Johnson, The Zeckendorf–Wythoff array applied to counting the number of representations of N as sums of distinct Fibonacci numbers, in: F. T. Howard (Eds.), *Applications of Fibonacci Numbers,* (1999) 53–60.

[11] E. Boros, V. Gurvich, V. Oudalov, A polynomial algorithm for a two-parameter extension of Wythoff Nim based on the Perron–Frobenius theory, *Int. J. Game Theory* 42 (2013), no. 4, 891–915.

[12] M. Boshernitzan, A. S. Fraenkel, Nonhomogeneous spectra of numbers. *Discrete Math.* 34 (1981), no. 3, 325–327.

[13] M. Boshernitzan, A.S. Fraenkel, A linear algorithm for nonhomogeneous spectra of numbers. *J. Algorithms* 5 (1984), no. 2, 187–198.

[14] M. A. Berger, A. Felzenbaum and A. S. Fraenkel, A nonanalytic proof of the Newman–Znám result for disjoint covering systems, *Combinatorica* 6 (1986), no. 3, 235–243.

[15] U. Blass, A. S. Fraenkel, The Sprague–Grundy function for Wythoff's game, *Theoret. Comput. Sci.* 75 (1990), no. 3, 311–333.

[16] U. Blass, A. S. Fraenkel, R. Guelman, How far can nim in diguise be stretched? *J. Combin. Theory Ser. A* 84 (1998) 145–156.

[17] N. Brauner, V. Jost, Small deviations, JIT sequencing and symmetric case of Fraenkel's conjecture, *Discrete Math.* 308 (2008) no. 11, 2319–2324.

[18] N. Calkin, H. S. Wilf, Recounting the rationals, *Amer. Math. Monthly* 107 (2000), 360–363.

[19] J. H. Conway, *On numbers and games*, Academic Press, London 1976, 2nd ed., A K Peters, Wellesley, MA 2001.

[20] Ian G. Connell, A generalization of Wythoff's game, *Canad. Math. Bull.* 2 (1959) 181–190.

[21] M. Cook, U. Larsson, T. Neary, A cellular automaton for blocking queen games, cellular automata and discrete complex systems, *21st IFIP WG 1.5 International Workshop, Automata 2015, Turku, Finland, June 8–10, Proceedings* J. Kari, (Eds.) LNCS 9099 (2015) 71–84. Also an extended version in *Nat. Comput.* (2016) DOI 10.1007/s11047-016-9581-2.

[22] C. Berge, The theory of graphs and its applications (1962) Wiley, New York.

[23] C. Cooper, The k-Zeckendorff array, *Aportaciones Matemáticas, Investigación* 20 (2011) 79–90.

[24] H. S. M. Coxeter, The golden section, Phyllotaxis and Wythoff's game, *Scripta Math.* 19 (1953) 135–143.

[25] A. Dress, A. Flammenkamp, N. Pink, Additive periodicity of the Sprague–Grundy function of certain Nim games, *Adv. in Appl. Math.* 22 (1999) 249–270.

[26] E. Duchêne, A. S. Fraenkel, S. Gravier, R. J. Nowakowski, Another bridge between NIM and WYTHOFF NIM' *Australas. J. Combin.* 44 (2009) 43–56.

[27] E. Duchêne, A. S. Fraenkel, R. J. Nowakowski, M. Rigo, Extensions and restrictions of Wythoff's game preserving its P-positions. *J. Combinat. Theory Ser. A* 117 (2010) 545–567.

[28] E. Duchêne, S. Gravier, Geometrical extensions of Wythoff's game, *Discrete Math.* 309 (2009) 3595–3608.

[29] E. Duchêne, M. Rigo, A morphic approach to combinatorial games: the Tribonacci case, *Theor. Inform. Appl.* 42 (2008) no. 2, 375–393.

[30] E. Duchêne, M. Rigo, Cubic Pisot unit combinatorial games, *Monatshefte für Mathematik* 155 (2008) 217–249.

[31] E. Duchêne, M. Rigo, Invariant games, *Theoret. Comput. Sci.* 411 (2010) 3169–3180.

[32] E. Duchêne, A. Parreau, M. Rigo, Deciding game invariance, http://arxiv.org/abs/1408.5274.

[33] P. Erdös, On a problem concerning congruence systems (Hungarian; Russian, English summary) *Mat. Lapok* 3 (1952) 122–128.

[34] P. Erdös, R. L. Graham, Old and new problems and results in combinatorial number theory, *Monographies de L'Enseignement Mathématique* [Monographs of L'Enseignement Mathématique] 28, Université de Genève, L'Enseignement Mathématique, Geneva (1980) 128 pp.

[35] L. Ericksen, P. G. Anderson, Patterns in differences between rows in k-Zeckendorf arrays, *Fibonacci Quart.* 50 (2012) 11–18.

[36] A. S. Fraenkel, The bracket function and complementary sets of integers, *Canad. J. Math.* 21 (1969) 6–27.

[37] A. S. Fraenkel, Complementing and exactly covering sequences. *J. Combin. Theory Ser. A* 14 (1973) 8–20.

[38] A. S. Fraenkel, How to beat your Wythoff games' opponent on three fronts, *Amer. Math. Monthly* 89 (1982) 353–361.

[39] A. S. Fraenkel, Wythoff games, continued fractions, cedar trees and Fibonacci searches, *Theoret. Comput. Sci.* 29 (1984) 49–73.

[40] A. S. Fraenkel, Scenic trails ascending from sea-level Nim to alpine Chess, in: R. J. Nowakowski (Eds.), *Games of no chance*, MSRI publications (Berkeley, Ca), Vol. 29 (1996) 13–42.

[41] A. S. Fraenkel, Complexity, appeal, and challenges of combinatorial games, *Theoret. Comput. Sci.* 313 (2004) 393–415.

[42] A. S. Fraenkel, New games related to old and new sequences, *Integers* 4 (2004) #G6, 18 pp.

[43] A. S. Fraenkel, Euclid and Wythoff games, *Discrete Math.* 304 (2005) 65–68.

[44] A. S. Fraenkel, Games played by Boole and Galois, *Discrete Appl. Math.* 156 (2008), 420–427.

[45] A. S. Fraenkel, Complementary iterated floor words and the flora game, *SIAM J. Discrete Math.* 24 (2010), 570–588.

[46] A. S. Fraenkel, From enmity to amity, *Amer. Math. Monthly* 117 (2010) 646–648.

[47] A. S. Fraenkel, RATWYT, *College Math. J.* 43 (2012), 160–164.

[48] A. S. Fraenkel, The rat game and the mouse game, in: R. J. Nowakowski (Eds.) *Games of no chance* 4, Proc. BIRS Workshop on Combinatorial Games, 2008, Banff, Alberta, Canada, MSRI Publ. Vol. 63, Cambridge University Press, Cambridge (2015), 95–113.

[49] A. S. Fraenkel, I. Borosh, A generalization of Wythoff's game. *J. Combin. Theory Ser. A* 15 (1973) 175–191.

[50] A. S. Fraenkel, C. Kimberling, Generalized Wythoff arrays, shuffles and interspersions, *Discrete Math.* 126 (1994) 137–149.

[51] A. S. Fraenkel, N. B. Ho, When are translations of P-positions of WYTHOFF NIM's game P-positions? *Integers* 15A #A4, Proc. Integers 2013 Conference, The Erdös Centennial Conference, 2015.

[52] A. S. Fraenkel, D. Krieger, The structure of complementary sets of integers: a 3-shift theorem, *Int. J. Pure Appl. Math.* 10 (2004) 1–49.

[53] A. S. Fraenkel, M. Lorberbom, Nimhoff games, *J. Combin. Theory Ser. A* 58 (1991) 1–25.

[54] A. S. Fraenkel, M.Ozery, Adjoining to Wythoff's game its P-positions as moves, *Theoret. Comput. Sci.* 205 (1998) 283–296.

[55] A. S. Fraenkel, U. Peled, Harnessing the unwieldy MEX function, in: R. J. Nowakowski (Eds.) *Games of No Chance* 4, Proc. BIRS Workshop on Combinatorial Games, 2008, Banff, Alberta, Canada, MSRI Publ., Vol. 63, Cambridge University Press, Cambridge, 77–94, 2015.

[56] A. S. Fraenkel, H. Porta, K. B. Stolarsky, Some arithmetical semigroups. *Analytic number theory (Allerton Park, IL, 1989)*, 255–264, Prog. Math., 85, Birkhauser Boston, MA, 1990.

[57] A. S. Fraenkel, E. Reisner, The game of End-Wythoff, in: M. H. Albert, R. J. Nowakowski (Eds.) *Games of No Chance* 3, Proc. BIRS Workshop on Combinatorial Games, July 2005, Banff, Alberta, Canada, MSRI Publ. Vol. 56, Cambridge University Press, Cambridge, 329–347, 2009.

[58] A. S. Fraenkel, Y. Tanny, A class of Wythoff-like games, *Integers* 12B (2013) #7, (Proc. of the Integers Conference 2011).

[59] A. S. Fraenkel, D. Zusman, A new heap game, Computer and games (Tsukuba, 1998), *Theoret. Comput. Sci.* 252 (2001) 5–12.

[60] M. Gardner, Wythoff Õs nim in Penrose tiles to trapdoor ciphers, (Freeman New York 1989), chapter 8.

[61] H. Gavel, P. Strimling, Nim with a modular Muller twist, *Integers* 4 (2004) #G04.

[62] L. Goldberg, A. S. Fraenkel, Extensions of Wythoff's game, Preprint.

[63] S. W. Golomb, A mathematical investigation of games of "take-away". *J. Combin. Theory* 1 (1966) 443–458.

[64] R. L. Graham, On a theorem by Uspensky, *Amer. Math. Monthly* 70 (1963) 407–409.

[65] R. L. Graham, Covering the positive integers by disjoint sets of the form $\{\lfloor n\alpha + \beta \rfloor : n = 1, 2, \ldots\}$, *J. Combin. Theory Ser. A* 15 (1973) 354–358.

[66] R. L. Graham, C-S. Lin, Spectra of numbers, *Mathematics Magazine* 51 (1978) 174–176.

[67] R. Graham and K. O'Bryant, A discrete Fourier kernel and Fraenkel's tiling conjecture, *Acta Arith.* 118 (2005) 283–304.

[68] J. P. Grossman and A. Siegel, Reductions of partizan games, *Games of No Chance* 3, Cambridge Univ. Press, 2009, 437–456.

[69] P. M. Grundy, Mathematics and games, *Eureka* **2** (1939), 6–8.

[70] V. Gurvich, On the misère version of game Euclid and miserable games, *Discrete Math.*, 307 (2007) 1199–1204.

[71] V. Gurvich, A four parametric generalization of the Wythoff Nim and its recursive solution, *RUTCOR Research Report* 18–2010, Rutgers University.

[72] V. Gurvich, Miserable and strongly miserable impartial games, *RUTCOR Research Report* 18–2011, Rutgers University.

[73] V. Gurvich, Further generalizations of the Wythoff game and the minimum excludant, *Discrete Appl. Math.* 160 (2012) 941–947.

[74] V. Gurvich, N. B. Ho, On tame, pet, domestic, and miserable impartial games, *RUTCOR Research Report* 4–2015, Rutgers University: http://arxiv.org/abs/1508.06343, *Discrete Appl. Math.*, to appear.

[75] R. K. Guy, R. J. Nowakowski, Unsolved problems in combinatorial games, problem 53, in: R. J. Nowakowski (Eds.) *More games of no chance*, Cambridge University Press, Cambridge, 457–473, 2002.

[76] P. Hegarty, U. Larsson, Permutations of the natural numbers with prescribed difference multisets, *Integers* 6 (2006) #A03 1–25.

[77] N. B. Ho, Two variants of Wythoff's game preserving its P-positions, *J. Combin. Theory Ser. A* 119 (2012) 1302–1314.

[78] V. E. Hoggatt, M. Bicknell-Johnsson and R. Sarsfield, A generalization of Wythoff's game, *Fib. Quart.* 17 (1979), 198–211.

[79] J. C. Holladay, Some generalizations of Wythoff's game and other related games, *Math. Magazine* 41 (1968) 7–13.

[80] A. Holshouser, H. Reiter, Blocking Nim, problem 714, *College Mathematics Journal*, 32 (2001) 5.

[81] Y. Jiao, On the Sprague–Grundy values of the F-WYTHOFF game, *Electron. J. Combin.* 20 (2013) no. 1 #P14.

[82] J. Kappraff, D. Blackmore, G. Adamson, Phyllotaxis as a dynamical system: a study in number (Section 6 Wythoff's Game) Chapter 17, in R. V. Jean and D. Barabé (Eds.) *Symmetry in Plants*, Series in Mathematical Biology and Medicine, Vol. 4, World Scientific, 1998.

[83] C. Kimberling, Ordering of the set of all positive Fibonacci sequences, in G. E. Bergum et al. (Eds.), *Applications of Fibonacci Numbers* 5 (1993) 405–416.

[84] C. Kimberling, Interspersions and dispersions, *Proc. Amer. Math. Soc.* 117 (1993) 313–321.

[85] C. Kimberling, The first column of an interspersion, *Fibonacci Quart.* 32 (1994) 301–314.

[86] C. Kimberling, Stolarsky interspersions, *Ars Combin.* 39 (1995) 129–138.

[87] C. Kimberling, The Zeckendorf array equals the Wythoff array, *Fibonacci Quart.* 33 (1995) 3–8.

[88] C. Kimberling, Numeration systems and fractal sequences, *Acta Arith.* 73 (1995) 103–117.

[89] C. Kimberling, Fractal sequences and interspersions, *Ars Combin.* 45 (1997) 157–168.

[90] C. Kimberling, A self-generating set and the golden mean, *J. Integer Seq.* 3 (2000) 1–6.

[91] C. Kimberling, Complementary equations, *J. Integer Seq.*, 10 (2007), Article 07.1.4.

[92] C. Kimberling, Complementary equations and Wythoff sequences, *J. Integer Seq.* 11 (2008) 1–8.

[93] C. Kimberling, The Wythoff difference array, *Congr. Numer.* 194 (2009) 153–158.

[94] C. Kimberling, Beatty sequences generated several ways, *Congr. Numer.* 201 (2010) 23–33.

[95] C. Kimberling, P. J. C. Moses, Complementary equations and Zeckendorf arrays, *Congr. Numer.* 201 (2010) 161–178.

[96] C. Kimberling, Beatty sequences and Wythoff sequences, generalized, *Fibonacci Quart.* 49 (2011) 195–200.

[97] C. Kimberling, The Wythoff triangle and unique representations of positive integers, *Aportaciones Matemáticas, Investigación* 20 (2011) 155–169.

[98] C. Kimberling, Beatty sequences and trigonometric functions, *Integers* 16 (2016) #A15.

[99] C. Kimberling, P. J. C. Moses, The infinite Fibonacci tree and other trees generated by rules, *Fibonacci Quart.* 52 (2014) 136–149.

[100] C. Kimberling, K. Stolarsky, Slow Beatty sequences, devious convergence, and partitional divergence, *Amer. Math. Monthly.* 123 (2016) 267–273.

[101] R. Knott, S. Northshield, Problem 14 and solution, in problem proposals, C. Kimberling, (Eds.) *Proceedings of the sixteenth conference on Fibonacci numbers and their applications, Fibonacci Quart.*, 2014.

[102] J. Lambek and L. Moser, On some two way classifications of integers, *Canad. Math. Bull.* 2 (1959) 85–89.

[103] H. Landman, A simple FSM-based proof of the additive periodicity of the Sprague–Grundy function of Wythoff's game, in: R. J. Nowakowski (Eds.), *More games of no chance*, Cambridge University Press, Cambridge (2002) 383–386.

[104] W. Lang, The Wythoff and the Zeckendorf representations of numbers are equivalent, in G. E. Bergum et al. (Eds.) *Applications of Fibonacci numbers,* Vol. 6, Kluwer, Dordrecht (1996) 319–337.

[105] U. Larsson, 2-pile Nim with a restricted number of move-size imitations. With an appendix by Peter Hegarty. *Integers* 9 (2009) #G04 671–690.

[106] U. Larsson, *Sequences and games generalizing the combinatorial game of Wythoff Nim*, Lic. Thesis, Göteborg University, October 2009.

[107] U. Larsson, Blocking Wythoff Nim, *Electron. J. Combin.* 18 (2011), Paper 120.

[108] U. Larsson, A Generalized Diagonal Wythoff Nim, *Integers* 12 (2012) #G02.

[109] U. Larsson, The $\star$-operator and invariant subtraction games, *Theoret. Comput. Sci.*, 422 (2012) 52–58.

[110] U. Larsson, Splitting sequences and Wythoff Nim extensions, *J. Integer Seq.*, Vol. 17 (2014) Article 14.5.7.

[111] U. Larsson, Restrictions of m-Wythoff Nim and p-complementary Beatty sequences, in: R. J. Nowakowski (Eds.) *Games of no chance* 4, Proc. BIRS Workshop on Combinatorial Games, 2008, Banff, Alberta, Canada, MSRI Publ. Vol. 63, Cambridge University Press, Cambridge, pp. 137–160, 2015.

[112] U. Larsson, Permutation games and the star-operator, (in preparation). See also preprint of "The $\star$-operator and invariant subtraction games": https://arxiv.org/abs/1009.4220.

[113] U. Larsson, Converging dictionaries and Wythoff Nim extensions, (in preparation). See also an extended preprint of Maharaja Nim: http://arxiv.org/abs/1207.0765.

[114] U. Larsson, M. Fisher, Chromatic Nim, in this issue.

[115] U. Larsson, A. S. Fraenkel, Take-away games on Beatty's theorem and the notion of k-invariance, in this issue.

[116] U. Larsson, P. Hegarty, A. S. Fraenkel, Invariant and dual subtraction games resolving the Duchêne–Rigo conjecture, *Theoret. Comput. Sci.* 412 (2011) 729–735.

[117] U. Larsson, S. Rubinstein-Salzedo, Grundy values of Fibonacci Nim, *Int. J. Game Theory*, 45 (2016) 617–625.

[118] U. Larsson, S. Rubinstein-Salzedo, Global Fibonacci Nim, *Int. J. Game Theory* (2017).

[119] U. Larsson, M. Weimerskirch, Impartial games whose rulesets produce given continued fractions, in this issue.

[120] U. Larsson, J. Wästlund, Maharaja Nim: Wythoff's Queen meets the Knight, *Integers* (2014) G05.

[121] C. G. Lekkerkerker, Voorstelling van natuurlijke getallen door een som van getallen van Fibonacci, *Simon Stevin* 29 (1952) 19.

[122] T. Lengyel, A Nim-type game and continued fractions, *Fibonacci Quart.* (2003) 310–320.

[123] J. Y-T. Leung (Ed), *Handbook of scheduling, algorithms, models, and performance analysis*, Chapter 19: Wieslaw Kubiak, Fair sequences; 19.5: Small deviations, balanced words and Fraenkel's conjecture, Chapman and Hall/CRC, 2004.

[124] C. D. Meyer, Matrix analysis and applied linear algebra, SIAM, Philadelphia, PA, 2000.

[125] N. A. McKay, U. Larsson, R. J. Nowakowski, A. A. Siegel, Finding golden nuggets by reduction, preprint, https://arxiv.org/abs/1510.07155.

[126] R. Morikawa, On eventually covering families generated by the bracket function, *Bull. Fac. Liberal Arts Nagasaki Univ.* 23 (1982/83) 17–22.

[127] R. Morikawa, Disjoint sequences generated by the bracket function, *Bull. Fac. Liberal Arts Nagasaki Univ.*, 26 (1985), 1–13.

[128] D. R. Morrison, A Stolarsky array of Wythoff pairs, in: V. E. Hoggatt, Jr., M. Bicknell-Johnson (Eds.), *A collection of manucripts related to the Fibonacci Sequence*, The Fibonacci Association (1980) 134–136.

[129] G. Nivasch, The Sprague–Grundy function of the game Euclid, *Discrete Math.* 306 (2006) 2798–2800.

[130] G. Nivasch, More on the Sprague–Grundy function for Wythoff's game, in: M. H. Albert, R. J. Nowakowski (Eds.), *Games of no chance* 3, MSRI Publications 56, Cambridge University Press (2009) 377–410.

[131] I. Niven, Diophantine approximations, Hedrick Lectures, Mathematical Association of America, Interscience Tracts in Pure and Applied Mathematics 14, Wiley/Interscience, New York, 1963.

[132] *Online encyclopedia of integer sequences,* http://oeis.org/.

[133] K. O'Bryant, Fraenkel's partition and Brown's decomposition, *Integers* 3 (2003) A11.

[134] A. Ostrowski, Bemerkungen zur Theorie der Diophantischen Approximationen, *Abh. Math. Sem. Hamburg* 1 (1922), 77–98.

[135] H. Porta and K. B. Stolarsky, Half-silvered mirrors and Wythoff's game, Canad. Math. Bull. 33 (1990), 119–125.

[136] V. V. Prasolov, Polynomials, in *Algorithms and computation in mathematics* 11, Springer, Berlin-Heidelberg, 2010.

[137] G. Rauzy, Nombres algébriques et substitutions, *Bull. Soc. Math. France* 110 (1982), 147–178.

[138] J. W. S. Rayleigh, *The Theory of Sound*, MacMillan, London, 1st ed. 1877, 2nd ed. 1894, reprinted 1928, 1929, 1937. (There is also a Dover 1945 publication, which is a reprint of the 2nd. The "sound" equivalent of Beatty's theorem, is stated, without proof, in §92a, pp. 122–123, in the 1937 reprint of the 2nd ed., the only one we saw. Kevin O'Bryant told us that he checked the 1st ed. and the statement is not there.)

[139] J. N. Ridley, Fourth order linear recurrences satisfied by Wythoff pairs, *Ramanujan J.* 5 (2001), 159–165.

[140] R. Silber, A Fibonacci property of Wythoff pairs, *The Fibonacci Quart.*, 14 (1976) 380–384.

[141] R. J. Simpson, Exact coverings of the integers by arithmetic progressions, *Discrete Math.* 59 (1986) 181–190.

[142] J. Simpson, Disjoint Beatty sequences, *Integers* 4 (2004) A12.

[143] Th. Skolem, On certain distributions of integers in pairs with given differences, *Math. Scand.* 5 (1957) 57–68.

[144] Th. Skolem, Über einige Eigenschaften der Zahlenmengen $\lfloor n\alpha + \beta \rfloor$ bei irrationalem α mit einleitenden Bemerkungen über einige kombinatorische Probleme, *Norske Vid. Selsk. Forh., Trondheim* 30 (1957) 118–125.

[145] F. Smith, P. Stănică, Comply/constrain games or games with a Muller twist, *Integers*, 2 (2002).

[146] C. P. Spears, M. Bicknell-Johnson, J. J. Yan, Fibonacci phyllotaxis by asymmetric cell division: Zeckendorf and Wythoff trees, *Congr. Numer.* 201 (2010) 257–271.

[147] R. P. Sprague, Über mathematische Kampfspiele, *Tôhoku Mathematical Journal* **41** (1935), 438–444.

[148] J. Steiner, Combinatorishe aufgabe, *J. reine angew. Math.* 45 (1853), 181–182 (= Gesammelte Were II, Berlin, 1882, 437–438).

[149] K. B. Stolarsky, Beatty sequences, continued fractions, and certain shift operators, *Canadian Math. Bull.* 19 (1976) 473–482.

[150] K. B. Stolarsky, Fourth order linearly recurrent Wythoff pairs, *Ramanujan J.* 2(1998), 441–448.

[151] X. Sun, Wythoff's sequence and N-heap Wythoff's conjectures, *Discrete Math.* 300 (2005) 180–195.

[152] X. Sun, D. Zeilberger, On Fraenkel's N-heap Wythoff's conjectures, *Ann. Comb.* 8 (2004) 225–238.

[153] R. Tijdeman, Exact covers of balanced sequences and Fraenkel's conjecture, *Algebraic number theory and Diophantine analysis* (Proceedings of the International Conference held in Graz, 1998), 467–483, de Gruyter, Berlin, 2000.

[154] R. Tijdeman, Fraenkel's conjecture for six sequences, *Discrete Math.* 222 (2000) no. 1-3, 223–234.

[155] J. V. Uspensky, On a problem arising out of the theory of a certain game, *Amer. Math. Monthly*, 34 (1927) 516–521.

[156] Vangelis Th. Paschos (ed), *Application of combinatorial optimization*, Chapter 4, Wiley-ISTE, Hoboken, NJ, 2010.

[157] M. J. Whinihan, Fibonacci nim, *Fibonacci Quart.* 1 (1963), 9–13.

[158] W. A. Wythoff, A modification of the game of Nim, *Nieuw Arch. Wisk.* 7 (1907) 199–202.

[159] E. Zeckendorf, Représentation des nombres naturels par une somme de nombres de Fibonacci ou de nombres de Lucas, *Bull. Soc. R. Sci. Liège* 41 (1972) 179–182.

eric.duchene@univ-lyon1.fr — *Laboratoire LIRIS, Université Claude Bernard Lyon 1, Villeurbanne, France*

aviezri.fraenkel@weizmann.ac.il — *Department of Computer Science and Applied Mathematics, Weizmann Institute of Science, Rehovot, Israel*

vgurvich@hse.ru — *Department of Computer Sciences, National Research University, Higher School of Economics, Moscow, Russia*

nhan.ho@latrobe.edu.au — *Department of Mathematics and Statistics, La Trobe University, Melbourne VIC, Australia*

ck6@evansville.edu — *Department of Mathematics, University of Evansville, Evansville, IN, United States*

urban031@gmail.com — *Department of Industrial Engineering and Management, Technion - Israel Institute of Technology, Haifa, Israel*

Games of No Chance 5
MSRI Publications
Volume **70**, 2017

Scoring games: the state of play

URBAN LARSSON, RICHARD J. NOWAKOWSKI
AND CARLOS PEREIRA DOS SANTOS

We survey scoring-play combinatorial game theory, and reflect upon similarities and differences with normal- and misère-play. We illustrate the theory by using new and old scoring rulesets, and we conclude with a survey of scoring games that originate from graph theory.

1. Introduction

Recent progress in scoring-play combinatorial game theory motivates a survey on the subject. There are similarities with classical settings in normal- and misère-play, but the subject is richer than both those combined. This survey has a three-fold purpose: first to survey the combinatorial game theory (CGT) work in the area (such as disjunctive sum, game comparison, game reduction and game values), secondly to point at some important ideas about scoring rulesets (in relation with normal- and misère-play), and at last we show that existing literature includes many scoring combinatorial games which have not yet been studied in the broader CGT context. Although CGT was first developed in positional (scoring) games by Milnor (inspired by game decomposition in the game of Go), the field took off only with the advances in normal-play during the 1970-80s, and recently via successes in understanding misère games.

1.1. ***Normal- and misère-play.*** The family of combinatorial games consists of two-player games with perfect information (no hidden information as in some card games), no chance moves (no dice), and where the two players move alternately. We primarily consider games in which the positions decompose into independent subpositions. This class of games has been called *additive* to distinguish it from maker-maker and maker-breaker positional games [5], such as HEX.

The first author was partially supported by the Killam Trust.
This work was partially funded by Fundação para a Ciência e a Tecnologia through the project UID/MAT/04721/2013.
MSC2010: 91A46.
Keywords: all-normal, guaranteed games, konane variations, maximizer-minimizer games, misère-play, normal-play, reversibility, scoring ruleset, take-small, combinatorial game theory, scoring games on graphs, scoring-play.

When a player has no more moves (in *any* game component, which is called the "long" disjunctive sum) the game ends and some convention is required to be able to determine the outcome. There are two natural conventions that tie together the last move and the outcome:

(1) *Normal-play convention*: last player wins;

(2) *Misère-play convention*: last player loses.

The obtained body of results considering these conventions is what we call *classical combinatorial game theory*. See [16; 7; 1; 47] for background, [25] for a survey, and [39] for a list of open problems.

The theory of normal-play was developed first. The analysis of NIM, by Charles Bouton, was published in the early twentieth century [10]. After this, the Sprague–Grundy theory for impartial combinatorial games[1] [49; 27] appeared, played with normal-play convention. This was a fundamental step for the subsequent establishment of combinatorial game theory; the important concept of *disjunctive sum* of games was established, and this idea was already present in the game of NIM, where, of course, the most central concept is that of "nim-sum". The next big step was in the 1950s, by Milnor, when the notion of game comparison first appeared [38] in so-called "positional games", and this theory was inspired by the famous eastern game of GO. The game appears to decompose into components during play; the longer play continues, the more independent the regions become. By the end of play, the game board has decomposed into a finite number of regions, and only the scoring part remains, where the difference of the total numbers of captured pieces together with the "captured territories" determines the final result. We will return to Milnor's class of games in Section 2. Later, in the 1970s, Conway [16] was able to expand the theory to include partizan game theory[2] followed up in the 1980s with Conway, Berlekamp and Guy's "Winning Ways" [7].

It is important to note that *normal-play convention* is a very special case, with many nice properties. (The notation for a normal-play game is $G = \{G^{\mathcal{L}} \mid G^{\mathcal{R}}\}$, where $G^{\mathcal{L}}$ is the set of G's Left options, and similar for Right. Also, G^L and G^R are elements of $G^{\mathcal{L}}$, $G^{\mathcal{R}}$, respectively.) A "ten commandments list" is given next. The properties of all the other conventions are compared to this list.

Properties of normal-play. N1 Combinatorial games played with normal-play convention, together with the disjunctive sum, are an ordered, abelian group.

N2 The inverse of G is obtained recursively by $-G = \{-G^{\mathcal{R}} \mid -G^{\mathcal{L}}\}$.

---

[1]A impartial combinatorial ruleset is a ruleset in which the allowable moves depend only on the positions and not on which of the two players is currently moving. In other words, from any subposition, the options are exactly the same for both players.

[2]There are two players, Left and Right, with usually different move options.

N3 To check if $G \succcurlyeq H$, it is only necessary to see if Left wins $G - H$ playing second.

N4 The empty set is the worst possible set of options.

N5 There are two fundamental reductions: domination and reversibility. The second happens when there is a Right response to a Left play in G such that $G^{LR} \preccurlyeq G$ (or the mirror image, from Right's point of view). In that case, G^L may be replaced by $G^{LR\mathcal{L}}$. This can be done even if $G^{LR\mathcal{L}} = \varnothing$ (atomic reversibility) and in this case, G^L is erased. In other words, G^L is replaced by the empty set!

N6 Given a game G, there is a unique simplest game equivalent to G. That game is obtained by making all the possible reductions in G and its followers.

N7 There is a bijection between the outcome classes and the order:

$$G \succ 0 \Leftrightarrow G \in \mathcal{L}; \quad G = 0 \Leftrightarrow G \in \mathcal{P}; \quad G \| 0 \Leftrightarrow G \in \mathcal{N}; \quad G \prec 0 \Leftrightarrow G \in \mathcal{R}.$$

N8 The disjunctive sum table of the outcome classes is not particularly complex:

$+$	$\mathcal{P}$	$\mathcal{N}$	$\mathcal{R}$	$\mathcal{L}$
$\mathcal{P}$	$\mathcal{P}$	$\mathcal{N}$	$\mathcal{R}$	$\mathcal{L}$
$\mathcal{N}$	$\mathcal{N}$	all	$\mathcal{N} \cup \mathcal{R}$	$\mathcal{N} \cup \mathcal{L}$
$\mathcal{R}$	$\mathcal{R}$	$\mathcal{N} \cup \mathcal{R}$	$\mathcal{R}$	all
$\mathcal{L}$	$\mathcal{L}$	$\mathcal{N} \cup \mathcal{L}$	all	$\mathcal{L}$

N9 If $LS(G) < RS(G)$ ("the number of move advantage is worse for Left if Left plays first") then G is a *zugzwang*. In the normal-play convention, there is no canonical zugzwangs because this inequality always corresponds to a number.

N10 If $LS(G) > RS(G)$ then G is *hot*. If a game is hot then $G^{\mathcal{L}} \neq \varnothing$ and $G^{\mathcal{R}} \neq \varnothing$, because otherwise G would be a number, and numbers are cold.

All these properties fail in misère universes: for **N1**, misère games have only monoid structures; for instance, considering impartial misère games, $*2$ has no inverse. Also, it is possible to have invertible elements that don't satisfy **N2** [37; 40]. In misère, $\{\,|\,*\} > 0$ and Left does not win going second in $\{\,|\,*\}$, therefore **N3** fails. About **N4**, in misère, an empty set of options may be a good thing; in $\{\,|\,*\}$, Left wins going first (the worst thing is an infinite string of moves, which only exists in "loopy" games). In [21], the authors proposed nice reductions for dicot misère games; for reversibility, in some situations, if $G^{LR\mathcal{L}} = \varnothing$ then G^L should be replaced by $*$. Hence, **N5** also fails. The failure of **N6** will be discussed in the next section. **N7** fails because, as we have seen, $\{\,|\,*\} > 0$ and $\{\,|\,*\} \in \mathcal{N}$. For **N8**, the algebraic table of the outcomes is

$+$	$\mathcal{P}$	$\mathcal{N}$	$\mathcal{R}$	$\mathcal{L}$
$\mathcal{P}$	all	all	all	all
$\mathcal{N}$	all	all	all	all
$\mathcal{R}$	all	all	all	all
$\mathcal{L}$	all	all	all	all

Finally, in misère, **N9** and **N10** fail because we don't have numbers to define stops.

2. Scoring combinatorial game theory

In scoring-play there is no direct correspondence between the last player and the outcome. When a player has no more moves, the game ends, but an evaluation about the outcome is still needed. In common language, the player wants to get as many points as possible irrespective of who moves last. Therefore, the information that the set of options is empty ($G^{\mathcal{L}} = \varnothing$ or $G^{\mathcal{R}} = \varnothing$) is not enough. It reveals the end of the game, but not the outcome of the game. It is necessary to include that information in the description of the game forms.

In [50], Fraser Stewart introduced a notation based in triples, instead of pairs. Considering a totally ordered group of results $\mathbb{B}$, in Stewart's notation, a game is a triple $\{G^{\mathcal{L}} \mid b \in \mathbb{B} \mid G^{\mathcal{R}}\}$ where, as usual, $G^{\mathcal{L}}$ and $G^{\mathcal{R}}$ are sets of games. For instance, the game of day 0,

$$\{\,|\,3\,|\,\} \underbrace{=}_{\text{definition}} 3$$

is a game where the players have no options, but the last player gets 3 points (which is usually bad for Right, who wants a negative number of points). Another example is the game $\{\,|\,2\,|\,-3\}$; if it is Left's turn, she has no options, the game ends, and she gets 2 points; if it is Right's turn, he has an option, that is $-3 = \{\,|\,-3\,|\,\}$. After that, the game ends and the final result is -3.

Later, in [34], the authors proposed an alternative notation with pairs instead of triples. Basically, the different empty sets are adorned with elements of $\mathbb{B}$. This allows a representation of the result when a player has no more moves. Using the new notation, $\{\,|\,3\,|\,\}$ becomes $\langle \varnothing^3 \mid \varnothing^3 \rangle$ and $\{\,|\,2\,|\,-3\}$ is represented by $\langle \varnothing^2 \mid -3 \rangle$. The notation with adorns has two fundamental advantages: first, it is more suitable for asymmetrical forms like $\langle \varnothing^3 \mid \varnothing^4 \rangle$, which is particularly interesting when we discuss reductions to simplest form! Second, and the most important, it has practical advantages when we want to relate classical combinatorial game theory to scoring combinatorial game theory. On the other hand, Stewart's notation may be useful when analyzing operators different than the disjunctive sum. A different type of bracket is used to distinguish scoring-play from the classical context. See Figure 1 for how this relates to game trees.

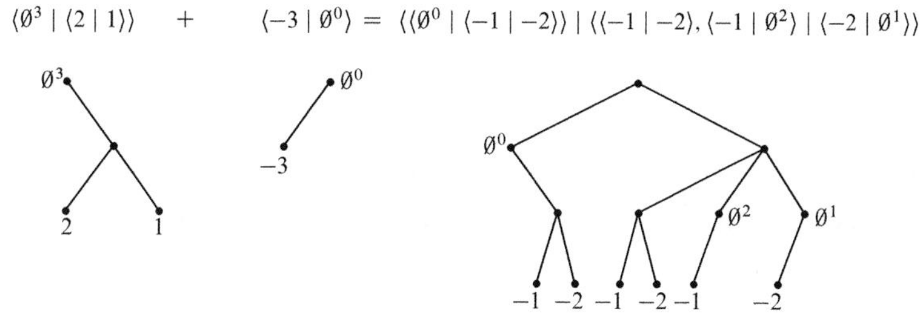

Figure 1. The disjunctive sum of two game trees.

There is a problem with inverses, so we define the *conjugate* as $\sim G = \langle \sim G^{\mathcal{R}} \mid \sim G^{\mathcal{L}} \rangle$. The game $G = \langle \langle 1 \mid -1 \rangle \mid \langle 1 \mid -1 \rangle \rangle$ is a zugzwang; if Left starts, the final result is -1, if Right starts, the final result is 1. The conjugate of that form is exactly the same, and so, $G + (\sim G) = \langle \langle 1 \mid -1 \rangle \mid \langle 1 \mid -1 \rangle \rangle + \langle \langle 1 \mid -1 \rangle \mid \langle 1 \mid -1 \rangle \rangle$. We may observe that $G + (\sim G)$ is not zero, but a new zugzwang. In conclusion, $\sim G$ is not the inverse of G. In [23], Mark Ettinger proved that G is not invertible. Therefore, as in misère-play, scoring-play often leads to monoid structures, instead of group structures.

The research done on scoring combinatorial game theory always considered short games, requiring that the sets $G^{\mathcal{L}}$ and $G^{\mathcal{R}}$ be finite. Also, in some works, only *dicot* forms were considered — from any subposition, a player can move if and only if the opponent also can.

The first mathematical approach to scoring-play was done by John Milnor and Olof Hanner with the publication of the papers [38; 29]. In the first, Milnor restricted the studied universe to nonzugzwang dicot forms. That means that every Milnor dicot form G satisfies $Ls(G) \geq Rs(G)$ so these games still satisfy **N9** and **N10**. (We write Ls, Left score, in scoring-play instead of LS, which is the Left stop, in normal-play.) In fact, Milnor's universe of scoring games is very well behaved, constituting a group structure. Following Milnor's approach, Olof Hanner proved the existence of a mean value of a game; a kind of "average value" for G [29]. This result was very important, inspiring the modern temperature theory [16; 7; 1; 47]. Because of that, in a game theorist's mind, this work is fundamentally important for the establishment of CGT, more than a first contribution for scoring-play analysis[3].

The next stride forward is in the 1990s when Mark Ettinger published his visionary PhD thesis ([23]; see also [22]). Ettinger studied dicot scoring games

[3]In their terminology they used a pay-off function instead of outcome function/Left stop/Left score and so on, probably inspired by the novelties in economic game theory regarding Nash equilibrium, mixed strategies, and so on. However, since then the fields ECG and CGT seem to have developed in completely different directions. A big open question is if there is a merging theory, but the authors are not yet aware of any.

without the restriction $Ls(G) \geq Rs(G)$, allowing zugzwang positions as $G = \langle\langle 1 \mid -1\rangle \mid \langle 1 \mid -1\rangle\rangle$. Even with the dicot restriction, Ettinger was able to prove useful reductions in forms and the conjugation property **N2**. Also, he proved a very useful theorem for checking if a player should prefer a game G or some number of points r ([23], Ettinger's theorem, page 20). We used the adjective "visionary" because his ideas, already in the 1990s, are very similar to those used 15 years later in misère theory. In particular, with different words, he proved the "downlink concept" ([23], page 22), used later by Aaron Siegel [48] and a "carousel" procedure for proving the conjugate property that is very useful in other contexts [33]. Finally, in Ettinger's universe, there are reversibility situations with $G^{LR\mathcal{L}} = \varnothing$. In some of them, the reversible option *cannot be erased*, but replaced by a well-chosen representative (atomic reversibility). Ettinger was the first who understood this particular situation of reversibility, filling in the cases that do not occur in the normal-play convention.

Fraser Stewart [50] analyzed scoring games almost without restrictions (only asymmetrical empty trees like $\langle\varnothing^3 \mid \varnothing^4\rangle$ were not allowed). The absence of restrictions brought the possibility of non-well-behaved forms like $\langle\varnothing^1 \mid -1\rangle$; the problematic forms are those hot games without options (*hot atomic games*). Left wants to have the right to move, even knowing that there are no available moves (seemingly resembling local misère-play)! To understand the interesting mathematical implications of these forms, we need to clarify first the different nature of the concepts r-score and r moves. For instance, 1-score is the game $1 = \langle\varnothing^1 \mid \varnothing^1\rangle$ whose tree is empty; on the other hand, one move is the game $\langle\langle\varnothing^0 \mid \varnothing^0\rangle \mid \varnothing^0\rangle$, whose tree is analogous to the tree of 1 in classical normal-play (a hat is used for moves, trees adorned only with zeros; $\hat{1} = \langle\langle\varnothing^0 \mid \varnothing^0\rangle \mid \varnothing^0\rangle$). In the second, Left does not win points, but the extra move may be very useful in presence of zugzwang components. In the symbiosis between scores and moves lies the soul of scoring combinatorial game theory.

2.1. ***Normal-play behavior in scoring games.*** In misère- or scoring-play, if we consider the game forms just like trees without scores, we can think about their behavior under normal-play convention. For example, in scoring, if we replace all the adorns by zero, we essentially obtain moves. Only the shape of the trees is relevant. The relation between a specific convention (sc) and the normal-play convention (nc) is very interesting. Often, we have an order preserving relation, that is,

$$G \succcurlyeq_{\text{sc}} H \Rightarrow G \succcurlyeq_{\text{nc}} H.$$

To give some intuition, consider $G \succcurlyeq H$ in misère-play and suppose that $G \not\succcurlyeq H$ in normal-play. For some game X, in normal-play, we have *Left loses playing first* $G+X$ *and Left wins playing first* $H+X$ or *Left loses playing second* $G+X$ *and*

Left wins playing second $H+X$. Without loss of generality, assume the first. Now, consider the form $\{\hat{n} \mid -\hat{n}\}$ where n is arbitrary large, $\hat{n}$ is a string of moves for Left, and $-\hat{n}$ a string of moves for Right. It is an easy check that, with misère convention, Left loses going first in $G+X+\{\hat{n} \mid -\hat{n}\}$ and Left wins going first in $H+X+\{\hat{n} \mid -\hat{n}\}$. This happens because $\{\hat{n} \mid -\hat{n}\}$ works like a zugzwang and zugzwangs promote normal-play (players want to make the last move in the disjunctive sum of the remaining components). Hence, this situation contradicts $G \succcurlyeq H$ in misère-play and we have the mentioned order-preserving. Of course, in scoring context, we can build an analogous argument with a "large zugzwang" $\langle -r \mid r\rangle$.

The one million dollar question is the opposite question: are extra moves always good? That is, is it true that $\hat{1} = \langle\langle\varnothing^0 \mid \varnothing^0\rangle \mid \varnothing^0\rangle \succcurlyeq 0$? In other words, under what circumstances do we have an order embedding? In Stewart's universe we do not have it. Consider $X = \langle\varnothing^1 \mid -1\rangle$: if Left plays first in $\hat{1}+X$, the final result is -1, if Left plays first in $0+X$, the final result is 1. Therefore, $\hat{1} \not\succcurlyeq 0$, and that happens because of the existence of hot atomic games.

It is also possible to prove that if $G = 0$ in Stewart's universe then the game tree of G is empty. Without loss of generality, consider $G^{\mathcal{L}} \neq \varnothing$ and r a negative real number less than all the atoms of G. Playing first, Left loses $G + \langle\varnothing^1 \mid r\rangle$ because she has a move in G. On the other hand, Left wins $0 + \langle\varnothing^1 \mid r\rangle$ with the final result 1. Hence G with options cannot be equal to 0. Again, that happens because of the existence of hot atomic games.

Finally, consider an hypothetical *atomic reversibility* such that $G^{LR\mathcal{L}} = \varnothing$. It cannot exist! If so, $G \succcurlyeq \langle\varnothing^a \mid G^{LR\mathcal{R}}\rangle$, for some real number a, and this inequality is contradictory with a distinguishing game like $X = \langle\varnothing^1 \mid r\rangle$, where r is an arbitrary small negative real number. Namely, $G+X \not\succcurlyeq \langle\varnothing^a \mid G^{LR\mathcal{R}}\rangle + X$. The importance of Stewart's contribution was to bring this whole range of unusual situations to scoring theory research.

In the 2010s, the authors and João Pedro Neto proposed the analysis of *guaranteed scoring games*, $\mathbb{GS}$, in which all the atoms in $G^{\mathcal{R}}$ of a game of the form $\langle\varnothing^r \mid G^{\mathcal{R}}\rangle$, have scores larger than or equal to r and analogously for $\langle G^{\mathcal{L}} \mid \varnothing^r\rangle$ [34; 33]. In the guaranteed scoring universe, in each game component, it is always good to continue playing instead of being without moves. More importantly, in that universe *there are no hot atomic games* and, because of that, an order-embedding of normal-play was proved [34]. Also, Ettinger's ideas were adapted in order to obtain comparison techniques, the conjugate property, reductions (Figure 2), and to propose useful canonical forms [33].

As in Ettinger's universe, the guaranteed universe allows for atomic reversibility. To better understand this concept, we make some preliminary observations. A *game* is a directed acyclic graph $(P, (M^{\mathcal{L}}, M^{\mathcal{R}}))$ where the set of nodes P contains the *game positions*, and the edges are partitioned into the distinguished sets $M^{\mathcal{L}}$ and $M^{\mathcal{R}}$, which are the *Left*- and *Right-moves* respectively. On the other hand, with a recursive construction and the definitions of disjunctive sum and order, the structure of *forms* $(\mathbb{GS}, +, \succcurlyeq)$ is obtained (game positions are described by forms) where the equivalence classes of the quotient $(\mathbb{GS}, +, \succcurlyeq)/{=}$ are what we call *game values*. In normal-play, because there is a unique simplest form in each equivalence class $[G]$, that is the natural choice for a representative — it is called the *canonical form*. However, in scoring context, we lose the uniqueness: there is more than one simplest form in an equivalence class $[G]$. Even so, the problem of canonical forms has an interesting approach [34]: an atomic reversible option G^L is replaced by $r - \widehat{n+1}$ where n is minimal such that $G \geq r - \hat{n}$ (and r is unique); this procedure constitutes a *choice*. With that choice, considering a form G, it is possible to prove that, after all reductions, a unique and simplest form is achieved, representing $[G]$. In conclusion, there is not a unique simplest form in the class $[G]$ but, *after a choice in the atomic reversibility procedure*, we get a unique simplest form (canonical form). The reversibility is illustrated in Figure 2.

	Milnor	Ettinger	Larsson et al.	Stewart
structure	group	monoid	monoid	monoid
existence of atomic reversible options	yes	yes	yes	no
existence of zugzwangs	no	yes	yes	yes
existence of hot empty games	no	no	no	yes
large equivalence class of zero	yes	yes	yes	no

This section would be incomplete without mentioning Will Johnson. His work imposes a big restriction on scoring games (which also restricts its relation to surrounding work). Johnson's universe consists of dicot games in which, for a given game G, the length of any play (distance to any leaf on the game tree) has the same parity [31]; see Section 4 for some graph games in this universe. The games are called *odd*- or *even-tempered* if all the lengths are odd or even, respectively. A game G is *inversive* if $Ls(G+X) \geq Rs(G+X)$ for every even-tempered game X. Johnson proved that the universe of inversive games is well behaved, having a group structure. Every inversive game has a canonical form and an additive inverse which is equal to its conjugate; moreover, $G \succcurlyeq H$ if G and H have the same "temper" and $Rs(G-H) \geq 0$.

3. Scoring rulesets

Many rulesets, usually played in normal (or misère) convention adapt easily to some scoring convention, and we may even mix rulesets in one and the same game; the scoring aspects may vary to create various effects such as zugzwangs or change the class of games for consideration to obtain more structure and nicer

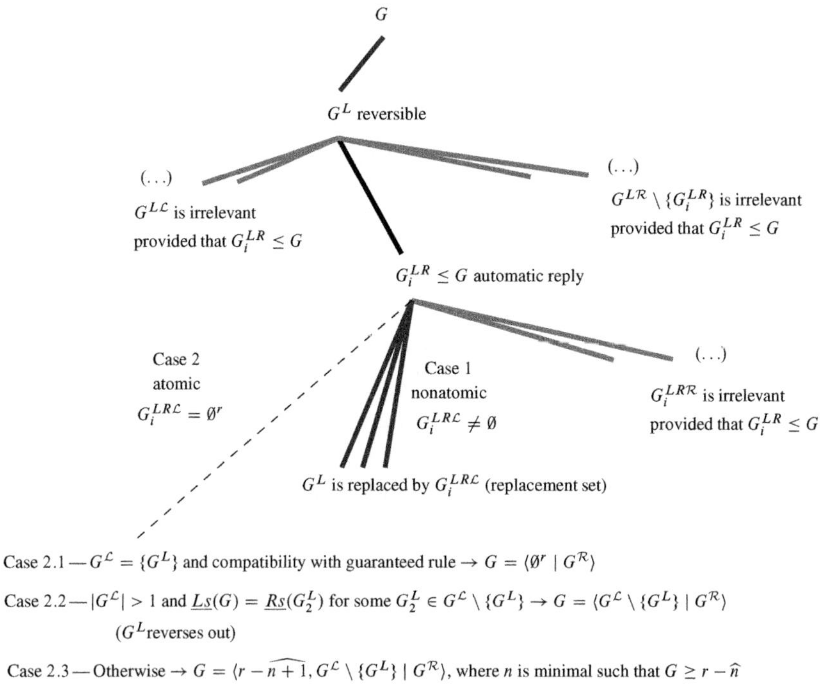

Figure 2. Reversibility of guaranteed games.

game reductions. We consider some games and properties in detail in this section and look briefly at many others in the next.

In scoring-play, as in normal- and misère-play, there are impartial rulesets, and there are partizan rulesets, and between them, there are dicot rulesets (the latter class is called "all-small" in normal-play). A scoring ruleset is *impartial* if all subpositions can be expressed as $r + G$, where the game tree of G is symmetric both in terms of moves and scores. A scoring ruleset is *partizan* if it is not impartial. A scoring ruleset is *dicot* if, from any nonterminal position, *both* players have moves; otherwise, the ruleset is *nondicot*.

In this section we also introduce another interesting classification for scoring rulesets:

(1) In an *all-normal* ruleset, making the last move is *never worse* than allowing the opponent to make the last move.

(2) In an *all-misère* ruleset, allowing the opponent to make the last move is *never worse* than making the last move.

(3) A *hybrid* ruleset is a ruleset that does not satisfy any of the above items.

We will return to this classification scheme, when we have defined some rulesets, but we mention that being a hybrid ruleset is richer than either of the other two.

3.1. DOTS-AND-BOXES. A classical impartial scoring ruleset is DOTS-AND-BOXES, a pencil and paper ruleset carefully studied in [7] and [6]. In DOTS-AND-BOXES, zugzwang positions are very common. The players alternate to connect nodes in an $n \times n$ grid. If you connect all four edges surrounding a cell, you get another move, until you are not able to capture any more cells. By the end of game, the captured number of cells is your score. Because of the nature of the game, near the end it decomposes very nicely into many nearly filled strips of cells, and the strategy depends only on parity. Therefore the concept of a "double-box" is one of the most interesting features of this game. An open-ended double-box, shown below, is a (scoring) zugzwang. Note however that in this example, the parity problem can be avoided (in case parity is already in your favor), by connecting the middle nodes.

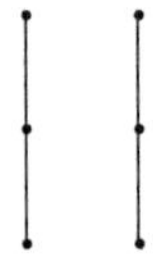

If we for the moment ignore the "extra moves", this position is the zugzwang $\langle\langle 2 \mid -2\rangle, \langle 2, \langle 2 \mid -2\rangle \mid -2, \langle 2 \mid -2\rangle\rangle \mid \langle 2 \mid -2\rangle, \langle 2, \langle 2 \mid -2\rangle \mid -2, \langle 2 \mid -2\rangle\rangle\rangle$. Because of the special rule in DOTS-AND-BOXES where you get an extra move (in another component) when you finish off a game component, disjunctive (long) sum is not exactly the right concept.

3.2. BRUSH. The game of BRUSH is an impartial scoring ruleset where the players alternate to place a brush on any node of a given finite graph. A vertex is "primed" if it has at least as many brushes as its degree. When a primed vertex v is "fired", a brush from v is placed on each adjacent vertex and v and all incident edges are erased — the edges have been "cleaned". If other vertices are primed then chose one and fire it and continue until there are no more primed vertices. (The order of firing is irrelevant.) For each move, the current player adds a number of points according to the number of "cleaned" edges. For example:

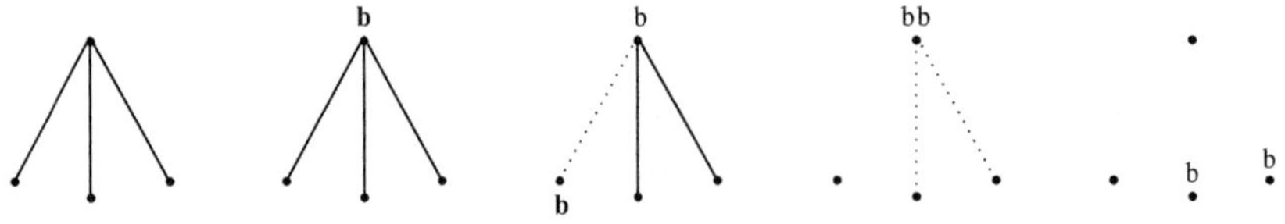

The left most picture represents the starting position, and the second picture represents the first move. When the second player plays the brush on the lower left vertex, as in the third picture, then a sequence of automatic cleaning finishes the game. As this happens, the second player obtains three points (before this move no point was awarded). If the player instead would have played on the upper vertex, then one more move would have been required to finish the game.

More generally, let $S(l, d)$ be the star with l leaves and needing d brushes for the center vertex to fire. It is easy to show

$$S(l, 1) = \langle l \mid -l \rangle \quad \text{and} \quad S(l, 2) = \langle l, \langle l \mid -l \rangle \mid -l, \langle l \mid -l \rangle \rangle.$$

When $d = 3$, the game is a zugzwang — the first play can take one edge but, regardless, the second player will take all the remaining edges — and $S(l, 3) = \langle \varnothing^{-l+1} \mid \varnothing^{l-1} \rangle$. The canonical forms become increasingly more complicated as d increases.

We describe some other variants under CLEANING GAMES in Section 4.

3.3. ***Variations of* KONANE.** Konane is a traditional Hawaiian normal-play game, played with black and white stones on an $n \times n$ grid. This game is convenient to adapt to scoring-play. In KONANE, the players alternate to capture the other player's pieces by jumping neighboring pieces, one at the time, and not changing direction in a single move. Played on the same game-board, the ruleset for normal-play CLOBBER is even simpler; by moving, a player captures one single neighboring piece by replacement. One interesting dicot scoring ruleset is KOBBER, mixing the rulesets KONANE and CLOBBER. In KOBBER, a player may choose between a capture as in KONANE or a capture as in CLOBBER. Each captured piece as in KONANE provides one point. Pieces captured by a CLOBBER move do not add to the score. KOBBER is dicot, but it is not impartial. (The convention is that Left plays black.)

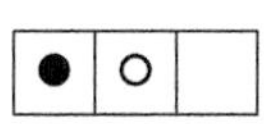

$$\langle 0, 1 \mid 0 \rangle = \langle 1 \mid 0 \rangle$$

In KOBBER, as in DOTS-AND-BOXES, there are zugzwang positions. This is one example:

3.4. ***Bonus/penalty rules.*** For many (nondicot) rulesets, one can adjoin a natural bonus/penalty at the end of play. When a player has no moves, he takes a final penalty equal to the number of own dead stones over the board. Consider TERMINAL KONANE: the rules are exactly as in KONANE except that each captured piece corresponds to one gained point. When a player has no moves he takes a final penalty equal to the number of his own stones still on the board. Next, a zugzwang position of TERMINAL KONANE:

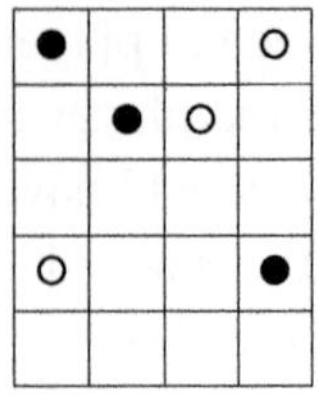

Another KONANE variant is DISKONNECT. In this ruleset a piece is *insecure* if it can be captured by the opponent with a well-chosen sequence of moves (ignoring the alternating-move condition). Otherwise, the piece is *safe*. When a player has no more moves he takes a final penalty equal to the number of own *insecure* stones remaining on the board (this is a GO-type rule).

3.5. ***Examples of all-normal and hybrid games.*** These examples of partizan scoring games fit well into our all-normal, all-misère, or hybrid classification.

In TERMINAL KONANE, if a player makes the last move, then he takes any remaining opponent's pieces. Therefore, to make the last move is *never worse* than to allow the opponent to make the last move; it is all-normal. Played in a disjunctive sum, the strategy is identical to that of normal-play.

DISKONNECT is a more interesting ruleset. In the following example, Left can make the last move. However, in the only winning line for Left, Right makes the last move. Therefore DISKONNECT is not all-normal. It is also easy to check that it is not all-misère, and so it is hybrid.

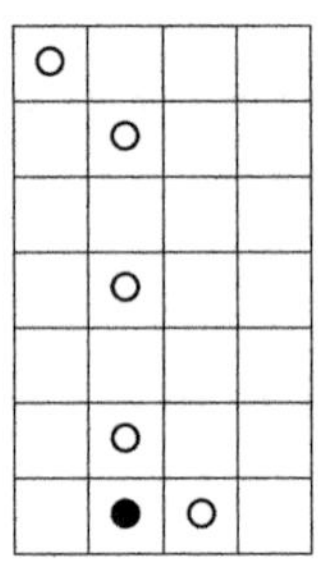

actual score $= -1$

Observe that classical normal-play theory is enough to analyze good TERMINAL KONANE strategies. Although the last move is not guaranteed to win, if you cannot win by making the last move, then you cannot win at all. On the other hand, because DISKONNECT is hybrid (in a sense, it has both normal-play and misère-play behavior), for that game we need scoring combinatorial game theory. Note that in DISKONNECT, as in TERMINAL KONANE, players always prefer to keep playing rather than having no moves. These rulesets satisfy the *guaranteed property*, with the accompanying theory [34]. If a player, say Left, runs out of moves in a component game, then the score of this component cannot decrease (in case of Right opening the component for more play). For instance, consider the following DISCONNECT positions:

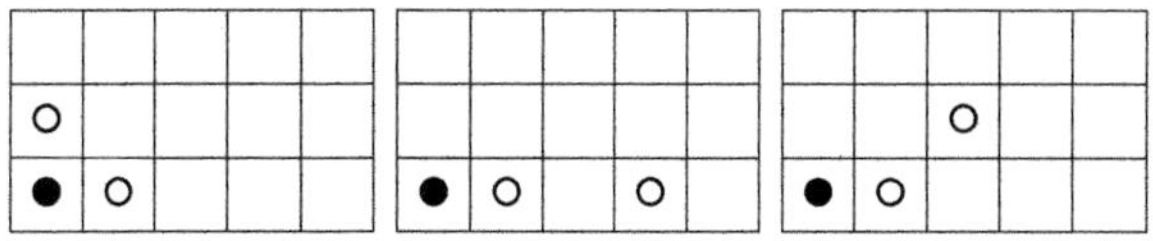

It is known, and not difficult to see, that the canonical forms of these positions are $\langle 1 \mid \varnothing^2 \rangle$, $\langle 2 \mid \varnothing^2 \rangle = 2 + \hat{1}$, and $\langle \langle 2 \mid \varnothing^2 \rangle \mid \varnothing^2 \rangle = 2 + \hat{2}$, respectively. (The middle picture needs some theory, but intuitively it is clear that Left will never jump just one stone: if she wants to play first in some other component, then she will do this directly, otherwise she captures two stones, and lets Right play first elsewhere. The penalty rule guarantees her 2 points in this component either way.) We can also compare

$$\langle 1 \mid \varnothing^2 \rangle < 2 + \hat{1} < 2 + \hat{2}.$$

Online software [33] is available for computations of reduced guaranteed forms, where more complicated positions may be analyzed. This module also contains the games TAKE-SMALL and TAKE-TALL. It turns out that the former is more interesting (because it is hybrid).

3.6. TAKE-SMALL. A nondicot scoring ruleset is TAKE-SMALL. It is played with a strip of a finite number of sticks of different (positive) integer length, possibly with empty spots between some of the sticks. Left chooses two adjacent sticks of lengths a and b where $a \geq b$ (the a-stick to the left). She captures the b-stick by removing it and putting the a-stick in its position. She is not allowed to move the stick over an empty spot. Right plays similarly in the opposite direction. If Right cannot move, Left will be rewarded a bonus: to complete all her possible moves. The reverse bonus is applied to Right if Left cannot move. The winner is the player who has accumulated the greater total length of sticks.

We'll represent a TAKE-SMALL position as a string of numbers where empty spots are indicated by $\cdot$ (dot). For example, in $(3, 2, \cdot, 1, 4)$ Left can move to the position $(\cdot, 3, \cdot, 4, 1)$ plus a capture of 2; Right can move to $(3, 2, \cdot, 4, \cdot)$ plus a capture of 1. Note that $(3, 2, \cdot, 1, 4)$ is actually the disjunctive sum $(3, 2) + (1, 4)$.

TAKE-SMALL is a good example of a partizan scoring ruleset with zugzwang positions that is not all-normal. A position like $(1, 0, 1) = \langle (1, 1) \mid (1, 1) \rangle = \langle \langle 1 \mid -1 \rangle \mid \langle 1 \mid -1 \rangle \rangle$ is a zugzwang. The canonical form for this position is $\langle \varnothing^{-1} \mid \varnothing^{1} \rangle$, therefore the position behaves like a *purely atomic game*, the empty tree with a Left-score and a Right-score.

Consider now the position $(1, 2, 2, 9, 8, 1) - 2$. It has an interesting feature. Left's options are

- $(1, \cdot, 2, 9, 8, 1)$;
- $(1, 2, 2, \cdot, 9, 1) + 6$;
- $(1, 2, 2, 9, \cdot, 8) - 1$.

Right's options are

- $(2, \cdot, 2, 9, 8, 1) - 3$;
- $(1, 2, \cdot, 9, 8, 1) - 4$;
- $(1, 2, 9, \cdot, 8, 1) - 4$.

The canonical form of this position is

$$\langle\langle\langle 11 \mid \varnothing^{11}\rangle \mid 6\rangle, \langle\langle 11 \mid \varnothing^{11}\rangle \mid \langle 1 \mid \varnothing^{1}\rangle\rangle \mid \langle \varnothing^{-4} \mid -4\rangle\rangle - 2.$$

Playing *without points*, as if it were normal-play, the option $\langle\langle 11 \mid \varnothing^{11}\rangle \mid 6\rangle$ reverses out. However, regarded as a scoring game (which it is), the left option $\langle\langle 11 \mid \varnothing^{11}\rangle \mid 6\rangle - 2$ is the only winning move. Left does not make the last move, but it does not matter. The existence of such positions in TAKE-SMALL explains why it is a hybrid ruleset.

3.7. ***Order in scoring-play and outcomes.*** As mentioned before, scoring order does not directly relate to outcomes. Consider the following DISKONNECT position G:

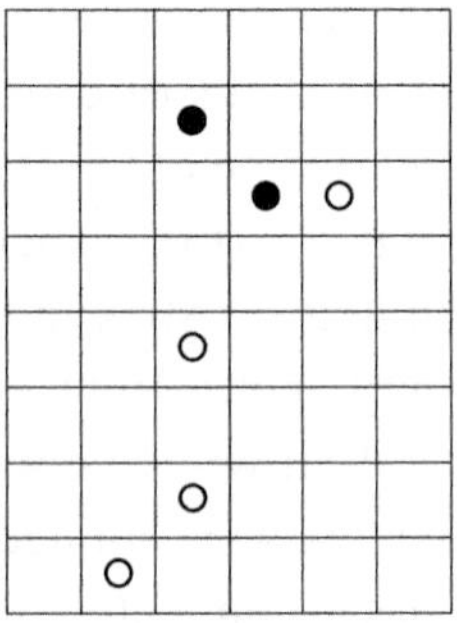

After reductions, the canonical form of G is $\langle 1 \mid 1\rangle$. Left always wins, playing first or second. However, $G \not\succcurlyeq 0$. In order to distinguish G from 0, consider the following zugzwang distinguishing game X:

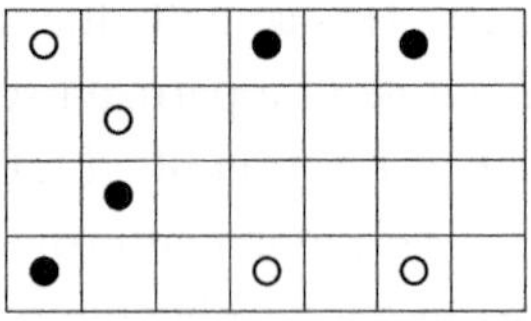

In fact, X reduces a lot! Its canonical form is $\langle \varnothing^{-2} \mid \varnothing^{2}\rangle$; it behaves like an empty tree. Hence, $G + X = \langle 1 \mid 1\rangle + \langle \varnothing^{-2} \mid \varnothing^{2}\rangle$ whose canonical form is $\langle\langle \varnothing^{-1} \mid \varnothing^{3}\rangle \mid \langle \varnothing^{-1} \mid \varnothing^{3}\rangle\rangle$.

Therefore, in X, Right loses going first and, in $G+X$, Right wins going first. Here we display the game $G+X$, as it might appear in a play-situation:

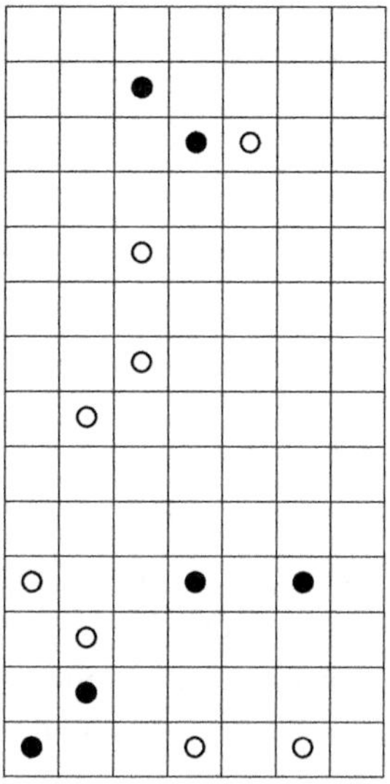

3.8. ***Order in different contexts.*** We recall that the behavior of a position may change depending on the context. Consider the following KOBBER position G:

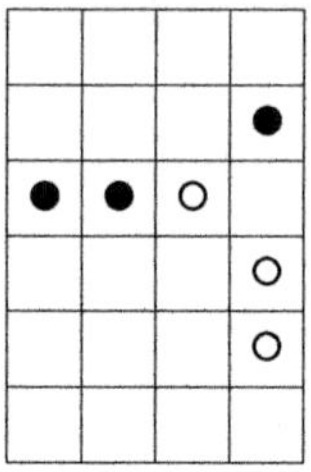

Without reductions, the literal form of G is

$$\langle 0, \langle\langle 2 \mid -1, \langle 1 \mid -1\rangle\rangle \mid \langle 3, \langle 3 \mid 1\rangle \mid 0\rangle\rangle \mid \langle 1 \mid 0\rangle\rangle,$$

where the only Right option is $\langle 1 \mid 0\rangle$ (the left-most white stone can capture a black one via a COBBER move). Left has two options, and in the dicot context, $G \succcurlyeq 1$, that is, in all situations Left *should prefer to have this component rather than add one point to her score* [23]. On the other hand, in the partizan context, there are situations where one point is better than having that form. Consider the position $G-\hat{2}$ (Right has two free moves); Left cannot finish with a final score better than 0.

3.9. ***A scoring games calculator.*** The algebra of games (disjunctive sums, canonical forms, and so on) is not a trivial task and cannot be done manually except for very simple positions. A computer program is required for more complex situations. The *scoring games calculator* [33] is such a program, suitable for the universe of guaranteed scoring games. It was implemented as a set of Haskell modules that run on an interpreter available in any Haskell distribution or embedded in a program that imports these modules.

4. Scoring games on graphs

Despite not having a theory of scoring-play, several scoring games have been studied in the literature. A typical approach, first identified in [43], is to approximate a graph parameter by having the players choose, for example, vertices or edges. When the game is over, an instance of the structure or set has been created. Left wants the cardinality of the structure or set as large as possible, i.e., Left is the *maximizer*, and Right wants it as small as possible, i.e., he is the *minimizer*. Left always wins because Right never scores points, therefore "this is boring for Right" is a valid comment. To convert to a game to be potentially interesting for both players, prelabel the connected components as either Left subgraphs and Right subgraphs. Left's score is then the sum of the cardinalities in the Left components, Right's score is the sum in the Right components and the score in the game is the difference. In all of the games *M2*, *O1*, *O3*, *O4*, and *O8*, the game finishes after a fixed number of moves, which is either the number of vertices or the number of edges. The techniques of Will Johnson [31] will be useful in analyzing these games.

We consider the maximizer/minimizer games first and then report on some others.

4.1. *Maximizer-minimizer games.*

M1: CLEANING GAMES [9]. The toppling number of a graph can be defined in the context of a zero-sum two-player game played on the graph as follows. The maximizer (who moves first in the original game of [9]) and the minimizer alternately place chips on the vertices of the graph. If the number of chips at a vertex is equal to its degree, it sends one chip to each neighbor vertex. The game ends if there exist an infinite sequence of vertices that send chips to their neighbors. Player 1 wants to maximize the number of chips played during the game, while player 2 tries to minimizes this.

M2: COMPETITION-REACHABILITY OF A GRAPH [44]. On a turn, players choose an undirected edge and give it an orientation. After all edges have been chosen in a subgraph H, the score is

$$|\{(x, y) \mid x, y \in V(H) \text{ and there is a directed path from } x \text{ to } y\}|.$$

If H is a Left subgraph then the score is added to Left's total, otherwise it is added to Right's total.

Let P_n be the path on n vertices. In [44], they give $Rs(P_n)$ and $Ls(P_n)$ and, in particular, show that if n is even the game is a zugzwang. This is an impartial game but not all-normal.

M3: DOMINATION GAME. Given a graph G, start with $S = \varnothing$. Players choose a vertex and add it to S but a new vertex of G must be dominated by S. The players are called Staller (Left) and Dominator (Right). This is an impartial game but not all-normal.

This is considered in several papers. In [11] it is shown that $|Ls(G) - Rs(G)|$ is at most 1 and that $\gamma(G) \leq Ls(G) \leq 2\gamma - 1$, where γ is the cardinality of the least dominating set. In [30] they conjecture that $Ls(G) \leq \frac{3}{4}|V(G)|$.

M4: GAME CHROMATIC NUMBER *and* GRUNDY NUMBERS. In the CHROMATIC NUMBER GAME, there is a graph G and a fixed number of colors k. On a turn a player chooses an uncolored vertex and colors it with one of the allowed colors, maintaining a proper coloring. Right wins if G can be colored and Left wins otherwise. Note that for a given k this is a *short disjunctive sum* since Left only has to win in one component. The game chromatic number of G is the least k such that Right can win going first. Despite this concept not coming directly from a scoring game there are several scoring games that have been used to approximate it. See [35] for a follow up and [4] for a survey.

GAME COLORING NUMBER: There are two variants.

(1) MARKING GAME: Given a graph G, a move consists of a player marking an unmarked vertex of G. The game ends when all vertices are marked. For $v \in V(G)$, let $b(x)$ be the number of neighbors of v that are marked before v is marked. The score of the game is $s = 1 + \max\{b(v), v \in V(G)\}$. By convention of this game, Right (the minimizer) starts. The game coloring number $\mathrm{col}_g(G)$ of G is $Rs(G)$. Some known results include

$$\mathrm{col}_g(G) \leq \begin{cases} 4 \text{ if } G \text{ is a forest, [24];} \\ 17 \text{ if } G \text{ is a planar, [55];} \\ 7 \text{ if } G \text{ is a outerplanar, [28; 32].} \end{cases}$$

(2) GAME GRUNDY NUMBER: The colors are the positive integers. Players choose uncolored vertices and it is colored the least integer not in its neighborhood, i.e., let $c(x)$ be the color of vertex x, so that $c(v) = \mathrm{mex}\{c(x) : x \sim v\}$.

M5: The GAME TRANSVERSAL NUMBER. Let $H = (V, E)$ be a hypergraph with vertex set V and edge set E of order $n(H) = |V|$ and size $m(H) = |E|$. A transversal in H is a subset of vertices in H that has a nonempty intersection with every edge of H. A vertex hits an edge if it belongs to that edge. In the transversal game played on H, Left and Right choose vertices where each vertex chosen must hit at least one edge not hit by the vertices previously chosen. The game ends when the set of vertices chosen becomes a transversal in H. This game was introduced in [12]. The game is impartial but not all-normal.

M6: GRAPH SATURATION GAME. Given a family of graphs $\{H_i\}$ and an integer n, build a graph by adding edges such that no copy of any H_i is built. The score is the number of edges.

The original game, a misère- play game was proposed in [13] and rediscovered by Hajnal. In both, the first player to create a triangle K_3 lost.

In [8] they show that $Ls(K_3, n) \leq (n \log n)/2 - 2n \log \log n + O(n)$ and report that Erdös claimed $n^2/5$ as the correct value. See also [45; 35].

The most general formulation of the game is given by Doug West [52].

Also related is the GRAPH MATCHING GAME [53]. Players choose independent edges, and the score is the number of edges. In [17] it is shown that $|Ls(G) - Rs(G)| \leq 1$ and that $Ls(G) \geq \frac{2}{3}m(G)$, where $m(G)$ is the cardinality of the largest matching of G.

4.2. *Other games.*

O1: EDGE-BALANCE INDEX GAME. Given a Graph $G = (V, E)$, Left colors edges blue and Right colors edges by red. Vertex v is colored as blue if it is incident with more blue than red edges and is colored red if it is incident with more red than blue edges and remains uncolored otherwise. The score is the difference between the blue and red vertices. The game is impartial but not all-normal.

The edge balance index is defined in [26] where the indications are that it will be difficult to find this number. In [15] the game is introduced. In addition, the edge-balance index with terminals is also introduced. Here, there are two specified vertices s and t. The play is the same but Left wins if there is a completely blue path (both edges and vertices) from s to t; Right wins if there is completely red path and there is a tie if there is one of each. This is not a scoring game but a normal-play partizan game, except a tie is possible. This becomes an impartial, not all-normal game by making the Left score is the number of blue paths and the Right score the number of red paths.

O2: BRIDGE BUILDING. Suppose a directed graph G, where each arc has an associated cost, and some $s, t \in V(G)$. The goal is to build a path from s to t. The first choice is an edge out of s and then, from each vertex, the next player selects the next vertex to be visited among all neighboring vertices of the current vertex — a path from s to t must still be constructible. The player has to pay the associated cost of the arc. The score is the difference between what the players pay. This is an impartial game but not all-normal, particularly if negative costs (a profit can be made) are allowed.

The game was introduced in [19] where the authors consider the "economic" game where each player minimizes its costs taking into account that also the other player acts in a selfish and rational way. They show that the decision problem associated with such a path is PSPACE-complete even for bipartite graphs. They also note that forcing the play around an even cycle is not a rational play. If we play the difference-of-scores then playing around an even cycle could be a good strategy.

In [20], the authors introduce a variant, SHORTEST CONNECTION GAME, where the two players start at different vertices, say s and t, and build their own paths. The game ends when the two paths first intersect say at vertex m. There are now two paths: L, which is Left's from s to m, and R, Right's from t to m. Left's score S_L is the sum of the edges of L and Right's score R_S is the sum of the edges of R and the game score is the $R_S - L_S$. They prove some complexity results; for example, SHORTEST CONNECTION GAME *is PSPACE-complete for directed bipartite graphs even if all costs are bounded by a constant.* This is not impartial nor all-normal.

O3: GRAPH GRABBING GAME. The vertices of the graph have coins, of possibly different denominations, at each vertex. On a turn, a player removes a noncut vertex and pockets the coins on the vertex. The score is the difference; that is, the winner is the player with the most coins. This is an impartial, not all-normal game.

This is a generalization of the original game played on a path [54]. The next version was played on trees [41] and generalized to the graphs in [36]. In [42], it is shown that if played on a tree T with an even number of vertices then $Ls(T)$ is at least half the total amount of money.

O4: MEDIAN GRAPH GAME. For a set of vertices X, set $d(v, X) = \sum_{x \in X} d(x, v)$ and $\operatorname{med}(X) = \min\{d(v, X) : v \in X\}$. The players choose vertices until all are chosen. Let L be the set chosen by Left and R that by Right. The Left and Right scores are $-\operatorname{med}(L)$ and $-\operatorname{med}(R)$ respectively and the alternating play score, assuming Left plays first, is $\mu(G) = \operatorname{med}(R) - \operatorname{med}(L)$. Compare this with the Weiner game *O8* .

The game was introduced in [14], where the authors give $\mu(G)$ for various G including a subset of trees and for complete bipartite graphs $K_{m,n}$, $m \geq n$, where

$$\mu(K_{m,n}) = \begin{cases} 1 \text{ if } m \neq \text{ and both are odd;} \\ 0 \text{ if } m = n \text{ and both are even;} \\ -1 \text{ if } m \text{ is odd and } n \text{ is even;} \\ -2 \text{ if } m \text{ is even and } n \text{ is odd.} \end{cases}$$

O5: GRAPH OCCUPATION GAME. Let G be a connected graph and let L and R be empty sets. On a turn, Left chooses a vertex and adds it to L, and Right adds a vertex to R. The extra condition is that both sets must be connected and

$L \cap R = \varnothing$. If a player cannot move the other player is allowed to take the rest of the (legal) vertices. The Left score is $|L|/|V(G)|$ and Right's is $|R|/|V(G)|$.

The game is introduced in [46]. The author notes a similarity to GO and shows that for each $\epsilon > 0$ there exists graphs where $Ls(G) \geq 1 - \epsilon$ but also that there exists graphs with $Ls(G) \leq \epsilon$.

O6: PIRATES AND TREASURE. Suppose a graph G with weights on some vertices (treasure) and some blue and red tokens on some vertices — there is no standard starting position. On a turn, Left must move a blue token to an adjacent vertex that contains treasure that they take and which is added to the player's score. The game is over when the player to move cannot and the player with the greatest amount of treasure wins.

The game is introduced in [51] where it is shown to be NP-hard to determine the final scores. In [2] it is shown to be PSPACE complete to determine if Right can win.

O7: SUBSET SUM GAME. Suppose an integer n and a set of weights $\{w_i\}$. The sum S starts at 0 and is not allowed to exceed n. On a turn, a player chooses a weight, removes from the set and adds it to the sum. A player's score is the sum of the weights they added to S.

The game, coming from the *knapsack* problem, is introduced in [18], where they consider two strategies: the greedy strategy and the one move look-ahead.

O8: WEINER INDEX GAME. Given a graph G, players choose vertices. The Left score is the sum of all the distances between the vertices Left chose. Similarly for Right. Since the winner is the player with the least sum, the score is defined as $Rs - Ls$. This is introduced in [3], where the authors require that $|V(G)|$ be even. It arises out of the study of the Weiner index from mathematical chemistry. They report that $|Ls(K_{1,2n-1}) - Rs(K_{1,2n-1})| = n - 1$.

References

[1] M. H. Albert, R. J. Nowakowski, and D. Wolfe. *Lessons in play*. A K Peters, Ltd., 2007.

[2] P. Arendas, Z. L. Blazsik, B. Bodor, and Cs. Szabo. On the complexity and topology of scoring games: of pirates and treasure. 2015, unpublished.

[3] K. Balakrishnan, B. Bresar, M. Changat, S. Klavzar, A. Vesel, and P. Z. Pletersek. Equal opportunity networks, distance-balanced graphs, and Wiener game. *Discrete Optim.*, 12:150–154, 2014.

[4] T. Bartnicki, J. Grytczuk, H. A. Kierstead, and X. Zhu. The map-coloring game. *Amer. Math. Monthly*, 114:793–803, 2007.

[5] J. Beck. *Combinatorial games: tic-tac-toe theory*. Series: Encyclopedia of Mathematics and its Applications (No. 114). Cambridge University Press, 2006.

[6] E. R. Berlekamp. Sums of $N \times 2$ amazons. In F.T. Bruss and L.M. Le Cam, editors, *Lecture Notes—Monograph Series*, volume 35, pages 1–34. Institute of Mathematical Statistics, Beechwood, Ohio, 2000. Papers in honor of Thomas S. Ferguson.

[7] E. R. Berlekamp, J. H. Conway, and R. K. Guy. *Winning ways for your mathematical plays*, volume 1. Academic Press Inc. [Harcourt Brace Jovanovich Publishers], London, 1982. Games in general.

[8] C. Biró, P. Horn, and D. J. Wildstrom. An upper bound on the extremal version of Hajnal's triangle-free game. *Discrete Appl. Math.*, 198:20–28, 2016.

[9] A. Bonato, William B. Kinnersley, and P. Prałat. Toppling numbers of complete and random graphs. *Discrete Math. Theor. Comput. Sci.*, 16(3):229–251, 2014.

[10] C. L. Bouton. Nim, a game with a complete mathematical theory. *Ann. of Math.*, 3(2):35–39, 1902.

[11] B. Bresarar, S. Klavzar, and D. F. Rall. Domination game and an imagination strategy. *SIAM J. Discrete Math.*, 24:979–991, 2010.

[12] Csilla Bujtas, Michael A. Henning, and Zsolt Tuza. Bounds on the game transversal number in hypergraphs. *European J. Combin.*, 59:34–50, 2017.

[13] S. C. Cater, F. Harary, and R. W. Robinson. One-color triangle avoidance game. *Congr. Numer.*, 153:211–221, 2001.

[14] Manoj Changat, Divya S. Lekha, Iztok Peterin, Ajitha R. Subhamathi, and Simon Spacapan. The median game. *Discrete Optim.*, 17:80–88, 2015.

[15] Ping-Tsai Chung. Edge-balance index games. *Appl. Math. Sci.*, 9:6691–6696, 2015.

[16] John H. Conway. *On numbers and games*. Academic Press [Harcourt Brace Jovanovich Publishers], London, 1976. London Mathematical Society Monographs, No. 6.

[17] Daniel W. Cranston, William B. Kinnersley, Suil O, and Douglas B. West. Game matching number of graphs. *Discrete Appl. Math.*, 161(13-14):1828–1836, 2013.

[18] Andreas Darmann, Gaia Nicosia, Ulrich Pferschy, and Joachim Schauer. The subset sum game. *European J. Oper. Res.*, 233(3):539–549, 2014.

[19] Andreas Darmann, Ulrich Pferschy, and Joachim Schauer. On the shortest path game. *Discrete Appl. Math.*, 2015.

[20] Andreas Darmann, Ulrich Pferschy, and Joachim Schauer. The shortest connection game. *Discrete Appl. Math.*, to appear.

[21] Paul Dorbec, Gabriel Renault, Aaron N. Siegel, and Eric Sopena. Dicots, and a taxonomic ranking for misère games. *J. Combin. Theory Ser. A*, 130:42–63, 2015.

[22] J. M. Ettinger. A metric for positional games. *Theoret. Comput. Sci.* (Math Games), 230:207–219, 2000.

[23] J. M. Ettinger. *Topics in combinatorial games*. PhD thesis, University of Wisconsin — Madison, 1996.

[24] U. Faigle, W. Kern, H. Kierstead, and W. T. Trotter. On the game chromatic number of some classes of graphs. *Ars Combin.*, 35:143–150, 1993.

[25] Aviezri S. Fraenkel. The vile, dopey, evil and odious game players. *Discrete Math.*, 312(1):42–46, 2012. Algebraic Graph Theory: a volume dedicated to Gert Sabidussi on the occasion of his 80th birthday.

[26] J. A. Gallian. A dynamic survey of graph labeling *Elect. J. Comb.* Dynamic Survey #DS6: Update of Dec. 29, (2014)-

[27] Patrick M. Grundy. Mathematics and games. *Eureka*, 2:6–8, 1939.

[28] D. J. Guan and X. Zhu. Game chromatic number of outerplanar graphs. *J. Graph Theory*, 30:67–70, 1999.

[29] Olof Hanner. Mean play of sums of positional games. *Pacific J. Math.*, 9:81–99, 1959.

[30] M. A. Henning and D. F. Rall. Progress towards the total domination game $\frac{3}{4}$-conjecture *Discrete Math.*, 339:2620–2627, 2016.

[31] Will Johnson. The combinatorial game theory of well-tempered scoring games. *Internat. J. Game Theory*, 43(2):415–438, 2014.

[32] H. A. Kierstead and Daqing Yang. Very asymmetric marking games. *Order*, 22(2):93–107 (2005).

[33] Urban Larsson, Richard J. Nowakowski, João P. Neto, and Carlos P. Santos. Guaranteed scoring games. *Electron. J. Combin.*, 23(3):Paper 3.27, 29, 2016.

[34] Urban Larsson, Richard J. Nowakowski, and Carlos P. Santos. When waiting moves you in scoring combinatorial games, 2015.

[35] N. Mehta and A. Seress. Connected, bounded degree, triangle avoidance games. *Elect. J. Comb.*, 18(1):1–37, 2011.

[36] Piotr Micek and Bartosz Walczak. A graph-grabbing game. *Combin. Probab. Comput.*, 20:623–629, 2011.

[37] Rebecca Milley. Partizan kayles and misère invertibility. *Integers*, 15(1309.1631), 2015.

[38] John Milnor. Sums of positional games. *Ann. of Math. Stud.* (Contributions to the Theory of Games, H. W. Kuhn and A. W. Tucker, eds.), Princeton, 2(28):291–301, 1953.

[39] Richard J. Nowakowski. Unsolved problems in combinatorial game theory. In *Games of No Chance* 4, pages 279–308. Cambridge University Press, 2015.

[40] Thane E. Plambeck and Aaron N. Siegel. Misere quotients for impartial games. *J. Combin. Theory Ser. A*, 115(4):593–622, 2008.

[41] Moshe Rosenfeld. A gold-grabbing game. http://garden.irmacs.sfu.ca/?q=op/a_gold_grabbing _game.

[42] Deborah E. Seacrest and Tyler Seacrest. Grabbing the gold. *Discrete Math.*, 312(10):1804–1806, 2012.

[43] Suk J. Seo and Peter J. Slater. Competition parameters of a graph. *AKCE Int. J. Graphs Comb.*, 4(2):183–190, 2007.

[44] Suk J. Seo and Peter J. Slater. Competition-reachability of a graph. *Discrete Math.*, 306(6): 580–590, 2006.

[45] Á. Seress. On Hajnal's triangle-free game. *Graphs Combin.*, 8:75–79, 1992.

[46] Alexander Shapovalov. Occupation games on graphs in which the second player takes almost all vertices. *Discrete Appl. Math.*, 159(15):1526–1527, 2011.

[47] Aaron N. Siegel. *Combinatorial game theory*. American Math. Society, 2013.

[48] Aaron N. Siegel. Misère canonical forms of partizan games. In *Games of No Chance* 4, pages 225–240. Cambridge University Press, 2015.

[49] Roland P. Sprague. Über mathematische Kampfspiele. *Tôhoku Mathematics Journal*, 13(290): 438–444, 1935.

[50] Fraser Stewart. *Scoring play combinatorial games*. PhD thesis, University of Dundee, 2011.

[51] Fraser Stweart. Pirates and treasure. *Integers*, 13(#G3):18 pages, 2013.

[52] Doug West. http://www.math.illinois.edu/dwest/regs/fsatgame.html

[53] Doug West. http://www.math.illinois.edu/dwest/regs/gamatch.html

[54] Peter Winkler. *Mathematical puzzles: a connoisseur's collection*. A K Peters, Ltd., 2004.

[55] Xuding Zhu. Refined activation strategy for the marking game. *J. Combin. Theory Ser. B*, 98(1):1–18, 2008.

urban031@gmail.com *Department of Industrial Engineering and Management, Technion - Israel Institute of Technology, Haifa, Israel*

rjn@mathstat.dal.ca *Department of Mathematics and Statistics, Dalhousie University, Halifax, NS, Canada*

cmfsantos@fc.ul.pt *Center for Functional Analysis, Linear Structures and Applications, University of Lisbon, Lisboa, Portugal*

Games of No Chance 5
MSRI Publications
Volume **70**, 2017

Restricted developments in partizan misère game theory

REBECCA MILLEY AND GABRIEL RENAULT

Misère games have excited new interest over the past decade with the introduction of an indistinguishability relation for analyzing positions modulo restricted subsets of games. We present a survey of recent progress in the theory of partizan misère games, including some results for general misère play, but focusing primarily on this *restricted* misère play. We discuss new and current work on game comparison and game inverses, as well as ongoing research around reversibility and canonical forms in restricted misère play. We also show how general results in each of these areas have been applied to specific games to find solutions under misère play.

1. Introduction

Most research in combinatorial game theory assumes normal play, where the first player unable to move loses, as opposed to misère play, where the first player unable to move wins. It is rather remarkable how much changes when we simply switch the goal from *getting* the last move to *avoiding* the last move. At first glance one might think misère play is merely the "opposite" of normal play, but this is not at all the case. There is actually no relationship between normal outcome and misère outcome: for every pair of (not necessarily distinct) outcomes $\mathcal{O}_1, \mathcal{O}_2 \in \{\mathcal{L}, \mathcal{N}, \mathcal{P}, \mathcal{R}\}$, there is a game with normal outcome $\mathcal{O}_1$ and misère outcome $\mathcal{O}_2$ [11]. Likewise, strategies from normal play are in general neither the same nor reversed for misère play. For example, in normal play, Left would always choose a move to $1 = \{0|\cdot\}$ over a move to $0 = \{\cdot|\cdot\}$, but in misère play there are situations[1] in which Left should choose 1 over 0 and others where Left prefers 0 over 1. This means that 0 and 1 are incomparable in misère play, which goes against our intuition that Left is trying to run out of moves before Right.

Renault was supported by the ANR-14-CE25-0006 project of the French National Research Agency.
MSC2010: 91A46.
Keywords: combinatorial game theory, misére games, partizan games.

[1] Left wins playing first on the single game 0 and loses playing first on the single game 1, but loses playing first on $0 + *$ and wins playing first on $1 + *$.

So we really are in a fog in misère play. We look to the elegant algebra of normal-play games and hope for some semblance of structure, but we are dismayed at every turn:

- Zero is trivial. In normal play we have the wonderful property that every previous-win game is equal to zero. In misère, the zero game is next-win, but our hopes that perhaps every next-win game is equal to zero are more than dashed; in fact, only the game $\{\cdot|\cdot\}$ is equal to zero [11]. In particular, for any game $G \neq 0$, the game $G - G$ is not equal to zero (a very troublesome fact indeed). Consequently, there are no nonzero inverses, and there is no longer an easy test for the equality and inequality of games.
- Equality (and inequality) is rare and difficult to prove. Partly due to the triviality of zero, equivalence classes induced by the equality relation are much smaller in misère play, and it is not often possible to simplify games. Inequality is likewise uncommon, resulting in unfortunate situations like the incomparability of 1 and 0.
- Addition is less intuitive. Disjunctive sum is defined in misère as in normal play, but much of our intuition for the interaction of games in a sum is lost. For example, the sum of two left-win games may be right-win! In fact, nothing can be said about the addition table of outcomes in miséré play: for any three outcomes $\mathcal{O}_1, \mathcal{O}_2, \mathcal{O}_3 \in \{\mathcal{L}, \mathcal{N}, \mathcal{P}, \mathcal{R}\}$, we can find positions G and H such that G has misère outcome $\mathcal{O}_1$, H has misère outcome $\mathcal{O}_2$, and $G + H$ has misère outcome $\mathcal{O}_3$ [11]. Other problems arise with sums, due to the lack of simplification under misère play: for example, the sum of a game with value $n \in \mathbb{Q}_2$ and a game with value $m \in \mathbb{Q}_2$ may not even be a number-valued game[2], let alone the game with value $n + m$.

For these reasons and others, the study of misère games was neglected for most of the 20th century. One chapter of *On Numbers and Games* presents an analysis of "How to Lose When You Must", and *Winning Ways* extends this work in their chapter "Survival in the Lost World", but both texts consider only impartial misère games. The *genus theory* developed in the latter allowed for the analysis of certain impartial misère games, but left most unsolvable [19]. A theory for partizan misère games seemed, if possible, even more elusive.

The fog began to lift when Thane Plambeck [18; 20] and Aaron Siegel [20] introduced a modified equality relation for games under misère play. Instead of requiring games to be interchangeable in *any* sum of games, two games will be considered *equivalent modulo* $\mathcal{U}$ if they can be interchanged in any sum of games from the set $\mathcal{U}$. For example, we might take $\mathcal{U}$ to be the set of all positions that

[2]This, along with the incomparability of number-valued games, demonstrates that the numerical value system developed for normal play is virtually meaningless in misère.

occur in some particular game, such as Domineering, and then two domineering positions are equivalent "modulo Domineering" if they are interchangeable in any sum of domineering positions. This is a natural and practical equivalence relation, and its introduction has encouraged renewed interest in the study of misère games.

Although initially designed only for impartial games, this "restricted" misère analysis works equally well for partizan games [21]. The study of restricted partizan misère games began with the doctoral theses of Paul Ottaway [16] and Meghan Allen [2], and has continued with a relative flurry of recent activity from a number of additional researchers. The present survey of partizan misère game theory highlights the most significant results from recent research, including canonical forms of partizan misère games, the invertibility of games under restricted misère play, and applications to specific partizan misère versions of Nim, Kayles, and Hackenbush. We begin with some prerequisite definitions.

2. Prerequisites

We use the notation $G = \{G^L \mid G^R\}$, where $G^L = \{G^{L_1}, G^{L_2}, \ldots\}$ is the set of left options from G and G^L is a particular left option. Any position which can be reached from a game G is called a *follower* of G.

The *outcome* of a game is $\mathscr{L}$ if Left wins playing first or second, $\mathscr{R}$ if Right wins playing first or second, $\mathscr{N}$ if either player can win going first, and $\mathscr{P}$ if neither player can win going first. These outcomes are partially ordered as in normal play; that is, $\mathscr{L} > \mathscr{P} > \mathscr{R}$, $\mathscr{L} > \mathscr{N} > \mathscr{R}$, and $\mathscr{P} \| \mathscr{N}$. We use the outcome function $o^-(G)$ to denote the misère outcome of G and $o^+(G)$ to denote the normal outcome of G. The *outcome classes* $\mathscr{L}^-, \mathscr{N}^-, \mathbb{R}^-, \mathscr{P}^-$ are the sets of all games with the indicated outcome under misére play, so that we can write $G \in \mathscr{L}^-$ when $o^-(G) = \mathscr{L}$.

In normal play, the *negative* of a game is defined recursively as $-G = \{-G^R | -G^L\}$, and is so-called because $G + (-G) = 0$ for all games G under normal play. As mentioned in the introduction, this property holds in misère play only if G is the zero game $\{\cdot \mid \cdot\}$ [11]. To avoid confusion and inappropriate cancellation, we generally write $\overline{G}$ instead of $-G$ and refer to this game as the *conjugate* of G.

Most other definitions from normal-play game theory are used without modification for misère games, including disjunctive sum, equality, and inequality. See [1] for an excellent introduction to normal play. In this paper, when equality and inequality relations are used, misère play is assumed unless otherwise stated. The equivalence relation developed by Plambeck and Siegel is formalized in Definition 2.1 below.

Definition 2.1. For games G and H and a set of games $\mathcal{U}$, the terms *equivalence* and *inequality, modulo* $\mathcal{U}$, are defined by

$$G \equiv H \pmod{\mathcal{U}} \iff o^-(G+X) = o^-(H+X) \text{ for all games } X \in \mathcal{U},$$
$$G \geqq H \pmod{\mathcal{U}} \iff o^-(G+X) \geq o^-(H+X) \text{ for all games } X \in \mathcal{U}.$$

The word *indistinguishable* is sometimes used instead of *equivalent*, and if $G \not\equiv H \pmod{\mathcal{U}}$ then G and H are said to be *distinguishable* modulo $\mathcal{U}$. In this case there must be a game $X \in \mathcal{U}$ such that $o^-(G+X) \neq o^-(H+X)$, and we say that X *distinguishes* G and H. The set $\mathcal{U}$ is called the *universe*. All universes in this survey are closed under followers and disjunctive sum, and most are also closed under conjugation. Although we usually assume G and H are games in $\mathcal{U}$, this stipulation is unnecessary, and it is sometimes useful to compare games modulo a universe $\mathcal{U}$ even when the games do not belong to $\mathcal{U}$.

Notice that $G \equiv H \pmod{\mathcal{U}}$ implies $G \equiv H \pmod{\mathcal{V}}$ for any subset $\mathcal{V} \subseteq \mathcal{U}$, but in general games can be equivalent in the smaller universe and distinguishable in the larger. Also note that this equivalence is actually a congruence relation with respect to disjunctive sum of games.

2A. ***Specific universes and properties.*** A number of specific game universes are discussed in the sections to follow, and we will define them here. Firstly, we identify games where one or both players have no move: a *left end* is a position with no first move for Left (that is, G with $G^L = \varnothing$), a *right end* is a position with no first move for Right ($G^R = \varnothing$), and an *end* is a position that is either a left end or a right end or both (the zero game).

A left (right) end is called *dead* if each of its followers is also a left (right) end. Games in which every end follower is a dead end are called *dead-ending*. Figures 1 and 2 provide examples to illustrate these definitions. By definition, in dead-ending games, if Left has no move at some point, then Left will never have a move again. This is a natural property held by well-studied games such as Hackenbush, Domineering, and other so-called *placement* games (where players move by placing pieces on a board). The set of all dead-ending games is denoted $\mathcal{E}$ and has proven to be rich in interesting results for misère play. The set of all dead ends and sums of dead ends is denoted $\mathcal{E}_e$.

Figure 1. A dead left end and a left end that is not dead.

Figure 2. A dead-ending game and a game that is not dead-ending.

Games in which the only end is zero — that is, where Left can move if and only if Right can move — are called *all-small* in normal play and *dicot* in misère. The set of all dicot games is denoted $\mathcal{D}$. Note that $\mathcal{D}$ is a proper subset of $\mathcal{E}$.

A position is called *alternating* if neither player can make consecutive moves; that is, if G^{LL} and G^{RR} are empty for all G^L and all G^R. This restriction allows for easier analysis under misère play. The set of all sums of alternating games is denoted $\mathcal{A}$, and the set of all alternating ends and their sums is denoted $\mathcal{A}_e$.

We end this section by mentioning two significant properties that universes may have. The following definitions appear in recent work by Larsson, Nowakowski, and Santos, as part of their new framework called "absolute game theory", in which they generalize the theories of normal, misère, and other types of play. Note that their universes are closed under conjugates. We give the definition of *density* specifically for the misère case, but it can be defined generally as well.

Definition 2.2 [8]. A universe $\mathcal{U}$ is *parental* if for any two nonempty sets $\mathcal{A}, \mathcal{B} \subseteq \mathcal{U}$, the position $\{\mathcal{A} | \mathcal{B}\}$ is also in $\mathcal{U}$.

Definition 2.3 [8]. A universe $\mathcal{U}$ is *dense* under misère play if, for all $G \in \mathcal{U}$ and any outcome $\mathcal{O}$ in $\{\mathcal{L}, \mathcal{R}, \mathcal{N}, \mathcal{P}\}$, there is an $H \in \mathcal{U}$ such that the misère outcome of $G + H$ is $\mathcal{O}$.

These properties are relevant to the current research areas of comparability and invertibility of misère games. These areas are discussed in Sections 3 and 4; Section 5 discusses the problems with and very recent solutions to the reversibility of misère games, and Section 6 discusses solutions to specific games under misère play. We begin with the comparability of misère games.

3. Comparability

In normal play, $G \geq H$ if and only if $G - H$ is previous-win, and so there is an easy test for inequality of games. In general misère play, we do not have this test, and so proving $G \geq H$ in misère play requires proving $o(G + X) \geq o(H + X)$ for all games X.

We do at least have a slightly modified *hand-tying* principle for misère games [11]. In normal-play, this principle says that if two games G and H differ only by the addition of one or more extra left options to G, then Left can do at least as well playing G as playing H ($G \geq H$, in normal play); at worst, Left can "tie her hand" and ignore the extra options, thereby essentially playing the game H instead of G. In misère play, the same argument holds, with one stipulation: the set H^L of left options cannot be empty. If it is, adding a left option is not always beneficial to Left, who is sometimes happy to have no move in a position. However, when there already exists at least one left option, Left can simply ignore any additional ones. This principle was used in [12] and [6] to classify day-2 and day-3 dicot games.

If we restrict ourselves to a particular universe of games $\mathcal{U}$, then we need only consider games X in $\mathcal{U}$, and so we may be able to show $G \geq H \pmod{\mathcal{U}}$ even if $G \not\geq H$ in general. When current research from absolute game theory [8] (see also [7]) is applied to misère games, we see that comparability of games G and H can be demonstrated without considering the sum of G and H with *all* $X \in \mathcal{U}$, provided certain conditions are met by $\mathcal{U}$. Specifically, if $\mathcal{U}$ is parental and dense, then the following result holds.

Theorem 3.1 [8]. *Let $\mathcal{U}$ be a universe that is conjugate-closed, parental, and dense. Then $G \geq H \pmod{\mathcal{U}}$ if and only if*

(i) *for all G^R there is $G^{RL} \geq H \pmod{\mathcal{U}}$ or $G^R \geq H^R \pmod{\mathcal{U}}$;*

(ii) *for all H^L there is $G^L \geq H^L \pmod{\mathcal{U}}$ or $G \geq H^{LR} \pmod{\mathcal{U}}$;*

(iii) *if H is a left end, then Left wins playing first in $G + X$ for any left end X in $\mathcal{U}$; and*

(iv) *if G is a right end, then Right wins playing first in $H + X$ for any right end X in $\mathcal{U}$.*

We will see this importance of ends in other areas of misère analysis, including invertibility, where we look next.

4. Invertibility

As stated in the introduction, no nonzero game has an additive inverse in general misère play. However, in a restricted universe $\mathcal{U}$, a game G may satisfy $G + \overline{G} \equiv 0 \pmod{\mathcal{U}}$ — or perhaps even $G + H \equiv 0 \pmod{\mathcal{U}}$ for $H \not\equiv \overline{G} \pmod{\mathcal{U}}$, as discussed in Section 4A — and then the game G is said to be *invertible* modulo $\mathcal{U}$. The first result of this kind was Meghan Allen's demonstration that $* + * \equiv 0$ in any universe of dicot games [3]. Allen's result is generalized in [10] with the following sufficient condition for invertibility in the universe of dicots.

Theorem 4.1 [10]. *If $G + \overline{G} \in \mathcal{N}^-$ and $H + \overline{H} \in \mathcal{N}^-$ for all followers H of G, then $G + \overline{G} \equiv 0 \pmod{\mathcal{D}}$.*

invertible position	universe of invertibility	reference
$*$	$\mathcal{D}$	[3]
$* : x$ for $x \in \mathbb{Q}_2$	$\mathcal{D}$	[10]
any dead end (e.g., $n \in \mathbb{Z}$)	$\mathcal{E}$	[15]
any alternating end	$\mathcal{A}$	[14]
any alternating game not in $\mathcal{P}^-$	$\mathcal{A}$	[12]

Table 1. Some known instances of invertibility in restricted misère play.

Theorem 4.1 was used to show that the ordinal sum of $*$ and a number[3], $* : x$, is invertible in the universe of dicots. This result and others are presented in Table 1, which lists some of the positions known to be invertible in the universes of alternating games ($\mathcal{A}$), dicots ($\mathcal{D}$), or dead-ending games ($\mathcal{E}$).

Many of these instances of invertibility were demonstrated using the following sufficiency condition for invertibility in restricted misère play. Generally, one proves $G + \overline{G} \equiv 0 \pmod{\mathcal{U}}$ by showing that the outcome of $G + \overline{G}$ is the same as the outcome of $G + \overline{G} + X$ for any X in $\mathcal{U}$. Theorem 4.2 essentially says that you need only check the X positions that are ends, an idea that is paralleled by the more recent result about comparability of misère games (Theorem 3.1).

Theorem 4.2 [15]. *Let U be a universe of games closed under followers, sum, and conjugation, and let $S \subseteq U$ be a set of games closed under followers. If $G + \overline{G} + X \in \mathcal{L}^- \cup \mathcal{N}^-$ for every game $G \in S$ and every left end $X \in \mathcal{U}$, then $G + \overline{G} \equiv 0 \pmod{\mathcal{U}}$ for every $G \in S$.*

4A. ***Nonconjugal invertibility.*** A bizarre property of restricted misère play is that a game G can have an additive inverse modulo some universe $\mathcal{U}$ without that inverse being the conjugate $\overline{G}$. The only known partizan result of this kind appears in [13], where the games $\{0|\cdot\}$ and $\{1|0\}$ sum to zero among the set of all *partizan Kayles*[4] positions, despite neither being equivalent to the conjugate of the other in this universe. This inverse pair is further remarkable for the fact that one position is right-win and the other is previous-win.

In the example from partizan Kayles, the actual conjugates of $\{0|\cdot\}$ and $\{1|0\}$ do not even belong to the universe. In [12] it was conjectured that being closed under conjugation would prevent such occurrences of nonconjugal invertibility; however, a counterexample can be seen in [20, Appendix 6] for a subset of impartial games.

[3]By *number* (*integer*) in misère play, we mean a game that is identical to the normal-play canonical form of a number (integer).

[4]The paper [13] solves a partizan version of the game Kayles, played on rows of pins, where Left can knock down a single pin and Right can knock down two adjacent pins.

This leads us to a pressing open question in misère theory: in what universes $\mathcal{U}$ do we have no nonconjugate inverses, so that $G + H \equiv 0 \pmod{\mathcal{U}}$ only if $H \equiv \overline{G} \pmod{\mathcal{U}}$? In-progress research suggests that adapting the proof of a similar result on scoring games [9] can prove that no nonconjugate inverses occur in dicot games, dead-ending games, or any universe that is parental, dense, and amenable to a type of "replacement" reversibility through ends, as discussed in the next section.

5. Reversibility and canonical forms

Given the relative lack of structure in misère play, it is perhaps surprising that we have canonical forms here just as in normal play, with precisely the same definitions of domination and reversibility (with inequality under misère play instead of normal play). This was shown in the collaborative paper of G. A. Mesdal [11] and subsequent work by Aaron Siegel [21]. The latter also demonstrated that, as in normal play, the simplified game obtained by removing dominated options and bypassing reversible ones is unique.

So canonical forms "work" in misère play; but in general the concept is less useful than in normal play, because it is so hard to find instances of domination or reversibility. If we restrict ourselves to a particular universe of games $\mathcal{U}$, then we may get domination or reversibility in the restricted universe that does not occur in general, due to inequalities of the form $G \geq H \pmod{\mathcal{U}}$. Consequently, a game could have different "restricted canonical forms" in different universes. However, the construction of a canonical form — specifically, how we deal with reversible options — is not quite the same when the universe is restricted in this way.

The problem of reversibility in restricted universes is related to the following result of [21], which is used in the construction of misère canonical forms.

Lemma 5.1 [21, Lemma 3.5]. *If H is a left end and G is not, then $G \not\geq H$.*

This result holds in the context of all misère games; however, it may be the case that a non-Left-end G can be greater than a left end H *modulo some universe* $\mathcal{U}$. For example, in the universe of dicot games $\mathcal{D}$, we have $\{0, *|*\} \geq 0 \pmod{\mathcal{D}}$ [6].

Why is this a problem? In general, Lemma 5.1 means we never have to worry about reversibility through an end; it cannot happen that $G \geq G^{LR}$ if G^{LR} is a left end, and so in such a case G^L could not be reversible. This fact is exploited in the proof that reversibility works in misère play: that $G' = G$ when G' is obtained from G by replacing a reversible option G^L with the left options of G^{LR} [21]. Since the same fact does not hold in restricted misère play, the result from [21] no longer applies, and so we cannot necessarily bypass all reversible options. In the example above from [6], even though $G = \{0, *|*\} \geq 0 \pmod{\mathcal{D}}$ and $0 = *^R = G^{LR}$, it is not the case that $\{0, *|*\} \equiv \{0|*\} \pmod{\mathcal{D}}$. Left's only good move in G is to $*$, so removing $*$ and replacing it with no options does not result in a game that is just as good for Left.

In [6], the proof from [21] of uniqueness of misère canonical forms was adapted to construct unique restricted canonical forms in the universe of dicot games. For dicots, the problem of reversibility through ends is dealt with as follows: if G^L is reversible through a left end, then replacing G^L with $*$ results in an equivalent game. This solution should be further adaptable to other restricted universes: we would just need to find a suitable "replacing game", that might depend on G, to replace an option G^L that is reversible through a left end. As the invertibility of $*$ (modulo dicots) is used in the proof of the uniqueness of the canonical forms for dicots, the replacing game in other universes will most likely have to be invertible. Solving the problem of reversibility in specific misère universes is an open area of research; notably, in-progress work from the authors of [8] suggests a solution for certain universes, including dead-ending games.

This completes our survey of the recent developments in misère theory, including comparability, invertibility, and reversibility of misère games. We next show how some of these advancements have been applied to solve specific games under misère play.

6. Applications to specific games

A number of specific partizan games have been successfully solved using the theory of restricted misère play. These solutions consider equivalence modulo the universe of all positions that occur under the specific game rule set, and take advantage of results for broader superset universes.

Penny Nim is a partizan variant of Nim played with stacks of coins. In each stack, coins are all heads up or all tails up, and the entire stack may be lying sideways. On her turn, Left chooses a stack with tails-up or sideways orientation, removes any number of coins from it, and turn it heads up. Right plays similarly on heads-up or sideways coins stacks, but leaves them tails-up. Notice that any position of this game is alternating, and the potential for sideways stacks means that not all components are initially ends. The game is solved in [12], using the analysis of the alternating universe $\mathscr{A}$, in which most "single-stack" positions are invertible. The solution involves first simplifying single stacks of coins, modulo $\mathscr{A}$, and then determining outcomes of sums of these finitely many simplified positions.

Partizan Kayles is a variant of Kayles, played on a row of pins, where Left can knock down a single pin and Right can knock down exactly two adjacent pins. Notice that any position of this game is dead-ending. The game is solved in [13]. The key is to see that Left should always take an isolated single pin when she can; this allows for removal of dominated options and decomposition of long rows of pins into shorter rows — into only isolated single pins and pairs of pins, in fact — and then all that remains is to see who wins on a sum of such positions. This is easily done once it is shown that an isolated single pin and an isolated pair of adjacent pins "cancel" (that is, they are additive inverses).

Hackenbush Sprigs is a particular case of the game of Hackenbush. The game can be seen as rows of blue, green and red dominoes where each row has exactly one green domino, which is the leftmost domino. A move of Left is to pick a blue or green domino and remove it with all dominoes of the same row to its right. Right plays similarly with red or green dominoes. Notice that any position of this game is dicot. The game is solved in [10]. The authors first show that all games are invertible by finding the canonical forms of all rows, modulo dicot games, and then finding the outcomes of sums of such positions. They end by showing that no other simplification can be made, completely solving the game.

7. Current and future directions

We conclude by highlighting two of the open problems that were introduced above.

(1) *Nonconjugal invertibility.* In what universes does $G+H \equiv 0$ imply $\overline{G} \equiv H$? Can we indeed prove that this is true for certain parental, dense universes, and if so, what known games naturally occur in such universes? Can we find other examples of universes in which this is *not* true, besides the one impartial and one partizan example that have been identified to date?

(2) *Reversibility through ends.* There is a solution for dicots, where options that are reversible through ends are replaced with $*$, and there is a suggestion that a similar solution will work in a few other specific universes. Can we solve reversibility in other universes, perhaps starting even with small ruleset-specific universes? Can we find a general process for constructing the necessary "replacement" games? Are there universes in which the problem of reversibility through ends does not even arise — that is, in which Lemma 5.1 holds?

References

[1] M.H. Albert, R.J. Nowakowski, D. Wolfe, *Lessons in Play*, A.K. Peters, Ltd., Massachusetts, 2007.

[2] M.R. Allen, An investigation of partizan misère games, PhD thesis, Dalhousie University, 2009.

[3] M.R. Allen, Peeking at partizan misère quotients, in R.J. Nowakowski (Ed.) *Games of No Chance* 4, MSRI Book Series **63**, Cambridge University Press, Massachusetts, 2015.

[4] E.R. Berlekamp, J.H. Conway, R.K. Guy, *Winning Ways for Your Mathematical Plays, Vol. 2.*, A.K. Peters Ltd., MA, 2001.

[5] J.H. Conway, *On Numbers and Games* (2nd ed.), A.K. Peters, Ltd., Massachusetts, 2001.

[6] P. Dorbec, G. Renault, A.N. Siegel, E. Sopena, Dicots, and a taxonomic ranking for misère games, *J. Combin. Theory Ser. A* **130** (2015), 42–63.

[7] J.M. Ettinger, Topics in combinatorial games, PhD Thesis, University of Wisconsin–Madison, 1996.

[8] U. Larsson, R.J. Nowakowski, C.P. Santos, Absolute combinatorial game theory, (2016), preprint.

[9] U. Larsson, R.J. Nowakowski, J.P. Neto, C.P. Santos, When game comparison becomes play: absolutely categorical game theory, (2016), preprint.

[10] N.A. McKay, R. Milley, R.J. Nowakowski, Misère-play hackenbush sprigs, *Internat. J. Game Theory* **45** (2015), 731–742.

[11] G.A. Mesdal, P. Ottaway, Simplification of partizan games in misère play, *Integers* **7** (2007), #G6.

[12] R. Milley, Restricted universes of partizan misère games, PhD thesis, Dalhousie University, 2013.

[13] R. Milley, Partizan kayles and misére invertibility, *Integers* **15** (2015), #G3.

[14] R. Milley, R.J. Nowakowski, P. Ottaway, The misère monoid of one-handed alternating games, *Integers* **12B** (2012), #A1.

[15] R. Milley, G. Renault, Dead ends in misère play: the misère monoid of canonical numbers, *Discrete Math.* **313** (2013), 2223–2231.

[16] P. Ottaway, Combinatorial games with restricted options under normal and misère play, PhD thesis, Dalhousie University, 2009.

[17] T.E. Plambeck, Daisies, kayles, and the Sibert–Conway decomposition in misère octal games, *Theoret. Comput. Sci.* **96** (1992), 361–388.

[18] T.E. Plambeck, Taming the wild in impartial combinatorial games, *Integers* **5** (2005), #G5.

[19] T.E. Plambeck, Advances in losing, in M.H. Albert and R.J. Nowakowski (Eds.) *Games of No Chance* 3, MSRI Book Series **56**, Cambridge University Press, Massachusetts, 2008.

[20] T.E. Plambeck, A.N. Siegel, Misère quotients for impartial games, *J. Combin. Theory Ser. A* **115** (2008), 593–622.

[21] A.N. Siegel, Misère canonical forms of partizan games, in R.J. Nowakowski (Ed.) *Games of No Chance* 4, MSRI Book Series **63**, Cambridge University Press, Massachusetts, 2015.

rmilley@grenfell.mun.ca *Computational Mathematics, Grenfell Campus, Memorial University of Newfoundland, Corner Brook, NL, Canada*

gabriel.renault@umons.ac.be *Université de Mons, Mons, Belgium*

Games of No Chance 5
MSRI Publications
Volume **70**, 2017

Unsolved problems in combinatorial games

RICHARD J. NOWAKOWSKI

During the recent development of combinatorial game theory many more problems have been suggested than solved. Here is a collection of problems that many people have found interesting. Included is a discussion of recent developments in these areas. This collection was started by Richard Guy in 1991, and has been updated in each *Games of No Chance* volume.

The problems are divided into five sections:

A. Taking and breaking

B. Pushing and placing pieces

C. Playing with pencil and paper

D. Disturbing and destroying

E. Theory of games

In the sections, the problems are identified by letters and numbers. Any number in parenthesis is the old number used in each of the lists of unsolved problems given on pp. 183–189 of AMS *Proc. Sympos. Appl. Math.* **43** (1991), called PSAM **43** below; on pp. 475–491 of *Games of No Chance*, hereafter referred to as GONC.

Previous versions of the unsolved problems can also be found on pp. 457–473 of *More Games of No Chance* (MGONC); pp. 475–500 of *Games of No Chance* 3 (GONC3); and pp. 279–308 of *Games of No Chance* 4 (GONC4).

Some problems have little more than the statement of the problem if there is nothing new to be added. References [#] are at the end of this article. Useful references for the rules and an introduction to many of the specific games mentioned below are:

- M. Albert, R. J. Nowakowski, and D. Wolfe, *Lessons in Play*: *An Introduction to the Combinatorial Theory of Games*, A K Peters, 2007 (*LIP*);
- Berlekamp, Conway, and Guy, *Winning Ways for your Mathematical Plays*, vol. 1–4, A K Peters, 2000–2004 (*WW*);
- Siegel, *Combinatorial Game Theory*, American Math. Society, 2013 (*CGT*).

The term "game" is ambiguous and needs to be clarified. If a board is not specified then the question pertains to all possible boards. Berlekamp noted that many games have a standard opening position, an empty board for example. However, some results include positions, such as Garden-of-Eden positions, that could never occur by any sequence of moves from a standard opening position, and ones which would never occur in decent play. Indeed, Plambeck has asked at several meetings whether there is any difference in the complexity results for all positions and those that would occur in decent play.

Common usage suggests the term "game-name" as meaning the ruleset without a specified opening board. For example, Drummond-Cole's $*2$ in DOMINEERING has "holes" which could not be formed starting from the empty board. The term *restricted* will be used if the positions are (or must be) obtainable from a standard opening board.

Acknowledgement. I have had help in compiling this collection from all those mentioned, and from others. I would especially like to mention Elwyn Berlekamp, Gabriel Drummond-Cole, Alex Fink, Aviezri Fraenkel, Urban Larsson, Richard Low, Neil McKay, Carlos Santos, Simon Rubinstein-Salzedo, Alex Sierra Ballarin, and Eric Sopena.

A. Taking and breaking games

A1. (1) Subtraction games with finite subtraction sets are known to have periodic nim-sequences. Investigate the relationship between the subtraction set and the length and structure of the period. The same question can be asked about *partizan* subtraction games, in which each player is assigned an individual subtraction set. See Fraenkel and Kotzig [36].

(A move in the game $S(s_1, s_2, s_3, \ldots)$ is to take a number of beans from a heap, provided that number is a member of the *subtraction-set*, $\{s_1, s_2, s_3, \ldots\}$. Analysis of such a game and of many other heap games is conveniently recorded by a *nim-sequence*,

$$n_0 n_1 n_2 n_3 \ldots,$$

meaning that the nim-value of a heap of h beans is n_h; i.e., that the value of a heap of h beans in this particular game is the *nimber* $*n_h$.)

For examples, see Table 2 in Section 4 on p. 67 of "Impartial games" in GONC.

It would now seem feasible to give the complete analysis for games whose subtraction sets have just three members, though this has so far eluded us. Several people, including Mark Paulhus and Alex Fink, have given a complete analysis for all sets $\{1, b, c\}$ and for sets $\{a, b, c\}$ with $a < b < c < 32$. Mark Daniel Ward has pushed this to $a < b < c < 4096$ and has conjectured [127]:

Case 1: Suppose $c = a + b$. Let $j \equiv b - a \mod 2a$, $0 \leq j < 2a$. Then

$$p = \begin{cases} 2b + a - j & \text{if } 0 \leq j < a, \\ a(2b - a + j)/\gcd(a, 2a - j) & \text{if } a \leq j < 2a. \end{cases}$$

Case 2: Suppose $c \neq a + b$; then p is a divisor of at least one of $a + b$, $a + c$, and $b + c$. Moreover, if G is the subset of these terms which p divides then $p = \gcd_{(x+y)\in G}(x + y)$.

In general, period lengths can be surprisingly long, and it has been suggested that they could be superpolynomial in terms of the size of the subtraction set. However, Guy conjectures that they are bounded by polynomials of degree at most $\binom{n}{2}$ in s_n, the largest member of a subtraction set of cardinality n. It would also be of interest to characterize the subtraction sets which yield a purely periodic nim-sequence, i.e., for which there is no preperiod. Carlos Santos [101] reduced this upper bound slightly by using a dynamical system approach.

Angela Siegel [109] (and *LIP*) considered infinite subtraction sets which are the complement of finite ones and showed that the nim-sequences are always arithmetic periodic. That is, the nim-values belong to a finite set of arithmetic progressions with the same common difference. The number of progressions is the period and their common difference is called the *saltus*. For instance, the game $S\{\hat{4}, \hat{9}, \hat{26}, \hat{30}\}$ (in which a player may take any number of beans except 4, 9, 26, or 30) has a preperiod of length 243, period-length of 13014, and saltus of 4702. Marla Clusky and Danny Sleator have proved her conjecture that the class of games $S\{\hat{a}, \hat{b}, \widehat{a+b}\}$ is purely periodic with period length $3(a + b)$.

For infinite subtraction games in general there are corresponding questions about the length and purity of the period.

Suppose the elements of the finite subtraction sets are not constants. It was shown by Fraenkel [35] that the game with subtraction set

$$\{1, 2, \ldots, t - 1, \lfloor n/t \rfloor\},$$

where $t \geq 2$ is a fixed integer and n is the pile size, has an *aperiodic nim-sequence*. See also Fraenkel [35], Guo [47]. Sopena [112] has a variant, I-MARK(S, D), where S contains the subtraction, and D the division elements so that a move from n is one of $n - s$ for $s \in S$, and $\lfloor n/d \rfloor$ for $d \in D$. He shows that I-MARK($\{1\}$, $\{2\}$) has an aperiodic nim-sequence but a periodic outcome sequence. The nim-sequence has a "period" except for one element which is nonperiodic and the values are taken from $\{0, 1, 2\}$. Larsson and Fox [70] show that the subtraction game with $S = \{F_{2n+1} - 1\}$ has a nonperiodic nim-sequence with values from $\{0, 1, 2\}$.

It is possible to get a subtraction-division game with nonperiodic nim-sequence with values from {0, 1}. Write n in binary and a move is to remove a leading 1, but not if n is a power of 2, or remove a trailing 0. The number of moves is fixed so this is a version of SHE-LOVES-ME-SHE-LOVES-ME-NOT. Can this be done with a subtraction game set?

An old, voiced but not written, question asks for an infinite subtraction set where the associated outcome-sequence is periodic but the nim-sequence is aperiodic. These are easy to find in octal games.

A2. (2) Are all finite *octal games* ultimately periodic? (If the binary expansion of the k-th code digit in the game with code $\mathbf{d}_0{\cdot}\mathbf{d}_1\mathbf{d}_2\mathbf{d}_3\ldots$ is

$$\mathbf{d}_k = 2^{a_k} + 2^{b_k} + 2^{c_k} + \cdots,$$

where $0 \le a_k < b_k < c_k < \ldots$, then it is legal to remove k beans from a heap, provided that the rest of the heap is left in exactly a_k or b_k or c_k or ... nonempty heaps. See [*WW*, pp. 81–115]. Some specimen games are exhibited in Table 3 of Section 5 of "Impartial games" in GONC.)

Resolve any number of outstanding particular cases, e.g., **·6** (OFFICERS), **·04**, **·06**, **·14**, **·36**, **·37**, **·64**, **·74**, **·76**, **·004**, **·005**, **·006**, **·007**, **·014**, **·015**, **·016**, **·024**, **·026**, **·034**, **·064**, **·114**, **·125**, **·126**, **·135**, **·136**, **·142**, **·143**, **·146**, **·162**, **·163**, **·164**, **·166**, **·167**, **·172**, **·174**, **·204**, **·205**, **·206**, **·207**, **·224**, **·244**, **·245**, **·264**, **·324**, **·334**, **·336**, **·342**, **·344**, **·346**, **·362**, **·364**, **·366**, **·371**, **·374**, **·404**, **·414**, **·416**, **·444**, **·564**, **·604**, **·606**, **·744**, **·764**, **·774**, **·776**, and GRUNDY'S GAME (split a heap into two unequal heaps [*WW*, pp. 96–97, 111–112; *LIP*, p. 142]), which has been analyzed, first by Dan Hoey, and later by Achim Flammenkamp, as far as heaps of 2^{35} beans.

In this volume, J. P. Grossman, using a new approach based on the sparse space phenomenon, has analyzed **·6** up to heaps of size 2^{47} and has found no periodicity.

Perhaps the most notorious and deserving of attention is the game **·007**, one-dimensional TIC-TAC-TOE, or TREBLECROSS, which Flammenkamp has pushed to 2^{25}. Figure 1 shows the first 11 million nim-values, a small proportion of which are ≥ 1024; the largest, $\mathcal{G}(6193903) = 1401$ is shown circled. Will 2048 ever be reached?

Flammenkamp has settled **·106**: it has the remarkable period and preperiod lengths of 328226140474 and 465384263797. For information on the current status of each of these games, we refer the reader to Flammenkamp's web page at uni-bielefeld.de/~achim/octal.html.

A game similar to GRUNDY'S GAME, and which is also unsolved, is John Conway's COUPLES-ARE-FOREVER [*LIP*, p. 142], where a move is to split any

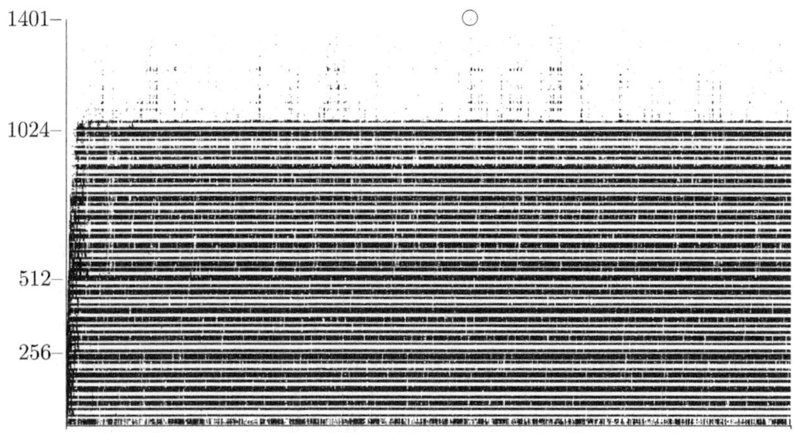

Figure 1. Plot of 11000000 nim-values of the octal game ·**007**.

heap except a heap of two. The first 50 million nim-values haven't displayed any periodicity. See Caines et al. [14]. More generally, Bill Pulleyblank suggests looking at splitting games in which you may only split heaps of size $> h$, so that $h = 1$ is SHE-LOVES-ME-SHE-LOVES-ME-NOT and $h = 2$ is COUPLES-ARE-FOREVER. David Singmaster suggested a similar generalization — you may split a heap provided that the resulting two heaps each contain at least k beans: $k = 1$ is the same as $h = 1$, while $k = 2$ is the third cousin of DAWSON'S CHESS.

Explain the structure of the periods of games known to be periodic.

In *Discrete Math.*, 44 (1983), pp. 331–334, Problem 38, Fraenkel raised questions concerning the computational complexity (see Section E1 below) of octal games. In Problem 39, he and Kotzig define *partizan octal games* in which distinct octals are assigned to the two players. Mesdal [GONC3, pp. 447–461] shows that in many cases, if the game is "all-small" [*WW*, pp. 229–262; *LIP*, pp. 183–207], then the atomic weights are arithmetic periodic. (See Section E13 for an explanation of the new name "dicot" games.) In Problem 40, Fraenkel introduces *poset games*, played on a partially ordered set of heaps, each player in turn selecting a heap and then removing a nonnegative number of beans from this heap and from each heap above it in the ordering, with at least one heap being reduced in size. For posets of height one, new regularities in the nim-sequence can occur. Tanya Khovanova and the author propose the following simple version. Given n nim-heaps of sizes $a_1, a_2, \ldots, a_n$, a move is to choose some heap i and some number $1 \le x \le a_i$ and for each $j \ge i$ replace a_j by $\max\{a_j - x, 0\}$. The solution for three heaps involves the continued fraction expansion of $\sqrt{2}$.

Note that this includes, as particular cases, SUBSET TAKEAWAY, CHOMP or DIVISORS, and GREEN HACKENBUSH forests. Compare Problems A3, A13, D1, and D2 below.

A3. (3) Hexadecimal games have code digits $\mathbf{d}_k$ in the interval from 0 to f (= 15), so that there are options splitting a heap into three heaps. See [*WW*, pp. 116–117].

Such games may be arithmetically periodic. Nowakowski has calculated the first 100000 nim-values for each of the 1-, 2-, and 3-digit games. Richard Austin's Theorem 6.8 in his thesis [4] and the generalization by Howse and Nowakowski [59] suffice to confirm the arithmetic periodicity of several of these games.

Some interesting specimens are $\cdot$**28** = $\cdot$**29**, which have period 53 and saltus 16, the only exceptional value being $\mathcal{G}(0) = 0$; $\cdot$**9c**, which has period 36, preperiod 28, and saltus 16; and $\cdot$**f6** with period 43 and saltus 32, but its apparent preperiod of 604 and failure to satisfy one of the conditions of the theorem prevent us from verifying the ultimate periodicity. The game $\cdot$**205200c** is arithmetic periodic with preperiod length 4, period length 40, and saltus 16, except that $40k + 19$ has nim-value 6 and $40k + 39$ has nim-value 14. This regularity (which also seems to be exhibited by $\cdot$**660060008** with a period length of approximately 300000) was first reported in Horrocks and Nowakowski [57] (see Problem A2). Grossman and Nowakowski [GONC4] have shown that the nim-sequences for $\cdot$**200...0048**, with an odd number of zero code digits, exhibit "ruler function" patterns. The game $\cdot$**9** has not so far yielded its complete analysis, but, as far as analyzed (to heaps of size 100000), exhibits a remarkable fractal-like set of nim-values. See Howse and Nowakowski [59]. Also of special interest are $\cdot$**e**; $\cdot$**7f** (which has a strong tendency to period 8, saltus 4, but, for $n \leq 100000$, has 14 exceptional values, the largest being $\mathcal{G}(94156) = 26614$); $\cdot$**b6** (which "looks octal"); $\cdot$**b33b** (where a heap of size n has nim-value n except for 27 heap sizes which appear to be random).

Other unsolved hexadecimal games are

$\cdot$**1x**,	x ∈ {8, 9, c, d, e, f},	$\cdot$**2x**,	a ≤ x ≤ f,
$\cdot$**3x**,	8 ≤ x ≤ e,	$\cdot$**4x**,	x ∈ {9, b, d, f},
$\cdot$**5x**,	8 ≤ x ≤ f,	$\cdot$**6x**,	8 ≤ x ≤ f,
$\cdot$**7x**,	8 ≤ x ≤ f,	$\cdot$**9x**,	1 ≤ x ≤ a,
$\cdot$**9d**,		$\cdot$**bx**,	x ∈ {6, 9, d},
$\cdot$**dx**,	1 ≤ x ≤ f,	$\cdot$**fx**,	x ∈ {4, 6, 7}.

A4. (53) *N*-HEAP WYTHOFF GAME. We are given $N \geq 2$ heaps of finitely many tokens, whose sizes are $p_1, \ldots, p_N$ with $p_1 \leq \cdots \leq p_N$. Players take turns removing any positive number of tokens from a *single* heap or removing $(a_1, \ldots, a_N)$ from *all* the heaps — a_i from the i-th heap — subject to the conditions:

(i) $0 \leq a_i \leq p_i$ for each i,

(ii) $\sum_{i=1}^{N} a_i > 0$,

(iii) $a_1 \oplus \cdots \oplus a_N = 0$, where $\oplus$ is nim-addition.

The player making the last move wins and the opponent loses. Note that the classical Wythoff game is the case $N = 2$.

For $N \geq 3$, Fraenkel makes the following conjectures.

Conjecture 1. For every fixed set $K := (A^1, \ldots, A^{N-2})$ there exists an integer $m = m(K)$ (i.e., m depends only on K) such that

$$(A^1, \ldots, A^{N-2}, A_n^{N-1}, A_n^N), \quad A^{N-2} \leq A_n^{N-1} \leq A_n^N,$$

with $A_n^{N-1} < A_{n+1}^{N-1}$ for all $n \geq 1$, is the n-th $\mathcal{P}$-position, and

$$A_n^{N-1} = \operatorname{mex}(\{A_i^{N-1}, A_i^N : 0 \leq i < n\} \cup T), \quad A_n^N = A_n^{N-1} + n,$$

for all $n \geq m$, where $T = T(K)$ is a (small) set of integers.

That is, if you fix $N - 2$ of the heaps, the $\mathcal{P}$-positions resemble those for the classical Wythoff game. For example, for $N = 3$ and $A^1 = 1$, we have $T = \{2, 17, 22\}$ and $m = 23$.

Conjecture 2. For every fixed K there exist integers $a = a(K)$ and $M = M(K)$ such that

$$A_n^{N-1} = \lfloor n\phi \rfloor + \varepsilon_n + a \quad \text{and} \quad A_n^N = A_n^{N-1} + n,$$

for all $n \geq M$, where $\phi = \frac{1}{2}(1 + \sqrt{5})$ is the golden section, and $\varepsilon_n \in \{-1, 0, 1\}$.

In Fraenkel and Krieger [37] the following was shown, inter alia: Let $t \in \mathbb{Z}_{\geq 1}$, $\alpha = \frac{1}{2}(2 - t + \sqrt{t^2 + 4})$, where $\alpha = \phi$ for $t = 1$, $T \subset \mathbb{Z}_{\geq 0}$ a finite set, and $A_n = \operatorname{mex}(\{A_i, B_i : 0 \leq i < n\} \cup T)$, where $B_n = A_n + nt$. Let $s_n := \lfloor n\alpha \rfloor - A_n$. Then there exist $a \in \mathbb{Z}$ and $m \in \mathbb{Z}_{\geq 1}$ such that for all $n \geq m$, either $s_n = a$, or $s_n = a + \varepsilon_n$, $\varepsilon_n \in \{-1, 0, 1\}$. If $\varepsilon_n \neq 0$, then $\varepsilon_{n-1} = \varepsilon_{n+1} = 0$. Also the general structure of the ε_n was characterized succinctly.

This result was then applied to the N-heap WYTHOFF GAME. In particular, for $N = 3$ (such that $K = A^1$) it was proved that $A_n^2 = \operatorname{mex}(\{A_i^2, A_i^3 : 0 \leq i < n\} \cup T)$, where

$$T = \{x \geq K : \text{there exists } 0 \leq k < K \text{ such that } (k, K, x) \text{ is a } \mathcal{P}\text{-position}\} \cup \{0, \ldots, K - 1\}.$$

The following upper bound for A_n^3 was established: $A_n^3 \leq (K + 3)A_n^2 + 2K + 2$. It was also proved that Conjecture 1 implies Conjecture 2.

In Sun and Zeilberger [117], a sufficient condition for the conjectures to hold was given. It was then proved that the conjectures are true for the case $N = 3$, where the first heap has up to ten tokens. For those ten cases, the parameter values m, M, a, T were listed in a table.

Sun [116] obtained results similar to those in Fraenkel and Krieger [37], but the proofs are different. In that work, it was also proved that Conjecture 1 implies Conjecture 2. A method was given to compute a in terms of certain indexes of the A_i and B_j.

A5. (23) BURNING-THE-CANDLE-AT-BOTH-ENDS. Conway and Fraenkel ask us to analyze NIM played with a row of heaps. A move may only be made in the leftmost or in the rightmost heap. When a heap becomes empty, then its neighbor becomes the end heap.

Albert and Nowakowski [2] have determined the outcome classes in impartial and partizan versions (called END-NIM [*LIP*, pp. 210, 263]) with finite heaps, and Duffy, Kolpin, and Wolfe [GONC3, pp. 419–425] extend the partizan case to infinite ordinal heaps. Wolfe asks for the actual values.

Nowakowski suggested to analyze impartial and partizan END-WYTHOFF: take from either end-pile, or the *same* number from both ends. The impartial game is solved by Fraenkel and Reisner [GONC3, pp. 329–347]. Fraenkel [32] asks a similar question about a generalized Wythoff game: take from either end-pile or take $k > 0$ from one end-pile and $\ell > 0$ from the other, subject to $|k - \ell| < a$, where a is a fixed integer parameter ($a = 1$ is END-WYTHOFF).

There is also HUB-AND-SPOKE NIM, proposed by Fraenkel. One heap is the hub and the others are arranged in rows forming spokes radiating from the hub. Albert notes that this game can be generalized to playing on a forest, i.e., a graph where each of whose components is a tree. The most natural variant is that beans may only be taken from a leaf (valence 1) or isolated vertex (valence 0).

The partizan game of RED-BLUE CHERRIES is played on an arbitrary graph; see McCurdy [81]. A player picks an appropriately colored cherry from a vertex of minimum degree, which disappears at the same time. Albert, Grossman, McCurdy, Nowakowski, and Wolfe [1] (mistakenly) show that if the graph has a leaf, then the value is an integer. They ask this: Is every RED-BLUE CHERRIES position an integer? Matthew Bardoe and Scott Herman (personal communication) disproved both by finding a 7-cycle with two leaves that has value $\frac{1}{2}$.

A6. (17) Extend the analysis of KOTZIG'S NIM [*WW*, pp. 515–517]. Is the game eventually periodic in terms of the length of the circle for every finite move set? Analyze the misère version of KOTZIG'S NIM.

Let $\Gamma(S; n)$ be the outcome of the game where S lists all the moves and n is the size of the circle. Tan and Ward [118] gave more evidence of the periodicity by showing the following: If $n \in \{1, 3, 5, 7, 15\}$, or if $n \equiv 3 \pmod 5$ and $n \geq 23$, then $\Gamma(1, 4; n) \in \mathcal{P}$; otherwise, $\Gamma(1, 4; n) \in \mathcal{N}$. They also give several conjectures including these: If n is odd and $n \notin \{9, 11, 17\}$ then $\Gamma(3,5;n) \in \mathcal{P}$ and is in $\mathcal{N}$ otherwise. Their evidence bolsters the periodicity conjecture but also indicates that there is likely to be a lot of noise when n is small.

A7. (18) Obtain asymptotic estimates for the proportions of $\mathcal{N}$-, $\mathcal{O}$- and $\mathcal{P}$-positions in Epstein's PUT-OR-TAKE-A-SQUARE game [*WW*, pp. 518–520].

A8. GALE'S NIM. This is NIM played with four heaps, but the game ends when three of the heaps have vanished, so that there is a single heap left. Brouwer and Guy have independently given a partial analysis, but the situation where the four heaps have distinct sizes greater than 2 is open. An obvious generalization is to play with h heaps and play finishes when k of them have vanished.

A9. EUCLID'S NIM is played with a pair of positive integers, a move being to diminish the larger by any multiple of the smaller. The winner is the player who reduces a number to zero. Analyses have been given by Cole and Davie [19], Spitznagel [114], Lengyel [75], Collins [20], Fraenkel [34], and Nivasch [91]. Lengyel [75] reports that Schwartz first found that (a, b) is the sequential sum of nim-heaps given by the normal continued fraction of a/b, where the last number in the continued fraction depends on the presence or absence of Fibonacci numbers. Gurvich [49] shows that the nim-value $g^+(a, b)$ for the pair (a, b) in normal play is the same as the misère nim-value $g^-(a, b)$ except for $(a, b) = (kF_i, kF_{i+1})$, where $k > 0$ and F_i is the i-th Fibonacci number. In this case, $g^+(kF_i, kF_{i+1}) = 0$ and $g^-(kF_i, kF_{i+1}) = 1$ if i is even and the values are reversed if i is odd. Aaron Siegel notes that this can be restated as this: EUCLID is tame, and the fickle positions have the form (kF_i, kF_{i+1}). The latter have genus 0^1 when i is even and 1^0 when i is odd.

Collins and Lengyel [21] play EUCLID'S NIM with three integers and solve some special cases.

In PARTIZAN EUCLID, from the position (p, q), where $p = kq + r$, $0 \le r < q$, Left moves to (q, r) and Right to $(q, (k+1)q - p)$. The outcome classes are investigated in [82]. Determining the values seems hard. The atomic weights may be easier — see Figure 2, where the *mean atomic weights* for $0 < q < p \le 100$ are given, where the coordinates (x, y) are $x = q/p$ and y is the mean atomic weight of (p, q).

A10. (20) DUDENEY [*WW*, pp. 521–523] is NIM, but with an upper bound Y on the number of beans that may be taken, and with the restriction that a player may not repeat his opponent's last move. If Y is even, the analysis is easy. Some advance in the analysis, when Y is odd, has been made by Marc Wallace, Alex Fink and Kevin Saff.

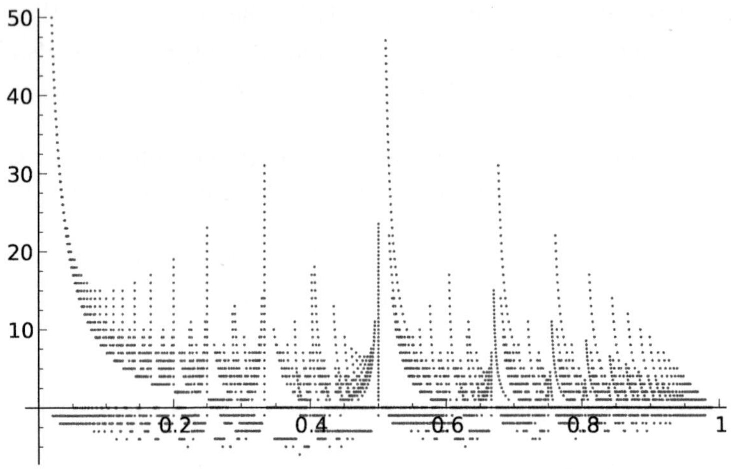

Figure 2. PARTIZAN EUCLID: mean atomic weights for $0 < q < p < 101$.

We can, for example, extend the table of strings of pearls given in [*WW*, p. 523], with the following values of Y which have the pure periods shown, where $\mathrm{D} = \mathrm{Y}+2$ and $\mathrm{E} = \mathrm{Y}+1$. The first entry corrects an error of $128r+31$ in *WW*.

$256r+31$	DEE	$512r+153$	DEE	$1024r+415$	DEE
$512r+97$	DDEDDDE	$512r+159$	DEE	$512r+425$	DE
$1024r+103$	DE	$512r+225$	DDE	$512r+487$	DEE
$128r+119$	DEE	$512r+255$	E	$1024r+521$	DDDE
$1024r+127$	DEEE	$512r+257$	DDDDE	$1024r+607$	DDE
$512r+151$	DDDEE	$512r+297$	DDEDEDE	$1024r+735$	DEEE

It seems likely that the string for $Y = 2^{2k+1} + 2^{2k} - 1$ has the simple period E for all values of k. But there is some evidence to suggest that an analysis will never be complete. Indeed, consider the following table showing the fractions, among 2^k cases, that remain undetermined:

$k =$	3	5	6	7	8	9	10	11	12	13	14	15	16	17
	$\frac{1}{2}$	$\frac{5}{16}$	$\frac{9}{32}$	$\frac{11}{64}$	$\frac{21}{128}$	$\frac{33}{256}$	$\frac{60}{512}$	$\frac{97}{1024}$	$\frac{177}{2048}$	$\frac{304}{4096}$	$\frac{556}{8192}$	$\frac{974}{16384}$	$\frac{1576}{32768}$	$\frac{2763}{65536}$

Moreover, the periods of the pearl-strings appear to become arbitrarily long.

A11. (21) SCHUHSTRINGS is the same as DUDENEY, except that a deduction of zero is also allowed, but cannot be immediately repeated [*WW*, pp. 523–524]. In *WW*, it was stated that it was not known whether there is any SCHUHSTRING game in which three or more strings terminate simultaneously. Kevin Saff has found three such strings (when the maximum deduction is $Y = 3430$, the three strings are multiples of 2793, 3059, 3381 terminate simultaneously) and he conjectures that there can be arbitrarily many such simultaneous terminations.

A12. (22) Analyze **DUDE**, i.e., unbounded DUDENEY, or NIM in which you are not allowed to repeat your opponent's last move.

Let $[h_1, h_2, \ldots, h_k; m]$, $h_i \le h_{i+1}$, be the game with heaps of size h_1 through h_k, where m is the move just made and $m = 0$ denotes a starting position. Then, the $\mathcal{P}$-positions are [22]

$$[(2s+1)2^{2j}; (2s+1)2^{2j}] \quad \text{for } k = 1,$$
$$[(2s+1)2^{2j}, (2s+1)2^{2j}; 1] \quad \text{for } k = 2,$$

and for $k \ge 3$ the heap sizes are arbitrary, the only condition being that the previous move was 1. The nim-values do not seem to show an easily described pattern.

A13. Games with pass moves. NIM WITH PASS. David Gale suggested an analysis of NIM played with the option of a single pass by either of the players, which may be made at any time up to the penultimate move. It may not be made at the end of the game. Once a player has passed, the game is as in ordinary NIM. The game ends when all heaps have vanished. Morrison, Friedman and Landsberg [87] have looked at this game with their renormalization techniques and suggest that obtaining a full solution may be intractable. Despite that warning, Low and Chan [79] made some progress. They show that:

(1) for two heaps, $[2m+1, 2m+2]$ are the only $\mathcal{P}$-positions;

(2) for three heaps, if one is of size 1, then $[1, m, m]$, $m \ne 2$, are the only $\mathcal{P}$-positions.

They have several other results; for example: Suppose $p + 2q + 3r \ge 10$; then the position with p heaps of size 1, q heaps of size 2, and r heaps of size 3 is a $\mathcal{P}$-position if and only if p is odd and q, r are even, or p is even and q, r are odd.

Chan et al. [18] have extended the analysis to 4 heaps.

CHOCOLATE BAR with a pass [86]. CHOCOLATE BAR is a generalization of CHOMP; see D1. Fix a natural number k and a nonnegative integer s. For nonnegative integers y and z such that $y \le \lfloor (z+s)/c \rfloor$, the chocolate bar will consist of $z+1$ columns where the zero column is the bitter square and the height of the i-th column is $t(i) = \min(y, \lfloor (i+s)/k \rfloor + 1)$. A move, as in CHOMP, is to take a square and remove it and every square in the upper-right quadrant based on this square. Let $CB(s, k, y, z, p)$ be the CHOCOLATE BAR game where there is a single pass move which can be used by either player but not at a terminal position. Here, $p = 1$ if the pass move is available and $p = 0$ otherwise. For $s < k$, they solve the game for s odd and ask for a solution when s is even.

In octal games, with a single pass, Horrocks and Nowakowski [57] ask for the regularities that can occur in the $\mathcal{G}$-sequences. They exhibit sequences which are periodic, arithmetic-periodic and split periodic, arithmetic-periodic.

Now what happens if more than one pass is allowed?

Given a game, the *Oslo* (one-sided-loopy) version gives Left the option to pass in any position except in the terminal positions. Angela Siegel [109] shows that the lattice of Oslo games born by day n is a distributive lattice.

The Oslo version of some games becomes almost trivial. OSLO WYTHOFF has the values $W(x, y) = 0$ if $x = y = 0$; $W(x, y) = \mathbf{upon}*$ if $x = y > 0$ or one of x and $y = 0$; and $W(x, y) = 2.\mathbf{upon}$ otherwise. (See E5 and E7.)

A14. Games with a Muller twist. In such games, each player specifies a condition on the set of options available to their opponent on their next move.

In ODD-OR-EVEN NIM, for example, each player specifies the parity of the opponent's next move. This game was analyzed by Smith and Stănică [111], who propose several other such games which are still open (see also Gavel and Strimling [45]).

The game of BLOCKING NIM proceeds in exactly the same way as ordinary NIM with N heaps, except that before a given player takes his turn, his opponent is allowed to announce a *block*, $(a_1, \ldots a_N)$; i.e., for each pile of counters, he has the option of specifying a positive number of counters which may not be removed from that pile. Flammenkamp, Holshouser and Reiter [56; 55] give the $\mathcal{P}$-positions for three-heap BLOCKING NIM with an incomplete block containing only one number, and ask for an analysis of this game with a block on just two heaps, or on all three. There are corresponding questions for games with more than three heaps. Larsson [69] shows a game with two piles of counters where at most $k-1$ moves, for some fixed k, may be blocked off at each stage. Then the $\mathcal{P}$-positions are of the form $\{x, y\}$, where either $|y-x| < k$ and $y - x \equiv k-1 \pmod 2$ or $x + y < k$.

Let S be a set of positive integers. The *complementary subtraction game* $\hat{S}$ is played on a heap where the last act of a move is to say whether the next subtraction is to be a number from S or from its complement. Horrocks and Trenton [58] introduced this variant. They analyzed heaps up to size 8000, the case where S is the Fibonacci numbers, without finding periodicity although there appears to be regularity. They also ask about the case $S = \{x \mid x \equiv a \text{ or } b \pmod c\}$. They report that the nim-values appear chaotic. The set of $(a_n, a_n + n)$ where $a_1 = 1$ and $a_n = \text{mex}\{a_i : i < n\}$ form the $\mathcal{P}$-positions of Wythoff's game. We ask what happens in this game if $S = \{a_n : n + 1, 2, 3, \ldots\}$.

A15. (13) Misère quaternary and octal games. Misère analysis has been revolutionized by Thane Plambeck and Aaron Siegel [97] with their concept of the *misère quotient* of a game, though the number of unsolved problems continues to increase. See Section E13 for more theoretical questions and an explanation of some of the concepts mentioned here.

Plambeck and Siegel ask these specific questions:

(1) The misère quotient of ·**07** (DAWSON'S KAYLES) has order 638 at heap size 33. Is it infinite at heap size 34? Even if the misère quotient is infinite at heap size 34, then, by Rédei's theorem [46, p. 142; 99], it must be isomorphic to a finitely presented commutative monoid. Call this monoid D_{34}. Exhibit a monoid presentation of D_{34}, and having done that, exhibit D_{35}, D_{36}, etc., and explain what is going on in general. Given a set of games $\mathcal{A}$, describe an algorithm to determine whether the misère quotient of $\mathcal{A}$ is infinite. Much harder: if the quotient is infinite, give an algorithm to compute a presentation for it.

(2) Give complete misère analyses for any of the (normal-play periodic) octal games that show "algebraic-periodicity" in misère play. Some examples are ·**54**, ·**261**, ·**355**, ·**357**, ·**516**, and ·**724**. Give a precise definition of algebraic periodicity and describe an algorithm for detecting and generalizing it. This is a huge question: if such an algorithm exists, it would likely instantly give solutions to at least a half-dozen unsolved 2- and 3-digit octals.

Plambeck offers prizes of US\$500.00 for complete analysis of DAWSON'S CHESS, ·**137** (alias DAWSON'S KAYLES, ·**07**); US \$200.00 for the "wild quaternary game" (·**3102**); and US \$25.00 each for ·**3122**, ·**3123**, and ·**3312**.

The website miseregames.org contains thousands of misère quotients for octal games.

Siegel notes that Dawson first proposed his problem in 1935, making it perhaps the oldest open problem in combinatorial game theory. (Michael Albert offers the alternative "Is chess a first player win?".) It may be of historical interest to note that Dawson showed the problem to Richard K. Guy around 1947. Fortunately, he forgot that Dawson proposed it as a losing game, was able to analyze the normal play version, rediscover the Sprague–Grundy theory, and get Conway interested in games.

A16. Now called E12.

A17. MEM and MNEM. MEM is played with heaps of tokens. Remove any number of tokens from any one heap. The number of tokens removed must be at least as large as the number that were removed on the previous move from that heap. Equivalently, a "heap" is a pair of integers (n, k), and a move is to any pair $(n-i, i)$, where $k \leq i \leq n$.

MNEM is exactly the same as MEM, with these additional options: Either player may add tokens to a heap instead of removing them. If adding tokens, the number added must be strictly less than the number of tokens added or removed on the previous move. If removing tokens, the number must be at least as large as the number added or removed on the previous move. So a move from (n, k) is to $(n-i, i)$ for $k \le i \le n$, as in MEM, or to $(n+i, i)$ for $1 \le i < k$.

Conway and (Aaron) Siegel have investigated this game. They conjecture this: *Every position in* MNEM *has finite nim value.* They verified this experimentally up to $n = 10000$. They also conjecture the following: *For both* MEM *and* MNEM, *if* $k^2 \ge n$, *then* $(n, k) = \lfloor n/k \rfloor$. Siegel reports that they have no idea how to prove either, so these may be very difficult problems. Rubinstein-Salzedo reports that, for MEM, he has a proof.

A18. FIBONACCI NIM. We are given $\alpha > 0$, and a heap of n counters. The first move is to take any number except n. On the $(m+1)$-st move, $m > 0$, the next player can take up to αq_m, where q_m is the number taken on the m-th move. Whinihan [128] attributes this game to R. E. Gaskell, and he solves the case $\alpha = 12$ by showing that the $\mathcal{P}$-positions are Fibonacci numbers. If $\alpha = 1$ then the $\mathcal{P}$-positions are powers of 2. Schwenk [105] solves the general case using enumeration schemes. Let P_α be the set of $\mathcal{P}$-positions for a given α. Extending Schwenk's result, Rubinstein-Salzedo and Sarkar claim that as α increases, then there are intervals on which the $\mathcal{P}_\alpha$ is constant. For example, for $1 \le \alpha < 2$, $\mathcal{P}_\alpha$ are the powers of 2 together with 0, and for $2 \le \alpha < 2.5$, $\mathcal{P}_\alpha$ is the set of Fibonacci numbers.

Larsson and Rubinstein-Salzedo [74] define the multiheap game, $\alpha = 2$, in which the first move can remove all of a heap and the restriction on subsequent moves remains regardless of which heap the next player chooses. For two heaps, they characterize the $\mathcal{P}$-positions, again in terms of Fibonacci numbers. The cases for different α and for more heaps are open.

See Levine [76] for the game where there is an upper and/or lower bound on the number that can be taken, where the bounds depend upon the size of the heap.

A19. Self-referential subtraction games. There several games under this heading, but the best-known game under this heading is EUCLID with its variants; see A9. A recent partizan variant is SUBVERSION [31]; see E16. (SUBVERSION is played with a pair of nonnegative integers, (a, b). If $a = 0$ or $b = 0$ then the game is over. Otherwise, if $a \ge b$ then Left can move to $(a, 0)$, if $a < b$ then Left can move to $(a, b-a)$; if $a \le b$ then Right can move to $(0, b)$, and if $a > b$ then Right can move to $(a, b-a)$.)

Two games, SUSEN and ALBUS, were developed by Larsson and Nowakowski. In SUSEN (SUbtraction Set ENcoded), the nonzero heap sizes form the subtraction set. For example, $(2, 3)$ is a $\mathcal{P}$-position — the moves are to (2) or (3), both followed by $()$; and to $(2, 1)$ followed by $(1, 1)$. The move to $(2, 2)$ is not available.

If there is only one heap, the next player wins. For two heaps, removing a heap is a losing move. If the players never remove a heap, the length of the game (a, b) until they reach (c, c) for some c, is fixed. Call this $L(a, b)$. Then (a, b) is a $\mathcal{P}$-position if and only if $L(a, b)$ is even. Alex Fink gives a proof that (a, b, c) is in $\mathcal{P}$ if and only if (a, b), (a, c), and (b, c) are all in $\mathcal{N}$. Extend the analysis to more heaps.

In ALBUS (ALl BUt set Sizes), the heap sizes are the illegal moves. For one heap, reduce the heap to 1 counter and win. The case of two heaps is still unsolved.

B. Pushing and placing pieces

B1. (5) **GO** is of particular interest, partly because of the loopiness induced by the "ko" rule, and many problems involve general theory; see E4 and E5.

Elwyn writes:

> I attach one region that has been studied intermittently over the past several years. The region occurs in the southeast corner of the board (Figure 3). At move 85, Black takes the ko at L6. What then is the temperature at N4? This position is copied from the game Jiang and Rui played at MSRI in July 2000. In 2001, Bill Spight and I worked out a purported solution by hand, assuming either Black komaster or White komaster.

Elwyn also writes:

> Nakamura [GONC3] has shown how capturing races in GO can be analyzed by treating liberties as combinatorial games. Like atomic weights, when the values are integers, each player's best move reduces

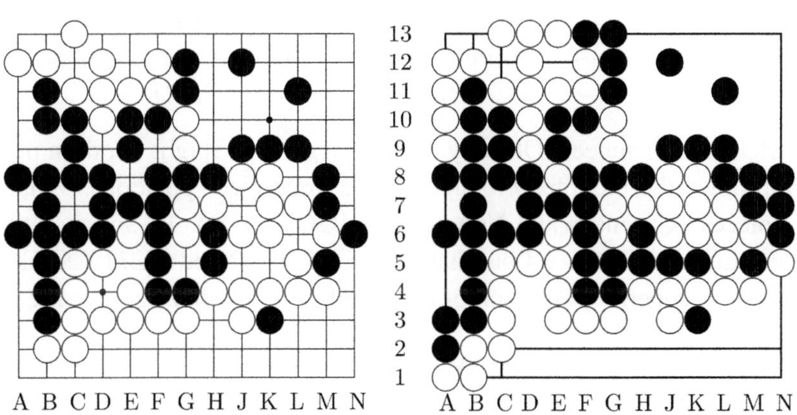

Figure 3. A position from Jiang vs. Rui at MSRI in July 2000.

his opponent's resources by one. The similarities between atomic weights and Nakamura's liberties are striking.

Theoretical problem: Either find a common formulation which includes much or all of atomic weight theory and Nakamura's theory of liberties, *or* find some significant differences.

Important practical applied problem: Extend Nakamura's theory to include other complications which often arise in GO, such as simple kos, either internal and/or external.

B2. WOODPUSH (see [LIP, pp. 214, 275]) is a game that involves ko but is simpler than GO. WOODPUSH is played on a finite strip of squares. Each square is empty or occupied by a black or white piece. A piece of the current player's color retreats — Left retreats to the left and Right to the right — to the next empty square, or off the board if there is no empty square; except, if there is a contiguous string containing an opponent's piece, then it can move in the opposite direction, *pushing* the string ahead of it. Pieces can be moved off the end of the strip. Immediate repetition of a global board positions is not allowed. A "ko-threat" must be played first. For example:

$$LRR\square \xrightarrow{\text{Left}} \square LRR \xrightarrow{\text{Right}} LR\square R \xrightarrow{\text{Left}} \text{ko-threat} \xrightarrow{\text{Right}} R\square\square R$$

Note that Right's first move to $LRR\square$ is illegal because it repeats the immediately prior board position and Left's second move to $\square LRR$ is also illegal so he must play a ko-threat. Also note that in $\square LRR\square$, Right never has to play a ko-threat since he can always push with either of his two pieces — with Left moving first,

$$\begin{array}{cccccc} & \text{Left} & & \text{Right} & \\ \square LRR\square & \rightarrow & \square\square LRR & \rightarrow & \square LR\square R \\ & \rightarrow & \text{ko-threat} & \rightarrow & \text{Right answers ko-threat} \\ & \rightarrow & \square\square LRR & \rightarrow & \square LRR\square \end{array}$$

Berlekamp, Plambeck, Ottaway, Aaron Siegel and Spight (work in progress) use top-down thermography to analyze the three-piece positions. What about more pieces? Cazenave and Nowakowski [GONC4, pp. 47–56] show that the position $L\square L\square R\square R$ is ± 4 but that $L\square L\square\square\square R\square R$ and $L\square L\square\square\square\square\square R\square R$ are draws by superko (repetition of the board position after more than 2 moves).

B3. (40) CHESS. Noam Elkies [MGONC, pp. 61–78] has examined DAWSON'S CHESS, but played under usual Chess rules, so that capture is not obligatory.

He would still welcome progress with his conjecture that the value $*k$ occurs for all k in (ordinary chess) pawn endings on sufficiently large chessboards.

Thea van Roode has suggested IMPARTIAL CHESS, in which the players may make moves of either color. Checks need not be responded to and kings may be captured. The winner could be the first to promote a pawn.

INFINITE CHESS is played with a regular set of chess pieces but on an infinite board. Evans and Hamkins [28] observe that for each finite number n, there is a position in INFINITE CHESS having value n. They show that positions with values ω, ω^2, $\omega^2 \times k$, and ω^3 occur, and recently, that ω^4 occurs. They ask about 3-dimensional CHESS.

B4. (30) CHESS: king and rook. Low and Stamp [80] have given a strategy in which White wins the king and rook vs. king problem within an 11×9 region. Kanungo and Low [64] show that with the initial position WK(1, 1), WR(x, y), and BK(a, b), where $1 < x < a$, White has a winning strategy on the $(a+b+3) \times (a+b+5)$ board.

With the same pieces but on a quarter-infinite board, spanning the first quadrant of the xy-plane, ENTREPRENEURIAL CHESS or ECHESS, introduced by Berlekamp in 2005; see [7], gives Black the additional option of "cashing out", which removes the board and converts the position into the integer $x + y$, where (x, y) are the coordinates of his king's position when he decides to cash out. Berlekamp and Low conjecture that the boiling point of ECHESS is $5\frac{5}{8}$.

B5. NONATTACKING QUEENS. Noon and Van Brummelen [92] alternately place queens on an $n \times n$ chessboard so that no queen attacks another. The winner is the last queen placer. They give nim-values for boards of sizes $1 \leq n \leq 10$ as 1121312310 and ask for the values of larger boards.

B6. (55) AMAZONS. Müller [88] has shown that the 5×5 game is a first player win and asks about the 6×6 game.

B7. PHUTBALL, more properly, Conway's PHILOSOPHER'S FOOTBALL, is usually played on a GO board with positions (i, j), $-9 \leq i, j \leq 9$ and the ball starting at (0, 0). For the rules, see [*WW*, pp. 752–755]. The game is loopy (see Section E5 below), and Nowakowski, Ottaway, and Siegel (see [108]) discovered positions that contained tame cycles, i.e., cycles with only two strings, one each of Left and Right moves. Aaron Siegel asks if there are positions in such combinatorial games which are stoppers but contain a *wild cycle*, i.e., one which contains more than one alternation between Left and Right moves. Sarkar (in this volume) has shown that draws can occur on the 19×19 board but it is not clear if they occur without collusion of the players. Demaine, Demaine, and Eppstein [MGONC, pp. 351–360] show that it is NP-complete to decide if a player can win on the next move. Loosen (see DIRECTIONAL PHUTBALL below) raises the question of whether there is any $\mathcal{P}$-position.

PHLAG PHUTBALL is a variant played on an $n \times n$ board with the initial position of the ball at $(0, 0)$ except that now only the ball may occupy the positions $(2i, 2j)$ with both coordinates even. This eliminates "tackling", and is an extension of one-dimensional ODDISH PHUTBALL, analyzed in Grossman and Nowakowski [MGONC, pp. 361–367]. The $(3, 2n+1)$ board (i.e., (i, j), $i = 0, 1, 2$ and $-n \le j \le n$) is already interesting and requires a different strategy from that appropriate to ODDISH PHUTBALL.

Loosen [78] introduces DIRECTIONAL PHUTBALL, a nonloopy version, which is also played on a grid. The ball starts in the bottom-left corner, where Left's goal-line is the right edge and Right's is the top edge. Players can only jump toward their opponent's goal-line, and win by jumping over that goal-line. Men can only be placed ahead of the ball (i.e., in the positive quadrant with origin on the ball). She notes that there is no $\mathcal{P}$-position in this game and that making an off-parity move ("poultry") can be good. She asks for a complete analysis of the $2 \times n$ board. She shows that the atomic weights of the $m \times n$ board for $2 \le m, n \le 4$ is 0 and is 0, 1, 1, 2 for $2 \times n$ for $n = 5, 6, 7, 8$, respectively.

B8. HEX. [*LIP*, pp. 264–265]. Nash's strategy-stealing argument shows that HEX is a first-player win but few quantitative results are known. Arneson et al. [131] report that 8×8 is solved, as are most openings on the 9×9.

Garikai Campbell [15] asks:

(1) For each n, what is the shortest path on an $n \times n$ board with which the first player can guarantee a win?

(2) What is the least number of moves that guarantees the first player a win? Campbell showed that this is at least n on an $n \times n$ board. Peng et al. [96] show that it is 7 on the 5×5 board.

B9. (54) FOX AND GEESE. Berlekamp and Siegel [108, Chapter 2] and [*WW* pp. 669–710], "analyzed the game fairly completely, relying in part on results obtained using CGSuite". In [*WW*, p. 710], the following open problems are given:

(1) Define a position's *span* as the maximum occupied row-rank minus its minimum occupied row-rank. Then quantify and prove an assertion such as the following: If the backfield is sufficiently large, and the span is sufficiently large, and if the separation is sufficiently small, and if the Fox is neither already trapped in a daggered position along the side of the board, nor immediately about to be so trapped, then the Fox can escape and the value is *off*.

(2) Show that any formation of three Geese near the center of a very tall board has a "critical rank" with the following property: If the northern Goose is far above, and the Fox is far below, then the value of the position is either positive,

HOT, or *off*, according as to whether the northern Goose is closer, equidistant, or further from the critical rank than the Fox, respectively.

(3) Welton asks what happens if the Fox is empowered to retreat like a bishop, going back several squares at a time in a straight line? More generally, suppose his straight-line retreating moves are confined to some specific set of sizes. Does $\{1, 3\}$, which maintains parity, give him more or less advantage than $\{1, 2\}$?

(4) What happens if the number of Geese and board widths are changed?

In Aaron Siegel's thesis there are several other questions:

(5) In the critical position — with Geese at (we use the algebraic CHESS notation of a, b, c, d, ... for the files and $1, 2, 3, \ldots, n$ for the ranks) (b, n), (d, n), (e, $n-1$), and (h, $n-1$), and the Fox at (c, $n-1$), which has value $1+2^{-(n-8)}$ on an $n \times 8$ board with $n \geq 8$ in the usual game — is the value $-2n+11$ for all $n \geq 6$ when played with "Ceylonese rules"? (Fox allowed two moves at each turn.)

(6) On an $n \times 4$ board with $n \geq 5$ and Geese at (b, n) and (c, $n-1$), do all Fox positions have value *over*? With the Geese on (b, n) and (d, n) are the only other values 0 at (c, $n-1$) and $\{\mathit{over} \mid 0\}$ at (b, $n-2$) and (d, $n-2$)?

(7) On an $n \times 6$ board with $n \geq 8$ and Geese at (b, n), (d, n), and (e, $n-1$) do the positions (a, $n-2k+1$), (c, $n-2k+1$), and (e, $n-2k+1$) all have value 0, and those at (b, $n-2k$), (d, $n-2k$), and (f, $n-2k$) all have value Star? And if the Geese are at (b, n), (d, n), and (f, n), are the zeroes and Stars interchanged?

B10. HARE AND HOUNDS. Aaron Siegel asks if the sequences of positions of increasing board length shown in Figure 4, left, are increasingly hot, and in Figure 4, right, if they have arbitrarily large negative atomic weight. He also conjectures that the starting position on a $6n+5 \times 3$ board, for $n > 0$, has value

$$-(n-1) + \left\{ b, c \,|0\|0\Big\|0\Big\|0 \ldots \Big\|0 \right\},$$

where there are $2n+4$ zeroes and slashes and $b = \big\{0, a\|0, \{0 \mid \mathit{off}\}\big\}$,

$$c = \big\{0 \,\|\, \downarrow_{[2]}*|0\|0\big\} \quad \text{and} \quad a = \{0, \downarrow_{[2]}* \mid 0, \downarrow_{[2]}*\}.$$

B11. (4) DOMINEERING. [*WW*, pp. 119–122, 138–142; *LIP* pp. 1–7, 260]. Extend the analysis.

(Left and Right take turns to place dominoes on a checker-board. Left orients her dominoes North–South and Right orients his East–West. Each domino exactly covers two squares of the board and no two dominoes overlap. A player unable to play loses.)

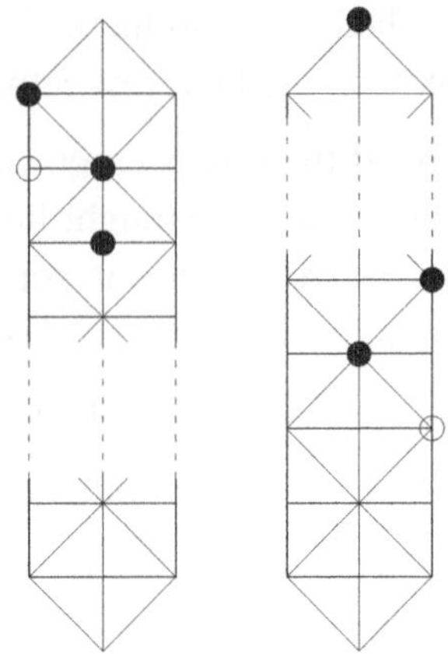

Figure 4. Sequences of hare and hounds positions.

See Berlekamp [5] and the second edition of [*WW*, pp. 138–142], where some new values are given. For example, David Wolfe and Dan Calistrate have found the values (to within "-ish", i.e., infinitesimally shifted) of 4×8, 5×6 and 6×6 boards. The value for a 5×7 board is

$$\left\{ \tfrac{3}{2} \,\middle|\, \left\{\tfrac{5}{4} \middle| -\tfrac{1}{2}\right\}, \left\{ \tfrac{3}{2} \,\middle|\, -\tfrac{1}{2}, \left\{\tfrac{3}{2} \middle| -1\right\} \,\middle\|\, -1 \middle| -3 \right\} \,\middle\|\, -1, \left\{ \tfrac{3}{2} \,\middle|\, -\tfrac{1}{2} \middle\| -1 \right\} \middle| -3 \right\}$$

Lachmann, Moore, and Rapaport [MGONC, pp. 307–315] discover who wins on rectangular, toroidal, and cylindrical boards of widths 2, 3, 5, and 7, but do not find their values. Bullock [11, p. 84] showed that 19×4, 21×4, 14×6, and 10×8 are wins for Left and that 10×10 is a first-player win. Uiterwijk [124] using a new DOMINEERING solver, shows that Left going first wins on 8×15, 11×9, 11×11, 12×8, 12×15, 14×8, and 17×6 boards and loses on 8×12, 9×11, and 11×10 boards. He also shows the following: Left wins on the board 6×15, Right wins on the boards 11×14 and 11×18 [120]; Left wins on the boards 6×19, 10×15, 14×15, and 18×15 [121]; and, in [122] proves that:

(1) all $2k \times n$ boards for $n = 3, 5, 7, 9, 11$ and $2k \times 13$ with $k \geq 3$ are Left wins, and

(2) all $m \times 2k$ boards for $m = 5, 9, 13$, and all $m \times 2k$ boards for $m = 3, 7$ with $k \geq 2$, and all $11 \times 2k$ boards with $k \geq 6$ are Right wins.

He also shows [123] how to recursively construct large positions with given values. Drummond-Cole [25] gives outcome classes for many sets of positions. For example, the boards $6 \times n$ is a Left win for n odd, $n \geq 41$, and $n \neq 43, 45, 51, 59$.

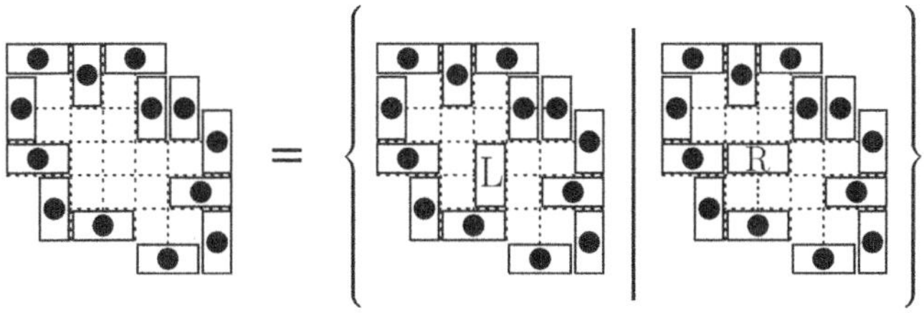

Figure 5. A DOMINEERING position of value $\pm 2*$.

Berlekamp notes that the value of a $2 \times n$ board, for n even, is only known to within "ish", and that there are problems on $3 \times n$ and $4 \times n$ boards that are still open.

Berlekamp asks, as a hard problem, to characterize all hot DOMINEERING positions to within "ish". As a possibly easier problem he asks for a DOMINEERING position with a new temperature, i.e., one not occurring in [GONC, Table 1, p. 477]. In an on-going investigation, Gabriel Drummond-Cole found values with temperatures between 1.5 and 2. Figure 5 shows a position of value $\pm 2*$ and temperature 2 [24].

Shankar and Sridharan [106] have found many DOMINEERING positions with temperatures other than those shown in [GONC, Table 1, p. 477]. Blanco and Fraenkel [8] have obtained partial results for the game of TROMINEERING, played with trominoes in place of (or, alternatively, in addition to) dominoes.

B12. NOGO can be found under the name "Anti-Atari Go" at the Sensei Library (see senseis.xmp.net/?AntiAtariGo) and was invented independently by Neil McKay. On a GO board (or on any graph) pieces are placed as in GO; the only restriction is that every connected group of each player must be adjacent to at least one empty intersection. Burke and Hearn [12] have shown that NOGO is NP-hard on a graph. McKay, Nowakowski and (Angela) Siegel found positions of value 1, $\frac{1}{2}$, $\frac{1}{4}$, and $\frac{1}{8}$ on a GO board. Using CGSuite, one can find the nimbers $*$, $*2$, $*4$, and$*8$ on a one-dimensional board. The position $\bullet \cdot \circ \cdot \bullet \cdot \circ \ldots$, where $\bullet$ is a black piece, $\circ$ is a white piece and "$\cdot$" is an empty space, is equivalent to the octal game **.6** (see Section A2), which is not known to be periodic. We ask for the nim-dimension of both the one- and two-dimensional board. Is there a limit to the denominator of the fractions found on the two-dimensional board?

C. Playing with pencil and paper

C1. (51) DOTS-AND-BOXES. Sierra [110] reports that 3×3 is a first-player win [*WW*], 4×4 (Wilson) and 4×5 (Barker and Korf) are draws, and that William Fraser reports that 5×5 is a first-player win. See Berlekamp's book [6] for more

Figure 6. Starting position for "Faroese" 1×6 Dots-and-Boxes and Dots-and-Triangles

problems about this popular children's (and adult's) game and see also [*WW*, pp. 541–584; *LIP*, pp. 21–28, 260].

Several other opening positions have been contemplated. "Open" is the standard position where there are no edges drawn, "Swedish" has all boundary lines drawn, "Faroese" has three, and "Icelandic" has 2 (see Figure 6). Jobson et al. [62] show that:

(1) Faroese $1 \times n$, $n \neq 2$, DOTS-AND-TRIANGLES is a first-player win, and

(2) the first player can at least tie on both Open and Faroese DOTS-AND-BOXES $1 \times n$ boards.

For the Faroese, they also conjecture that the first player wins by 1 for all $n \geq 9$ and odd. Sierra has also checked this (using David Wilson's analyzer) for Faroese $1 \times n$, $n \leq 19$ and odd (which he calls "staples", S_n), and adds to the conjecture that a winning move for the first player is a vertical move that splits the staple into two staples $S_2 + S_{n-2}$.

Sierra also conjectures for Icelandic DOTS-AND-BOXES with n odd that the first player loses by 1. Again, this is true for $n \leq 19$.

For misère play, Collette et al. [GONC4] show that the open $1 \times n$ position is a first-player win. They also conjecture that *the outcome of misère* $1 \times n$ *Swedish* DOTS-AND-BOXES *is eventually periodic, starting at* $n = 22$ *and with a period of* 10.

Elwyn Berlekamp asks for a complete theory of the impartial Icelandic $1 \times n$ game. Here, the player who completes the last box loses since they cannot complete their move by drawing another edge.

C2. (25) SPROUTS. Extend the analysis of this Conway–Paterson game in either the normal or misère form [*WW*, pp. 564–568].

(A move joins two spots, or a spot to itself by a curve which doesn't meet any other spot or previously drawn curve. When a curve is drawn, a new spot must be placed on it. The valence of any spot must not exceed three.)

C3. (26) SYLVER COINAGE. [*WW*, pp. 575–597]. Extend the analysis.

(Players alternately name distinct positive integers, but may not name a number which is a sum, with repetitions allowed, of previously named integers. Whoever names 1 loses.) Sicherman [107] contains the most recent information.

C4. (28) Extend Úlehla's [126] or Berlekamp's [*WW*, pp. 570–572] analysis of VON NEUMANN'S GAME from directed forests to directed acyclic graphs.

(VON NEUMANN'S GAME, or HACKENDOT, is played on one or more rooted trees. The roots induce a direction, towards the root, on each edge. A move is to delete a node, together with all nodes on the path to the root, and all edges incident with those nodes. Any remaining subtrees are rooted by the nodes that were adjacent to deleted nodes.)

C5. (43) INVERTING HACKENBUSH. Thea van Roode has written a thesis [100] investigating both this and REVERSING HACKENBUSH, but there is plenty of room for further analysis of both games.

In INVERTING HACKENBUSH, when a player deletes an edge from a component, the remainder of the component is replanted with the new root being the pruning point of the deleted edge. In REVERSING HACKENBUSH, the colors of the edges are all changed after each deletion. Both games are hot, in contrast to BLUE-RED HACKENBUSH [*WW*, pp. 1–7; *LIP*, pp. 82, 88, 111–112, 212, 266] which is cold, and GREEN HACKENBUSH [*WW*, pp. 189–196], which is tepid.

C6. (42) BEANSTALK and BEANS-DON'T-TALK are games invented respectively by John Isbell and John Conway. See Guy in [GONC]. BEANSTALK is played between Jack and the Giant. The Giant chooses a positive integer, n_0. Then Jack and the Giant play alternately $n_1, n_2, n_3, \ldots$, according to the rule

$$n_{i+1} = \begin{cases} \frac{1}{2}n_i & \text{if } n_i \text{ is even,} \\ 3n_i \pm 1 & \text{if } n_i \text{ is odd;} \end{cases}$$

that is, if n_i is even, there's only one option, while if n_i is odd there are just two. The winner is the person moving to 1.

We still don't know if there are any $\mathcal{O}$*-positions* (positions of infinite remoteness).

C7. (63) THE ERDŐS–SZEKERES SEQUENCE GAME [26] (and see Schensted [103]) was introduced by Harary et al. [52]. From a deck of cards labeled from 1 through n, Alexander and Bridget alternately choose a card and append it to a sequence of cards. The game ends when there is an ascending subsequence of a cards or a descending subsequence of d cards.

The game appears to have a strong bias towards the first player. Albert et al. [GONC3] show that for $d = 2$ and $a \le n$ the outcome is $\mathcal{N}$ or $\mathcal{P}$ accordingly as n is odd or even, and is $\mathcal{O}$ (drawn) if $n < a$. They conjecture that for $a \ge d \ge 3$ and all sufficiently large n, it is $\mathcal{N}$ with both normal and misère play, and also with normal play when played with the rationals in place of the first n integers.

They also suggest investigating the form of the game in which players take turns naming pairs (i, π_i) subject to the constraint that the chosen values form part of the graph of some permutation of $\{1, 2, \ldots, n\}$.

C8. THINNING THICKETS. This game, played on trees consisting of a stalk with at most one leaf at each vertex, is introduced in [60]. They show that the game has infinite nim-dimension and infinite boiling-point. The trees are very thin (a stalk with maybe a leaf at each node) and either all the arcs are green or just blue and red. They ask to extend the analysis to general trees.

D. Disturbing and destroying

D1. (27) Extend the analysis of CHOMP [*WW*, pp. 598–599; *LIP*, pp. 19, 46, 216].

David Gale asks for the solution of the infinite 3D version where the board is the set of all triples (x, y, z) of nonnegative integers, that is, the lattice points in the positive octant of $\mathbb{R}^3$. The problem is to decide whether it is a win for the first or second player.

CHOMP (Gale [43]) is equivalent to DIVISORS (Schuh [104]). CHOMP is easily solved for $2 \times n$ arrays, Sun [115], and indeed a recent result by Byrnes [13] shows that any poset game eventually displays periodic behavior if it has two rows, and a fixed finite number of other elements. See also the Fraenkel poset games mentioned near the end of A2.

Thus, most of the work in recent years has been on three-rowed CHOMP. The situation becomes quite complicated when a third row is added; see Zeilberger [129] and Brouwer et al. [10]. A novel approach (renormalization) is taken by Friedman and Landsberg [GONC3]. They demonstrate that three-rowed CHOMP exhibits certain scaling and self-similarity patterns similar to chaotic systems. Is there a deterministic proof that there is a unique winning move from a $3 \times n$ rectangle? The renormalization approach is based on nonlinear dynamics techniques from physics; its results are highly suggestive but as of yet not fully mathematically rigorous.

TRANSFINITE CHOMP has been investigated by Huddleston and Shurman [MGONC]. An open question is to calculate the nim-value of the position $\omega \times 4$ — they conjecture this to be $\omega \cdot 2$, but it could be as low as 46, or even uncomputable! Perhaps the most fascinating open question in TRANSFINITE CHOMP is their *stratification conjecture*, which states that if the number of elements taken in a move is $< \omega^i$, then the change in the nim-value is also $< \omega^i$.

Andries Brouwer has a CHOMP site: www.win.tue.nl/~aeb/games/chomp.html.

Nakamura and Miyadera [90] introduce a new board. Instead of the columns nonincreasing in size going out from the poisoned square, they are nondecreasing.

They also add a strip going off to the left. Dozier and Perry [23] introduce more poisoned squares and generalize to ideals and Grobner bases.

D2. (33) SUBSET TAKE-AWAY. Given a finite set, players alternately choose proper subsets subject to the rule that once a subset has been chosen no proper subset can be removed. Last player to move wins.

Many people play the dual, that is, a nonempty subset must be chosen and no proper superset of this can be chosen. We discuss this version of the game which now can be considered a poset game with the sets ordered by inclusion.

The $(n;\, k)$ SUBSET TAKE-AWAY game is played using all subsets of sizes 1 through k of a n-element set. In the $(n;\, n)$ game, one has the whole set (i.e., the set of size n) as an option, so a strategy-stealing argument shows this must be a first-player win.

Gale and Neyman [44], in their original paper on the game, conjectured that the winning move in the $(n;\, n)$ game is to remove just the whole set. This is equivalent to the statement that the $(n;\, n-1)$ game is a second-player win, which has been verified only for $n \leq 5$. A stronger conjecture states that $(n;\, k)$ is a second-player win if and only if $k+1$ divides n. This was proved in the original paper only for $k = 1$ or 2.

Andries Brouwer and J. Daniel Christensen [9] show that these are false. The first player loses in (7; 7) if they take the top set. Instead, they win by taking a set of cardinality 4. The second player wins (7; 3), but 4 does not divide 7.

See also Fraenkel and Scheinerman [41].

D3. (39) Sowing or MANCALA games. There appears to have been no advance on the papers mentioned in MGONC, although we feel that this should be a fruitful field of investigation at several different levels.

D4. Annihilation games. *k*-ANNIHILATION. Initially place tokens on some of the vertices of a finite digraph. Denote by $\rho_{\text{out}}(u)$ the outvalence of a vertex u. A move consists of removing a token from some vertex u, and "complementing" $t := \min(k, \rho_{\text{out}}(u))$ (immediate) followers of u, say $v_1, \ldots, v_t$: if there is a token on v_j, remove it; if there is no token there, put one on it. The player making the last move wins. If there is no last move, the outcome is a draw. For $k = 1$, there is an $O(n^6)$ algorithm for deciding whether any given position is in $\mathcal{P}$, $\mathcal{N}$, or $\mathcal{O}$; and for computing an optimal next move in the last 2 cases (Fraenkel and Yesha [42]). Fraenkel asks: Is there a polynomial algorithm for $k > 1$? For an application of k-annihilation games to lexicodes, see Fraenkel and Rahat [40].

D5. TOPPLING DOMINOES [*LIP*, pp. 110–112, 274] is played with a row of vertical dominoes, each of which is either blue or red. A player topples one of his/her dominoes to the left or to the right.

we have stacks	0	1&7	2&5	3	4	6
which can be relabeled	0	1&3	1&2	1	2	3

Table 1. Strings of "pearls" in DUDENEY for values of Y of various forms.

See Fink et al. [GONC4, pp. 65–76] for the proofs that every number occurs exactly once and is a palindrome; there are exactly n positions with value $*n$. Several conjectures are listed but the most intriguing seems to be this: *If G is a palindrome then G's value appears uniquely.*

There are several variants of TOPPLING DOMINOES. If all the dominoes must be toppled in the same direction then this is a HACKENBUSH string. TIMBER [95] is an impartial version of TOPPLING DOMINOES played on a directed graph. The dominoes are on the edges. A player chooses a domino which is toppled in the direction of the edge. The dominoes on incident edges are then toppled (regardless of the underlying edge) and the process is iterated. If only outcomes are required then only trees are interesting. In [95], an algorithm is given to determine the outcome class of a tree, which surprisingly requires nim-values. Values, however, appear to be difficult to determine. The normal and misère versions on a path are related to Dyck paths and Catalan numbers; see Section E15 for more details. The partizan version, imaginatively called PARTIZAN TIMBER, has Left and Right dominoes placed on the edges of a directed graph. Any TOPPLING DOMINOES position can be transformed into a PARTIZAN TIMBER position by subdividing each edge and directing the edge toward the new vertex. A similar algorithm to that of TIMBER can be used to determine the outcome class of a tree. The uniqueness of numbers doesn't hold even on a path. In the other direction, are there any values that do not occur in the game?

D6. TOWER OF HANOI. Chappelon et al. [17] consider the TOWER OF HANOI as a two-player game. The first player can win when there are 3 pegs and it is a draw with 4 or more pegs — parity of moves! They introduce a scoring version (see E14) where there are weights on the edges. Almost all are first-player wins. Are there any variations with second-player wins?

HANOI STICK-UP is played with the disks of the TOWERS OF HANOI puzzle, starting with each disk in a separate stack. A move is to place one stack on top of another such that the size of the bottom of the first stack is less than the size of the top of the second; the two stacks then fuse (&) into one. The only relevant information about a stack are its top and bottom sizes, and it's often possible to collapse the labeling of positions: so, for instance, starting with 8 disks and fusing 1&7 and 2&5, in which the legal moves are still the same. John Conway, Alex Fink, and others have found that the $\mathcal{P}$-positions of height ≤ 3 in normal HANOI STICKUP are exactly those which, after collapsing, are of the form $0^a\ 01^b\ 1^c\ 12^d\ 2^e$ with $\min(a+b+c, c+d+e, a+e)$ even, except that if $a+e \le a+b+c$ and $a+e \le c+d+e$ then both a and e must be even (02 can't be involved in a legal move so can be dropped).

They also found the normal and misère outcomes of all positions with up to six stacks, but there is more to be discovered.

D7. (56) Are there any draws in BEGGAR-MY-NEIGHBOR? Marc Paulhus showed that there are no cycles when using a half-deck of two suits, but the problem for the whole deck (one of Conway's "anti-Hilbert" problems) is still open. Lakshtanov and Aleksenko [68] showed that the expected number of moves is finite in the modified game where each pile is shuffled when being returned to the deck and that the player to start a new pile is chosen by a Bernoulli trial with $p \in (0, 1)$.

E. Theory of games

E1. (49) Fraenkel updates Berlekamp's questions on computational complexity as follows:

Demaine, Demaine, and Eppstein [MGONC] proved that deciding whether a player can win in a *single* move in PHUTBALL [*WW*, pp. 752–755; *LIP*, p. 212] is NP-complete. Grossman and Nowakowski [MGONC] gave constructive partial strategies for one-dimensional PHUTBALL. Thus, these papers do not show that PHUTBALL has the required properties.

Perhaps NIMANIA (Fraenkel and Nešetřil [39]) and MULTIVISION (Fraenkel [33]) satisfy the requirements. NIMANIA begins with a single positive integer, but after a while there is a multiset of positive integers on the table. At move k, a copy of an existing integer m is selected, and 1 is subtracted from it. If $m = 1$, the copy is deleted. Otherwise, k copies of $m-1$ are adjoined to the copy $m-1$. The player first unable to move loses and the opponent wins. It was proved that:

(i) the game terminates, and

(ii) Player 1 can win.

In Fraenkel, Loebl and Nešetřil [38], it was shown that the max number of moves in NIMANIA is an Ackermann function, and the min number satisfies $2^{2^{n-2}} \le \mathrm{Min}(n) \le 2^{2^{n-1}}$.

The game is thus intractable simply because of the length of its play. This is a *provable* intractability, much stronger than NP-hardness, which is normally only a *conditional* intractability. One of the requirements for the tractability of a game is that a winner can consummate a win in at most $O(c^n)$ moves, where

$c > 1$ is a constant, and n a sufficiently succinct encoding of the input (this much is needed for nim on 2 equal heaps of size n).

To consummate a win in NIMANIA, player 1 can play randomly most of the time, but near the end of play, a winning strategy is needed, given explicitly. Whether or not this is an "intricate" solution depends on the beholder. But it seems that it's of even greater interest to construct a game with a very *simple* strategy which still has high complexity!

Also every play of MULTIVISION terminates, the winner can be determined in linear time, and the winning moves can be computed linearly. But the length of play can be arbitrarily long. So let's ask the following: Is there a game which has

1. simple, playable rules,
2. a simple explicit strategy,
3. length of play at most exponential, and
4. is NP-hard or harder.

Tung [119] proved the following:

Theorem. *Given a polynomial* $P(x, y) \in \mathbb{Z}[x, y]$, *the problem of deciding whether for all* x *there exists* $y[P(x, y) = 0]$ *holds over* $\mathbb{Z}_{\geq 0}$, *is co-NP-complete.*

Define the following game of length 2: Player 1 picks $x \in \mathbb{Z}_{\geq 0}$, Player 2 picks $y \in \mathbb{Z}_{\geq 0}$. Player 1 wins if $P(x, y) \neq 0$, otherwise Player 2 wins. For winning, Player 2 has only to compute y such that $P(x, y) = 0$, given x, and there are many algorithms for doing so.

Also Jones and Fraenkel [63] produced games, with small length of play, which satisfy these conditions.

So we are led to the following reformulation of Berlekamp's question: Is there a game which has

1. simple, playable rules,
2. a finite set of options at every move,
3. a simple explicit strategy,
4. length of play at most exponential, and
5. is NP-hard or harder.

E2. Complexity closure. Aviezri Fraenkel asks: Are there partizan games G_1, G_2, G_3 such that:

(i) G_1, G_2, G_3, $G_1 + G_2$, $G_2 + G_3$, and all their options have polynomial-time strategies, and

(ii) $G_1 + G_3$ is NP-hard?

E3. Sums of switch games. David Wolfe considers a sum of games G, each of the form $a\|b|c$ or $a|b\|c$, where a, b, and c are integers specified in unary. Is there a polynomial time algorithm to determine who wins in G, or is the problem NP-hard?

E4. (52) How does one play *sums* of games with varied overheating operators?

Sentestrat and top-down thermography [*LIP*, p. 214]:

David Wolfe would like to see a formal proof that sentestrat works, an algorithm for top-down thermography, and conditions for which top-down thermography is computationally efficient.

Aaron Siegel asks the following generalized thermography questions:

(1) Show that the Left scaffold of a dogmatic (neutral ko-threat environment [*LIP*, p. 215]) thermograph is decreasing as function of t. (Note, this is *not* true for komaster thermographs.) (Dogmatic thermography was invented by Berlekamp and Spight. See [113] for a good introduction.)

(2) Develop the machinery for computing dogmatic thermographs of double kos (multiple alternating 2-cycles joined at a single node).

In the same vein as (2):

(3) Develop a temperature theory that applies to all loopy games.

Siegel thinks that (3) is among the most important open problems in combinatorial game theory. The temperature theory of GO appears radically different from the classical combinatorial theory of loopy games (where infinite plays are draws). It would be a huge step forward if these could be reconciled into a "grand unified temperature theory". Problem (2) seems to be the obvious next step toward (3).

Conway asks for a natural set of conditions under which the mapping $G \mapsto \int^* G$ is the *unique* homomorphism that annihilates all infinitesimals.

E5. Loopy games [*WW*, pp. 334–377; *LIP*, pp. 213–214] are partizan games that do not satisfy the ending condition. A *stopper* is a game that, when played on its own, has no ultimately alternating, Left and Right, infinite sequence of legal moves. Aaron Siegel reminds us of [*WW*, 2nd ed., p. 369], where the authors tried hard to prove that every loopy game had stoppers, until Clive Bach found the CAROUSEL counterexample. Is there an alternative notion of simplest form that works for *all* finite loopy games, and, in particular, for the CAROUSEL? The simplest form theorem for stoppers is in [*WW*, p. 351].

Siegel conjectures that, if Q is an arbitrary cycle of Left and Right moves that contains at least two moves for each player, and is not strictly alternating, then there is a stopper consisting of a single cycle that matches Q, together

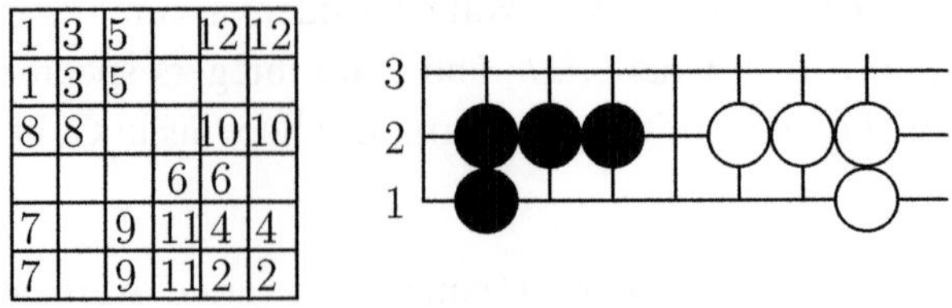

Figure 7. A DOMINEERING position and a chilled GO position of value $*2$.

with various exits to enders, i.e., games which end in a finite, though possibly unbounded, number of moves. (Note that games normally have Left and Right playing alternately, but if the game is a sum, then play in one component can have arbitrary sequences of Left and Right moves, not just alternating ones.)

A long cycle is *tame* if it alternates just once between Left and Right, otherwise it is *wild*. Aaron Siegel writes:

> I can produce wild cycles "in the laboratory," by specifying their game graphs explicitly. So the question is to detect one "in nature", i.e., in an actual game with (reasonably) playable rules such as PHUTBALL (Problem B7).

Siegel also asks under what conditions does a given infinitesimal have a well-defined atomic weight, and asks to specify an algorithm to calculate the atomic weight of an infinitesimal stopper g. The algorithm should succeed whenever the atomic weight is well-defined, i.e., whenever g can be sandwiched between loopfree all-smalls of equal atomic weight.

E6. (45) Elwyn Berlekamp asks for the *habitat* of $*2$, where $*2 = \{0, * \mid 0, *\}$.

In DOMINEERING, Uiterwijk and Barton [125] improve on Drummond-Cole's original $*2$ in that the position is reachable from a standard board (no initial holes) and also exhibits a connected $*3$ position. See, for example, Figure 7, which also shows a GO position, found by Nakamura and Berlekamp [89], whose chilled value is $*2$. The Black and White groups are both connected to life via unshown connections emanating upwards from the second row. Either player can move to $*$ by placing a stone at E2, or to 0 by going to E1. Given a game, let n be the smallest nonnegative integer n such that if a position of the game has value $*k$ then $k < 2^n$.

E7. Partial ordering of games. David Wolfe lets $g(n)$ be the number of games born by day n, notes that an upper bound is given by $g(n+1) \leq g(n) + 2^{g(n)} + 2$, and a lower bound for each $\alpha < 0$ is given by $g(n+1) \geq 2^{g(n)^{\alpha}}$, for n sufficiently large, and asks us to tighten these bounds.

He also asks what group is generated by the all-small games (or — much harder — of all games) born by day 3. Describe the partial order of games born by day 3, identifying all the largest "hypercubes" (Boolean sublattices) and how they are interconnected. These questions have been answered for day 2 [GONC3, pp. 125–130].

Berlekamp suggests other possible definitions for games born by day n, $\mathcal{G}_n$, depending on how one defines $\mathcal{G}_0$. Our definition is 0-based, as $\mathcal{G}_0 = \{0\}$. Other natural definitions are integer-based (where $\mathcal{G}_0$ are integers) or number-based. These two alternatives do not form a lattice, for if G_1 and G_2 are born by day k, then the games

$$H_n := \bigl\{G_1, G_2 \,\|\, G_1, \{G_2 \mid -n\}\bigr\}$$

form a decreasing sequence of games born by day $k+2$ exceeding any game $G \geq G_1, G_2$, and the day $k+2$ join of G_1 and G_2 cannot exist. What is the structure of the partial order given by one of these alternative definitions of birthday?

The set of all short games does not form a lattice, but Calistrate, Paulhus, and Wolfe [MGONC] have shown that the games born by day n form a distributive lattice $\mathcal{L}_n$ under the usual partial order. They ask for a description of the exact structure of $\mathcal{L}_3$. Siegel describes $\mathcal{L}_4$ as "truly gigantic and exceedingly difficult to penetrate" but suggests that it may be possible to find its dimension and the maximum *longitude*, $\mathrm{long}_4(G)$, of a game in $\mathcal{L}_4$, which he defines as

$$\mathrm{long}_n(G) = \mathrm{rank}_n(G \vee G^\bullet) - \mathrm{rank}_n(G),$$

where $\mathrm{rank}_n(G)$ is the rank of G in $\mathcal{L}_n$ and $G^\bullet$ is the *companion* of G,

$$G^\bullet = \begin{cases} * & \text{if} \quad G = 0, \\ \{0, (G^L)^\bullet \mid (G^R)^\bullet\} & \text{if} \quad G > 0, \\ \{(G^L)^\bullet \mid 0, (G^R)^\bullet\} & \text{if} \quad G < 0, \\ \{(G^L)^\bullet \mid (G^R)^\bullet\} & \text{if} \quad G \,\|\, 0. \end{cases}$$

Albert and Nowakowski [3] show that starting with any set of games, instead of just $0 = \{\cdot|\cdot\}$, then the games born on the next day form a complete but not necessarily distributive lattice. They ask which sets of games will give a distributive lattice? Is there a set which will give a nondistributive but modular lattice? Carvalho, dos Santos, Dias, Coelho, Neto, Nowakowski, and Vinagre [16] answered this by showing that for any lattice L, there is a set of games which generates L on the next day.

The set of all-small games does not form a lattice, but Siegel forms a lattice $\mathcal{L}_n^0$ by adjoining the least and greatest elements $\triangle$ and $\triangledown$ and asks: Do the elements of $\mathcal{L}_n^0$ have an intrinsic "handedness" that distinguishes, say, $(n-1)\cdot\uparrow$ from $(n-1)\cdot\uparrow + *$?

A game is *option-closed* [94] (now called hereditarily transitive) if, recursively, each G^{LL} is also $\mathcal{G}^L$ and the same for Right. For example, HACKENBUSH strings are option-closed. McKay, Nowakowski, and (Angela) Siegel [83] show that the option-closed games born on day n form a planar, nondistributive lattice, but the question of *how many?* remains unanswered. Are there other natural families of games that form planar lattices?

E8. Aaron Siegel asks, given a group or monoid, $\mathcal{K}$, of games, to specify a technique for calculating the simplest game in each $\mathcal{K}$-equivalence class.

He notes that some restriction on $\mathcal{K}$ might be needed; for example, $\mathcal{K}$ might be the monoid of games absorbed by a given idempotent.

E9. Siegel also would like to investigate how search methods might be integrated with a canonical-form engine.

E10. **(9)** Develop a *misère theory* for unions of partizan games [*WW*, p. 312].

E11. Four-outcome games. Guy [50] has given a brute force analysis of a parity subtraction game (start with a subtraction set and an odd number of tokens, the winner is the player who removes an even number of tokens). The Sprague–Grundy theory does not apply because the game wasn't impartial, nor the Conway theory, because it was not last-player-winning. Is there a class of games in which there are four outcomes, $\mathcal{N}$ext, $\mathcal{P}$revious, $\mathcal{L}$eft and $\mathcal{R}$ight, and for which a general theory can be given? The discussion in [73, §2] hints that there will not be an extension of normal- or misère-theory that will solve this type of game.

E12. Impartial misère analysis. See A16. From the works of Plambeck and Siegel: Let $\mathcal{A}$ be some set of games (the *universe*) played under misère rules. Typically, $\mathcal{A}$ is the set of positions that arise in a particular game, such as DAWSON'S CHESS. Games $H, K \in \mathcal{A}$ are said to be equivalent, denoted by $H \equiv K \pmod{\mathcal{A}}$, if $H + X$ and $K + X$ have the same outcome for all games $X \in \mathcal{A}$. The relation $\equiv$ is an equivalence relation, and a set of representatives, one from each equivalence class, forms the *misère quotient*, $\mathcal{Q} = \mathcal{A}/\equiv$. A *quotient map* $\Phi : \mathcal{A} \to \mathcal{Q}$ is defined, for $G \in \mathcal{A}$, by $\Phi : \mathcal{G} = [G]_{\equiv}$.

(1) A quotient map $\Phi : \mathcal{A} \to \mathcal{Q}$ is said to be *faithful* if, whenever $\Phi(G) = \Phi(H)$, then G and H have the same normal-play Grundy value. Is every quotient map faithful?

(2) Let $(\mathcal{Q}, \mathcal{P})$ be a quotient and $\mathcal{S}$ a maximal subgroup of $\mathcal{Q}$. Must $\mathcal{S} \cap \mathcal{P}$ be nonempty? (Note: it's easy to get a "yes" answer in the special case when $\mathcal{S}$ is the kernel.)

(3) Extend the classification of impartial misère quotients. We have preliminary results on the number of quotients of order $n \leq 18$ but believe that this can be pushed far higher.

(4) Exhibit an impartial misère quotient with a period-5 element. Same question for period 8, etc. We've detected quotients with elements of periods 1, 2, 3, 4, 6, and infinity, and we conjecture that there is no restriction on the periods of quotient elements.

(5) In the flavor of both (3) and (4): What is the smallest quotient containing a period 4 (or 3 or 6) element?

E13. Inverses in partizan misère games. Siegel [Misère canonical forms of partizan games, GONC4] and Allen [GONC4] extend the misère monoid concepts to partizan games. In normal play, games in which both players have a move or neither does are called *all-small* because their values are infinitesimal. This is not true in misère play and the term *dicotic* has been coined to refer to these games in all play conventions. (See Section E14 as well.)

Recall that $-G$ is obtained from the game G by reversing the roles of Left and Right — "turning the board around". In normal play, $G - G = 0$ which is decidedly not true in misère play. To avoid confusion, when dealing with misère play $-G$ is frequently written as $\overline{G}$ and is called the *conjugate* of G.

(1) Allen asks, in the dicotic universe $\mathcal{D}$, when is it true that $G+\overline{G} \equiv 0 \pmod{\mathcal{D}}$? This was shown to hold for all games in [71].

(2) Rebecca Milley asks, in a universe $\mathcal{U}$, closed under sums, conjugation, and subpositions, if $G + H \equiv 0 \pmod{\mathcal{U}}$, is it necessarily true that $H = \overline{G}$? Jason Brown had asked the question about arbitrary universes. Milley [84] has the counterexample: on a finite strip of squares, Left places a 1×1 piece and Right places a 1×2 (domino). In the universe of sums of strips a strip of length 1 plus a strip of length 2 is equivalent to 0. This leads to a different question (see [Milley and Renault, in this volume]): in what universes U do we have no nonconjugate inverses?

A universe $\mathcal{U}$ is *reverse-friendly* if for every $G = \{A \| \{\cdot | B\}\}$, there is $H = \{A' \| \{\cdot | B\}\}$, where H is invertible modulo $\mathcal{U}$ and $G \geq \{\cdot | B\} \geq \{\cdot | C\}$. Adapting the proofs from [73], it appears that nonconjugate inverses cannot occur in any parental, dense, and reverse-friendly universes.

E14. Scoring games. Instead of "the last play determines the winner", another natural way is to have "whichever player has the higher score wins". Both DOTS-AND-BOXES and GO, for example, are scoring games.

Milnor looked at dicotic scoring games with nonnegative incentives. Milnor [85] defines a "positional game" to be what we would call a dicot (Section E13) scoring game, but only looks at positional games with nonnegative incentives. Hanner [51] looked at the same class of games as Milnor, and invented thermography. Ettinger [27] looked at dicotic scoring games (possibly with negative incentives), which he calls "positional games" following Milnor. Johnson [61] looked at the dicot scoring games that are "well-tempered" in some sense (a subset of Ettinger's games). Stewart [in this volume] looks at general scoring games. Since the games are no longer dicotic, a choice has to be made about how to handle the situation where the current player can't move but their opponent can.

Larsson, Nowakowski, and Santos [72; 71] have developed the theory of guaranteed scoring games. At the end of the game, the score is indicated by $\varnothing^a$ — the empty set to indicate there is no move and a is the score. A Left atomic game has the form $\langle \varnothing^\ell \mid A, B, \ldots \rangle$ — Left has no move and $A, B, \ldots$ are the Right options. In a Right atomic game, it is Right who has no move. A scoring game G is guaranteed if for every Left atomic game $\langle \varnothing^\ell \mid A, B, \ldots \rangle$ in G if $\varnothing^r$ occurs in one of the Right options $A, B, \ldots$ then $\ell \leq r$ and the reverse needs to be true for a Right atomic game. Normal play games can be embedded in scoring games by replacing every occurrence of $\varnothing$ by $\varnothing^0$. This mapping is an order-embedding into the set of guaranteed scoring games. They ask: Is the set of guaranteed scoring games the largest subset of scoring games (closed under disjunctive sum and conjugation) in which normal play games can be order-embedded? This subset of scoring games cannot contain games of the form $\langle \varnothing^\ell \mid \varnothing^r \rangle$ where $\ell > r$.

They ask for a solution of BRUSHES; see ["Scoring games: the state of play", in this volume] for the rules.

E15. Find a formula for the number of $\mathcal{P}$-positions for NIM played with $2n$ tokens. There are none with an odd number of tokens. The On-line Encyclopedia of Integer Sequences (OEIS) has the value for small values of n. Khovanova and Xiong [66] have formulas for when:

(i) the number of heaps is restricted and

(ii) the number in a heap is restricted.

In general, given a game, enumerate the number of $\mathcal{P}$-positions of a given size. We only know of TIMBER (Section D5) where the number of $\mathcal{P}$-positions, both normal and misère, are related to Dyck paths with certain properties (also to Catalan numbers and Fine numbers) and Hetyei [53; 54] which relates the number of $\mathcal{P}$-positions to the Bernoulli numbers of the second kind.

E16. Miller and Guo (see [48] for an overview) consider games played on lattice points in a polyhedra (often a Cartesian grid in 2 or more dimensions). There are finitely many terminal points, and finitely many vectors which indicate the moves. They originally conjectured that all lattice games have strategies that can be computed efficiently. Fink [30] disproved this. In *square-free lattice games*, the vectors only have zeroes and ones as entries. These generalize octal, hexadecimal and many other heap games. Miller and Guo conjecture [48] that there are efficient algorithms for determining the winning moves from any $\mathcal{N}$-position.

E17. A game has *nim-dimension n* if it contains a position $*2^{n-1}$ but not $*2^n$ (Santos [102]). A game has infinite nim-dimension if all the nimbers can be constructed. It has null, or $\varnothing$, nim-dimension if $*$ cannot be constructed. This was motivated by E6. (See also THINNING THICKETS [60].)

One can ask for the nim-dimension for each game one encounters. TOPPLING DOMINOES (Section C5), THINNING THICKETS [60], and GENERALIZED KONANE [Carvalho and Santos, in this volume] all have infinite nim-dimension, COL has nim-dimension 1 [*WW*], it is conjectured that DOMINEERING (Sections B11, E6) has nim-dimension = 1; CLOBBER has nim-dimension = 1; and AMAZONS (Section B6) has nim-dimension 2 (only $*4$ is known).

E18. Atomic weight calculus and nim-dimension. SUBVERSION (see A19) is a dicotic game in which both players have exactly one move and, in all positions, at least one of the players can end the game. The nim-dimension is 0 since the only nimbers that can occur are 0 and $*$. Fisher et al. [31] show that the exceptional case in the atomic weight calculations involve only a finite number of cases. They ask if this is true for any game with bounded nim-dimension. Of course, if true then the number of cases will probably increase with the nim-dimension.

E19. Universal partizan games. NIM is *universal* for impartial games in that for every impartial value there is (at least) one NIM of that value. Carvalho and Santos show that GENERALIZED KONANE [in this volume] is universal for the short Conway's group. Are there any other universal games? (Other than sliding a coin down a game-tree.) TOPPLING DOMINOES (D5) is a candidate but dealing with reversible options has proven difficult.

A related question: are there any general techniques that, given a game, show that certain values do not occur?

E20. Temperature. Nowakowski and Santos define the *boiling point* of a rule set R as $BP(R) = \limsup_G \operatorname{temp}(G)$. (See B11 and E17.)

As with nim-dimension, one can ask for the boiling point for each game one encounters. TOPPLING DOMINOES (D5), THINNING THICKETS [60], and

GENERALIZED KONANE [Carvalho and Santos, in this volume] all have infinite boiling point. It is conjectured that the boiling point of DOMINEERING (B11, E6) is 2. See Berlekamp [in this volume].

Find feasible methods that give good upper bounds on the temperatures. Nowakowski and Santos show that if the length of the confusion intervals are bounded by k then the boiling point is no more than $3k/2$.

E21. Joins of games. When playing game A, a position may be a valid position for another game B.

This has been utilized in *conjoined games*; see Huggan and Nowakowski [in this volume]. Start playing A and when a terminal position is reached, start playing B. The two games must be compatible in some not-well-defined sense. This is a possible mechanism to generate a starting position for games without a prescribed start.

Heinrich (with Duchene, Larsson, and Parreau) makes it an option for a player to switch games (from A to B) at any point. Tennenhouse makes it an option to switch from normal to misère (or the reverse depending on the rules of the original game). Conway [*WW*, p. 534] introduced WHIM, which is NIM with the extra option, to be only be exercised once, of deciding if the outcome convention is normal or misère.

Farr and Ho [29] explore another possible join.

E22. Multiplayer games. (Propp [98] and [67] are good places to start.) If a player cannot win but can make either of the other two the winner, how should play proceed?

One way is to invoke a preference. Nowakowski, Santos, and Silva consider the podium rule (each player wants to be highest on the podium). They consider THREE-PLAYER NIM and determine the equivalence classes of the associated monoid. Koki Suetsugu gives each player a preference ordering. An intriguing aspect is that a player doesn't necessarily want themselves to win.

Using the podium rule, misère THREE-PLAYER NIM with one pile might be easier to analyze. Both Kelly [65] and Zhao and Liu [130] consider this game. Liu and Duan [77] finally solve the one heap and also the case when the number of players is at least as large as the number of heaps. The other case is still open. They ask for an analysis under other preferences.

We can consider cooperative two- or more player games such that all players want that one of the players make the last move. For two-player games this is a "puzzle" rather than a game. However, disjunctive sums still occur and the understanding about the recursive construction, comparison procedures, and canonical forms is an open field.

References

[1] M. H. Albert, J. P. Grossman, S. McCurdy, R. J. Nowakowski and D. Wolfe. Cherries. Preprint, 2005. [Problem A5]

[2] M. H. Albert and R. J. Nowakowski. The game of end-nim. *Electr. J. Combin.*, 8(2):#R1, 12pp., 2001. [Problem A5]

[3] M. H. Albert and R. J. Nowakowski. Lattices of games. *Order*, 29:75–84, 2011. [Problem E7]

[4] R. B. Austin. Impartial and partizan games. Master's thesis, The University of Calgary, 1976. [Problem A3]

[5] E. R. Berlekamp. Blockbusting and domineering. *J. Combin. Theory* (*Ser. A*), 49:67–116, 1988. An earlier version, entitled "Introduction to blockbusting and domineering", appeared in *The Lighter Side of Mathematics*, Proc. E. Strens Memorial Conf. on Recr. Math. and its History, Calgary, 1986, Spectrum Series (R. K. Guy and R. E. Woodrow, eds.), Math. Assoc. of America, Washington, DC, 1994, pp. 137–148. [Problems B11, E6]

[6] E. Berlekamp. *The dots and boxes game: sophisticated child's play.* A K Peters, Ltd., 2000. [Problem C1]

[7] E. Berlekamp and R. Low. *Enterpreneurial chess*, to appear in *International Journal of Game Theory.*

[8] S. A. Blanco and A. S. Fraenkel. Tromineering, tridomineering and L-tridomineering. Preprint, August 2006. [Problem B11]

[9] A. E. Brouwer and J. D. Christensen. Counterexamples to conjectures about subset takeaway and counting linear extensions of a boolean lattice. Preprint, 2017. [Problem D2]

[10] A. E. Brouwer, G. Horváth, I. Molnár-Sáska and C. Szabó. On three-rowed chomp. *INTEGERS*, 5:#G07, 11pp., 2005. [Problem D2]

[11] N. Bullock. Domineering: solving large combinatorial search spaces. Master's thesis, University of Alberta, 2002. [Problem B11]

[12] Kyle Burke and Bob Hearn. Pspace-complete two-color placement games. Preprint, 2016. [Problem B12]

[13] S. Byrnes. Poset game periodicity. *INTEGERS*, 3:#G3, 12pp., 2003. [Problem D1]

[14] I. Caines, C. Gates, R. K. Guy and R. J. Nowakowski. Periods in taking and splitting games. *American Math Monthly,* 106:359–361, 1999. [Problem A2]

[15] G. Campbell. On optimal play in the game of hex. *INTEGERS*, 4:#G2, 23pp., 2004. [Problem B4]

[16] A. Carvalho, C. P. dos Santos, C. Dias, F. Coelho, J. P. Neto, R. J. Nowakowski and S. Vinagre. On lattices from combinatorial game theory, modularity and a representation theorem: finite case. *Theor. Comput. Sci.*, 527:37-49, 2014. [Problem B4]

[17] J. Chappelon, U. Larsson and A. Matsuura. Two-player tower of hanoi. Preprint 2015. [Problem C1]

[18] W. H. Chan, R. M. Low, S. C. Locke and O. L. Wong. The P-positions in "nim with a pass" played on heap sizes of at most four. Preprint, 2017. [Problem A13]

[19] A. J. Cole and A. J. T. Davie. A game based on the euclidean algorithm and a winning strategy for it. *Math. Gaz.*, 53:354–357, 1969. [Problem A9]

[20] D. Collins. Variations on a theme of Euclid. *INTEGERS*, 5:#G3, 12pp., 2005. [Problem A9]

[21] D. Collins and T. Lengyel. The game of 3-Euclid. *Discrete Math.*, 308:1130–1136, 2008. [Problem A9]

[22] N. Comeau, J. Cullis, R. J. Nowakowski and J. Paek, personal communication (class project). [Problem A12]

[23] H. Dozier and J. Perry. Androids armed with poisoned chocolate squares: ideal Nim and its relatives. *Math. Mag.*, 89(4):235–250, 2016. [Problem D1]

[24] G. C. Drummond-Cole. Temperature 2 in domineering. https://cgp.ibs.re.kr/~gabriel/2star.pdf [Problems B11, E6]

[25] G. C. Drummond-Cole. An update on domineering on rectangular boards. *INTEGERS*, 14:#G3, 13pp., 2014. [Problem B11]

[26] P. Erdős and G. Szekeres, A combinatorial problem in geometry, *Compositio Math.*, 2:464–470, 1935. [Problem C7]

[27] J. M. Ettinger. A metric for positional games. *Theoret. Comput. Sci.* (*Math Games*), 230:207–219, 2000. [Problem E14]

[28] C. D. A. Evans and J. D. Hamkins. Transfinite game values in infinite chess. *INTEGERS*, 14:#G2, 2014. [Problem B3]

[29] G. Farr and N. Bao Ho. The Sprague–Grundy function for some nearly disjunctive sums of nim and silver dollar games. Preprint, 2017. [Problem E21]

[30] A. Fink. Lattice games without rational strategies. *J. Combin. Theory* (*Ser. A*), 119(2):450–459, 2012. [Problem E16]

[31] M. Fisher, N. McKay, R. J. Nowakowski, P. Ottaway and C. Santos. Atomic weight calculus of subversion. Preprint, 2017. [Problems A19, E17]

[32] A. S. Fraenkel. How to beat your Wythoff games' opponent on three fronts. *Amer. Math. Monthly*, 89:353–361, 1982. [Problem A5]

[33] A. S. Fraenkel. Multivision: an intractable impartial game with a linear winning strategy. *Amer. Math. Monthly*, 105:923–928, 1998. [Problem E1]

[34] A. S. Fraenkel. Euclid and Wythoff games. *Discrete Math.*, 304:65–68, 2005. [Problem A9]

[35] A. S. Fraenkel. Aperiodic subtraction games. *Electr. J. Combin.*, 18:P19, 2011. [Problem A1]

[36] A. S. Fraenkel and A. Kotzig. Partizan octal games: partizan subtraction games. *Internat. J. Game Theory*, 16:145–154, 1987. [Problem A1]

[37] A. S. Fraenkel and D. Krieger. The structure of complementary sets of integers: a 3-shift theorem. *Int. J. Pure and Appl. Math.*, 10:1–49, 2004. [Problem A4]

[38] A. S. Fraenkel, M. Loebl and J. Nešetřil. Epidemiography II, games with a dozing yet winning player. *J. Combin. Theory (Ser. A)*, 49:129–144, 1988. [Problem E1]

[39] A. S. Fraenkel and J. Nešetřil. Epidemiography. *Pacific J. Math.*, 118:369–381, 1985. [Problem E1]

[40] A. S. Fraenkel and O. Rahat. Complexity of error-correcting codes derived from combinatorial games. (Y. Björnsson, M. Müller, and J. Schaeffer, eds.), in *Proc. Intern. Conference on Computers and Games CG'*2002, Edmonton, Alberta, Canada, July 2002, Lecture Notes in Computer Science, Springer, 2883:201–221, 2003. [Problem D4]

[41] A. S. Fraenkel and E. R. Scheinerman. A deletion game on hypergraphs. *Discrete Appl. Math.,* 30:155–162, 1991. [Problem D2]

[42] A. S. Fraenkel and Y. Yesha. Theory of annihilation games, I. *J. Combin. Theory (Ser. B)*, 33:60–86, 1982. [Problem D4]

[43] D. Gale. A curious nim-type game. *Amer. Math. Monthly,* 81:876–879, 1974.

[44] D. Gale and A. Neyman. Nim-type games. *Int. J. Game Theory,* 11:17–20, 1982. [Problem D2]

[45] H. Gavel and P. Strimling. Nim with a modular muller twist. *INTEGERS*, 4:#G4, 9pp., 2004. [Problem A14]

[46] P. A. Grillet. *Commutative Semigroups, Advances in Mathematics*, 2. Springer 2001. [Problem A15]

[47] A. Guo. Winning strategies for aperiodic subtraction games. *Theoret. Comput. Sci.*, 421:70–73, 2012. [Problem A1]

[48] A. Guo and E. Miller. Algorithms for lattice games. *Int. J. Game Theory*, 42(4):777–788, 2013. [Problem E16]

[49] V. A. Gurvich. On the misere version of game Euclid and miserable games. *Disc. Math*, 307:1199–1204, 2007. [Problem A9]

[50] R. K. Guy. A parity subtraction game. *Crux Mathematicorum*, 33:37–39, 2004. [Problem E11]

[51] O. Hanner. Mean play of sums of positional games. *Pacific J. Math.*, 9:81–99, 1959. [Problem E14]

[52] F. Harary, B. Sagan, and D. West. Computer-aided analysis of monotonic sequence games. *Atti Accad. Peloritana Pericolanti Cl. Sci. Fis. Mat. Natur.*, 61:67–78, 1985. [Problem E7]

[53] G. Hetyei. Enumeration by kernel positions. *Adv. Appl. Math.,* 42:445–470, 2009. [Problem E15]

[54] G. Hetyei. Enumeration by kernel positions for strongly Bernoulli type truncation games on words. *J. Combin. Theory (Ser. A)*, 117:1107–1126, 2010. [Problem E15]

[55] A. Holshouser, H. Reiter, and J. Rudzinski. Pilesize dynamic one-pile nim and beatty's theorem. *INTEGERS*, 4:#G3, 13pp., 2004. [Problem A14]

[56] A. Holshouser, H. Reiter and J. Rudzinski. Dynamic one-pile nim. *Fibonacci Quart.*, 41(3):253–262, 2003. [Problem A14]

[57] D. G. Horrocks and R. J. Nowakowski. Regularity in the g-sequences of octal games with a pass. *INTEGERS*, 3:#G1, 10 pp., 2003. [Problems A3, A13]

[58] D. G. Horrocks and M. A. Trenton. Subtraction with a muller twist. *INTEGERS*, 8:#G05, 2008. [Problem A14]

[59] S. Howse and R. J. Nowakowski. Periodicity and arithmetic-periodicity in hexadecimal games. *Theoret. Comput. Sci.*, 313(3):463–472, 2004. [Problem A3]

[60] M. Huggan and R. J. Nowakowski. Thinning Thickets. To appear in *INTEGERS*. [Problem C8]

[61] W. Johnson. Well-tempered scoring games. *Int. J. Game Theory*, 43:415–438, 2014. [Problem E14]

[62] A. Jobson, L. Sledd, S. C. White and D. J. Wildstrom. Variations on narrow dots-and-boxes and dots-and-triangles. *INTEGERS*, 2017. [Problem C1]

[63] J. P. Jones and A. S. Fraenkel. Complexities of winning strategies in diophantine games. *J. Complexity*, 11:435–455, 1995. [Problem E1]

[64] S. Kanungo and R. M. Low. Further analysis on the "king and rook vs. king on a quarter-infinite board" problem. *INTEGERS*, 7:#G9, 2007. [Problem B4]

[65] R. Kelly Annela. One pile misère nim for three or more players. *Int. J. Math. Math. Sci.*, 2006:1–8, 2006. [Problem E22]

[66] T. Khovanova and J. Xiong. Nim fractals, *J. Integer Seq.*, 17(7):article 14.7.8, 17pp., 2014. [Problem E15]

[67] W. O. Krawec. Analyzing n-player impartial games. *Int. J. Game Theory*, 41(2):345–367, 2011. [Problem E22]

[68] E. L. Lakshtanov and A. I. Aleksenko. Finiteness in the beggar-my-neighbor card game. *Probl. Inf. Transm.*, 49(2):163–166, 2013. [Problem D7]

[69] U. Larsson. Blocking Wythoff nim. *Electron. J. Combin.*, 18(1):paper 120, 18pp., 2011. [Problem A14]

[70] U. Larsson and N. Fox. An aperiodic subtraction game of nim-dimension two. *J. Integer Seq.*, 18(7):article 15.7.4, 6pp., 2015. [Problem A1]

[71] U. Larsson, R. J. Nowakowski, J. P. Neto and C. P. Santos. Guaranteed scoring games. *Electron. J. Combin.*, 23(3):paper 3.27, 29pp., 2016. [Problem E14]

[72] U. Larsson, R. J. Nowakowski and C. P. Santos. When waiting moves you in scoring combinatorial games. Preprint, 2015. [Problem E14]

[73] U. Larsson, R. J. Nowakowski and C. P. Santos. When game comparison becomes play: Absolutely Categorical Game Theory, arXiv:1609.02764 [Problem E14]

[74] U. Larsson and S. Rubinstein-Salzedo. Global fibonacci nim. 2015. [Problem A18]

[75] T. Lengyel. A nim-type game and continued fractions. *Fibonacci Quart.*, 41:310–320, 2003. [Problem A9]

[76] L. Levine. Fractal sequences and restricted nim. *Ars Combinatoria*, LXXX:113–128, 2006. [Problem A18]

[77] W. A. Liu and J. W. Duan. Misère nim with multi-player. *Discrete Appl. Math.*, 219:40–50, 2017. [Problem E22]

[78] A. M. Loosen. Directional Phutball. *Atl. Electron. J. Math.*, 3:30–45, 2008. [Problem B7]

[79] R. M. Low and W. H. Chan. An atlas of n- and p-positions in 'nim with a pass'. *INTEGERS*, 15:#G2, 2015. [Problem A13]

[80] R. M. Low and M. Stamp. King and rook vs. king on a quarter infinite board. *INTEGERS*, 6:#G03, 2006. [Problem B4]

[81] S. McCurdy, *Two combinatorial games*, MSc thesis, Dalhousie University, 2004. [Problem A5]

[82] N. A. McKay and R. J. Nowakowski. Outcomes of Partizan Euclid, *INTEGERS*, 12B:#A9, 2012. [Problem A9]

[83] R. J. Nowakowski and A. A. Siegel. *The lattices of option-closed games*. 2014. Preprint. [Problem E7]

[84] R. Milley. *Restricted universes of partizan misère games*. PhD thesis, Dalhousie University, 2013. [Problem E13]

[85] J. Milnor. Sums of positional games. *Ann. of Math. Stud.* (Contributions to the Theory of Games, H. W. Kuhn and A. W. Tucker, eds.), Princeton, 2(28):291–301, 1953. [Problem E14]

[86] R. Miyadera, M. Inoue and M. Fukui. Impartial chocolate bar games with a pass. *INTEGERS*, 16:#G5, 2016. [Problem A13]

[87] R. E. Morrison, E. J. Friedman and A. S. Landsberg. Combinatorial games with a pass: a dynamical systems approach. *Chaos* 21:043108, 21pp., 2011. [Problem A13]

[88] M. Müller. Solving 5×5 Amazons. In *The 6th Game Programming Workshop* 2001, **14** in IPSJ Symposium Series Vol. 2001, pp. 64-71, Hakone, Japan, 2001. [Problem B6]

[89] T. Nakamura and E. Berlekamp. Analysis of composite corridors. (Y. Björnsson, M. Müller, and J. Schaeffer, eds.), In *Proc. Intern. Conference on Computers and Games CG*2002, Edmonton, Alberta, Canada, July 2002, Lecture Notes in Computer Science, Springer, 2883:213–222, 2003. [Problem E6]

[90] S. Nakamura and R. Miyadera. Impartial chocolate bar games. *INTEGERS*, 15:#G4, 2015. [Problem D1]

[91] G. Nivasch. The Sprague–Grundy function of the game Euclid. *Discrete Math.*, 306(21):2798–2800, 2006. [Problem A9]

[92] H. Noon and G. Van Brummelen. The non-attacking queens game. *Coll. Math. J.*, 37:223–227, 2006. [Problem B5]

[93] N. A. McKay, R. Milley and R. J. Nowakowski, Misère-play Hackenbush Sprigs, *Int. J. Game Theory,* 45: 731–742, 2016. [Problem E13]

[94] R. J. Nowakowski and P. Ottaway. Option-closed games. *Contrib. Discrete Math.*, 6(1):142–153, 2011. [Problem E7]

[95] R. J. Nowakowski, G. Renault, E. Lamoureux, S. Mellon and T. Miller. Timber. *J. Combin. Math. Combin. Comput.*, 85:213–225, 2013. [Problem D5]

[96] Y. Peng, X. dong Xu, H. peng Lou and X. feng Cui. New approach on optimal play in hex game. *Appl. Res. Comput.*, 2:024, 2010. [Problem B8]

[97] T. E. Plambeck and A. N. Siegel. Misere quotients for impartial games. *J. Combin. Game Theory,* (*Ser. A*), 115(4):593–622, 2008. [Problem A15]

[98] J. Propp. Three-player impartial games. *Theoret. Comput. Sci.* (*Math Games*), 233:263–278, 2000. [Problem E22]

[99] L. Rédei. *The Theory of Finitely Generated Commutative Semigroups*, Pergamon, 1965. [Problem A15]

[100] T. van Roode. *Partizan forms of Hackenbush*, MSc thesis, University of Calgary, 2002. [Problem C5]

[101] C. dos Santos. *Alguns resultados sobre jogos imparciais e dimensao nim* (Results on NIM dimension), PhD Thesis, University of Lisbon, 2010. [Problem A1]

[102] C. dos Santos. Embedding processes in combinatorial game theory. *Discrete Appl. Math.*, 159(8):675–682, 2011. [Problem E17]

[103] C. Schensted. Longest increasing and decreasing subsequences, *Canad. J. Math.*, 13:179–191, 1961. *MR* 22 #12047. [Problem C7]

[104] F. Schuh. Spel van delers (The game of divisors). *Nieuw Tijdschrift voor Wiskunde,* 39:299–304, 1952. [Problem D1]

[105] A. J. Schwenk. Take-away games. *Fibonacci Quart.*, 8:225–234, 1970. [Problem A18]

[106] A. Shankar and M. Sridharan. New temperatures in Domineering. *INTEGERS*, 5:#G04, 13pp., 2005. [Problem B11]

[107] G. Sicherman. Theory and practice of Sylver coinage. *INTEGERS*, 2:#G2, 2002. [Problem C6]

[108] A. N. Siegel, *Loopy games and computation*. PhD dissertation, University of California, Berkeley, Spring 2005. [Problems B7, B9]

[109] A. A. Siegel. *On the structure of games and their posets*. PhD thesis, Dalhousie University, 2011. [Problems A1, A13]

[110] A. Sierra Ballarín. *Combinatorial game theory: the dots-and-boxes game.* Master's Thesis, Universtitat Politecnica Catalyuna, 2015. [Problem C1]

[111] F. Smith and P. Stănică. Comply/constrain games or games with a Muller twist. *INTEGERS*, 2:#G3, 10pp., 2002. [Problem A14]

[112] E. Sopena. i-MARK: A new Subtraction Division game. *Theoret. Comput. Sci.*, 627:90–101, 2016. arXiv:1509.04199. [Problem A1]

[113] W. L. Spight, Evaluating kos in a neutral threat environment: preliminary results. (J. Schaeffer, M. Müller and Y. Björnsson, eds.), In *Computers and Games: Third Internat. Conf., CG*2002, Lecture Notes in Computer Science, Springer, Berlin, 2883:413–428, 2003. [Problem E4]

[114] E. R. Spitznagel Jr. Properties of a game based on Euclid's algorithm. *Math. Mag.*, 46:87–92, 1973. [Problem A9]

[115] X. Sun. Improvements on Chomp. *INTEGERS*, 2:#G1, 2002. [Problem D1]

[116] X. Sun. Wythoff's sequence and n-heap Wythoff's conjectures. *Discrete Math.*, 300:180–195, 2005. [Problem A4]

[117] X. Sun and D. Zeilberger. On Fraenkel's N-heap Wythoff's conjectures. *Ann. Comb.*, 8(2):225–238, 2004. [Problem A4]

[118] X. L. Tan and M. D. Ward. On Kotzig's Nim. *INTEGERS*, 14:#G06, 27pp., 2014. [Problem A6]

[119] S. P. Tung. Computational complexities of Diophantine equations with parameters. *J. Algorithms* 8:324–336, 1987. [Problem E1]

[120] J. W. H. M. Uiterwijk. The impact of safe moves on perfectly solving Domineering boards, 1: analysis of and experiments with 1-step safe moves. *ICGA Journal* 37(2):97–105, 2014. [Problem B11]

[121] J. W. H. M. Uiterwijk. The impact of safe moves on perfectly solving Domineering boards, 2: analysis of and experiments with 1-step safe moves. *ICGA Journal* 37(3):144–160, 2014. [Problem B11]

[122] J. W. H. M. Uiterwijk. The impact of safe moves on perfectly solving Domineering boards, 3: analysis of and experiments with 1-step safe moves. *ICGA Journal* 37(4):207–213, 2014. [Problem B11]

[123] J. W. H. M. Uiterwijk. Polymerization and crystallization of Snowflake molecules in Domineering. *Theoret. Comput. Sci.*, 644:143–158, 2016. [Problem B11]

[124] J. W. H. M. Uiterwijk. 11×11 Domineering is solved: the first player wins. (A. Plaat, W. Kosters and H.J. van den Herik, eds.) *Computers and Games: 9th International Conference, CG*2016, revised selected papers, Lecture Notes in Computer Science, Springer, 10068:129–136, 2016. [Problem B11]

[125] J. W. H. M. Uiterwijk and M. Barton. New results for Domineering from combinatorial game theory endgame databases. *Theoret. Comput. Sci.*, 592:72–86, 2015. [Problem B11]

[126] J. Úlehla. A complete analysis of von Neumann's Hackendot. *Int. J. Game Theory*, 9:107–113, 1980. [Problem C4]

[127] M. D. Ward. A conjecture about periods in subtraction games, arXiv:1606.04029. [Problem A1]

[128] M. J. Whinihan. Fibonacci nim. *Fibonacci Quart.*, 1(4):9–13, 1963. [Problem A18]

[129] D. Zeilberger. Three-rowed Chomp. *Adv. Appl. Math.*, 26:168–179, 2001. [Problem D1]

[130] X. Zhao and W. A. Liu. One pile misère bounded nim with two alliances. *Discrete Appl. Math.*, 214:16–33, 2016. [Problem E22]

[131] B. Arneson, R. Hayward, and P. Henderson. "Solving Hex: Beyond Humans," in Computers and Games, vol. 6515, H. van den Herik, H. Iida, and A. Plaat, Eds. Springer Berlin/Heidelberg, 2011, pp. 1–10. [Problem B8]

rjn@mathstat.dal.ca

Department of Mathematics and Statistics,
Dalhousie University, Halifax, NS B3H 3J5, Canada

Games of No Chance 5
MSRI Publications
Volume **70**, 2017

Misère games and misère quotients

AARON N. SIEGEL

These notes are based on a short course offered at the Weizmann Institute of Science in Rehovot, Israel, in November 2006. The notes include an introduction to impartial games, starting from the beginning; the basic misère quotient construction; a proof of the periodicity theorem; and statements of some recent results and open problems in the subject.

Introduction

This course is concerned with *impartial combinatorial games*, and in particular with misère play of such games. Loosely speaking, a *combinatorial game* is a two-player game with no hidden information and no chance elements. We usually impose one of two winning conditions on a combinatorial game: under *normal play*, the player who makes the last move wins, and under *misère play*, the player who makes the last move loses. We will shortly give more precise definitions.

The study of combinatorial games began in 1902, with C. L. Bouton's published solution to the game of NIM [2]. Further progress was sporadic until the 1930s, when R. P. Sprague [17; 18] and P. M. Grundy [6] independently generalized Bouton's result to obtain a complete theory for normal-play impartial games.

In a seminal 1956 paper [8], R. K. Guy and C. A. B. Smith introduced a wide class of impartial games known as *octal games*, together with some general techniques for analyzing them in normal play. Guy and Smith's techniques proved to be enormously powerful in finding normal-play solutions for such games, and they are still in active use today [4].

At exactly the same time (and, in fact, in exactly the same issue of the *Proceedings of the Cambridge Philosophical Society*), Grundy and Smith published a paper on misère games [7]. They noted that misère play appears to be quite difficult, in sharp contrast to the great success of the Guy–Smith techniques.

Despite these complications, Grundy remained optimistic that the Sprague–Grundy theory could be generalized in a meaningful way to misère play. These hopes were dashed in the 1970s, when Conway [3] showed that the Grundy–Smith

MSC2010: 91A46.
Keywords: games.

complications are intrinsic. Conway's result shows that the most natural misère-play generalization of the Sprague–Grundy theory is hopelessly complicated, and is therefore essentially useless in all but a few simple cases.[1]

The next major advance occurred in 2004, when Thane Plambeck [10] recovered a tractable theory by *localizing* the Sprague–Grundy theory to various restricted sets of misère games. Such localizations are known as *misère quotients*, and they will be the focus of this course. While some of the ideas behind the quotient construction are present in Conway's work of the 1970s, it was Plambeck who recognized that the construction can be made systematic — in particular, he showed that the Guy–Smith *periodicity theorem* can be generalized to the local setting.

This course is a complete introduction to the theory of misère quotients, starting with the basic definitions of combinatorial game theory and a proof of the Sprague–Grundy theorem. We include a full proof of the periodicity theorem and many motivating examples. The final lecture includes a discussion of major open problems and promising directions for future research.

Lecture 1. Normal play

November 26, 2006 Scribes: Leah Nutman and Dan Kushnir

Impartial combinatorial games — a few examples. A combinatorial game is a two player game with no hidden information (i.e., both players have full information of the game's position) and no chance elements (given a player's move, the next position of the game is completely determined). Let us demonstrate this notion with a few useful examples.

Example (NIM). A position of NIM consists of several strips, each containing several boxes. A move consists of removing one or more boxes from a *single* strip. Whoever takes the last box (from the last remaining strip) wins. A sample game of NIM is illustrated in Figure 1.

Example (KAYLES). A position of KAYLES consists of several strips, each containing several boxes, as in NIM. A move consists of removing one or two *adjacent* boxes from a single strip. If the player takes a box (or two) from the middle of a strip then this strip is split into two *separate* strips. In particular, no future move can affect both sides of the original strip. (See Figure 2 for an illustration of one such move.)

Whoever takes the last box (from the last remaining strip) wins.

[1]Despite its apparent uselessness, Conway's theory is actually quite interesting from a theoretical point of view. We will not say much about it in this course, but it is well worth exploring; see [3] for discussion.

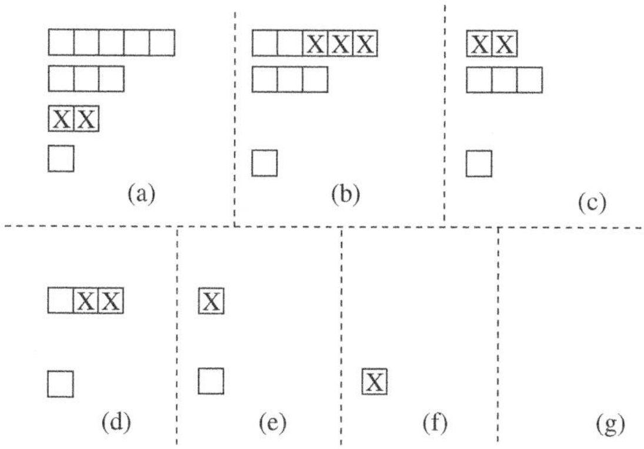

Figure 1. The seven subfigures represent seven consecutive positions in a play of a game of NIM. The last position is the empty one (with no boxes left and thus no more possible moves). The first six minifigures also indicate the move taken next (which transforms the current position into the next position): a box marked with "X" is a box that was selected to be taken by the player whose turn it is to play.

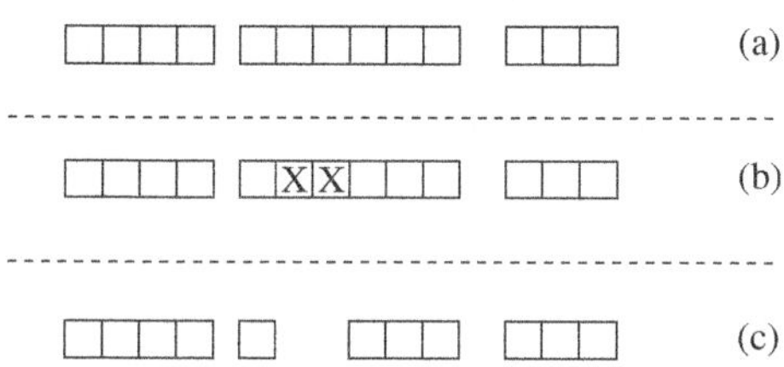

Figure 2. (a) The position before the move, consisting of three strips. (b) The move: the selected boxes are marked with "X". (c) In the new position, the middle strip was split, leaving four strips.

Example (DAWSON'S KAYLES). This game is identical to KAYLES up to two differences:

(1) A move consists of removing *exactly two* adjacent boxes from a single strip.

(2) The winning condition is flipped: whoever makes the last move *loses*.

Winning conditions and the difficulty of a game. All three examples above share some common properties. They are:

Finite: For any given first position, there are only finitely many possible positions that the game may take (throughout its execution).

Loopfree: No position can occur twice in an execution of a game. Once we leave a position, this position will never repeat itself.

Impartial: Both players have the same moves available at all times.

All of the games we will consider in this course have these three properties. As we will further discuss below, the first two properties (finite and loopfree) imply that one of the players must have a perfect winning strategy — that is, a strategy that guarantees a win no matter what his opponent does.

Main Goal. Given a combinatorial game Γ, find an efficient winning strategy for Γ.[2]

We will consider in this course two possible winning conditions for our games:

Normal play: Whoever makes the last move wins.

Misère play: Whoever makes the last move loses.

The different winning conditions of the aforementioned games turn out to have a great effect on their difficulty. NIM was solved in 1902 and KAYLES was solved in 1956. By contrast, the solution to DAWSON'S KAYLES remains an open problem after 70 years. (That is, we still do not know an efficient winning strategy for it.)

What makes DAWSON'S KAYLES so much harder? It is exactly the fact that the last player to move loses. In general, games with misère play tend to be vastly more difficult. The themes for this course are the following:

(1) Why is misère play more difficult?

(2) How can we tackle this difficulty?

Game representations and outcomes. We have mentioned that our goal is to obtain efficient winning strategies for impartial combinatorial games. We will in fact be even more concerned with the structure of individual positions. Therefore, by a "game" G, we will usually mean an individual position in a combinatorial game.

Sometimes we will shamelessly abuse terminology and use the term "game" to refer to a system of rules. It will (hopefully) always be clear from the context which meaning is intended. To help minimize confusion, we will always denote individual positions by roman letters ($G, H, \ldots$) and systems of rules by Γ.

One way to formally represent a game is as a tree. For example, the NIM position G which contains three boxes in a single strip can be represented by the tree given in Figure 3. Here 0 represents a strip with zero boxes, from which there are no possible moves.

Definition 1.1. Two games G and H are identical (isomorphic) if they have isomorphic trees. If G and H are identical, we write $G \cong H$.

[2]More precisely, we seek a winning strategy that can be computed in polynomial time (measured against the size of a *succinct description* of a game position). In general, any use of the word "efficient" in this course can be safely interpreted to mean "polynomial-time," though we will be intentionally vague about issues of complexity.

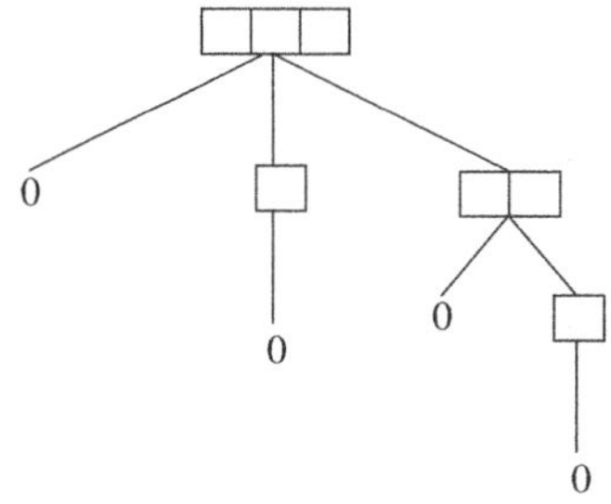

Figure 3. Tree representation of the NIM position G which contains three boxes in a single strip.

We can also think of the NIM position G from Figure 3 as a *set*: $\square\square\square = \{0, \square, \square\square\}$. We call the positions we can move to directly from a game G the *options* of G. So we are identifying G with the set of its options; for example, 0 is identified with the empty set $\{\}$.

We will now introduce some notation that will make it easier to discuss the value of any given position of a game and in particular, the values of NIM positions.

Definition 1.2. For every $n \geq 0$ we denote by $*n$ a strip in NIM of length n. We write 0 and $*$ as shorthand for $*0$ and $*1$, respectively. Formally, we have

$$*n = \{0, *, *2, *3, \ldots, *(n-1)\}.$$

As we mentioned above, every game with the properties we have specified has a well-defined outcome (indicating who will win when both players play perfectly). Assuming both players play perfectly, either

(1) the first player has a winning move, or

(2) any move the first player may make will move to a position where he loses. In this case the second player can win.

Definition 1.3. Let G be a game. The *normal outcome* $o^+(G)$ is defined by

- $o^+(G) = \mathscr{P}$ if second player can win G, assuming normal play;
- $o^+(G) = \mathscr{N}$ if first player can win G, assuming normal play.

Likewise, the *misère outcome* $o^-(G)$ is defined by

- $o^-(G) = \mathscr{P}$ if second player can win G, assuming misère play;
- $o^-(G) = \mathscr{N}$ if first player can win G, assuming misère play.

We say G is a *normal $\mathscr{P}$-position* if $o^+(G) = \mathscr{P}$, etc.

Note that o^+ and o^- have simple recursive descriptions:

$$o^+(G) = \mathscr{P} \iff o^+(G') = \mathscr{N} \text{ for every option } G' \text{ of } G.$$
$$o^-(G) = \mathscr{P} \iff G \neq 0 \text{ and } o^-(G') = \mathscr{N} \text{ for every option } G' \text{ of } G.$$

$\mathscr{P}$ and $\mathscr{N}$ are short for *p*revious player and *n*ext player, respectively.

For example, we can consider NIM played with a single strip and see which positions are $\mathscr{P}$-positions and which are $\mathscr{N}$-positions:

- $o^+(0) = \mathscr{P}$: If there are no more boxes, then the previous move was the winning move (the previous player took the last box).
- $o^+(*n) = \mathscr{N}$ for every $n > 0$: When there is only one strip left, the next player can take all the remaining boxes and thus win.

What about the misère outcomes?

- $o^-(0) = \mathscr{N}$: If there are no more boxes, then the previous player took the last box and lost. So the next player is the winning one.
- $o^-(*) = \mathscr{P}$: When there is only one box left, the next player must take it and lose, so the previous player is the winning one.
- $o^-(*n) = \mathscr{N}$ for every $n > 1$: Here the winning move is to take all boxes but one.

We now revise our main goal.

Main Goal (revised). Given a position G in a combinatorial game, find an efficient way to compute the outcome of G.

In all the examples we consider in this course, the two goals are equivalent: efficient methods for computing the outcomes of positions will instantly yield efficient winning strategies.

Disjunctive sums. The positions in each of our examples naturally decompose. In NIM, no single move may affect more than one strip, so each strip is effectively independent. Both KAYLES and DAWSON'S KAYLES exhibit an even stronger form of decomposition: a typical move cuts a strip into two components, and since the components are no longer adjacent, no subsequent move can affect them both.

These observations motivate the following definition.

Definition 1.4. Let G and H be games. The *(disjunctive) sum* of G and H, denoted $G + H$, is the game played as follows. Place copies of G and H side-by-side. A move consists of choosing exactly one component and making a move in that component.

Formally, we can define $G+H$ as the direct sum of the trees for G and H. Or, thinking in terms of sets,

$$G+H=\{G'+H : G' \text{ is an option of } G\} \cup \{G+H' : H' \text{ is an option of } H\}.$$

In combinatorial game theory, it is customary to be lazy in our use of notation and write simply

$$G+H=\{G'+H, G+H'\}.$$

***The strategy for* NIM.** Here is the strategy for NIM, assuming normal play: write the size of each strip in binary, and then do a bitwise XOR. G is a $\mathcal{P}$-position if and only if the result is identically 0. For example, the starting position of Figure 1(a) has strips of sizes 5, 3, 2 and 1, so we can write

$$\begin{array}{rr} & 101=5 \\ \oplus & 11=3 \\ \oplus & 10=2 \\ \oplus & 1=1 \\ \hline & 101 \end{array}$$

The result is nonzero, so Figure 1(a) is an $\mathcal{N}$-position (in normal play).

We will shortly prove a stronger statement that implies this strategy.

Equivalence. We would like to regard two games as equivalent if they behave the same way in any disjunctive sum. For now assume normal play.

Definition 1.5. We say G and H are *equal*, and write $G=H$, if and only if

$$o^+(G+X)=o^+(H+X) \text{ for every combinatorial game } X.$$

Note that if $G \cong H$, then necessarily $G=H$, but we will see that nonisomorphic games can be equal.

Proposition 1.6. *$G+0=G$ for any game G.*

Proof. Adding 0 does not change the structure of G at all. (In fact, $G+0 \cong G$.) □

Proposition 1.7. *$G+G=0$ for any game G.*

Proof. We need to show that X and $G+G+X$ have the same outcome, for any X.

First suppose $o^+(X)=\mathcal{P}$. Second player can win $G+G+X$ as follows. Whenever first player moves on X, just use the winning strategy there. If first player ever moves on one of the copies of G, make the identical move on the other copy. Second player will get the last move on X because she is following the winning strategy there, and she will get the last move on $G+G$ by symmetry.

Conversely, if $o^+(X) = \mathcal{N}$, then on $G + G + X$, just make a winning move on X and proceed as before. □

Example. Here is a simple example to show how disjunctive sums can be useful for studying combinatorial games. Consider a NIM position with strips of sizes 19, 23, 16, 45, 23 and 19. By the previous argument, the two strips of size 19 together equal 0, as do the two strips of size 23. So this is equivalent to NIM with strips of sizes 16 and 45.

Exactly the same argument works for KAYLES or DAWSON'S KAYLES.

Proposition 1.8. (a) $=$ *is an equivalence relation.*

(b) *If* $G = H$, *then* $G + K = H + K$.

Proof. Since equality of outcomes is an equivalence relation, (a) is immediate. For (b), if $G = H$ then

$$o^+(G + X) = o^+(H + X) \text{ for all } X,$$

so in particular

$$o^+(G + (K + X)) = o^+(H + (K + X)) \text{ for all } X.$$

Disjunctive sum is associative, so $G + K = H + K$. □

Proposition 1.9. *The following are equivalent, for games* G, H:

(i) $G = H$.

(ii) $o^+(G + H) = \mathcal{P}$.

Proof. (i) ⇒ (ii): If $G = H$, then $G + G = G + H$. But $G + G = 0$, so $o^+(G + H) = o^+(0) = \mathcal{P}$.

(ii) ⇒ (i): By a symmetry argument (just like Proposition 1.7), X and $G+H+X$ have the same outcome, for all X. Therefore $G + H = 0$, so $G + H + H = H$. But $H + H = 0$. □

The Sprague–Grundy theorem.

Theorem 1.10 (Sprague–Grundy). *For any game* G, *there is some* m *such that* $G = *m$.

We will in fact prove the following stronger statement.

Definition 1.11. Let S be a finite set of nonnegative integers. We define the *minimal excludant* of S, denoted $\operatorname{mex}(S)$, to be the least integer not in S.

Theorem 1.12 (mex rule). *Suppose* $G \cong \{*a_1, \ldots, *a_k\}$. *Then* $G = *m$, *where*

$$m = \operatorname{mex}\{a_1, \ldots, a_k\}.$$

Proof. By Proposition 1.9, it suffices to show that $G + *m$ is a $\mathcal{P}$-position. There are two cases.

Case 1: First player moves in G. This leaves the position $*a + *m$, where $*a$ is some option of G. Since $m \notin \{a_1, \ldots, a_k\}$, we necessarily have $a \neq m$. If $a > m$, second player can move to $*m + *m$; if $a < m$, she can move to $*a + *a$. In either case, she leaves a $\mathcal{P}$-position.

Case 2: First player moves in $*m$. This leaves $G + *a$, for some $a < m$. Since m is the *minimal* excludant of $\{a_1, \ldots, a_k\}$, we must have $a = a_i$ for some i. Therefore second player can move to $*a + *a$, a $\mathcal{P}$-position. □

The Sprague–Grundy theorem follows from one more ingredient.

Exercise (prove the *replacement lemma*). Suppose $G = \{G_1, \ldots, G_k\}$ and suppose $G_1 = H$ for some H. Then

$$G = \{H, G_2, \ldots, G_k\}.$$

Proof of Sprague–Grundy theorem. Write $G = \{G_1, \ldots, G_k\}$. Inductively, we may assume that $G_1 = *a_1, \ldots, G_k = *a_k$. By the replacement lemma, $G = \{*a_1, \ldots, *a_k\}$, and by the mex rule we are done. □

Lecture 2. Octal games and misère play

November 27, 2006 Scribes: Omer Kadmiel and Shai Lubliner

We introduce a broad class of games known as *octal games*, and then give the definition of misère quotient.

NIM *values.* In the previous lecture we showed the following:

- Assuming *normal* play, if G is any impartial combinatorial game, then $G = *m$ for some m. Moreover, if $G = \{*a_1, \ldots, *a_k\}$ then $m = \text{mex}\{a_1, \ldots, a_k\}$.
- For any impartial G, H, $o^+(G + H) = \mathcal{P}$ if and only if $G = H$.

We denote by $\mathcal{G}(G)$ the unique integer m such that $G = *m$ in normal play. $\mathcal{G}(G)$ is called the *nim value* of G.

***XOR and a winning strategy for (normal-play)* NIM.** If m, n integers then $m \oplus n$ denotes the binary XOR of m and n.

Theorem 2.1. *Let a, b, c be integers. Then*

$$o^+(*a + *b + *c) = \mathcal{P} \Longleftrightarrow a \oplus b \oplus c = 0.$$

Proof (by example). Consider the following example:

$$\begin{array}{rl} 11101001 & a \\ \oplus\ 01101111 & b \\ \oplus\ 00000111 & c \\ \hline 10000001 & \end{array}$$

As the XOR of these values $\neq 0$, we must show that this is an $\mathcal{N}$-position. The first player simply finds the most significant bit marked 1 in the XOR and chooses any component in which this bit is a 1. In this example, that component is a. He then makes an appropriate move in a that switches the most significant bit to 0, and sets all lower-order bits as needed to make the sum equal 0. Here the winning move is from a to $a' = 01101000$, changing just the first and last bits; $a' \oplus b \oplus c = 0$, so by induction it is a $\mathcal{P}$-position. □

Corollary 2.2. $$*a + *b = *(a \oplus b).$$

Proof. We know $a \oplus b \oplus (a \oplus b) = 0$, so $*a + *b + *(a \oplus b) = 0$. □

Example (DAWSON'S KAYLES). Recall that in DAWSON'S KAYLES, a move consists of removing *exactly* two adjacent boxes. We defined DAWSON'S KAYLES as a misère-play game, but we can just as easily consider it in normal play. Denote by H_n a single strip of length n. Then the moves from H_n are to $H_a + H_{n-2-a}$, where $1 \leq a \leq n-2$.

We can use the Sprague–Grundy theorem and the NIM addition rule to compute normal-play values of H_n easily:

$$H_0 = \{\} = 0,$$
$$H_1 = \{\} = 0,$$
$$H_2 = \{H_0\} = \{0\} = *,$$
$$H_3 = \{H_1\} = \{0\} = *,$$
$$H_4 = \{H_1 + H_1,\ H_2 + H_0\} = \{0+0,\ *+0\} = \{0,\ *\} = *2,$$
$$H_5 = \{H_2 + H_1,\ H_3 + H_0\} = \{*+0,\ *+0\} = \{*,\ *\} = 0,$$
$$H_6 = \{H_2 + H_2,\ H_3 + H_1,\ H_4 + H_0\} = \{*+*,\ *+0,\ *2+0\} = \{0,\ *,\ *2\} = *3.$$

This rapidly becomes tedious, and it's easily implemented on a computer. The results of a computer calculation are shown in Figure 4. Each row represents a block of 34 NIM values: the first row shows $\mathcal{G}(H_0)$ through $\mathcal{G}(H_{33})$; the next row shows $\mathcal{G}(H_{34})$ through $\mathcal{G}(H_{67})$; etc. The number 34 was obviously not chosen by accident; after a few initial anomalies, a strong regularity quickly emerges with period 34. We now prove a theorem that shows, for a wide class of games,

	0										1										2										3			
	0	1	2	3	4	5	6	7	8	9	0	1	2	3	4	5	6	7	8	9	0	1	2	3	4	5	6	7	8	9	0	1	2	3
0+	0	0	1	1	2	0	3	1	1	0	3	3	2	2	4	0	5	2	2	3	3	0	1	1	3	0	2	1	1	0	4	5	2	7
34+	4	0	1	1	2	0	3	1	1	0	3	3	2	2	4	4	5	5	2	3	3	0	1	1	3	0	2	1	1	0	4	5	3	7
68+	4	8	1	1	2	0	3	1	1	0	3	3	2	2	4	4	5	5	9	3	3	0	1	1	3	0	2	1	1	0	4	5	3	7
102+	4	8	1	1	2	0	3	1	1	0	3	3	2	2	4	4	5	5	9	3	3	0	1	1	3	0	2	1	1	0	4	5	3	7
136+	4	8	1	1	2	0	3	1	1	0	3	3	2	2	4	4	5	5	9	3	3	0	1	1	3	0	2	1	1	0	4	5	3	7
170+	4	8	1	1	2	0	3	1	1	0	3	3	2	2	4	4	5	5	9	3	3	0	1	1	3	0	2	1	1	0	4	5	3	7

Figure 4. NIM values of DAWSON'S KAYLES in normal play.

that if such periodicity is observed for "sufficiently long" (in a sense to be made precise) then it must continue forever.

Octal games and octal codes.

Definition 2.3. An *octal code* is a sequence of digits $0.d_1d_2d_3\ldots$ where $0 \leq d_i < 8$ for all i.

An octal code specifies the rules for a particular *octal game*. An octal game is played with strips of boxes, and the code describes how many boxes may be removed and under what circumstances. The digit d_k specifies the conditions under which k boxes may be removed.

Let us consider the bit representation of each d_k: denote $d_k = \epsilon_0 + \epsilon_1 \cdot 2 + \epsilon_2 \cdot 4$, where each $\epsilon_i = 0$ or 1.

- We can remove an entire strip of length k if and only if $\epsilon_0 = 1$.
- We can remove k boxes from the *end* of a strip (leaving at least one box) if and only if $\epsilon_1 = 1$.
- We can remove k boxes from the *middle* of a strip (leaving at least one box on each end) if and only if $\epsilon_2 = 1$.

Therefore,

- DAWSON'S KAYLES is represented by 0.07 as you have to remove exactly two blocks every time from anywhere in the strip, and you can remove an entire strip of length 2.
- KAYLES is represented by 0.77 as you can remove one or two boxes from a single strip.
- NIM is represented by the infinite sequence $0.3333333\ldots$ as you are allowed to take any number of boxes from the end or to take an entire strip of any length (but you are not allowed to separate the original strip into two strips).

Guy–Smith periodicity theorem.

Theorem 2.4 (Guy–Smith periodicity theorem). *Consider an octal game with* finitely many *nonzero code digits, and let k be largest with $d_k \neq 0$. Denote by H_n a strip of length n. Suppose that for some $n_0 > 0$ and $p > 0$ we have*

$$\mathcal{G}(H_{n+p}) = \mathcal{G}(H_n) \text{ for every } n \text{ with } n_0 \leq n < 2n_0 + p + k.$$

Then

$$\mathcal{G}(H_{n+p}) = \mathcal{G}(H_n) \text{ for all } n \geq n_0.$$

Proof. Note that a move from H_n is always to $H_a + H_b$, where $n - k \leq a + b < n$. (In taking a whole strip, or from the end of the strip, we may take one or both of a, b to be 0.)

Now proceed by induction on n. The base case $n < 2n_0 + p + k$ is given by hypothesis, so assume $n \geq 2n_0 + p + k$. A move from H_{n+p} is to $H_a + H_b$ where $a + b \geq n + p - k$.

Since $n \geq 2n_0 + p + k$, we have $n + p - k \geq 2n_0 + 2p$, so without loss of generality $b \geq n_0 + p$. (Since the sum $a + b$ is greater than or equal to $2(n_0 + p)$, at least one of the elements must be at least $n_0 + p$.) By induction $\mathcal{G}(H_{b-p}) = \mathcal{G}(H_b)$, so

$$\mathcal{G}(H_a + H_{b-p}) = \mathcal{G}(H_a) \oplus \mathcal{G}(H_{b-p}) = \mathcal{G}(H_a) \oplus \mathcal{G}(H_b) = \mathcal{G}(H_a + H_b).$$

Here is the picture:

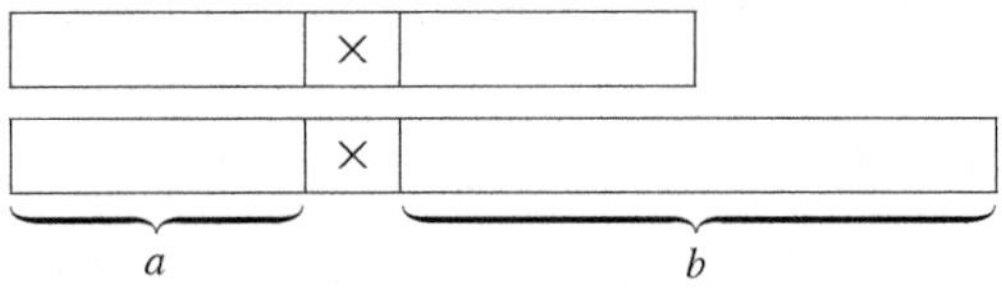

$$H_n \to H_a + H_{b-p}$$

$$H_{n+p} \to H_a + H_b$$

Now $H_a + H_{b-p}$ is an option of H_n, so we conclude that the options of H_{n+p} have exactly the same $\mathcal{G}$-values as those of H_n. Since the $\mathcal{G}$-values of H_{n+p} and H_n both observe the mex rule, we have

$$\mathcal{G}(H_{n+p}) = \mathcal{G}(H_n). \qquad \square$$

When p and n_0 are as small as possible, we say that Γ has (normal-play) *period* p and *preperiod* n_0.

Examples. In normal play:

- KAYLES (0.77) has period 12.
- DAWSON'S KAYLES (0.07) has period 34.
- 0.106 has period 328226140474. (See [4].)
- 0.007 is not known to be periodic.

Open Problem. Does there exist a finite octal code (i.e., an octal code with finitely many nonzero digits) that yields an aperiodic game?

***Misère* NIM.** We now consider NIM in misère play. It is not hard to show the following. If G consists of heaps of sizes $a_1, \ldots, a_k$, then

$$o^-(G) = \mathcal{P} \iff a_1 \oplus a_2 \oplus \cdots \oplus a_k = 0,$$

unless every a_i is equal to 0 or 1. In that case, $o^-(G) = \mathcal{P} \iff a_1 \oplus \cdots \oplus a_k = 1$.

So the strategy for misère NIM is: play exactly like in normal NIM, unless your move would leave only heaps of size 0 or 1. In that case, play to leave an odd number of heaps of size 1.

Misère equality. We now make the exact same definition of equality as before (see Definition 1.5), this time assuming misère play.

Definition 2.5. $G = H \iff o^-(G+X) = o^-(H+X)$ for all X.

Recall that in normal play any two $\mathcal{P}$-positions are equal (and in particular, any $\mathcal{P}$-position is equal to 0). We shall see that this is not the case in misère play.

In misère play:

- 0 is an $\mathcal{N}$-position.
- $*$ is a $\mathcal{P}$-position.
- $*2$ is an $\mathcal{N}$-position.

This we have already seen. Note that $*2 + *2$ is also a $\mathcal{P}$-position. No matter what first player does, second player can always respond by moving to $*$:

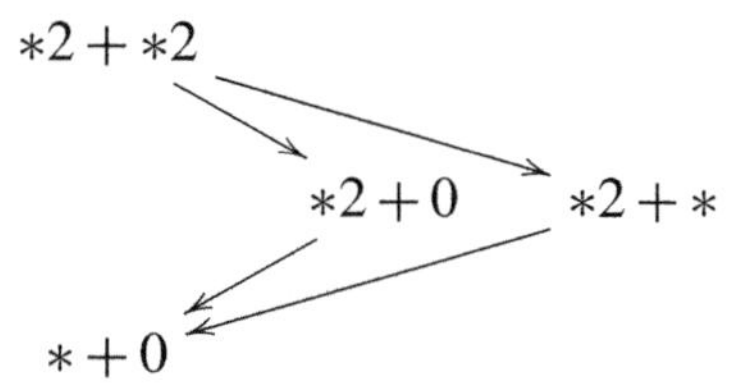

This immediately shows that $*2 + *2 \neq 0$, since $*2 + *2$ is a $\mathcal{P}$-position but 0 is an $\mathcal{N}$-position. In fact, we will now show that $*2 + *2 \neq *$, thus exhibiting two distinct $\mathcal{P}$-positions.

Proposition 2.6. $$* + * = 0.$$

Proof. Whoever can win X can also win $X + * + *$: he follows the winning strategy on X, and if his opponent ever moves on one copy of $*$, he responds by moving on the other. This guarantees that his opponent will make the last move on X, leaving either 0 or $* + *$. But both of these are $\mathcal{N}$-positions. □

Now $* + *2 + *2$ is an $\mathcal{N}$-position, since it has a move to $*2 + *2$. The following proposition therefore shows that $* \neq *2 + *2$.

Proposition 2.7. *$*2 + *2 + *2 + *2$ is a $\mathcal{P}$-position.*

Proof. The options are $*2 + *2 + *2 + 0$ and $*2 + *2 + *2 + *$. But these have moves to $*2 + *2 + 0 + 0$ and $*2 + *2 + * + *$, respectively. By the previous proposition, both of these are equal to $*2 + *2$, a $\mathcal{P}$-position. □

In fact, it is possible to show that $*2 + *2 \neq *m$ for any m. So even among sums of NIM-heaps, we have games that are not equivalent to any NIM-heap. This contrasts sharply with the situation in normal play, where *every* game is equivalent to a NIM-heap.

We have seen that $* + * = 0$. There are very few other identities we can establish in misère play. Here are really the only two:

Exercise (misère mex rule). Suppose $G \cong \{*a_1, \ldots, *a_k\}$. Then $G = *m$, where

$$m = \text{mex}\{a_1, \ldots, a_k\},$$

provided that at least one $a_i = 0$ or 1. (See Theorem 1.12.)

Exercise. For any m, we have $*m + * = *(m \oplus 1)$. (See Corollary 2.2.)

The misère mex rule is spectacularly false if every $a_i \geq 2$. For example, let

$$G = \{*2\},$$

the game whose only option is $*2$. (G is sometimes called $*2_\#$.) G is a $\mathcal{P}$-position, so right away we have $G \neq 0$. As an exercise, show that G is not equal to any $*m$. In fact, it is possible to show that G is not equal to any *sum* of NIM-heaps, but we won't do that in this course.

Birthdays. Clearly, things are more complicated in misère play than in normal play. We now state a result that shows just how much worse they are.

Definition 2.8. The *birthday* of a game G is the height of its game tree.

In *normal* play there are just six games with birthday ≤ 5 (modulo equality): 0, $*$, $*2$, $*3$, $*4$, and $*5$. In misère play, there are 4171780. On day 6 there are more than $2^{4171779}$.

The theory of misère games modulo $=$ is beautiful and fascinating, but these results suggest that it is not terribly useful: we very quickly run into seemingly intractable complications. We will not say much more about this "global theory" in this course; the interested reader is referred to [3].

Misère quotients. If $G = H$, then $G + X$ and $H + X$ have the same outcomes, for any game X. As we've just observed, this equality relation gives rise to a virtually intractable theory. The problem is that $G = H$ is too strong a relation — we are requiring that G and H behave identically in any context, which is asking a bit too much.

Key Idea. Suppose we just want to know how to play KAYLES (for example). We just need to specify how a KAYLES position G interacts with other positions that actually occur in KAYLES.

With this in mind, fix a set $\mathscr{A}$ of games (usually, $\mathscr{A}$ will be the set of positions that occur in some octal game). Assume that $\mathscr{A}$ is closed under addition.

Definition 2.9. Let $\mathscr{A}$ be a set of games, closed under addition. Then for $G, H \in \mathscr{A}$,

$$G \equiv_{\mathscr{A}} H \iff o^-(G + X) = o^-(H + X) \text{ for all } X \in \mathscr{A}.$$

Compare this to Definition 2.5: we are restricting the domain of games that can be used to distinguish G from H. This coarsens the equivalence and allows us to recover a tractable theory. Very often, the set of equivalence classes modulo $\equiv_{\mathscr{A}}$ is finite, even when $\mathscr{A}$ is infinite. (It is trivial to see that $\equiv_{\mathscr{A}}$ is an equivalence relation, since outcome-equality is an equivalence relation.)

Now, think of normal-play NIM values as elements of the group

$$\mathcal{D} = \bigoplus_{\mathbb{N}} \mathbb{Z}_2,$$

a (countably) infinite direct sum of copies of $\mathbb{Z}_2$ (one for each binary digit). The Sprague–Grundy theory maps each game G to an element of $\mathcal{D}$, thus representing the normal-play structure of G in terms of the group structure of $\mathcal{D}$. We will show that the equivalence classes modulo $\equiv_{\mathscr{A}}$ function as a *localized* misère analogue of the Sprague–Grundy theory.

We will make a slightly stronger assumption on $\mathscr{A}$ than closure under addition.

Definition 2.10. A set of games $\mathscr{A}$ is *hereditarily closed* if, for any $G \in \mathscr{A}$ and any option G' of G, we also have $G' \in \mathscr{A}$.

Definition 2.11. $\mathscr{A}$ is *closed* if it is both hereditarily closed and closed under addition.

Note that if $\mathscr{A}$ is the set of positions that occur in an octal game, then $\mathscr{A}$ is closed. In fact, virtually all sets of games that are interesting to us are closed, so there is little harm in making this assumption.

Example. Let $\mathscr{A} = \{\text{all sums of } * \text{ and } {*2}\}$, that is,

$$\mathscr{A} = \{m \cdot * + n \cdot {*2} : m, n \in \mathbb{N}\}.$$

Let's compute the equivalence classes modulo $\equiv_{\mathscr{A}}$:

- $* \not\equiv_{\mathscr{A}} 0$, since $*$ is a $\mathscr{P}$-position and 0 is an $\mathscr{N}$-position.
- Likewise, $*2 \not\equiv_{\mathscr{A}} *$ since $*2$ is an $\mathscr{N}$-position. Further, $*2 \not\equiv_{\mathscr{A}} 0$: let $X = *2$; then $*2 + X = *2 + *2$ is a $\mathscr{P}$-position, but $0 + X = *2$ is an $\mathscr{N}$-position.
- Finally, $*2 + * \not\equiv_{\mathscr{A}} *$ since it's an $\mathscr{N}$-position; $*2 + * \not\equiv_{\mathscr{A}} 0$, since they're distinguished by $X = *$; and $*2 + * \not\equiv_{\mathscr{A}} *2$, since they're distinguished by $X = *2$.

This gives four equivalence classes:

$[0]$	$[*]$	$[*2]$	$[*2+*]$
$\mathscr{N}$	$\mathscr{P}$	$\mathscr{N}$	$\mathscr{N}$

Are there others? Yes! $*2 + *2$ is a $\mathscr{P}$-position, so it's either equivalent to $*$, or a new equivalence class. But,

- $* + (*2 + *2)$ is an $\mathscr{N}$-position, since it has a move to $*2 + *2$, which is $\mathscr{P}$;
- $*2 + *2 + (*2 + *2)$ is a $\mathscr{P}$-position (Proposition 2.7).

Therefore $* \not\equiv_{\mathscr{A}} *2 + *2$. Similar reasoning shows that $*2 + *2 + *$ gives yet another equivalence class.

So we have six equivalence classes total:

$[0]$	$[*]$	$[*2]$	$[*2+*]$	$[*2+*2]$	$[*2+*2+*]$
$\mathscr{N}$	$\mathscr{P}$	$\mathscr{N}$	$\mathscr{N}$	$\mathscr{P}$	$\mathscr{N}$

We now show that these are the only six.

Lemma 2.12. *Let $n \geq 1$. Then $n \cdot *2$ is a $\mathscr{P}$-position if and only if n is even.*

Proof. If n is even, then second player's strategy is to cancel out copies of $*2$ (using the fact that $* + * = 0$) until we get down to $*2 + *2$, which is known to be a $\mathscr{P}$-position.

If n is odd, $n \geq 3$, then first player can win by moving to $(n-1) \cdot *2$.

Finally, if $n = 1$, then first player simply moves to $*$. □

Lemma 2.13. *Let $n \geq 1$. Then $n \cdot *2 + *$ is always an $\mathscr{N}$-position.*

Proof. If n is even, then the winning move is to $n \cdot *2$, which is a $\mathscr{P}$-position by the previous lemma.

If n is odd, $n \geq 3$, then the winning move is to $(n-1) \cdot *2 + * + *$, which again is a $\mathscr{P}$-position, since $* + * = 0$.

Finally, if $n = 1$, then the winning move is to $0 + *$. □

Corollary 2.14. *Suppose $G = m \cdot * + n \cdot *2$ and $X = m' \cdot * + n' \cdot *2$. If $n \geq 1$, then the outcome of $G + X$ depends only on the parities of $m + m'$ and $n + n'$.*

Proof. Follows immediately from the previous two lemmas and the fact that $* + * = 0$. □

Corollary 2.15. *Let* $G = m \cdot * + n \cdot *2$ *and* $H = m' \cdot * + n' \cdot *2$. *If* $n, n' \geq 1$, $m \equiv m' \pmod 2$, *and* $n \equiv n' \pmod 2$, *then* $G \equiv_{\mathscr{A}} H$.

Proof. Follows immediately from the previous corollary. □

Corollary 2.16. *There are exactly six equivalence classes modulo* $\equiv_{\mathscr{A}}$.

Proof. By the previous corollary, every $G \in \mathscr{A}$ is equivalent to $m \cdot * + n \cdot *2$, for some $m < 2$ and $n < 3$. There are only six such possibilities, and we've already shown that all six are mutually inequivalent. □

Warning. We've just shown that $*2 + *2 + *2 \equiv_{\mathscr{A}} *2$. However, equality does *not* hold:

Exercise. Show that $*2 + *2 + *2 \neq *2$. (Hint: try $X = *2_{\#}1$, defined by $*2_{\#}1 = \{*2_{\#}, *\} = \{\{*2\}, *\}$.)

This shows that the equivalence $\equiv_{\mathscr{A}}$ is a genuine coarsening of equality. There exist unequal games that are equivalent modulo $\mathscr{A}$.

This finishes our example. We now return to the general context.

Lemma 2.17. *Let* $\mathscr{A}$ *be any closed set of games and* $G, H \in \mathscr{A}$. *If* $G \equiv_{\mathscr{A}} H$ *and* $K \in \mathscr{A}$, *then* $G + K \equiv_{\mathscr{A}} H + K$.

Proof. For $X \in \mathscr{A}$, we have

$$o^-((G+K)+X) = o^-(G+(K+X)) \text{ and } o^-(H+(K+X)) = o^-((H+K)+X).$$

But $\mathscr{A}$ is closed, so $K + X \in \mathscr{A}$. Since $G \equiv_{\mathscr{A}} H$, we have $o^-(G + (K + X)) = o^-(H + (K + X))$, as needed. □

Moreover, since $\mathscr{A}$ is hereditarily closed, we have $0 \in \mathscr{A}$. So the equivalence class of 0 is an identity, and in fact we have a monoid.

Definition 2.18. A *semigroup* is a set S equipped with an associative binary operation $\cdot$. That is,

- if $x, y \in S$, then $x \cdot y \in S$;
- if $x, y, z \in S$, then $x \cdot (y \cdot z) = (x \cdot y) \cdot z$.

A semigroup S is a *monoid* if it has an identity, and *commutative* if its operation is commutative.

We've shown that the equivalence classes of $\mathscr{A}$ modulo $\equiv_{\mathscr{A}}$ form a commutative monoid $\mathbb{Q}$:

$$\mathbb{Q} = \{[G]_{\equiv_{\mathscr{A}}} : G \in \mathscr{A}\}.$$

Furthermore, if $G \equiv_{\mathscr{A}} H$, then since $o^-(G+0) = o^-(H+0)$, we have

$$G \text{ is a } \mathscr{P}\text{-position} \iff H \text{ is a } \mathscr{P}\text{-position}.$$

So we can define a subset $\mathcal{P} \subset \mathbb{Q}$ by

$$\mathcal{P} = \{[G]_{\equiv_{\mathscr{A}}} : G \in \mathscr{A} \text{ is a } \mathscr{P}\text{-position}\}.$$

Definition 2.19. The structure $(\mathbb{Q}, \mathcal{P})$ is the *misère quotient* of $\mathscr{A}$, and we denote it by $\mathbb{Q}(\mathscr{A})$.

We'll continue to use uppercase letters $G, H, \ldots$ to denote games in the set $\mathscr{A}$, and lowercase letters $a, b, \ldots$ to denote elements of the quotient $\mathbb{Q}(\mathscr{A})$. Likewise, disjunctive sums of games will always be written additively (for example, $G + 2 \cdot H$), whereas the quotient operation will always be written multiplicatively (for example, ab^2).

Example. Let's sketch the structure of $\mathbb{Q}(\mathscr{A})$ for our example:

$$\mathscr{A} = \{\text{sums of } * \text{ and } *2\}.$$

Denote by $\Phi : \mathscr{A} \to \mathbb{Q}$ the quotient map

$$\Phi(G) = [G]_{\equiv_{\mathscr{A}}}.$$

Now $\mathscr{A}$ is generated (as a monoid) by $*$ and $*2$. Put

$$1 = \Phi(0) = [0], \quad a = \Phi(*) = [*], \quad b = \Phi(*2) = [*2].$$

We know that $* + * = 0$, so in fact $a^2 = 1$. Furthermore, we've seen that $*2 + *2 + *2 \equiv_{\mathscr{A}} *2$, so we have $b^3 = b$. But we also know that the six elements

$$\begin{array}{lllllll} \mathscr{A} & [0] & [*] & [*2] & [*2+*] & [*2+*2] & [*2+*2+*] \\ \downarrow & & & & & & \\ \mathbb{Q} & 1 & a & b & ab & b^2 & ab^2 \end{array}$$

are all distinct. Thus $\mathbb{Q} = \{1, a, b, ab, b^2, ab^2\}$ and we have the presentation

$$\mathbb{Q} \cong \langle a, b \mid a^2 = 1,\ b^3 = b \rangle.$$

Since $*$ and $*2 + *2$ are the only $\mathscr{P}$-positions (up to equivalence), we also have $\mathcal{P} = \{a, b^2\}$. This misère quotient is called $\mathcal{T}_2$, and it is the first of many that we will see.

Lecture 3. The periodicity theorem

November 28, 2006 Scribes: Amir Kantor and Gideon Amir

Definition 3.1 (Definitions). Let $\mathscr{A}$ be any set of games. Define

$$\mathrm{hcl}(\mathscr{A}) = \{\text{subpositions of all games in } \mathscr{A}\},$$

$$\mathrm{cl}(\mathscr{A}) = \text{Closure under addition of } \mathrm{hcl}(\mathscr{A}).$$

Remark. To see that $\mathrm{cl}(\mathscr{A})$ is hereditarily closed, let $G = G_1 + G_2 + \cdots + G_k$, where $G_i \in \mathrm{hcl}(\mathscr{A})$. Without loss of generality, $G' = G_1' + G_2 + \cdots + G_k$. We know that $G_1' \in \mathrm{hcl}(\mathscr{A})$ since the latter is hereditarily closed.

Example. $\mathrm{cl}(\{*2\}) = \{\text{sums of } *, *2\} = \{i \cdot * + j \cdot *2 : i, j \in \mathbb{N}\}$.

Exercise.

- If $\mathscr{A} \subseteq \mathscr{B}$ and $\mathscr{B}$ is closed, then $\mathrm{cl}(\mathscr{A}) \subseteq \mathscr{B}$.
- $\mathrm{cl}(\mathrm{cl}(\mathscr{A})) = \mathrm{cl}(\mathscr{A})$.

Definition 3.2. If $\mathscr{A}$ is not closed, $\mathbb{Q}(\mathscr{A}) \triangleq \mathbb{Q}(\mathrm{cl}(\mathscr{A}))$. We sometimes write $\mathbb{Q}(G) \triangleq \mathbb{Q}(\mathrm{cl}(\{G\}))$.

Example. $$\mathcal{T}_2 \cong \mathbb{Q}(*2).$$

Quotients of octal games. Let's consider the context of a specific octal game, such as KAYLES. Denote by H_n a KAYLES heap of size n and let $\mathscr{A}$ be the set of all KAYLES positions; that is,

$$\mathscr{A} = \mathrm{cl}(H_0, H_1, H_2, H_3, \ldots).$$

Let $(\mathbb{Q}, \mathcal{P})$ be the misère quotient for KAYLES and consider the quotient map $\Phi : \mathscr{A} \to \mathbb{Q}$.

Remark. If we know $\Phi(H_n)$ for all n, then if $G = H_{n_1} + \cdots + H_{n_k}$ we can easily compute $\Phi(G) = \Phi(H_{n_1}) \cdots \Phi(H_{n_k})$. So, in order to specify Φ, it suffices to specify the single-heap values $\Phi(H_n)$.

The *main point* is this:

> Suppose we know $\mathbb{Q}(\mathscr{A})$, together with $\Phi(H_n)$ for all n. If we want to know $o^-(G)$ for $G \in \mathscr{A}$, we can write $G = H_{n_1} + \cdots + H_{n_k}$, compute $\Phi(G) = \Phi(H_{n_1}) \cdots \Phi(H_{n_k})$, and simply look up whether $\Phi(G) \in \mathcal{P}$. If the quotient is finite, we've reduced the problem of finding $o^-(G)$ to a small number of operations on a finite multiplication table. This yields an efficient way to compute $o^-(G)$.

So we direct our energies at computing the values of $\Phi(H_n)$ for all n. In practice, we can construct good algorithms for computing quotients of a finite number of heaps. (We won't have time to discuss them in this course; see [13,

Appendix C].) If we run these algorithms on KAYLES to heap 120, we get the result shown in Figure 5.

Now examine the Φ-values $\Phi(H_n) \in \mathbb{Q}$. We observe that

$$\Phi(H_{n+12}) = \Phi(H_n), \text{ for } 71 \leqslant n \leqslant 120 - 12.$$

This situation is much like the periodicity of $\mathcal{G}$-values that we observed in normal play.

The following notation will be very useful; it applies to KAYLES as well as to an arbitrary octal game Γ. Denote by

- $\mathcal{A}$ the set of all positions, $\mathcal{A} = \text{cl}(H_0, H_1, \ldots)$;
- $\mathbb{Q}(\Gamma) = \mathbb{Q}(\mathcal{A})$;
- $\mathcal{A}_n$ the set of all positions with no heap larger than n, $\mathcal{A}_n = \text{cl}(H_0, \ldots, H_n)$;
- $\mathbb{Q}_n(\Gamma) = \mathbb{Q}(\mathcal{A}_n)$, the n-th *partial quotient* for Γ.

$$\begin{aligned}\mathbb{Q}(H_0, H_1, H_2, \ldots, H_{120})& \\ \cong \langle a, b, c, d, e, f, g \mid{}& a^2 = 1,\ b^3 = b,\ bc^2 = b,\ c^3 = c,\ bd = bc, \\ & cd = b^2,\ d^3 = d,\ be = bc,\ ce = b^2,\ e^2 = de, \\ & bf = ab,\ cf = ab^2c,\ d^2f = f,\ f^2 = b^2, \\ & b^2g = g,\ c^2g = g,\ dg = cg, \\ & eg = cg,\ fg = ag,\ g^2 = b^2\rangle,\end{aligned}$$

$$\mathcal{P} = \{a, b^2, ac, ac^2, d, ad^2, e, ade, adf\}.$$

	1	2	3	4	5	6	7	8	9	10	11	12
0+	a	b	ab	a	c	ab	b	ab^2	d	b	bc	e
12+	ab^2	b	abc	ab^2	d^2e	ab	b	ade	b^2c	bc	abc	b^2c
24+	f	b	g	ab^2c	b^2c	abc	b	ab^2	g	bc	abc	b^2c
36+	ab^2	b	ab	ab^2	b^2c	abc	b	ab^2	g	b	abc	b^2c
48+	ab^2	b	g	ab^2	b^2c	abc	b	ab^2	b^2c	b	abc	b^2c
60+	ab^2	b	g	ab^2	b^2c	abc	b	ab^2	g	bc	abc	b^2c
72+	ab^2	b	g	ab^2	b^2c	abc	b	ab^2	g	b	abc	b^2c
84+	ab^2	b	g	ab^2	b^2c	abc	b	ab^2	g	b	abc	b^2c
96+	ab^2	b	g	ab^2	b^2c	abc	b	ab^2	g	b	abc	b^2c
108+	ab^2	b	g	ab^2	b^2c	abc	b	ab^2	g	b	abc	b^2c

Figure 5. Quotient presentation and pretending function for misère KAYLES to heap 120.

We've computed $\mathbb{Q}_{120}(\text{KAYLES})$, and the quotient map $\Phi_{120} : \mathscr{A}_{120} \longrightarrow \mathbb{Q}_{120}$, and found that it is periodic past a certain point.

A brief digression. In a moment we will state a misère version of the periodicity theorem. We first pause to consider some potential difficulties.

Remark. Suppose we computed $\mathbb{Q}_n$. Now we throw H_{n+1} into the quotient. There might be games $G, K \in \mathscr{A}_n$ such that $G \equiv_{\mathscr{A}_n} K$ but are distinguished by H_{n+1}. When this happens, we have $\Phi_n(G) = \Phi_n(K)$, but $\Phi_{n+1}(G) \neq \Phi_{n+1}(K)$.

This remark shows that we must be careful not to confuse the partial quotients of Γ with its full quotient.

Note that in normal play, there is no such concern. Given a set of games $\mathscr{A}$, it is possible to define *normal equivalence modulo* $\mathscr{A}$ in exactly the same way we've defined misère equivalence modulo $\mathscr{A}$. However, in normal play it will always be the case that $G \equiv_{\mathscr{A}} K$ if and only if $G = K$. That is, in normal play, local and global equivalence coincide. (To see this, observe that if $G \neq K$ in normal play, then G and K must have different NIM values, so $G + G$ and $G + K$ have different outcomes. So if G and K are distinguished by anything, then they must be distinguished locally, by G itself.) So, although the sorts of localizations we're discussing are perfectly applicable to normal play, they don't provide any further resolution (and in a sense, they don't need to, because normal play is simple enough to begin with).

Let us consider another difference between normal play and misère play. Consider a finitely generated set $\mathscr{A}$. In *normal play*, there can be only finitely many $\mathscr{G}$-values represented. To see this, let $H_1, \ldots, H_n$ generate $\mathscr{A}$. Then the $\mathscr{G}$-values represented by $\mathscr{A}$ are bitwise exclusive-or's of $\mathscr{G}(H_1), \ldots, \mathscr{G}(H_n)$, but these are bounded.

What about misère play? Is $\mathbb{Q}(\mathscr{A})$ finite? The answer is: not in general. Later in this course we will see an example of an infinite, finitely generated quotient. Our picture of such quotients is still very hazy. In fact, the following question is still open.

Open Problem. Specify an algorithm to determine whether $\mathbb{Q}(\mathscr{A})$ is infinite, assuming $\mathscr{A}$ is finitely generated.

We'll say more about this later in the course. Finally, now is as good a time as any to interject the following remark:

Remark. All monoids we consider in this course are commutative. Sometimes I will slip and say "monoid" when I really mean "commutative monoid."

Periodicity. We now return to the setting of an octal game Γ with heaps H_n.

Recall the periodicity theorem for normal play:

Let Γ be an octal game with last nonzero code digit k. Suppose there are integers n_0, p such that $\mathcal{G}(H_{n+p}) = \mathcal{G}(H_n)$ for $n_0 \le n < 2n_0 + p + k$. Then in fact

$$\mathcal{G}(H_{n+p}) = \mathcal{G}(H_n) \text{ for all } n \ge n_0.$$

Theorem 3.3 (periodicity theorem for misère play). *Let Γ be an octal game with last nonzero code digit k. Fix n_0, p and let $M = 2n_0 + 2p + k$. Let $(\mathbb{Q}_M, \mathcal{P}_M) = \mathbb{Q}_M(\Gamma)$. Suppose that $\Phi_M : A_M \longrightarrow \mathbb{Q}_M$, and that $\Phi_M(H_{n+p}) = \Phi_M(H_n)$ for $n_0 \le n < 2n_0 + p + k$. Then in fact*

$$\mathbb{Q}(\Gamma) \cong \mathbb{Q}_M(\Gamma),$$

and

$$\Phi(H_{n+p}) = \Phi(H_n) \textit{ for all } n \ge n_0.$$

Proof. Recall the proof for normal play. By induction on n,

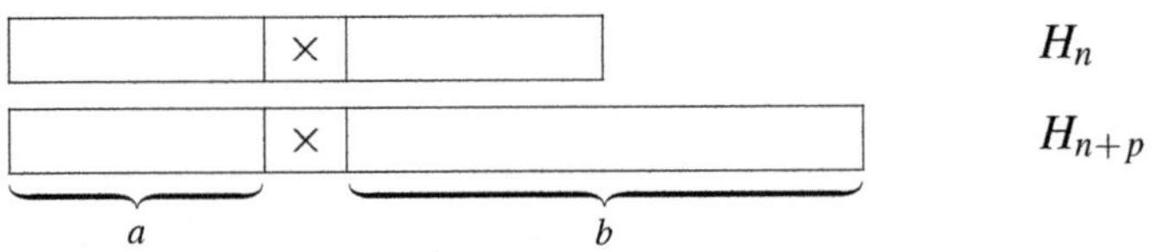

$H_{n+p} \longrightarrow H_a + H_b$ is a typical move from H_{n+p}. We chose the upper bound of our induction base case to be large enough that one of $a, b \ge n_0 + p$. Assume without loss of generality that it's b. But then $\mathcal{G}(H_{b-p}) = \mathcal{G}(H_b)$, so $\mathcal{G}(H_a + H_b) = \mathcal{G}(H_a + H_{b-p})$. We conclude that the options of H_n and H_{n+p} represent exactly the same $\mathcal{G}$-values. But $\mathcal{G}$-values are computed by the mex rule, so this implies $\mathcal{G}(H_{n+p}) = \mathcal{G}(H_n)$.

To prove the periodicity theorem for misère play, we can use exactly the same argument to show that the options of H_n, H_{n+p} represent exactly the same Φ_M-values. So the proof now depends only on the following lemma.

Lemma 3.4. *Suppose $\mathcal{A}$ is a closed set of games, and G is a game all of whose options are in $\mathcal{A}$. Assume that, for some $H \in \mathcal{A}$,*

$$\{\Phi(G') : G' \textit{ is an option of } G\} = \{\Phi(H') : H' \textit{ is an option of } H\}.$$

Then $\mathbb{Q}(\mathcal{A} \cup \{G\}) \cong \mathbb{Q}(\mathcal{A})$ and $\Phi(G) = \Phi(H)$.

Assuming Lemma 3.4, the proof of the periodicity theorem is complete, for we can go by induction to show that

$$Q_M(\Gamma) \cong Q_{M+1}(\Gamma) \cong Q_{M+2}(\Gamma) \cong \ldots,$$

and that the resulting Φ-values are periodic. □

Bipartite monoids. Although we could prove Lemma 3.4 directly, it will be easier after we introduce a suitable abstraction of the misère quotient construction. Since the abstract setting is also useful in other situations, this is worth the effort.

Definition 3.5. A *bipartite monoid* is a pair $(\mathbb{Q}, \mathcal{P})$ where $\mathbb{Q}$ is a commutative monoid, and $\mathcal{P} \subset \mathbb{Q}$ is some subset. We will usually write bm for bipartite monoid.

Definition 3.6. Let $(\mathbb{Q}, \mathcal{P})$ be a bm; $x, y \in \mathbb{Q}$ are said to be *indistinguishable* if, for all $z \in \mathbb{Q}$,

$$xz \in \mathcal{P} \Longleftrightarrow yz \in \mathcal{P}.$$

Definition 3.7. A bm $(\mathbb{Q}, \mathcal{P})$ is *reduced* if the elements of $\mathbb{Q}$ are pairwise distinguishable. We write rbm for reduced bipartite monoid.

Proposition 3.8. *Every misère quotient is a rbm.*

Proof. Suppose $[G]_{\equiv_{\mathscr{A}}}$ and $[H]_{\equiv_{\mathscr{A}}}$ are indistinguishable. Then for any $X \in \mathscr{A}$,

$$[G] + [X] \in \mathcal{P} \Longleftrightarrow [H] + [X] \in \mathcal{P}.$$

Therefore $o^-(G + X) = o^-(H + X)$ for all $X \in \mathscr{A}$, so $[G] = [H]$. □

Example. If $\mathscr{A}$ is a closed set of games, and $\mathscr{B}$ is the set of misère $\mathscr{P}$-positions of $\mathscr{A}$, then $(\mathscr{A}, \mathscr{B})$ is a bipartite monoid. The same is true if we take $\mathscr{B}$ to be the set of normal $\mathscr{P}$-positions of $\mathscr{A}$.

Definition 3.9. A function $f : (\mathbb{Q}, \mathcal{P}) \to (\mathcal{S}, \mathcal{R})$ is a *bipartite monoid homomorphism* if $f : \mathbb{Q} \to \mathcal{S}$ is a monoid homomorphism, and for every $x \in \mathbb{Q}$, we have $x \in \mathcal{P}$ if and only if $f(x) \in \mathcal{R}$.

Definition 3.10. Let $(\mathbb{Q}, \mathcal{P})$ and $(\mathcal{S}, \mathcal{R})$ be bipartite monoids. $(\mathcal{S}, \mathcal{R})$ is a *quotient* of $(\mathbb{Q}, \mathcal{P})$ if and only if there is a surjective homomorphism $f : (\mathbb{Q}, \mathcal{P}) \to (\mathcal{S}, \mathcal{R})$.

Definition 3.11. Let $(\mathbb{Q}, \mathcal{P})$ be a bm. Define a relation ρ on $\mathbb{Q}$ by $x \rho y$ if and only if x and y are indistinguishable.

Exercise. Show that ρ is an equivalence relation, and that the equivalence classes modulo ρ form a bipartite monoid.

Definition 3.12. The *reduction* of $(\mathbb{Q}, \mathcal{P})$ is the bipartite monoid of equivalence classes modulo ρ. We denote it by $(\mathbb{Q}', \mathcal{P}')$.

Exercise. Show that $(\mathbb{Q}', \mathcal{P}')$ is reduced and is a quotient of $(\mathbb{Q}, \mathcal{P})$.

Example. Let $\mathscr{A}$ be a closed set of games, and let $\mathscr{B}$ be the set of misère $\mathscr{P}$-positions in $\mathscr{A}$. Then the misère quotient $\mathbb{Q}(\mathscr{A})$ is the reduction of $(\mathscr{A}, \mathscr{B})$.

The following proposition is extremely useful.

Proposition 3.13. *Suppose* $(\mathbb{Q}, \mathcal{P})$ *is a bm with reduction* $(\mathbb{Q}', \mathcal{P}')$. *Let* $(\mathcal{S}, \mathcal{R})$ *be any quotient of* $(\mathbb{Q}, \mathcal{P})$, *via* $f : (\mathbb{Q}, \mathcal{P}) \to (\mathcal{S}, \mathcal{R})$, *and let* $(\mathcal{S}', \mathcal{R}')$ *be its reduction. Then there is an isomorphism* $i : (\mathbb{Q}', \mathcal{P}') \to (\mathcal{S}', \mathcal{R}')$ *making the following diagram commute*:

$$\begin{array}{ccc} \mathbb{Q} & \xrightarrow{f} & \mathcal{S} \\ \downarrow & & \downarrow \\ \mathbb{Q}' & \overset{i}{=\!=\!=} & \mathcal{S}' \end{array}$$

Proof. Let ρ be the reduction relation on $\mathbb{Q}$ ($\mathbb{Q}' = \mathbb{Q}/\rho$), and let τ be the reduction relation on $\mathcal{S}$ ($\mathcal{S}' = \mathcal{S}/\tau$).

Now for $x, y \in \mathbb{Q}$, we have

$$\begin{aligned} &[x]_\rho = [y]_\rho \\ &\qquad \Longleftrightarrow xz \in \mathcal{P} \Leftrightarrow yz \in \mathcal{P} \text{ for all } z \in \mathbb{Q}; \\ &\qquad \Longleftrightarrow f(xz) \in \mathcal{R} \Leftrightarrow f(yz) \in \mathcal{R} \text{ for all } z \in \mathbb{Q}; \\ &\qquad \Longleftrightarrow f(x)w \in \mathcal{R} \Leftrightarrow f(y)w \in \mathcal{R} \text{ for all } w \in \mathcal{S} \text{ (since } f \text{ is surjective)}; \\ &\qquad \Longleftrightarrow [f(x)]_\tau = [f(y)]_\tau. \end{aligned}$$

So we may define the map i by $i([x]_\rho) = [f(x)]_\tau$. We just showed that i is well defined and one-to-one. Since f is surjective, so is i, and it follows that i is an isomorphism. Commutativity of the diagram follows trivially from the definition of i. □

Corollary 3.14. *Every bipartite monoid has exactly one reduced quotient* (*up to isomorphism*).

Let us see why this is important. Let $\mathscr{A}$ be a closed set of games, and $\mathscr{B}$ the set of misère $\mathscr{P}$-positions in $\mathscr{A}$. Then the misère quotient $\mathbb{Q}(\mathscr{A})$ is the reduction of $(\mathscr{A}, \mathscr{B})$. Therefore, suppose we have some putative quotient $(\mathbb{Q}, \mathcal{P})$, and we want to assert that it is $\mathbb{Q}(\mathscr{A})$. We just need to show that

(a) $(\mathbb{Q}, \mathcal{P})$ is reduced; and

(b) $(\mathbb{Q}, \mathcal{P})$ is a quotient of $(\mathscr{A}, \mathscr{B})$.

By Proposition 3.13, these conditions imply that $(\mathbb{Q}, \mathcal{P}) \cong \mathbb{Q}(\mathscr{A})$. We can therefore avoid the exhaustive analysis used to construct $\mathcal{T}_2$ during the previous lecture.

Lecture 4. More examples

November 29, 2006 Scribes: Shai Lubliner and Ohad Manor

Proof of Lemma 3.4. We now prove Lemma 3.4, thus completing the proof of the periodicity theorem.

Definition 4.1. Suppose $\mathscr{A}$ is a set of games, and G is a game all of whose options are in $\mathscr{A}$. Define

$$\Phi''G = \{\Phi(G') : G' \text{ is an option of } G\}.$$

(This definition includes the case when $G \in \mathscr{A}$.)

Lemma 4.2. *Suppose $\mathscr{A}$ is a closed set of games and $(\mathcal{Q}, \mathcal{P})$ a rbm. The following are equivalent:*

(i) $(\mathcal{Q}, \mathcal{P}) \cong \mathcal{Q}(\mathscr{A})$;

(ii) *there exists a surjective monoid homomorphism $\Phi : \mathscr{A} \to \mathcal{Q}$, such that for all $G \in \mathscr{A}$,*

$$\Phi(G) \in \mathcal{P} \iff G \neq 0 \textit{ and } \Phi(G') \notin \mathcal{P} \textit{ for every option } G' \textit{ of } G.$$

Proof. (i) $\Rightarrow$ (ii): Let Φ be the quotient map $\mathscr{A} \to \mathcal{Q}(\mathscr{A})$. We know that, for all G,

$$G \text{ is a } \mathscr{P}\text{-position} \iff G \neq 0 \text{ and every } G' \text{ is an } \mathscr{N}\text{-position}.$$

But since Φ is a homomorphism of bipartite monoids, we have for all $X \in \mathscr{A}$,

$$X \text{ is a } \mathscr{P}\text{-position} \iff \Phi(X) \in \mathcal{P},$$

and the conclusion follows immediately.

(ii) $\Rightarrow$ (i): By Corollary 3.14, $\mathcal{Q}(\mathscr{A})$ is the unique reduced quotient of $(\mathscr{A}, \mathscr{B})$ (where $\mathscr{B}$ is the set of $\mathscr{P}$-positions in $\mathscr{A}$). Thus it suffices to show that Φ is a homomorphism of bipartite monoids, since this implies that $(\mathcal{Q}, \mathcal{P})$ is a quotient of $(\mathscr{A}, \mathscr{B})$. So we must prove the following, for all $G \in \mathscr{A}$:

$$G \text{ is a } \mathscr{P}\text{-position} \iff \Phi(G) \in \mathcal{P}.$$

Now by induction on G (i.e., on the height of the game tree of G), we may assume that

$$G' \text{ is a } \mathscr{P}\text{-position} \iff \Phi(G') \in \mathcal{P},$$

for all options G' of G. But now,

$$\begin{aligned} \Phi(G) \in \mathcal{P} &\iff G \neq 0 \text{ and } \Phi(G') \notin \mathcal{P} \text{ for all } G' \quad \text{(by assumption)} \\ &\iff G \neq 0 \text{ and every } G' \text{ is an } \mathscr{N}\text{-position} \quad \text{(by induction)} \\ &\iff G \text{ is a } \mathscr{P}\text{-position} \quad \text{(by definition of } \mathscr{P}\text{-position).} \end{aligned}$$

This proves the lemma. □

Proof of Lemma 3.4. Assume $\mathscr{A}$ is a closed set of games, all options of G are in $\mathscr{A}$ and $\Phi''G = \Phi''H$ for some $H \in \mathscr{A}$. We must show that $\mathcal{Q}(\mathscr{A} \cup \{G\}) \cong \mathcal{Q}(\mathscr{A})$ and $\Phi(G) = \Phi(H)$.

Define $\Phi^+ : \mathrm{cl}(\mathscr{A} \cup \{G\}) \to \mathcal{Q}$ by

- $\Phi^+(G) = \Phi(H)$;
- $\Phi^+(Y) = \Phi(Y)$ for all $Y \in \mathscr{A}$.

If we regard G as a free generator of the monoid $\mathrm{cl}(\mathscr{A} \cup \{G\})$ over $\mathscr{A}$, then this defines a monoid homomorphism. So we just need to show that Φ^+ satisfies condition (ii) of Lemma 4.2.

Fix $X \in \mathrm{cl}(\mathscr{A} \cup \{G\})$. We can write $X = n \cdot G + Y$ for some $n \geq 0$ and $Y \in \mathscr{A}$. The case $n = 0$ is already known, so we can assume $n \geq 1$. Let $W = n \cdot H + Y$; clearly $\Phi^+(X) = \Phi^+(W)$.

Now consider an option X' of X. We have $X' = n \cdot G + Y'$ or $(n-1) \cdot G + G' + Y$:

- If $X' = n \cdot G + Y'$, then $\Phi^+(X') = \Phi^+(n \cdot H + Y')$, which is an option of W.
- If $X' = (n-1) \cdot G + G' + Y$, then $\Phi^+(X') = \Phi^+((n-1) \cdot H + G' + Y)$. But since $\Phi''G = \Phi''H$, there must be some H' with $\Phi^+(H') = \Phi^+(G')$. So $\Phi^+(X') = \Phi^+((n-1) \cdot H + H' + Y)$, again an option of W'.

This shows that $(\Phi^+)''X \subset (\Phi^+)''W$, and an identical argument shows that $(\Phi^+)''W \subset (\Phi^+)''X$. But since $W \in \mathscr{A}$, we know that

$$\Phi(W) \in \mathcal{P} \Longleftrightarrow W \neq 0 \quad \text{and} \quad \Phi(W') \notin \mathcal{P} \text{ for all } W'.$$

Since $\Phi^+(X) = \Phi^+(W)$ and $(\Phi^+)''X = (\Phi^+)''W$, we have

$$\Phi^+(X) \in \mathcal{P} \Longleftrightarrow W \neq 0 \quad \text{and} \quad \Phi^+(X') \notin \mathcal{P} \text{ for all } X'.$$

This satisfies Lemma 4.2(ii) except for the condition $W \neq 0$. But if either of G, H is identically 0, then both must be, since $\Phi''G = \varnothing$ if and only if $\Phi''H = \varnothing$. Therefore $W \neq 0$ if and only if $X \neq 0$, and we are done. □

Further examples. The partial quotients of NIM are fundamental examples, and we denote them by $\mathcal{T}_n$:

- $\mathcal{T}_0 = \mathbb{Q}(0)$;
- $\mathcal{T}_1 = \mathbb{Q}(*)$;
- $\mathcal{T}_2 = \mathbb{Q}(*2)$;
- $\mathcal{T}_n = \mathbb{Q}(*2^{n-1})$.

Here are their presentations:

- $\mathcal{T}_0 = \{1\}$; $\mathcal{P} = \varnothing$.
- $\mathcal{T}_1 = \langle a \mid a^2 = 1 \rangle$; $\mathcal{P} = \{a\}$.
- $\mathcal{T}_2 = \langle a, b \mid a^2 = 1,\ b^3 = b \rangle$; $\mathcal{P} = \{a, b^2\}$.
- $\mathcal{T}_3 = \langle a, b, c \mid a^2 = 1,\ b^3 = b,\ c^3 = c,\ b^2 = c^2 \rangle$; $\mathcal{P} = \{a, b^2\}$.

- $\mathcal{T}_n = \langle a, b_1, b_2, \ldots, b_{n-1} \mid a^2 = 1,\ b_i^3 = b_i,\ b_1^2 = b_2^2 = \cdots = b_{n-1}^2 \rangle$; $\mathcal{P} = \{a, b_1^2\}$.

To find $\Phi(*m)$ (in any of the $\mathcal{T}_n$), write m in binary, as $\cdots \epsilon_3\epsilon_2\epsilon_1\epsilon_0$, and we have

$$\Phi(*m) = a^{\epsilon_0} \cdot b_1^{\epsilon_1} \cdot b_2^{\epsilon_2} \cdots b_n^{\epsilon_n}.$$

For example, in $\mathcal{T}_4$, we have

$$\Phi(*4)=b_2, \quad \Phi(*5)=ab_2, \quad \Phi(*6)=b_1b_2, \quad \Phi(*7)=ab_1b_2, \quad \Phi(*8)=b_3.$$

Notice that we always have

$$b_1^2 = b_2^2 = \cdots = b_{n-1}^2.$$

Denote this element by z; z represents the sum $*m + *m$, for any NIM-heap with $m \geq 2$. In fact, it represents any NIM position of $\mathcal{G}$-value 0, provided it has at least one heap of size ≥ 2.

The structure of $\mathcal{T}_n$. Let's write out the elements of $\mathcal{T}_3$:

$$\mathcal{T}_3 = \{1, a, b_1, ab_1, b_2, ab_2, b_1b_2, ab_1b_2, z, az\}.$$

Consider the subset

$$\mathcal{K} = \{b_1, ab_1, b_2, ab_2, b_1b_2, ab_1b_2, z, az\}.$$

Observe that $z \cdot z = z$, $z \cdot b_1 = b_1$, and $z \cdot b_2 = b_2$. Therefore z is an identity of $\mathcal{K}$ and $x^2 = z$ for all $x \in \mathcal{K}$. So $\mathcal{K}$ is a group, and we have

$$\mathcal{K} \cong \mathbb{Z}_2^3.$$

In fact $\mathcal{K}$ behaves just like normal play $\mathcal{G}$-values: it has eight elements, corresponding one-to-one with NIM positions of $\mathcal{G}$-value 0 through 7.

Recall the strategy for misère NIM: play exactly like in normal NIM, unless your move would leave only heaps of size 0 or 1. In that case, play to leave an odd number of heaps of size 1.

$\mathcal{K}$ corresponds to the "exactly like normal NIM" clause of this strategy: it is isomorphic to the normal-play quotient of $*4$. The two elements 1 and a correspond to the "unless": they represent positions with all heaps of size ≤ 1.

Note that every $\mathcal{T}_n$, for $n \geq 2$, can be written as $\mathcal{K} \cup \{1, a\}$, where $\mathcal{K} \cong \mathbb{Z}_2^n$. $\mathcal{K}$ is called the *kernel* of the monoid, and in the next lecture we will see how to generalize it.

In particular, we have

- $|\mathcal{T}_0| = 1$,
- $|\mathcal{T}_1| = 2$,

- $|\mathcal{T}_n| = 2^n + 2$ for all $n \geq 2$.

We can also define the full quotient of NIM:

$$\mathcal{T}_\infty = \mathbb{Q}(0, *, *2, *3, *4, \ldots)$$
$$\cong \langle a, b_1, b_2 \mid a^2 = 1,\ b_i^3 = b_i,\ b_1^2 = b_2^2 = \cdots \rangle, \quad \mathcal{P} = \{a, b_1^2\}.$$

Remember that normal-play $\mathcal{G}$-values look like

$$\bigoplus_{\mathbb{N}} \mathbb{Z}_2.$$

Well, we can write $\mathcal{T}_\infty = \mathcal{K}_\infty \cup \{1, a\}$ in exactly the same way, and we have $\mathcal{K}_\infty \cong \bigoplus_{\mathbb{N}} \mathbb{Z}_2$.

Tame and wild quotients.

Definition 4.3. A set $\mathcal{A}$ is *tame* if and only if $\mathbb{Q}(\mathcal{A}) \cong \mathcal{T}_n$ for some $n \in \mathbb{N} \cup \{\infty\}$. Otherwise it is *wild*.

Not all quotients are tame.

Example. Let $G = \mathbb{Q}(*2_\#320)$, where $*2_\#320 = \{0, *2, *3, *2_\#\}$ and $*2_\# = \{*2\}$. We have

$$\mathbb{Q}(G) \cong \langle a, b, t \mid a^2 = 1, b^3 = b, t^2 = b^2, bt = b \rangle; \quad \mathcal{P} = \{a, b^2\}.$$

This quotient is called $\mathcal{R}_8$. It is very common; many octal games have quotient $\mathcal{R}_8$, including (for example) 0.75. In fact, it can be shown that $\mathcal{R}_8$ is the smallest quotient except for $\mathcal{T}_0$, $\mathcal{T}_1$, $\mathcal{T}_2$ (see [16]). The quotient map is given by (writing $z = b^2$, as before)

$$\Phi(*) = a, \quad \Phi(*2) = b, \quad \Phi(*3) = ab, \quad \Phi(*2_\#) = z, \quad \Phi(G) = at.$$

Notice that $\mathcal{R}_8$ is just $\mathcal{T}_2$ with two extra elements:

$$\mathbb{Q} = \{\overbrace{1, a, \underbrace{b, ab, z, az}_{\mathcal{K}}}^{\mathcal{T}_2}, t, at\}.$$

Now $\mathcal{K} \cong \mathbb{Z}_2^2$, and $\{1, a\}$ is a (separate) isomorphic copy of $\mathbb{Z}_2$. But $\{t, at\}$ is not a group, because $t^2 = z \in \mathcal{K}$.

The right picture of $\mathcal{R}_8$ is this: it is the union

$$\mathcal{K} \cup \{1, a\} \cup \{t, at\},$$

where $\mathcal{K}$ and $\{1, a\}$ are two disjoint groups, and $\{t, at\}$ are two extra elements that are "associated" with $\mathcal{K}$. We'll say more about this in the next lecture.

General structure.

Lemma 4.4. *Suppose that $\mathscr{A}$ is hereditarily closed, $\mathscr{A} \neq \varnothing$, and $\mathscr{A} \neq \{0\}$. Then necessarily $* \in \mathscr{A}$.*

Proof. $*$ is the only game whose only option is 0. □

Proposition 4.5. *Let $(\mathbb{Q}, \mathcal{P})$ be any nontrivial misère quotient. Then for all $x \in \mathbb{Q}$, there is some $y \in \mathbb{Q}$ with $xy \in \mathcal{P}$.*

Proof. Write $(\mathbb{Q}, \mathcal{P}) = \mathbb{Q}(\mathscr{A})$ and choose $G \in \mathscr{A}$ with $\Phi(G) = x$. First suppose $G = 0$. Then $x = 1$. By the assumption of nontriviality, we have $\mathscr{A} \neq \{0\}$, so by the previous lemma $* \in \mathscr{A}$. But $\Phi(*) \in \mathcal{P}$ and $1 \notin \mathcal{P}$, so we can take $y = \Phi(*)$.

Now assume $G \neq 0$, and consider $G + G$. If it is a $\mathscr{P}$-position, then we are done, with $y = x$. Otherwise, some option of $G + G$ must be a $\mathscr{P}$-position, say $G + G'$. So we can take $y = \Phi(G')$. □

Proposition 4.6. *For any G and any option G', $\Phi(G) \neq \Phi(G')$.*

Proof. The proof is left as an exercise for the reader. (Hint: use the previous proposition.) □

Proposition 4.7. *If $\mathscr{A}$ is nontrivial and $G \in \mathscr{A}$, then $G \not\equiv_{\mathscr{A}} G + *$.*

Proof. By Proposition 4.5, there is a game $H \in A$ such that $G+H$ is a $\mathscr{P}$-position. But then $G + H + *$ is an $\mathscr{N}$-position, so H distinguishes G from $G + *$. □

Corollary 4.8. *Every nontrivial misère quotient has even order.*

Proof. The proof is left as an exercise for the reader. (Hint: consider the mapping $x \mapsto ax$.) □

In fact, one can prove the following facts:

- $\mathcal{T}_1$ is the only quotient of order 2. (Immediate from Lemma 4.4.)
- There are no quotients of order 4. (Proved in [13].)
- $\mathcal{T}_2$ is the only quotient of order 6. (Also proved in [13].)
- $\mathcal{R}_8$ is the only quotient of order 8. (Much harder to prove; see [16].)

Lecture 5. Further topics

November 30, 2006 Scribes: Shiri Chechik and Menachem Rosenfeld

In this lecture, we will discuss four interesting problems, most of which have not yet been solved completely. We will also discuss the structure of finite commutative monoids.

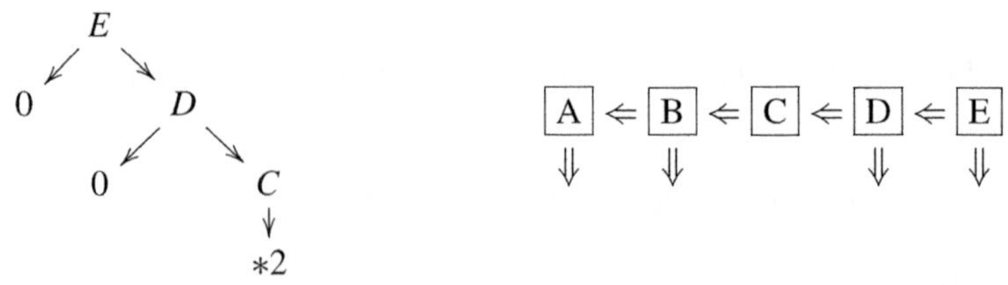

Figure 6. Two representations for the game $E = *(2_\#0)0$. Left: the game tree of E. Right: a visual representation of $\mathrm{cl}(E)$.

Four interesting problems.

1. Infinite quotients. We can think of infinite quotients as belonging to either one of two categories: those that are finitely generated, and those that are not. We have already seen one infinite quotient, $\mathcal{T}_\infty = \mathcal{Q}(0, *, *2, \ldots)$. It is not finitely generated. Every one of its finitely generated submonoids is finite, and it is built up from these finite quotients. It is therefore not an interesting quotient to study.

There also exist finitely generated infinite quotients. We can find an example of this by denoting

$$A = *,\ \ B = *2,\ \ C = \{B\} = *2_\#,\ \ D = *2_\#0 = \{C, 0\},\ \text{ and } E = \{D, 0\} = *(2_\#0)0.$$

Figure 6, left, shows the game tree of E.

Denoting $\mathscr{A} = \mathrm{cl}(E)$, a visual way to understand a game in $\mathscr{A}$ is suggested in Figure 6, right; for every game, there are several coins in every box, and a move consists of moving a coin along an arrow (either one step to the left, or from boxes other than C, outside the game board). The last player to move loses.

As it turns out, $|\mathbb{Q}(E)| = \infty$, but every game with a smaller tree has a finite quotient. So E is in some sense the *simplest* game that gives rise to an infinite quotient. To understand why the quotient is infinite, first note that every $X \in \mathscr{A}$ can be written as $X = iA + jB + kC + lD + mE$. In [13, §A.7], we compute the outcome of every such X. It turns out that when $k \geq 3$, the outcomes follow a simple rule: $o^-(X) = \mathscr{P} \iff i + l$ and $j + m$ are both even. However, when $k \leq 2$, the outcomes can be quite erratic. See Figure 7. Each table represents the outcomes for a particular choice of (i, j, k). Within each table, there is a dot at (row m, column l) if and only if $iA + jB + kC + lD + mE$ is a $\mathscr{P}$-position.

Inspecting this figure, we can see that the structure of the $\mathscr{P}$-positions is very complicated. For example, for $i = j = k = l = 0$, $X = mE$ is a $\mathscr{P}$-position $\iff$ $m \in \{1, 4, 7, 10, 12, 14, 16, \ldots\}$.

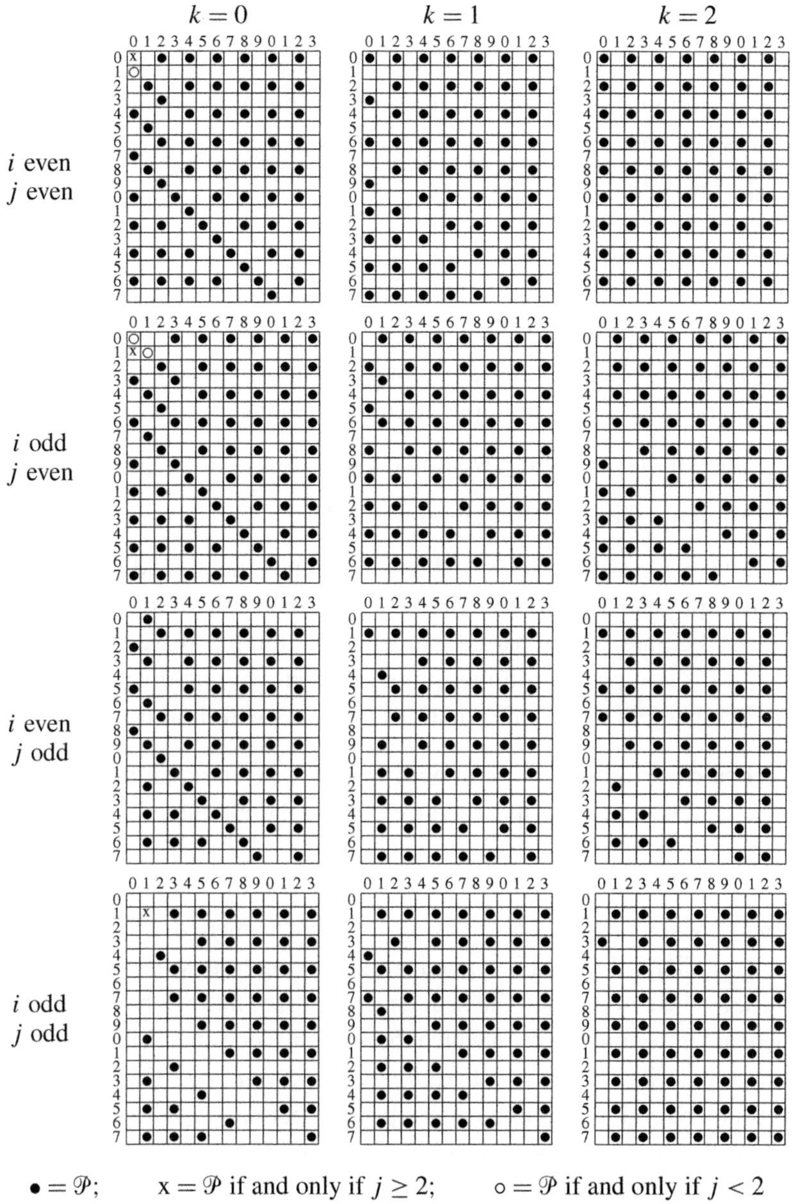

Figure 7. Schematic of the 𝒫-positions for cl$(*(2_{\#}0)0)$ with $k \leq 2$.

To see that the quotient is infinite, consider the case $i = j = k = 0$. For sufficiently large odd l, we have that $lD + mE$ is a 𝒫-position if and only if $m = l + 7$. This means that the lD's are pairwise distinguishable.

It was mentioned in a previous lecture that infinite quotients are still poorly understood. We still cannot solve the following problem.

Open Problem. Specify an algorithm to determine whether a quotient is infinite.

Of course, we'd really like to know much more about $\mathbb{Q}(\mathscr{A})$ than merely *whether* it's infinite. An old theorem about commutative semigroups guarantees that this is possible:

Theorem 5.1 (Rédei). *Every finitely generated commutative semigroup is finitely presented.*

We won't prove Rédei's theorem in this course; see [5; 14]. It makes the following question meaningful.

Open Problem. Specify an algorithm to compute the *presentation* of $\mathbb{Q}(\mathscr{A})$ (even if $\mathbb{Q}$ is infinite), assuming $\mathscr{A}$ is finitely generated.

In particular, the following would be a good start.

Open Problem. Give a presentation for $\mathscr{Q}(E)$.

Note: when we proved the periodicity theorem, at no point did we assume that the partial quotients are finite. Thus the periodicity theorem applies perfectly well to octal games whose partial quotients are infinite. If we could produce an algorithm for computing infinite quotients, then we could (in theory) use the periodicity theorem to provide solutions to games with infinite partial quotients.

2. Algebraic periodicity of octal games. Let Γ be an octal game. Then $\mathscr{Q}(\Gamma)$ is uniquely determined by its sequence of partial quotients,

$$\langle \mathbb{Q}_n(\Gamma) : n \in \mathbb{N} \rangle.$$

We can ask, when is it determined by only finitely many of these partial quotients?

The periodicity theorem is a good start in trying to answer this question — it happens, for instance, when the sequence stabilizes and we have periodicity.

There are intriguing cases in which the sequence does not stabilize but exhibits a strong regularity, which is called *algebraic periodicity*. This phenomenon is not yet understood well enough for a precise definition to be given. The term is derived from *arithmetic periodicity* in normal play, which means that the sequence is periodic but on each period we add a "saltus". For example, if the period is 5 and the saltus is 4, a possible sequence is

$$0, 4, 5, 3, 2, \quad 4, 8, 9, 7, 6, \quad 8, 12, 13, 11, 10, \ldots$$

Theorem 5.2 (see [1]). *No finite octal game* (*that is, one with finitely many nonzero digits*) *can be arithmetic periodic* (*with nontrivial saltus*) *in normal play.*

(Remark: NIM is a trivial example of a nonfinite octal game which is arithmetic periodic.)

order (n)	2	4	6	8	10	12	14	16	...
# of quotients	1	0	1	1	1	6	9	50	...

Table 1. Number of different quotients for every order.

However, algebraic periodicity is manifested in finite octal games with misère play. Several examples are presented in [13].

Several two-digit octal games for which the normal solution is known, have not yet been solved for misère play. Of these, 0.54 is the only one for which the solution seems to be in reach — because it appears to be algebraic periodic, which suggests a solution for it.

Open Problem. Prove the solution for 0.54 presented in [13].

Open Problem. Formulate a suitable general definition of "algebraic periodicity" and prove a theorem that states: if Γ is algebraic periodic for sufficiently long, then it continues this period, and we can compute $\mathcal{Q}(\Gamma)$.

Presumably, this would immediately provide a solution for 0.54, and probably six or eight three-digit octals as well.

3. Generalizations of the mex rule. Suppose we have a quotient map $\Phi : \mathscr{A} \to \mathcal{Q}$. Let G be a game all of whose options are in $\mathscr{A}$. Can we determine, based only on $\Phi''G$, whether $\mathcal{Q}(\mathscr{A} \cup \{G\}) \cong \mathcal{Q}(\mathscr{A})$? If they are isomorphic, can we determine $\Phi(G)$? (Recall that $\Phi''G$ is defined as $\{\Phi(G') : G' \text{ is an option of } G\}$.)

By asking these questions, we are essentially looking for a way to generalize the mex rule, which solves them for normal play.

The answer to both question is: yes! However, more information is needed than what it contained in $\mathcal{Q}$.

Recall that in the previous lecture, we proved a lemma that answers this question in case there is some $H \in \mathscr{A}$ such that $\Phi''G = \Phi''H$. It turns out that we can get a much stronger result. However, this result is beyond the scope of this lecture; see [13].

4. Classification. How many misère quotients are there of order n (up to isomorphism)? Table 1 displays some of what is known so far. The results for $n = 14$ and 16 are tentative.

A related question: can we identify other interesting classification results? Here is one such result.

It is possible to define the "tame extension" $\mathcal{T}(\mathcal{Q}, \mathcal{P})$ of an arbitrary quotient $(\mathcal{Q}, \mathcal{P})$. See [16] for a precise definition. It turns out that

$$(\mathcal{Q}, \mathcal{P}) \subsetneq \mathcal{T}(\mathcal{Q}, \mathcal{P}),$$

but $\mathcal{T}(\mathcal{Q}, \mathcal{P})$ adds no new $\mathscr{P}$-position types. Furthermore,

$$\mathcal{T}_{n+1} = \mathcal{T}(\mathcal{T}_n).$$

We therefore have two families of quotients,

$$\mathcal{T}_2, \mathcal{T}_3, \ldots, \mathcal{T}_\infty$$

and

$$\mathcal{R}_8, \mathcal{T}(\mathcal{R}_8), \mathcal{T}(\mathcal{T}(\mathcal{R}_8)), \ldots, \mathcal{T}^\infty(\mathcal{R}_8),$$

all of which have $|\mathcal{P}| = 2$. The following result is proved in [16].

Theorem 5.3. *Every quotient with $|\mathcal{P}| = 2$ is isomorphic to a quotient in one of these two families.*

So we have

$0, \mathcal{T}(0), \mathcal{T}(\mathcal{T}(0)), \ldots$	(normal play)
$\mathcal{T}_2, \mathcal{T}(\mathcal{T}_2), \ldots, \mathcal{T}^\infty(\mathcal{T}_2)$	(tame misère play)
$\mathcal{R}_8, \mathcal{T}(\mathcal{R}_8), \mathcal{T}^2(\mathcal{R}_8), \ldots, \mathcal{T}^\infty(\mathcal{R}_8)$	("almost tame" misère play)

Can we say anything else along these lines?

The structure of finite commutative monoids. Let $\mathcal{Q}$ be any finite commutative monoid, and let $x, y \in \mathcal{Q}$.

Definition 5.4. If $xz = y$ for some $z \in \mathcal{Q}$, x *divides* y. In this case, we write $x \mid y$.

Definition 5.5. If $x \mid y$ and $y \mid x$, x and y are *mutually divisible* (md).

Example. $$\mathcal{T}_2 = \langle a, b \mid a^2 = 1, b^3 = b \rangle = \{\underbrace{1, a}_{\text{md}}, \underbrace{b, ab, z, az}_{\text{md}}\}.$$

Exercise. Show that md is an equivalence relation.

Definition 5.6. The *mutual divisibility classes* of $\mathcal{Q}$ are the equivalence classes of $\mathcal{Q}$ under the relation md.

Example. The md classes of $\mathcal{R}_8 = \{1, a, b, ab, z, az, t, at\}$ are

$$\{1, a\}, \quad \{b, ab, z, az\} \quad \text{and} \quad \{t, at\}.$$

Definition 5.7. An element $x \in \mathcal{Q}$ is an *idempotent* if $x^2 = x$.

Example. In $\mathcal{T}_2$ (and also $\mathcal{R}_8$), 1 and z are the only idempotents.

Exercise. The md class of an idempotent x is a group with x for an identity.

Exercise. If S is a maximal subgroup of $\mathcal{Q}$ (that is, a group which is not contained in any larger subgroup of $\mathcal{Q}$) then it is the md class of its idempotent.

Let $z_1, z_2, \ldots, z_k$ be the idempotents of $\mathcal{Q}$ (since $\mathcal{Q}$ is finite, we can enumerate them all). We write $z = z_1 z_2 \ldots z_k$. We then have $z^2 = z_1^2 z_2^2 \ldots z_k^2 = z_1 z_2 \ldots z_k = z$ and $z z_i = z$.

Definition 5.8. The *kernel* of $\mathcal{Q}$ is the md class of z, and is denoted $\mathcal{K}$.

We will soon prove the following theorem.

Theorem 5.9. *The map $x \mapsto zx$ is a surjective homomorphism from $\mathcal{Q}$ onto $\mathcal{K}$.*

The kernel $\mathcal{K}$ can be characterized in two ways:

(1) It is the unique group such that there is a surjective homomorphism $f : \mathcal{Q} \to \mathcal{K}$ with the following property: if $g : \mathcal{Q} \to \mathcal{D}$ is a homomorphism onto a group $\mathcal{D}$, then there exists an $h : \mathcal{K} \to \mathcal{D}$ which makes the following diagram commute:

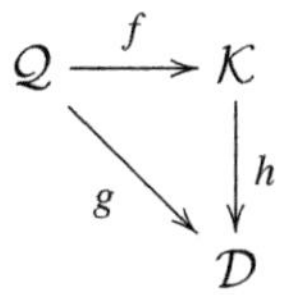

In other words, any map from $\mathcal{Q}$ onto a group $\mathcal{D}$ factors through f.

(2) $\mathcal{K}$ is the *group of fractions* of $\mathcal{Q}$, that is, it is the group obtained by adjoining formal inverses to $\mathcal{Q}$.

Lemma 5.10. *If $y \in \mathcal{Q}$ then for some $r > 0$, y^r is an idempotent.*

(Note: This does not hold for infinite monoids!)

Proof. Consider the sequence $y, y^2, y^3, y^4, \ldots$ Since $\mathcal{Q}$ is finite, there must be some $n > 0$ and some $k > 0$ such that $y^n = y^{n+k}$. We then have for every $t \geq 0$, $y^{n+tk} = y^n$. Let r be the unique integer such that $n \leq r < n+k$ and $r \equiv 0 \pmod{k}$. Then,

$$y^{2r} = y^{r+tk} = y^{n+tk} y^{r-n} = y^n y^{r-n} = y^r.$$

So y^r is an idempotent. □

Note that this idempotent is uniquely determined for any given y. Therefore, for any y, there is a unique idempotent x such that $y^n = x$ for some $n > 0$. This motivates the following definition:

Definition 5.11. For any idempotent $x \in \mathcal{Q}$, the *archimedean component* of x is $\{y \in \mathcal{Q} : \exists n (y^n = x)\}$.

What we have actually shown is that every $y \in \mathcal{Q}$ is a member of a unique archimedean component. Therefore, $\mathcal{Q}$ is partitioned into several archimedean components. For example, $\mathcal{R}_8$ is partitioned into $\{1, a\}$ and $\{b, ab, z, az, t, at\}$.

We complete the picture by defining a natural partial order on idempotents.

Definition 5.12. For idempotents $x, y \in \mathcal{Q}$, $x \leq y \Longleftrightarrow xy = x$.

Example. For any idempotent x, $z \leq x \leq 1$.

Theorem 5.13. *The idempotents of $\mathcal{Q}$ form a lattice with respect to the relation $\leq$.*

Exercise. Prove this theorem. *Hint*: Define

$$x \wedge y = xy$$

and

$$x \vee y = \prod \{w \in \mathcal{Q} : w \text{ is an idempotent and } w \geq x, y\}.$$

Examples. In these examples, $[a^b]$ denotes an archimedean component with a elements contained in the md class of the idempotent, and b additional elements.

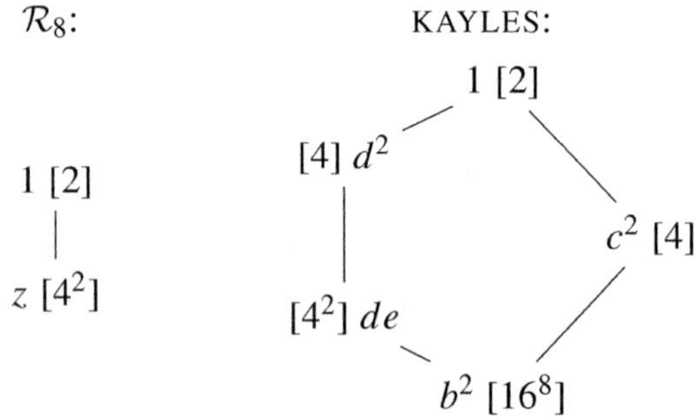

Theorem 5.14. *The map $x \mapsto zx$ is a surjective homomorphism $\mathcal{Q} \to \mathcal{K}$.*

Proof. We must show that $zx \in \mathcal{K}$ for all x; it follows easily that $x \mapsto zx$ is a surjective homomorphism.

Clearly $z \mid zx$, so we must show that $zx \mid z$. Let us take an $n > 0$ such that x^n is an idempotent. Then $zx^n = z$ by the definition of z, so $(zx)x^{n-1} = z$. □

Corollary 5.15. *If $x \in \mathcal{K}$, then $\forall y \in \mathcal{Q}$, $xy \in \mathcal{K}$ (because $x = zx$, so $xy = zxy$).*

Corollary 5.16. *$\mathcal{K} \cap \mathcal{P} \neq \varnothing$ (because by a previous lemma, $\exists x \in \mathcal{Q}$ such that $xz \in \mathcal{P}$, but $xz \in \mathcal{K}$ so $xz \in \mathcal{K} \cap \mathcal{P}$).*

Definition 5.17. $(\mathcal{Q}, \mathcal{P})$ is *normal* if $\mathcal{K} \cap \mathcal{P} = \{z\}$.

(Remark: the smallest known example of an abnormal quotient is of size 420.)

Definition 5.18. $(\mathcal{Q}, \mathcal{P})$ is *regular* if $|\mathcal{K} \cap \mathcal{P}| = 1$.

(Remark: the smallest known example of an irregular quotient is of size over 3000.)

Definition 5.19. A quotient map $\Phi : \mathcal{A} \to \mathcal{Q}$ is *faithful* if, for all $G, H \in \mathcal{A}$,

$$\Phi(G) = \Phi(H) \Rightarrow \mathcal{G}(G) = \mathcal{G}(H).$$

Open Problem. Is every quotient map faithful?

Theorem 5.20. *If* $(\mathcal{Q}, \mathcal{P})$ *is normal and* Φ *is faithful, then for all* $G, H \in \mathscr{A}$,

$$z\Phi(G) = z\Phi(H) \Longleftrightarrow \mathscr{G}(G) = \mathscr{G}(H).$$

There is therefore a one-to-one correspondence between elements of $\mathcal{K}$ and normal-play NIM values of games in $\mathscr{A}$. Furthermore, we can compute the mex function in the kernel. This gives us the following strategy for playing a misère octal game Γ: play as if you were playing normal Γ, unless your move would take you outside of $\mathcal{K}$. Then pay attention to the fine structure of the misère quotient.

Example. The octal game 0.414 has not yet been solved for normal play. Nevertheless, we can prove that $\Phi(H_n) \in \mathcal{K}$ for $n > 18$, and we can prove that its quotient is one of

$$\mathcal{Q}_{18}, \mathcal{T}(\mathcal{Q}_{18}), \mathcal{T}(\mathcal{T}(\mathcal{Q}_{18})), \ldots,$$

though we do not know which. The strategy for misère 0.414 is: play as if you were playing normal 0.414, unless your move would leave only heaps of size ≤ 18. Then pay attention to the fine structure of the misère quotient.

One last open problem:

Open Problem. Let $\mathcal{S}$ be an arbitrary maximal subgroup of $\mathcal{Q}$. Must $\mathcal{S} \cap \mathcal{P}$ be nonempty?

Further reading

Misère quotients for impartial games [13], by Plambeck and Siegel, includes most of the material presented in these notes, and a great deal else as well. It is the best resource both for additional examples of misère quotients and for a deeper look at the structure theory. Plambeck's original paper introducing misère quotients [10] includes a proof of the periodicity theorem that is somewhat different from the one presented here. His survey paper [11] provides a nice informal summary of much that is known about misère games. The forthcoming paper [16] dives much more deeply into the structure of misère quotients.

The most current source of information is the *Misére games* website [12], which includes Plambeck's misére games blog. See also [9].

The *canonical* theory is virtually useless in practice, but nonetheless absolutely fascinating. It is (essentially) the "quotient" obtained by taking $\mathscr{A}$ to be the universe of all misère games. See [3] for many results along these lines.

Finally, perhaps the best way to get acquainted with misère quotients is to download a copy of *MiséreSolver* [15] and start experimenting. It can easily reproduce all the examples in this paper, and of course many more as well.

Acknowledgements

First and foremost, I wish to thank the scribes for the course: Gideon Amir, Shiri Chechik, Omer Kadmiel, Amir Kantor, Dan Kushnir, Shai Lubliner, Ohad Manor, Leah Nutman, Menachem Rosenfeld, and Rivka Taub. I also wish to thank Professor Aviezri Fraenkel for inviting me to the Weizmann Institute and suggesting this course, and thereby making these notes possible. Finally, I wish to thank Thane Plambeck, for recognizing the importance of misère quotients and inventing this beautiful and fascinating theory.

References

[1] E. R. Berlekamp, J. H. Conway, and R. K. Guy. *Winning ways for your mathematical plays*. A. K. Peters, Ltd., Natick, MA, 2nd edition, 2001.

[2] C. L. Bouton. Nim, a game with a complete mathematical theory. *Ann. of Math.*, 3(2):35–39, 1902.

[3] J. H. Conway. *On numbers and games*. A. K. Peters, Ltd., Natick, MA, 2nd edition, 2001.

[4] A. Flammenkamp. Sprague–Grundy values of octal games. http://wwwhomes.uni-bielefeld.de /achim/octal.html.

[5] P. A. Grillet. *Commutative semigroups*. Number 2 in Advances in Mathematics. Springer, 2001.

[6] P. M. Grundy. Mathematics and games. *Eureka*, 2:6–8, 1939.

[7] P. M. Grundy and C. A. B. Smith. Disjunctive games with the last player losing. *Proc. Cambridge Philos. Soc.*, 52:527–533, 1956.

[8] R. K. Guy and C. A. B. Smith. The G-values of various games. *Proc. Cambridge Philos. Soc.*, 52:514–526, 1956.

[9] T. E. Plambeck. Misère games (web pages devoted to problems, computer software, and theoretical results).

[10] T. E. Plambeck. Taming the wild in impartial combinatorial games. *Integers*, 5(#G5), 2005.

[11] T. E. Plambeck. Advances in losing. In M. H. Albert and R. J. Nowakowski, editors, *Games of No Chance 3*, number 56 in MSRI Publications. Cambridge University Press, Cambridge, 2009.

[12] T. E. Plambeck and A. N. Siegel. Misère games on the web. http://www.miseregames.org/.

[13] T. E. Plambeck and A. N. Siegel. Misère quotients for impartial games. *J. Combin. Thy. Ser. A*, 115(5), 2008.

[14] L. Rédei. *The theory of finitely generated commutative semigroups*. Pergamon, 1965.

[15] A. N. Siegel. *MisereSolver*. http://www.cgsuite.org/.

[16] A. N. Siegel. The structure and classification of misère quotients. Forthcoming.

[17] R. P. Sprague. Über mathematische Kampfspiele. *Tôhoku Math. J.*, 41:438–444, 1935–36.

[18] R. P. Sprague. Über zwei Abarten von Nim. *Tôhoku Math. J.*, 43:351–359, 1937.

aaron.n.siegel@gmail.com *San Francisco, CA, United States*

Games of No Chance 5
MSRI Publications
Volume **70**, 2017

An historical tour of binary and tours

DAVID SINGMASTER

Recreational mathematics has an old and honourable history. We illustrate that history and perhaps a bit of the utility of recreational mathematics by discussing a number of recreations involving binary representations and paths on graphs.†

Leibniz's binary arithmetics

In the seventeenth century, Francis Bacon used binary 5-tuples as a code, but binary arithmetic as we currently understand it — doing actual arithmetic with binary numbers rather than just using binary representations — starts with Leibniz about 1679, though he didn't publicize it until the late 1600s. He heard about the Fu-Hsi ordering of the I-Ching hexagrams from Jesuit missionaries in China in 1701 and wrote a good deal about it thereafter (see Figures 1 and 2 and p. 219).

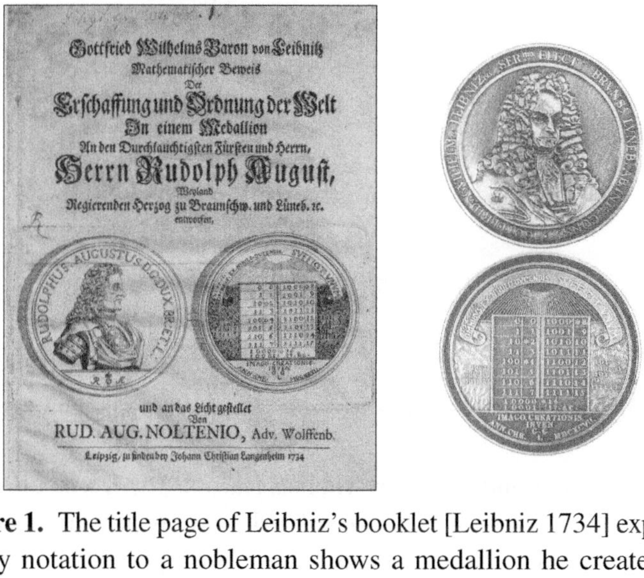

Figure 1. The title page of Leibniz's booklet [Leibniz 1734] explaining binary notation to a nobleman shows a medallion he created, later borrowed by the Stadtsparkasse of Hanover to honor Leibniz himself.

† This is an elaboration of notes for a talk at the First European Congress of Mathematics, Paris, July 1992, and given several times since. This material has been edited for *Games of No Chance* 5 by Urban Larsson (urban031@gmail.com). In a few occurences when uncertainty about historical

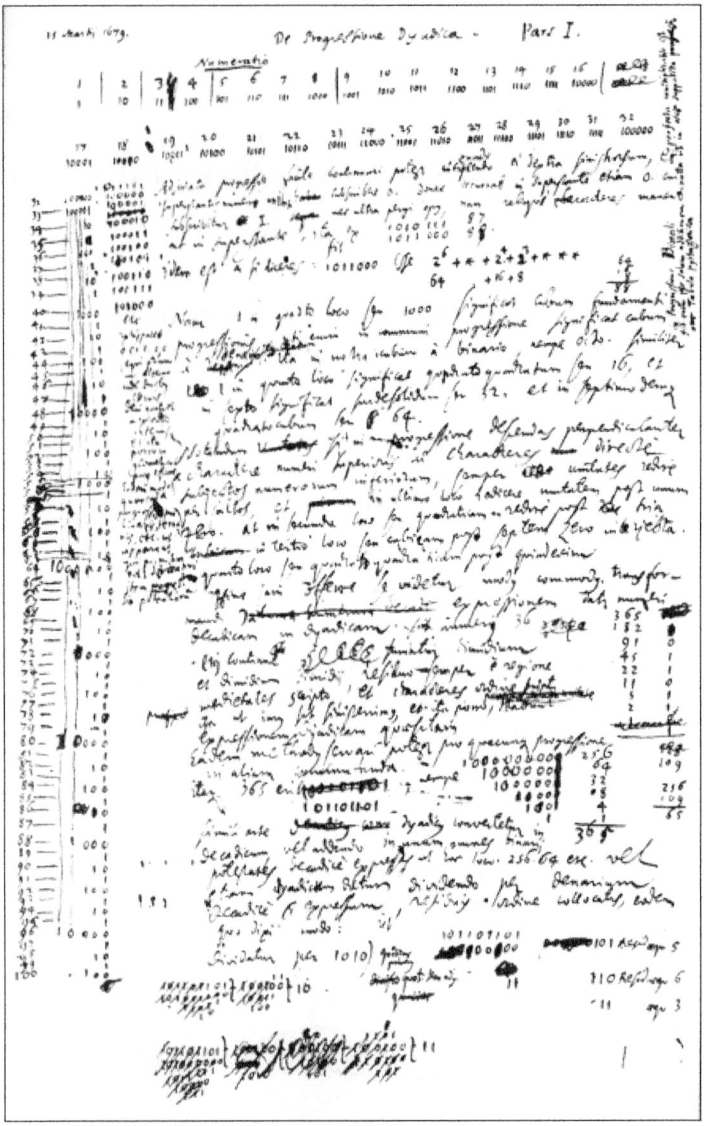

Figure 2. Leibniz's first writing on binary arithmetic, dated 11 March 1679.

However, Leibniz was anticipated by Thomas Harriot, 1604, who did not publish, and by John Napier, whose *Rabdologiæ* of 1617 gave binary arithmetic as far as computing square roots, but this seems to have been ignored.

references remained in the author's Word file, question marks have been erased in this version, following also suggestions from referees. The editor advanced this decision for readability, and would be greatful for any comments in the future to complement or assist the numerous findings in this work.

But binary ideas go much further back. Some simple counting systems are more or less base 2 and there are many instances of duality in nature — hands, sexes, etc. But we are interested in material that is somewhat more mathematical.

Binary multiplication

The earliest implicit use of binary representations occurs in ancient Egyptian mathematics. Figure 3 is Problem 30 of the Rhind Mathematical Papyrus, ca. 1700 B.C.E., computing $(\frac{2}{3}+\frac{1}{10})\times 13$. The problem is to solve $(\frac{2}{3}+\frac{1}{10})x=10$, which is being done by false position, using $x = 13$ as a trial.

Because of their complicated notation for numbers, especially fractions, they multiplied by repeatedly doubling, then adding the appropriate terms. For instance, to multiply a number by 13, they computed successively the double, the quadruple and the octuple of the number, then added the number to its quadruple and its octuple. This is also known as Russian peasant multiplication and was in use in Russia until the twentieth century. It was sufficiently common that duplication and mediation (halving) were reckoned among the basic rules of arithmetic in the middle ages and could be found in arithmetic books until the seventeenth century. It uses the fact that every integer is a sum of distinct powers of two, which is the same as saying that every integer has a binary representation.

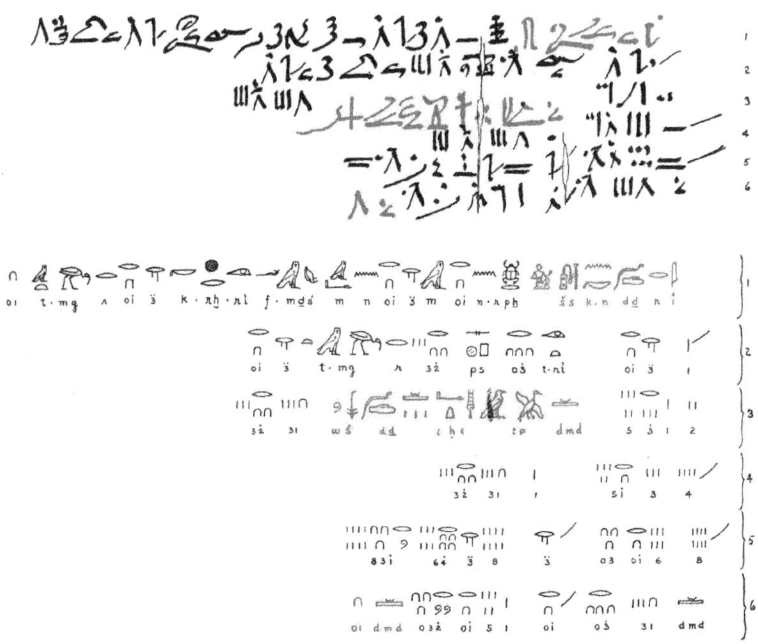

Figure 3. Problem 30 from the Rhind Mathematical Papyrus of ca. 1700 B.C.E.

Binary weights

A somewhat more explicit use of binary representation occurs in Arabic sources about the eleventh century. This is the use of weights 1, 2, 4, . . . to make all integer weights on a scale. (Actually, if we allow putting weights into the other side, one can use weights 1, 3, 9, . . .) This is often known as Bachet's weights problem, but it already appears in Fibonacci and several other European books before Bachet.

Chinese rings and Gray code

At about the same time, the puzzle known as the Chinese Rings (Figure 4) appears in China, though tradition attributes it back to the semi-legendary Hung Ming of about 200 C.E.

One Oriental name, *Lau kák ch'a*, translates as "delay-guest-instrument".

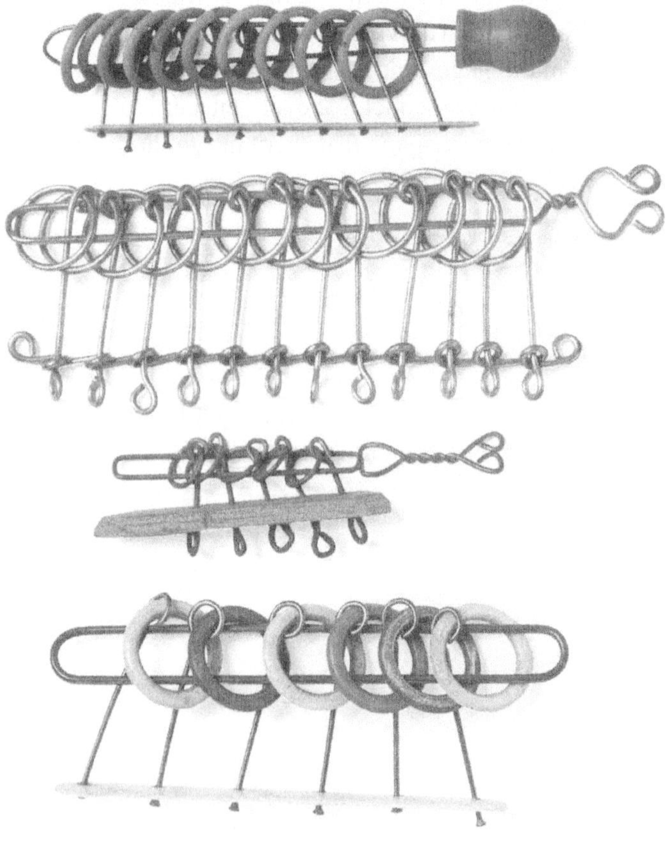

Figure 4. Four examples of Chinese Rings. From p. 107 of [Slocum and Botermans 1986].

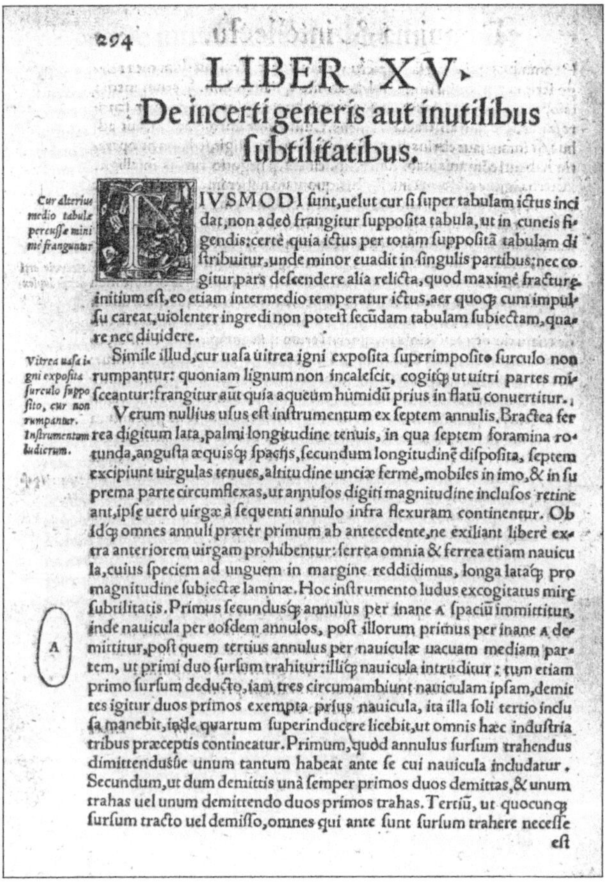

294

LIBER XV.

De incerti generis aut inutilibus ſubtilitatibus.

Cur alterius medio tabulæ percuſſæ minime franguntur

EIVSMODI ſunt, uelut cur ſi ſuper tabulam ictus incidat, non adeò frangitur ſuppoſita tabula, ut in cuneis figendis: certè quia ictus per totam ſuppoſitã tabulam diſtribuitur, unde minor euadit in ſingulis partibus; nec cogitur pars deſcendere alia relicta, quod maximè fracturæ initium eſt, eo etiam intermedio temperatur ictus, aer quoq; cum impulſu careat, uiolenter ingredi non poteſt ſecũdam tabulam ſubiectam, quare nec diuidere.

Vitrea uaſa igni expoſita ſurculo ſuppoſito, cur non rumpantur.

Simile illud, cur uaſa uitrea igni expoſita ſuperimpoſito ſurculo non rumpantur: quoniam lignum non incaleſcit, cogitq; ut uitri partes miſceantur: frangitur aũt quia aqueum humidũ prius in flatũ conuertitur.

Inſtrumentum ludicrum.

Verum nullius uſus eſt inſtrumentum ex ſeptem annulis. Bractea ferrea digitum lata, palmi longitudine tenuis, in qua ſeptem foramina rotunda, anguſta æquisq; ſpacijs, ſecundum longitudinẽ diſpoſita, ſeptem excipiunt uirgulas tenues, altitudine unciæ fermè, mobiles in imo, & in ſuprema parte circumflexas, ut annulos digiti magnitudine incluſos retineant, ipſę uerò uirgæ à ſequenti annulo infra flexuram continentur. Ob idq; omnes annuli præter primum ab antecedente, ne exiliant liberè extra anteriorem uirgam prohibentur: ferrea omnia & ferrea etiam nauicula, cuius ſpeciem ad unguem in margine reddidimus, longa lataq; pro magnitudine ſubiectæ laminæ. Hoc inſtrumento ludus excogitatus miræ ſubtilitatis. Primus ſecundusq; annulus per inane A ſpaciũ immittitur, inde nauicula per eoſdem annulos, poſt illorum primus per inane A demittitur, poſt quem tertius annulus per nauiculæ uacuam mediam partem, ut primi duo ſurſum trahitur: illicq; nauicula intruditur; tum etiam primo ſurſum deducto, iam tres circumambiunt nauiculam ipſam, demittes igitur duos primos exempta prius nauicula, ita illa ſoli tertio incluſa manebit, inde quartum ſuperinducere licebit, ut omnis hæc induſtria tribus præceptis contineatur. Primum, quòd annulus ſurſum trahendus dimittendusue unum tantum habeat ante ſe cui nauicula includatur. Secundum, ut dum demittis unà ſemper primos duos demittas, & unum trahas uel unum demittendo duos primos trahas. Tertiũ, ut quocunq; ſurſum tracto uel demiſſo, omnes qui ante ſunt ſurſum trahere neceſſe eſt

Figure 5. Early description of the Chinese rings, from [Cardan 1550].

It is first known in Europe in 1550 when Cardan describes and illustrates it in his *De subtilitatis*; see Figure 5. It has recently been recognised that Luca Pacioli, in his unpublished manuscript *De viribus quantitatis* of about 1500, describes the puzzle, with no diagram. Wallis explains it quite clearly in his *De algebra tractatus* of 1693 (Figure 6). An old English phrase for it is "tiring irons" or "tarrying irons", and these words are recorded in the Oxford English Dictionary as far back as 1601. Descriptive or picturesque names in various languages include *Chinese rings*, *Chainese rings*, *Cardan's rings*, *Ryou-kaik-tjyo*, *Lau kák ch'a*, *Kau tsz' lin wain* ('Nine-connected-rings'), *Tiring* or *Tarrying irons*, *Baguenaudier*, *Meleda*, *Zauberkette*, *Magische Ringspiel*, *Nürnberger Tand*, *Grillenspiel*, *Zankeisen*, *Nodi d'anelli*.

It is difficult to describe the puzzle and how to do it, and even finding good images is hard. Suffice it to say that it has some number, n, of rings attached to

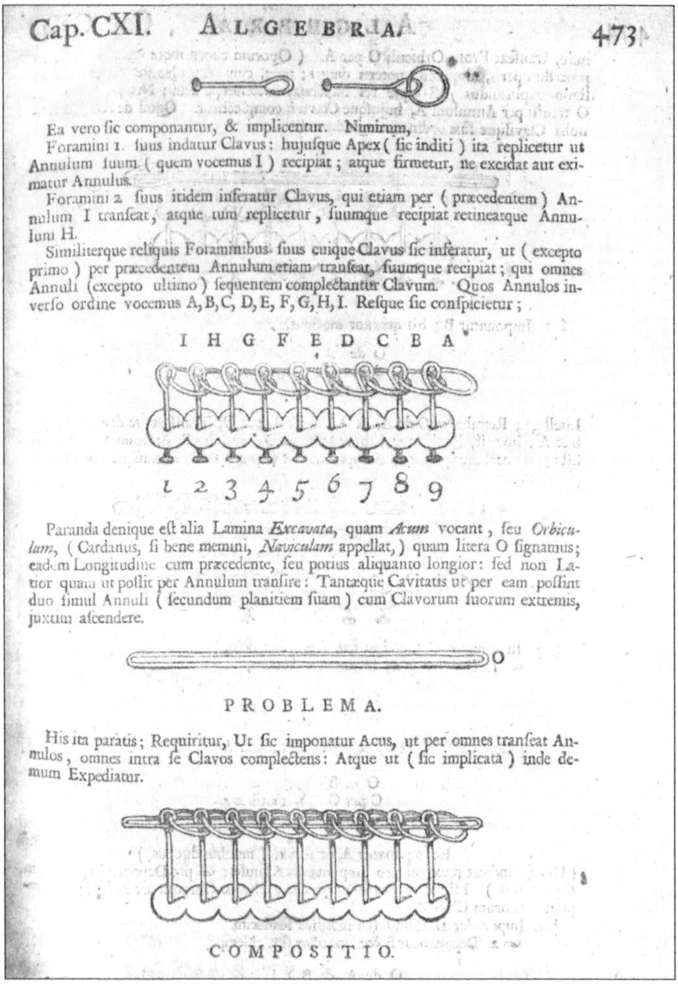
Cap. CXI. ALGEBRA. 473

Ea vero ſic componantur, & implicentur. Nimirum,

Foramini 1. ſuus indatur Clavus: hujuſque Apex (ſic inditi) ita replicetur ut Annulum ſuum (quem vocemus I) recipiat; atque firmetur, ne excidat aut eximatur Annulus.

Foramini 2. ſuus itidem inferatur Clavus, qui etiam per (præcedentem) Annulum I tranſeat, atque tum replicetur, ſuumque recipiat retineatque Annulum H.

Similiterque reliquis Foraminibus ſuus cuique Clavus ſic inſeratur, ut (excepto primo) per præcedentem Annulum etiam tranſeat, ſuumque recipiat; qui omnes Annuli (excepto ultimo) ſequentem complectantur Clavum. Quos Annulos inverſo ordine vocemus A, B, C, D, E, F, G, H, I. Reſque ſic conſpicietur;

I H G F E D C B A

1 2 3 4 5 6 7 8 9

Paranda denique eſt alia Lamina *Excavata*, quam *Acum* vocant, ſeu *Orbiculum*, (Cardanus, ſi bene memini, *Naviculam* appellat,) quam litera O ſignamus; eadem Longitudine cum præcedente, ſeu potius aliquanto longior: ſed non Latior quam ut poſſit per Annulum tranſire: Tantæque Cavitatis ut per eam poſſint duo ſimul Annuli (ſecundum planitiem ſuam) cum Clavorum ſuorum extremis, juxtim aſcendere.

O

PROBLEMA.

His ita paratis; Requiritur, Ut ſic imponatur Acus, ut per omnes tranſeat Annulos, omnes intra ſe Clavos complectens: Atque ut (ſic implicata) inde demum Expediatur.

COMPOSITIO.

Figure 6. First illustrated discussion of Chinese rings, from [Wallis 1685].

a bar by wires and systematically looped over a second bar in such a way that one can take off or put on only one ring at a time, which is either the end ring or the next to the last ring on the bar. Figure 7 shows the solution for four rings.

If we represent a state of the puzzle as a sequence of ons and offs, or better, 1s and 0s, then each position is a binary number and the movement changes such a binary number to another one which differs in just one place, which is either the last place or the place next to the last 1. Figure 8 (left) shows this for five rings, from Afriat, p. 31, with the binary patterns written in the image by me. Looking at the sequence of moves, Figure 8 (right) as it appears, one sees that starting from the position with all rings off is more interesting and the pattern

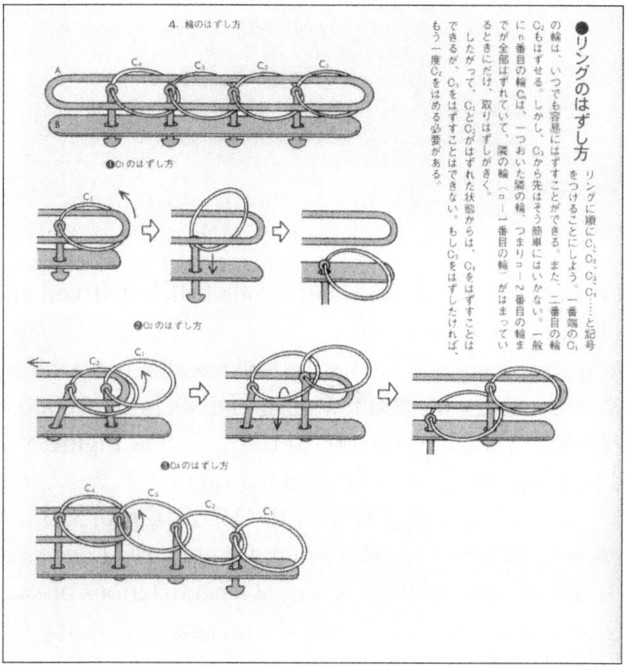

Figure 7. A 4-ring solution, from [Takagi 1982], p. 205.

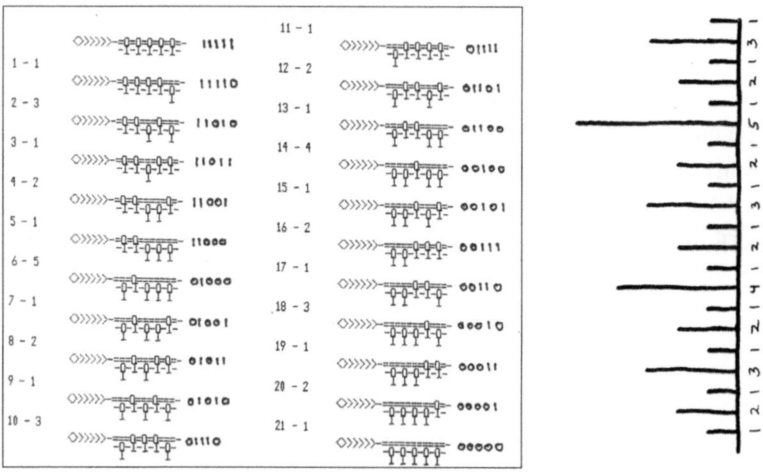

Figure 8. A 5-ring solution, from [Afriat 1982], with the corresponding Gray Code (right).

of moves is: 1, 2, 1, 3, 1, 2, 1, 4, 1, 2, 1, 3, 1, 2, 1, 5, . . . (Flip the picture, view Figure 8 from the side.)

You may recognize this pattern, especially if you are English or American and hence use rulers marked in halves, quarters, eighths, ... Figure 9 is a picture of

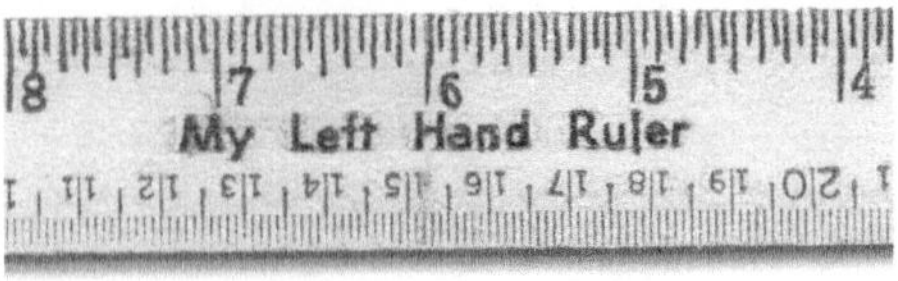

Figure 9. A genuine left-handed ruler.

a Left-Hand Ruler, obtainable from left-handed shops! If you are metric, then you may not be so familiar with this.

Consequently the binary number for the k-th position is not the binary representation of k. With five rings, the first few binary numbers are 00000, 00001, 00011, 00010, 00110, 00111, 00101, 00100, 01100, ... See Figure 8 (right). (Note that it would be easier to have numbered the rings 0, 1, 2, ... in the preceding discussion, so the sequence of moves is: 0, 1, 0, 2, 0, 1, 0, 3, 0, 1, 0, ...)

This pattern was rediscovered by Frank Gray in 1947 and patented by Bell Labs in 1953! See below for more history. Generalizations of such patterns are now called Gray codes, but we will call the present pattern the Gray Code.

Proposition 1 below is easy to obtain, but not as well known as it should be. (I first found it in 1970, but not clearly formulated.)

Let $n = \sum b_{n,i}2^i$ be the binary expansion of the integer n. Let $C = C_n$ be the number of consecutive 1s at the end of this binary expansion. We can also say that C_n is the largest i such that the i-th bit changes when we go from n to $n+1$ or that C_n is the power of 2 which exactly divides $n+1$, i.e. it is the number of 0s at the end of $n+1$. $C_n + 1$ gives the number of the ring moved at step $n+1$, if we start at one. Let G_n be the number of the n-th Gray Code value and let $\sum g_{n,i}2^i$ be its binary expansion. Then G_n has its C_n-th bit changed to produce G_{n+1}. Thus $|G_{n+1} - G_n| = 2^C$. By careful counting, one can show that

$$g_{n,i} \equiv b_{n,i} - b_{n,i+1} \equiv b_{n,i} + b_{n,i+1} \equiv b_{n',i+1}, \tag{1}$$

where $n' = n + 2^i$.

In other words, we can find the binary form of G_n as follows. Write n in binary as $\sum b_{n,i}2^i$. Shift it one place to the right, throwing away the rightmost bit; this produces the binary expression of $\lfloor n/2 \rfloor$ (with $\lfloor\ \rfloor$ = floor function). Then add the binary expressions for n and $\lfloor n/2 \rfloor$, but without carrying, which we denote by $\oplus$. This establishes the following.

Proposition 1. *Let $B(k)$ be the binary representation of k and let $G(k)$ be the k-th binary word in the Gray Code. Then $G(k) = B(k) \oplus B(\lfloor k/2 \rfloor)$. That is, we shift the binary representation of k to the right, losing the end digit, and do a Boolean addition (also known as addition mod 2, exclusive or, XOR) with $B(k)$.*

n	Binary	C	$C+1$	Gray	G
0	00000	0	1	00000	0
1	00001	1	2	00001	1
2	00010	0	1	00011	3
3	00011	2	3	00010	2
4	00100	0	1	00110	6
5	00101	1	2	00111	7
6	00110	0	1	00101	5
7	00111	3	4	00100	4
8	01000	0	1	01100	12
9	01001	1	2	01101	13
10	01010	0	1	01111	15
11	01011	2	3	01110	14
12	01100	0	1	01010	10
13	01101	1	2	01011	11
14	01110	0	1	01001	9
15	01111	4	5	01000	8
16	10000	0	1	11000	24
17	10001	1	2	11001	25
18	10010	0	1	11011	27
19	10011	2	3	11010	26
20	10100	0	1	11110	30
21	10101	1	2	11111	31

Figure 10. The relation between binary and Gray coding.

For example:

- For $k = 7$, we have $B(7) = 00111$, $B(\lfloor 7/2 \rfloor) = B(3) = 00011$, and $G(7) = 00111 \oplus 00011 = 00100 = 4$.
- For $k = 15$, we have $B(15) = 01111$, $B(\lfloor 15/2 \rfloor) = B(7) = 00111$, and $G(15) = 01111 \oplus 00111 = 01000 = 8$.

For the inverse process, we have

$$b_{n,j} = \sum_{i \geq j} g_{n,i}.$$

Otherwise stated, we have to sum all the shifts of G. That is, the binary value corresponding to a Gray code G is given by $G \oplus \lfloor G/2 \rfloor \oplus \lfloor G/4 \rfloor \oplus \lfloor G/8 \rfloor \oplus \cdots$

E.g., if $G = 15$, we have that the binary value of 15 is 1111, so we compute $1111 \oplus 0111 \oplus 0011 \oplus 0001 = 1010 = 10$, i.e. $G(10) = 15$.

Note that $n = 101010\ldots$ gives $G(n) = 111111\ldots$, which is the desired endpoint of the Chinese Rings process and the Rings are solved in n steps.

Suppose we have r rings. By considering the cases when r is odd and when it is even, we can determine n. In both cases, we can write $n = \lfloor \frac{2}{3} 2^r \rfloor$.

This and other basic results were (first) developed by Louis A. Gros, a notary of Lyon about whom little is known, in a very rare pamphlet titled *Théorie du Baguenodier*, Lyon, 1872. I have not been able to see this yet; a photocopy was made and given to the Radcliffe Science Library at Oxford by Afriat, but the library could not find it when I asked for it.

The function $G(k)$ can be viewed as a permutation of the binary n tuples— indeed of the binary numbers in $[2^{n-1}, 2^n - 1]$. Francis Clarke at Swansea has made some study of this permutation; in particular, he has determined its order.

In 1880, Lucas published a report that Dr. O.-J. Broch, former Minister and President of the Royal Norwegian Commission at the Universal Exposition of 1878, recently told him that country people in Norway still used the rings to close their chest and sacks. No one has ever confirmed this fact and in 1904 a Norwegian ethnographer said he had never heard of it.

Looking at the sequence of 5-tuples given above, we see that the right-hand triples are a sequence which goes through every binary triple and ends adjacent to where it started. If we depict a binary triple as a point on the 3-cube, we have a circuit along the edges of this cube which goes through every vertex just once and returns to its starting point. See Figure 11 (left).

Such circuits are generally known as Hamiltonian circuits, for reasons to be seen shortly, and the Hamiltonian circuits on the n-cube are the generalized Gray codes.

Exercise. Show that every Hamiltonian circuit on the 3-cube is equivalent to this circuit. (This is no longer true in higher dimensions.)

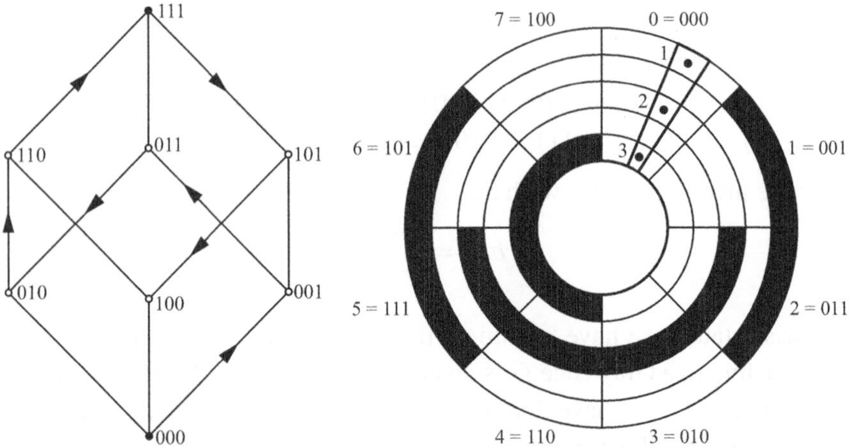

Figure 11. Left: a 3-cube showing Gray Code. Right: a switch, using Gray Code to minimize errors.

In 1947, Frank Gray of Bell Labs was using binary representations for coding and he found that a certain kind of error was minimized if the codings for adjacent numbers differed in just one binary bit. That is, he wanted to use a coding which was a Hamiltonian circuit on the n-cube.

Figure 11 (right) shows a switch for detecting the position of a central axis which has an arm containing three contact points marked $1, 2, 3$. The black areas are at some voltage and the white areas are insulation. As shown all three contacts have no voltage, so the signal on the arm is 000. (Alternatively, one can have the arm contacts having voltage and the black areas as detectors.) As we turn, the signal becomes 001, then 011, 010, ... When the arm is at one of the transition angles, it is possible for a contact to register 0 instead of 1 or vice versa. The Gray coding guarantees that there is never more than one contact point that can be in error and that the two possible signals give adjacent positions, so the effect of such an error is minimized. If one uses the ordinary binary coding, one can have all three contacts being subject to error and the position of the arm is completely undetermined.

Gray constructed his coding precisely by the pattern of the Chinese Rings, though he also went on to consider other Hamiltonian circuits on the n-cube. Bell Labs patented the idea in 1953 (Figure 12), but I believe the patent was cancelled as part of an anti-trust settlement.

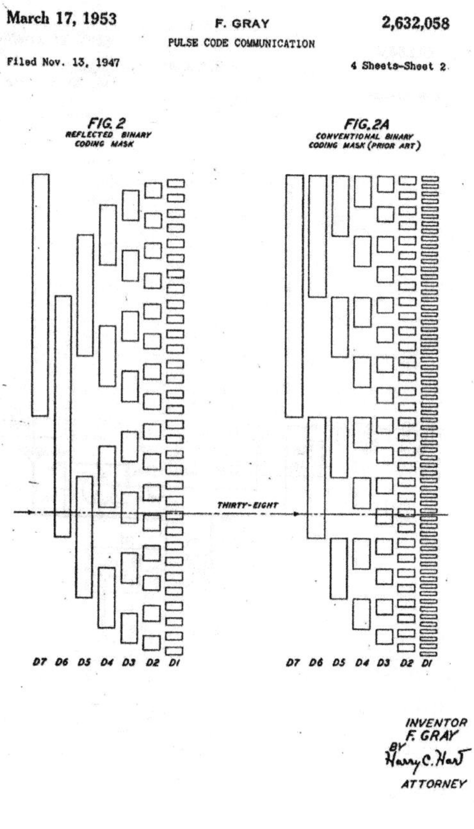

Figure 12. Patent for the Gray Code.

The number of Hamiltonian circuits or Gray codes on the n-cube remains an interesting problem, which leads to the question of when two circuits are equivalent. If we take both the symmetries of the circuit (i.e. starting at any point in either direction) and the symmetries of the cube as equivalences, the determination of the number of distinct circuits requires careful computation. For $n = 3$, there is essentially only one circuit and there are nine for the 4-cube. But already for the 5-cube, the counting which has been done failed to consider all the symmetries, so

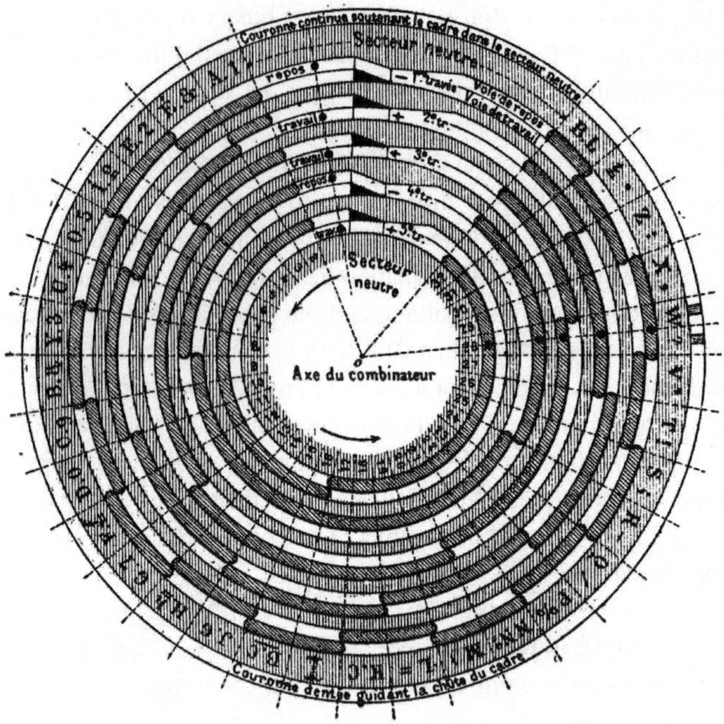

Figure 13. Baudot's use of the Gray Code in the 1870s. From [Heath 1972], p. 82.

we do not know how many inequivalent circuits there are, but it will be somewhere in the order of a million.

Surprisingly, the idea of a switch using the Gray code was discovered by the Belgian telegraph engineer Jean Émile Baudot (1845–1903) — the eponym of baud — in the 1870s, and utilised in his printing telegraphs. See Figure 13. I have also read that Stibitz used the same code in 1941, and Jack Good also recognised the pattern at Bletchley Park, 1943.

I-Ching hexagrams

Trying to keep in historical order, the ordinary binary representations of integers 0 to 63 occur in the 64 hexagrams of the I-Ching; Figure 14 (left) shows some objects decorated with the simpler 8 trigrams. The book itself goes back to ca. 600 B.C.E., but had the hexagrams in a traditional order attributed to King Wan, which has no known mathematical structure. Figure 15 (top) shows the 64 hexagrams in Wan's order, and Figure 14 (right) shows the fu-hsi ordering, which occurs in many places. The fu-hsi ordering only dates from the eleventh century and the binary pattern is clearly seen in the associated "segregation

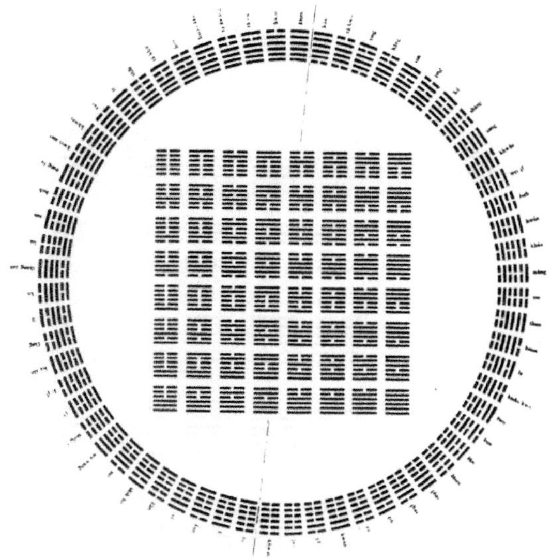

Figure 14. Left: I-Ching trigrams make a popular pattern. Right: the Fu-Hsi ordering (Yoshida Mitsuyoshi, *Jinkōki*, 1627; see [Sato 2013]).

table" (Figure 15, bottom), also occurring in many places. Leibniz learned about the later order from the Jesuit missionary Joachim Bouvet in 1701 and this inspired him to write a great deal more on binary arithmetic. He even produced a theological analogy that God was the One who created Everything out of Nothing; this pleased him so much that he had a special medal made to commemorate it (Figure 1, left). He expected this idea would convert the Emperor of China to Christianity.

Binary divination

Another explicit use of binary occurs in divination cards. These are 6 cards on which numbers are written. The subject, assumed to be less than 64 years old, says which cards his age occurs on and you add up the first numbers on these cards to determine his age (Figure 16, left). The first known European appearance of these is in the unpublished manuscript *De viribus quantitatis* by Luca Pacioli, about 1500. However, I have seen a Japanese description of two versions of the idea called Magic Cards and Magic Picture which says they have been known since at least the fourteenth century. Figure 16, right, shows versions from the early seventeenth century.

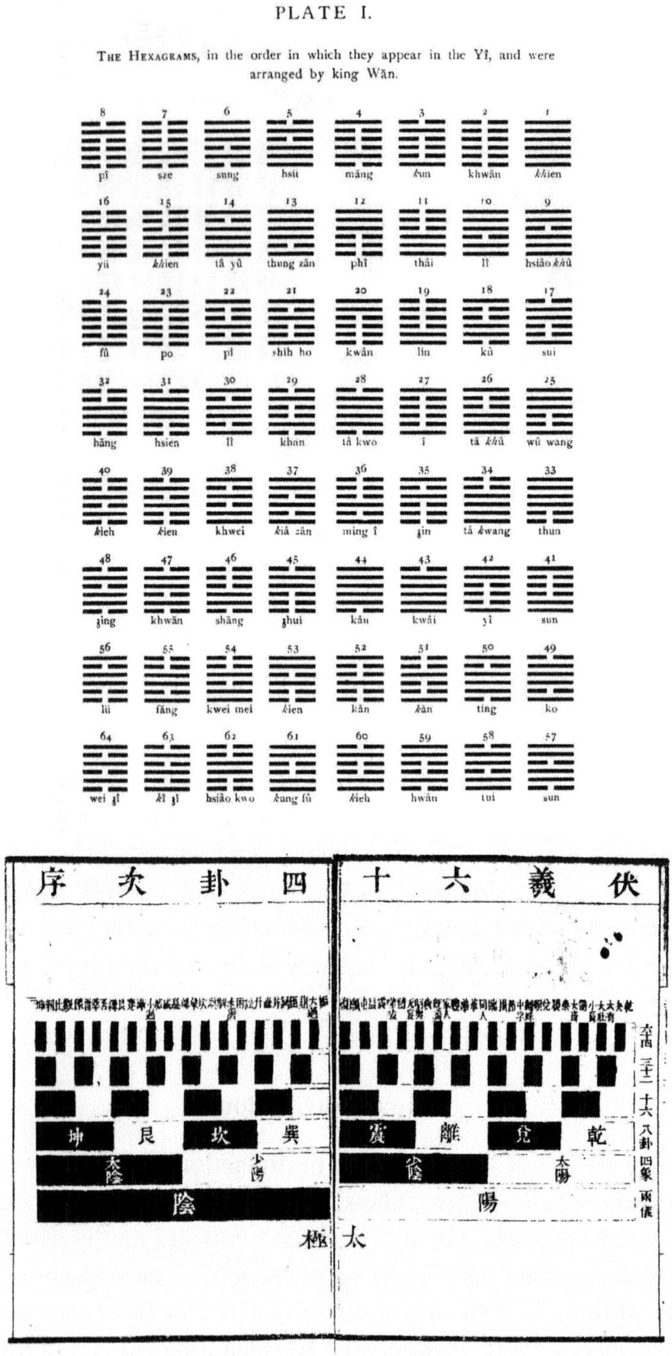

Figure 15. Top: the non-mathematical ordering of Wan [Legge 1899]. Bottom: the "segregation table" in Zhu Xi's *Zhouyi benyi* (twelfth century C.E.), reproduced in Hu Wei's *Yitu minghian* (1706).

1	3	5	7	9	11	13	15
17	19	21	23	25	27	29	31
33	35	37	39	41	43	45	47
49	51	53	55	57	59	61	63

2	3	6	7	10	11	14	15
18	19	22	23	26	27	30	31
34	35	38	39	42	43	46	47
50	51	54	55	58	59	62	63

4	5	6	7	12	13	14	15
20	21	22	23	28	29	30	31
36	37	38	39	44	45	46	47
52	53	54	55	60	61	62	63

8	9	10	11	12	13	14	15
24	25	26	27	28	29	30	31
40	41	42	43	44	45	46	47
56	57	58	59	60	61	62	63

16	17	18	19	20	21	22	23
24	25	26	27	28	29	30	31
48	49	50	51	52	53	54	55
56	57	58	59	60	61	62	63

32	33	34	35	36	37	38	39
40	41	42	43	44	45	46	47
48	49	50	51	52	53	54	55
56	57	58	59	60	61	62	63

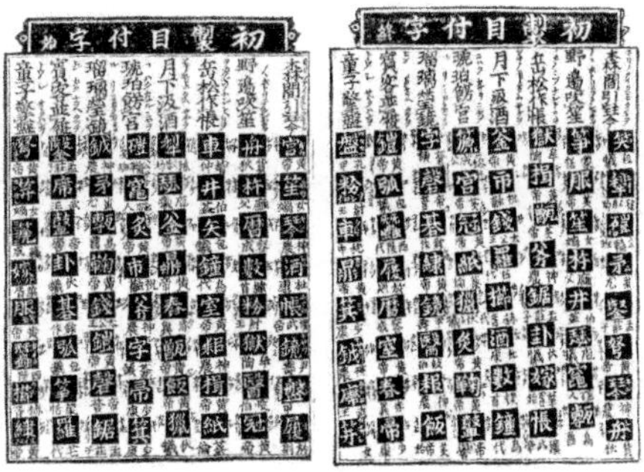

[Begin] [End]

[These figures are taken from another edition of 1631.]

Figure 16. Left: divination cards. Right: magic cards and a magic picture from the *Jinkōki* of Yoshida. (Note by editor: For more information about this early seventeenth-century work see [Sato 2013].)

Knight's tours

Now we go backwards a bit in time to find the beginning of our other main idea: Hamiltonian circuits and paths. The oldest Hamiltonian circuits and paths are knight's tours and paths on the chessboard. In the mathematical literature, knight's tours first appear in the 1723 edition of Ozanam's *Récréations mathématiques et physiques* (see Figure 17) and are first studied systematically by Euler (1759) and Vandermonde (1771). Euler describes it as "a curious question which does not submit to any analysis".

But if we examine the chess literature, we immediately discover tours going back to the dawn of chess. Murray's *History of Chess* describes a half-board path in the *Kāvyālaṅkāra* of Rudraṭa, c900 (Figure 18, left). These are given in poetic forms in the shapes of "wheel, sword, club, bow, spear, trident, and

I. *de M. de Montmort.*

1	38	31	44	3	46	29	42
32	35	2	39	30	43	4	47
37	8	33	26	45	6	41	28
34	25	36	7	40	27	48	5
9	60	17	56	11	52	19	50
24	57	10	63	18	49	12	53
61	16	59	22	55	14	51	20
58	23	62	15	64	21	54	13

II. *de M. Moivre.*

34	49	22	11	36	39	24	1
21	10	35	50	23	12	37	40
48	33	62	57	38	25	2	13
9	20	51	54	63	60	41	26
32	47	58	61	56	53	14	3
19	8	55	52	59	64	27	42
46	31	6	17	44	29	4	15
7	18	45	30	5	16	43	28

III. *de M. de Mairan.*

40	9	26	53	42	7	64	29
25	52	41	8	27	30	43	6
10	39	24	57	54	63	28	31
23	56	51	60	1	44	5	62
50	11	38	55	58	61	32	45
37	22	59	48	19	2	15	4
12	49	20	35	14	17	46	33
21	36	13	18	47	34	3	16

Figure 17. Knight's tours from [Ozanam 1723], pp. 261–262.

plough, which are to be read according to the chessboard squares of the chariot [= rook], horse [= knight], elephant [= bishop]." The poet placed syllables in the cells of a half chessboard so that it reads the same straight across as when following a piece's path. With help from the commentator Nami, of 1069, the rook's and knight's path's are reconstructed, and are given in Murray. Both are readily extended to full board paths, but not tours, by placing a second copy of

1	30	9	20	3	24	11	26
16	19	2	29	10	27	4	23
31	8	17	14	21	6	25	12
18	15	32	7	28	13	22	5

32	35	30	25	8	5	50	55
29	24	33	36	51	56	7	4
34	31	26	9	6	49	54	57
23	28	37	12	1	52	3	48
38	13	22	27	10	47	58	53
19	16	11	64	61	2	43	46
14	39	18	21	44	41	62	59
17	20	15	40	63	60	45	42

Figure 18. Rudrata, earliest half board path (left). Earliest knight's tour (right). From [Murray 1913], pp. 54 and 336.

the half board beneath the given copy and seeing that the first cell of the second board is connected to the last cell of the first board.

The next oldest known versions, which are full-board tours, appear in *Kitâb ash-shaṭranj mimma'l-lafahu'l-'Adli waṣ-Ṣûlî wa ghair-huma* (Book of the Chess; extracts from the works of al 'Adlî, aûlî and others), by an unknown author, copied by Abû Isḥâq Ibrâhîm ibn al Mubârak ibn 'Alî al Mudhahhab al Baghdâdî, in 1141. Murray gives two distinct tours. The solution of the first is a numbered diagram, Figure 18 (right), but the second is "solved" four times by acrostic poems, where the initial letters of the lines give the tour in an algebraic notation. There are also a knight/bishop tour and a knight/queen tour, where moves of the two types alternate.

A natural question arises: how many knight's tours are there? A little trial soon sends you to smaller boards, where two investigators found 9862 knight's tours on the 6×6 board in the 1970s.

This enumeration accounted for the symmetry group of a circuit, which is D_{36}, by taking a corner as the starting cell and one of the two cells adjacent to the corner as the second cell of the circuit. However, I don't believe anyone has examined these circuits to see how many have various symmetries of the board and thus to determine the number of inequivalent circuits. On the 8×8 board, some 75,000 tours were found having the same first 35 moves! In 1975, I made some crude estimates and predicted there are $10^{23\pm3}$ tours on the 8×8 board.

Martin Loebbing and Ingo Wegener, in "The number of knight's tours equals 33,439,123,484,294 – counting with binary decision diagrams" *Electronic J. Combinatorics* **3** (1996), R5, gives a somewhat vague description of a method for counting knight's tours — they speak of directed knight's tours, but it is not clear if they have properly accounted for the symmetries of a tour or of the board. Several people immediately pointed out that the number is incorrect because it has to be divisible by four. Two comments have appeared (ibid.) — on 15 May 1996, the authors admitted this and said they would redo the problem, but they have submitted no further comment as of Jan 2001. On 18 Feb 1997, Brendan McKay announced that he had done the computation another way and found 13,267,364,410,532.

In view of the difference between these values and my 1975 estimate, it might be worth explaining my reasoning. In 1964, Duby found 75, 000 tours with the same first 35 moves. The average valence for a knight on an 8×8 board is 5.25, but one cannot exit from a cell in the same direction as one entered, so we might estimate the number of ways that the first 35 moves can be made as $4.2535 = 9.9 \times 1021$. Multiplying by 75,000 then gives 7.4×1026. I think I assumed that some of the first moves had already been made, e.g. we only allow one move from the starting cell, and factored by 8 for the symmetries of the square, to get 2.2 × 1025. I can't find my original calculations, and I find the estimate 1025 in later papers, so I suppose I tried to reduce the effect of the 4.2535 some more. In retrospect, I had no knowledge of how many of these had already been tried. If about half of all moves from a cell had already been tried before any circuit was found, then the estimate would be more like $2.2534 \times 75,000 = 7.1 \times 1016$. If we divide the given number of circuits by 75,000 and take the 34th root, we get an average valence of 1.78 remaining, far less than I would have guessed.

I am grateful to Don Knuth for this reference. Neither he nor I expected to ever see this number calculated!

The Icosian Game

In the 1850s, Kirkman observed that there was only one inequivalent Hamiltonian circuit on the dodecahedron. Hamilton developed this idea into a board game called The Icosian Game (after the 20 vertices) and he also developed the mathematics into the first description of a group by means of generators and relations. See Figure 19.

Only one example of the board for The Icosian Game was known until recently. (Three more have been found in the last decade or so.) Lucas and Ahrens, writing about the turn of the century, describe a solid version of the game and Ahrens even gives an address of where to buy one, but no examples were known until about 2000.

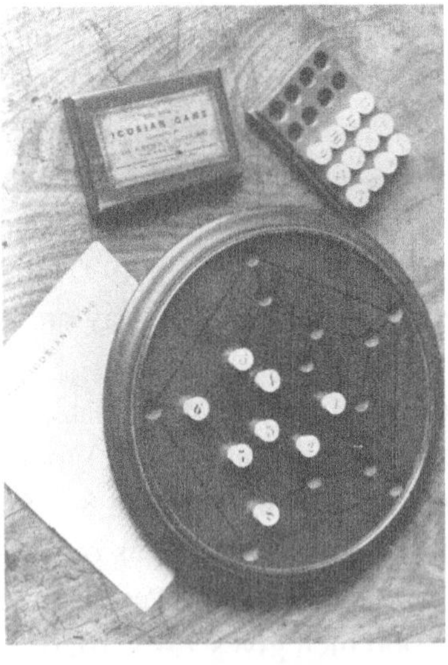

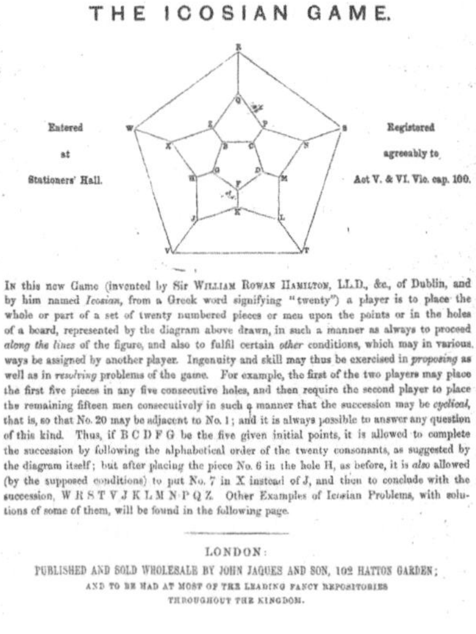

THE ICOSIAN GAME.

Entered at Stationers' Hall.

Registered agreeably to Act V. & VI. Vic. cap. 100.

In this new Game (invented by Sir William Rowan Hamilton, LL.D., &c., of Dublin, and by him named *Icosian*, from a Greek word signifying "twenty") a player is to place the whole or part of a set of twenty numbered pieces or men upon the points or in the holes of a board, represented by the diagram above drawn, in such a manner as always to proceed *along the lines* of the figure, and also to fulfil certain *other* conditions, which may in various ways be assigned by another player. Ingenuity and skill may thus be exercised in *proposing* as well as in *resolving* problems of the game. For example, the first of the two players may place the first five pieces in any five consecutive holes, and then require the second player to place the remaining fifteen men consecutively in such a manner that the succession may be *cyclical*, that is, so that No. 20 may be adjacent to No. 1; and it is always possible to answer any question of this kind. Thus, if B C D F G be the five given initial points, it is allowed to complete the succession by following the alphabetical order of the twenty consonants, as suggested by the diagram itself; but after placing the piece No. 6 in the hole H, as before, it is *also* allowed (by the supposed conditions) to put No. 7 in X instead of J, and then to conclude with the succession, W R S T V J K L M N P Q Z. Other Examples of Icosian Problems, with solutions of some of them, will be found in the following page.

LONDON:

PUBLISHED AND SOLD WHOLESALE BY JOHN JAQUES AND SON, 102 HATTON GARDEN;
AND TO BE HAD AT MOST OF THE LEADING FANCY REPOSITORIES
THROUGHOUT THE KINGDOM.

Figure 19. The Icosian Game: an exemplar in the Royal Irish Academy, from Robin Wilson, and its instructions.

Take, for instance, the object illustrated below this paragraph. For the sake of those who find pleasure in testing their wits, I ask: What is it?—hastening to add that archæologists themselves have yet to agree on an answer.

THE MYSTERIOUS DODECAHEDRON WITH THE PENTAGONAL ENDS.

Judging by its external appearance as shown in the picture, it is a bronze object shaped like a pentadodecahedron. Round openings of various sizes are found in the center of each face. The interior of the object is hollow. All specimens of this artifact have been found north of the Alps, which indicates a Roman origin.

One interpreter sees this mysterious thing as a mere toy; another as a die used in games of chance; a third as a model used in teaching the measurement of cylindrical bodies; a fourth as a candleholder.

What is it?

Since this book was first published, I have received over a hundred answers to this question from both experts and laymen all over the world. The experts' explanations tend to be quite authoritative in tone, though they contradict each other. The most probable solution—though far from established—is that we have here a musical instrument.

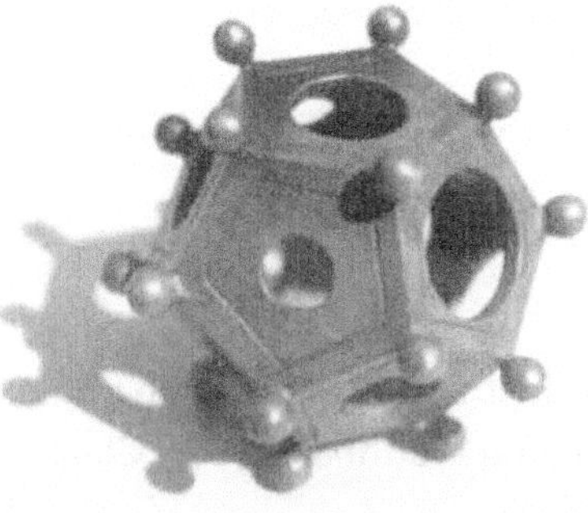

Figure 20. The Roman Dodecahedron and C. W. Ceram's discussion of it (*Gods, graves, and scholars*, 2nd ed., Gollancz, London, p. 25).

On the other hand, there are about 100 examples of Roman bronze hollow dodecahedra with knobs at the corners which look exactly like solid versions of the Icosian Game; Figure 20 shows a photo of a facsimile. Archaeologists are mystified as to what these objects are; there are hundreds on conjectures in the archaeological literature, now including mine that they might be early versions of Hamilton's game.

The Tower of Hanoi

The greatest of French recreational mathematicians was François-Édouard-Anatole Lucas (1842–1891), who died at the height of his powers from blood poisoning caused by a scratch from a plate dropped at a scientific banquet. In 1883, he brought out his Tower of Hanoi, which I will presume is known to all of you (see Figure 21). The story about the 64 discs in Benares appears in the original literature and was so widely spread that it appeared as truth in Robert Ripley's *Believe It Or Not!* (Figure 22).

When I gave this talk in 1993, Jean Brette, of the Palais de la Découverte in Paris, told me that there was an original example of the Tower of Hanoi in the Conservatoire National des Arts et Métiers – Musée National des Techniques. At the time, it was closed for refurbishment, but some years later, I wrote to ask about this and Elisabeth Lefevre sent details and photocopies of the box, instruction sheet and some other material. The bottom of the box has an ink

Fig. 2. — La tour d'Hanoï.

LA TOUR D'HANOÏ

VÉRITABLE CASSE-TÊTE ANNAMITE

JEU RAPPORTÉ DU TONKIN

PAR LE PROFESSEUR N. CLAUS (DE SIAM)

Mandarin du Collège Li-Sou-Stian!

Ce Jeu inédit a été trouvé, pour la première fois, dans les écrits de l'illustre Mandarin FER-FER-TAM-TAM, qui seront publiés, plus ou moins prochainement, suivant les ordres du Gouvernement chinois.

La **TOUR D'HANOÏ** se compose d'étages superposés et décroissants, en nombre variable, que nous avons représentés par huit pions en bois, percés à leur centre. Au Japon, en Chine, au Tonkin, on les fait en porcelaine.

Le Jeu consiste à démolir la tour, étage par étage, et à la reconstruire dans un lieu voisin conformément aux règles indiquées.

Amusant et instructif, facile à apprendre et à jouer, à la ville, à la campagne, en voyage, il a pour but la vulgarisation des sciences, comme tous les autres jeux curieux et inédits du professeur **N. CLAUS (DE SIAM)**.

Nous pourrions offrir une *prime de dix mille francs*, de *cent mille francs*, d'*un million de francs*, et plus encore, à celui qui réaliserait, à la main, le transport de la Tour d'Hanoï, à soixante-quatre étages, conformément aux règles du Jeu. Nous dirons, tout de suite, qu'il faudrait exécuter successivement le nombre de déplacements

18 446 744 073 709 551 615,

ce qui exigerait plus de *cinq milliards de siècles!*

D'après une vieille légende indienne, les brahmes se succèdent depuis bien longtemps, *sur* les marches de l'autel, dans le Temple de Bénarès, pour exécuter le déplacement de la *Tour Sacrée, de Brahma*, aux *soixante-quatre* étages en *or fin*, garnis de diamants de Golconde. Quand tout sera fini, la Tour et les brahmes tomberont, et ce sera la fin du monde!

PARIS, PÉKIN, YÉDO et SAÏGON

Chez les libraires et marchands de nouveautés.

1883

Tous droits réservés.

Figure 21. The original box cover and instructions (1883) for the Tower of Hanoi, where the game's introduction is attributed to N. Claus (de Siam), a pseudonym of Lucas (d'Amiens).

Figure 22. From Robert Ripley's *Believe It Or Not!*, book 2 (Simon & Schuster, 1931, and other editions).

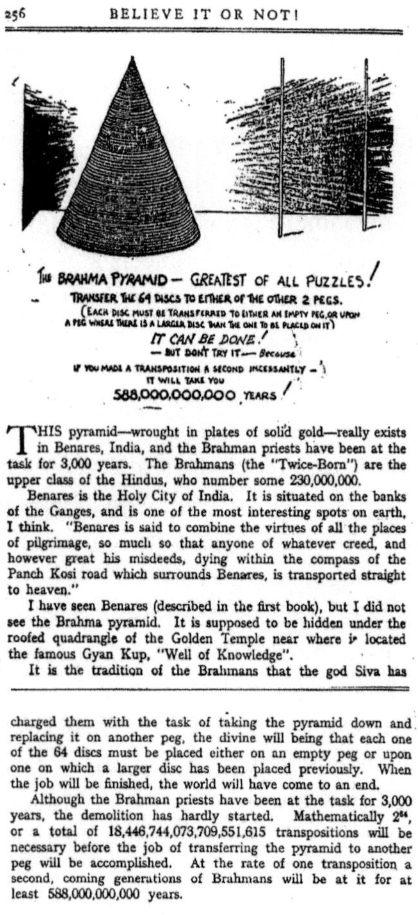

256 BELIEVE IT OR NOT!

THIS pyramid—wrought in plates of solid gold—really exists in Benares, India, and the Brahman priests have been at the task for 3,000 years. The Brahmans (the "Twice-Born") are the upper class of the Hindus, who number some 230,000,000.

Benares is the Holy City of India. It is situated on the banks of the Ganges, and is one of the most interesting spots on earth, I think. "Benares is said to combine the virtues of all the places of pilgrimage, so much so that anyone of whatever creed, and however great his misdeeds, dying within the compass of the Panch Kosi road which surrounds Benares, is transported straight to heaven."

I have seen Benares (described in the first book), but I did not see the Brahma pyramid. It is supposed to be hidden under the roofed quadrangle of the Golden Temple near where is located the famous Gyan Kup, "Well of Knowledge".

It is the tradition of the Brahmans that the god Siva has charged them with the task of taking the pyramid down and replacing it on another peg, the divine will being that each one of the 64 discs must be placed either on an empty peg or upon one on which a larger disc has been placed previously. When the job will be finished, the world will have come to an end.

Although the Brahman priests have been at the task for 3,000 years, the demolition has hardly started. Mathematically 2^{64}, or a total of 18,446,744,073,709,551,615 transpositions will be necessary before the job of transferring the pyramid to another peg will be accomplished. At the rate of one transposition a second, coming generations of Brahmans will be at it for at least 588,000,000,000 years.

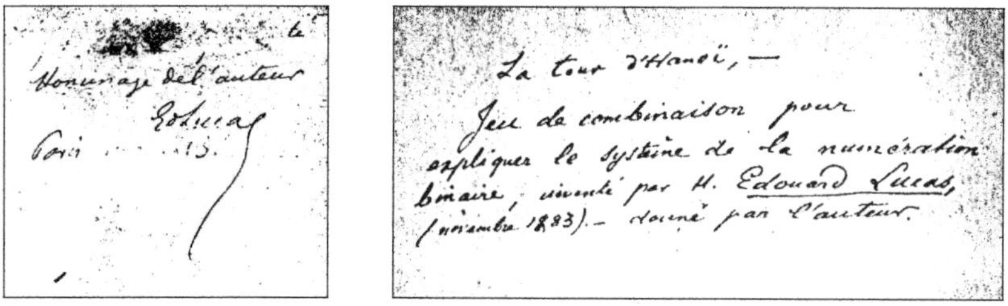

Figure 23. The bottom of the original Tower of Hanoi box (left); the inside of the lid of the box (right).

inscription: Hommage de l'auteur / Ed Lucas / Paris 1888 (Figure 23, left). The date is not clearly legible on the photocopy, but is known from the Museum's records. Inside the cover, apparently in the same hand (that is, in Lucas's writing), is an ink inscription (Figure 23, right); it translates to "The Tower of Hanoi: a combination game to explain the binary numbering system. Invented by Mr. Edouard Lucas, November 1883. Present of the author." These comments are very important historically in that Lucas never publicly admitted to inventing the game!

Despite its age, the Tower of Hanoi continues to surprise. In the late 1980s, I observed that the discs can be placed on the pegs in $3n$ ways and wondered if any position was more difficult to obtain from the initial position than the position with all discs on another peg. In fact there is not, but there are $2n$ positions which are just as difficult. The following analysis is based on work I did then, but this was improved by seeing the approach used by Daniele Parisse: "The Tower of Hanoi and the Stern-Brocot array", PhD Thesis at Fakultät für Mathematik, Ludwig-Maximilians-Universität München, 1997, under the direction of Andreas Hinz (an amended version was printed). In late 2000 and early 2001, I used this material as part of "The history of some combinatorial recreational problems", a chapter for *History of Combinatorics*, edited by Robin J. Wilson. In so doing, I found I needed some extra results; they eventually turned out to be pretty straightforward, but little of my original 1993 organization remains! The following is the revised material.

The first article on the puzzle, [Longchamps 1883], showed that it takes $2^n - 1$ moves to solve the problem when there are n discs. There has been some question as to whether this or any other early discussion actually showed that this number of moves is minimal, but if M_n is the minimal number of moves for n discs, then the basic argument clearly leads to

$$M_n = M_{n-1} + 1 + M_{n-1}. \tag{2}$$

In actually carrying out the minimal solution, one finds that the sequence of discs moved is precisely the same as the sequence of rings moved in the Chinese Rings: $1, 2, 1, 3, 1, 2, 1, 4, 1, 2, 1, 3, 1, 2, 1, 5, \ldots$, though there is not an analogous binary representation for the positions. Further each disc always moves through the pegs in the same cyclic order, with alternate discs moving in alternate directions. The Tower of Hanoi corresponds to a Hamiltonian circuit on the n-dimensional cube, while the Chinese Rings corresponds to a Hamiltonian path from one corner to a diametrically opposite corner—in both cases the route is the very particular one that we have called the Gray Code. This seems to have first been observed by Crowe [1956] (see also [Gardner 1957]). Because of the recursive nature of the solution, it has been a popular problem for testing programming techniques in recent years.

A position means a legal arrangement of the discs, i.e., no disc is on a smaller one. The order of discs on a peg is then determinate and we only have to say which discs are on which peg. Let us call the position, with all discs on one peg, an initial or perfect position.

To analyse my 1980s question of how many positions are maximally difficult, let $A(n, d)$ be the number of positions with n discs requiring d moves to obtain from the initial position. All the positions with the n-th disc still in its initial place can be viewed as positions on the first $n-1$ discs and so $A(n, d) = A(n-1, d)$ for $d < 2^{n-1}$. If the n-th disc has moved, there are two positions it can get to in $2^{n-1} - 1$ moves and the other discs are in a pile of $n-1$ at the other non-initial position. Hence $A(n, 2^{n-1} - 1 + d) = 2A(n-1, d)$, for $d < 2^{n-1}$.

If $0 \leq d < 2^n$ and we write d in its binary representation: $d = \sum d_i 2^i$ and set $S(d) = \sum d_i$—i.e., $S(d)$ is the number of ones in the binary representation of d—then one readily sees there are $2^{S(d)}$ positions that require d moves to achieve. The average value δ of d turns out to be precisely $\frac{2}{3}$ of the maximum number, $2^n - 1$, of moves. Andreas Hinz [1989a; 1992] found this same result and further determined the average number of moves between any two positions is asymptotic to $(466/885)2^n$.

The basic idea of Hinz is to examine the graph of positions connected by legal moves. This graph was already formulated by Scorer, Grundy and Smith [Scorer et al. 1944], though they didn't proceed as far as Hinz. This graph is in a triangular array with $2n$ points along each edge. It is constructed recursively and it is difficult to relate a given arrangement of the discs to its location in the graph. I have reformulated this and determined how to relate a position to its point in the graph. This process also yields the above results and some others. This takes a little notation. Recall we use $\oplus$ for Boolean addition or "exclusive or" and we will use $\odot$ for Boolean multiplication or "and". We count discs and pegs starting with 0.

The Disc-Peg Incidence matrix A is defined by: $A(i, j) = 1$ if disc i is on peg j, and $A(i, j) = 0$ otherwise. We will do two examples. If discs 0 and 1 are on peg 1 and disc 2 is on peg 0, then the matrix is the first shown below, while if discs 0 and 1 are on peg 1, disc 2 is on peg 0, and disc 3 is on peg 2, then the matrix is the second shown.

		peg 0 1 2	peg 0 1 2
	disc		
	0	0 1 0	0 1 0
$A =$	1	0 1 0	0 1 0
	2	1 0 0	1 0 0
	3		0 0 1

It is easier to view this matrix by its columns or its rows. The three columns give three binary n-tuples x_0, x_1, x_2 which indicate which discs are on the pegs. These "peg contents n-tuples" describe a partition of the discs, so $x_i \odot x_j = 00\ldots0$ for $i \neq j$ and $\sum x_i = 11\ldots1$. We can also interpret x_j as the integers given by the binary n-tuples, i.e. as $x_j = \sum_i A(i, j)2^i$ and we denote this vector as x. In our examples, we get $x = (4, 3, 0)$ and $x = (4, 3, 8)$. The fact that $\sum x_i = 2^n - 1$ tells us that we could plot the positions with triangular coordinates.

However, not all the points satisfying this condition correspond to a partition of the discs and adjacent points in the resulting plot are not connected by legal moves, so this does not lead to a useful graph. (One can see a connection with the Chinese Rings if one thinks of a Chinese Rings position as a pair of binary n-tuples, one recording the rings on the bar and the other recording the rings off the bar. Then these two n-tuples correspond to a partition of the rings.)

If we consider the rows of A, we can define the "disc location n tuples" or "disc location vector" as $p = (p_i)$, where $p_i = j$ if $A(i, j) = 1$, i.e. if disc i is on peg j. In our examples, we get $p = (1, 1, 0)$ and $p = (1, 1, 0, 2)$. Thus $p \in (\mathbb{Z}_3)^n$ and all such points occur, i. e. the puzzle has 3^n positions. Since each peg is adjacent to both others, $\mathbb{Z}_3$ behaves like a 3-point circle; when $n = 2$, we have a 9-point torus. We will tend to identify a position with its disc location vector. It is straightforward, but a bit messy, to find the mappings between these two descriptions and to describe the legal moves in each case.

Now we need some more notation.

Let $d(p, q)$ be the shortest number of moves between positions p and q. Since moves are reversible, we have $d(p, q) = d(q, p)$ and $d(\cdot, \cdot)$ is easily seen to be a metric.

Let $\boldsymbol{i}$ be the disc location vector $(i, i, \ldots, i)$, i.e., the perfect or initial position with all discs on peg i.

Let $i \circ j = -(i + j) \pmod 3$, so $i \circ j = i$ if $i = j$; $i \circ j \notin \{i, j\}$ if $i \neq j$.

We want to determine $d_i = d(\boldsymbol{i}, p)$ for some position p. Though the process is reversible, it seems easier to describe it as starting from $\boldsymbol{i}$. The largest disc, numbered $n-1$ since we start counting with zero, wants to go onto peg p_{n-1}.

If $p_{n-1} = j \neq i$, then we have to move the first $n-1$ discs from peg i onto the other disc, namely disc $i \circ j$. This takes $2^{n-1} - 1$ moves. Then we move disc $n-1$ onto peg j and we have reduced the problem by one disc, using 2^{n-1} moves. But our pile of $n-1$ discs is now on peg $i \circ p_{n-1}$, so our reduced situation starts from this peg and the roles of pegs $i = i \circ p_{n-1} \circ p_{n-1}$ and $i \circ j = i \circ p_{n-1}$ have been interchanged.

If $p_{n-1} = i$, then we don't have to carry out the 2^{n-1} moves and the reduced situation still starts from peg $i = i \circ p_{n-1}$.

This establishes the following.

Proposition 2. *For a position p, the value of $d = d_i = d(i, p)$ is determined by the following process.*

```
t = i
d = 0
FOR k = n − 1 TO 0 STEP − 1
    IF p_k ≠ t THEN d = d + 2^k
    t = t ∘ p_k
NEXT k
```

In our examples, the three distances are $d_0 = 0+2+1 = 3$, $d_1 = 4+2+1 = 7$, $d_3 = 4+0+0 = 4$ for the first and $d_0 = 8+4+2+1 = 15$, $d_1 = 8+0+2+1 = 11$, $d_3 = 0+4+0+0 = 4$ for the second.

Now consider computing d_0, d_1, d_2 in parallel as we will generally do. Observe that when (a, b, c) is a permutation of $(0, 1, 2)$, then so is $(a \circ p_k, b \circ p_k, c \circ p_k)$. Hence the three t values in the algorithm, which are originally $(0, 1, 2)$, always remain a permutation of $(0, 1, 2)$ at each stage. Hence we see:

Corollary 3. *Each binary place has the value* 1 *twice in the binary expansions of d_0, d_1, d_2.*

If we let $D_k(a)$ be the k-th digit of the binary representation of a, we can express the result of Corollary 3 as

$$\sum_i D_k(d_i) = 2 \text{ for each } k. \tag{3}$$

Hence we also have:

Corollary 4. $\sum_i d_i = 2(2^n - 1)$.

Since the situation is symmetric in the pegs, we readily see:

Corollary 5. *The average value of d_i over all p is $\frac{2}{3}(2^n - 1)$.*

Now considering the calculation of d_0, d_1, d_2 in parallel, we see that p uniquely determines (d_0, d_1, d_2). For if $q \neq p$, then the first place, counting down from $n-1$, where the vectors differ will give a different binary digit in two of the distances.

Proposition 6. *The set of positions in the Tower of Hanoi with n discs is in one-to-one correspondence with the set of triples of binary n-tuples,* (d_0, d_1, d_2), *satisfying* (3).

Proof. Proposition 2 gives a mapping from the set of positions to the distance triples and Corollary 3 says these triples satisfy (3). The above discussion shows the mapping is one-to-one. But there are precisely 3^n such triples and we already know there are 3^n positions of the puzzle, so the mapping must also be onto. □

One can determine $p = (p_k)$ from a triple (d_0, d_1, d_2) satisfying (3) by the following.

```
t(0) = 0; t(1) = 1; t(2) = 2
FOR k = n − 1 TO 0 STEP − 1
    FOR i = 0 TO 2
        IF D_k(d_i) = 0 THEN p_k = t(i)
    NEXT i
    FOR i = 0 TO 2
        t(i) = t(i) ∘ p_k
    NEXT i
NEXT k
```

We have that $d_0 + d_1 + d_2$ adds to a constant, $2(2^n - 1)$, which leads us to think of using triangular coordinates, but the sum is twice what it ought to be. But this is what happens when we take distances to the corners rather than the usual distances to the edges. This suggests that the natural coordinates are the complementary distances $d'_i = (2^n - 1) - d_i$. In our examples, these are (4, 0, 3) and (0, 4, 11). Then (3) becomes

$$\sum_i D_k(d'_i) = 1, \text{ for each } k. \tag{4}$$

Thus $\sum_i d'_i = 2^n - 1$, so the (d'_i) can be used as triangular coordinates for a graph of the positions within a triangle of edge $2^n - 1$, i.e. having 2^n points along each edge. Proposition 6 tells us which points in the triangle are legal positions of the puzzle.

In triangular coordinates, two points are adjacent if and only if they differ by one in two coordinates. E.g. (0, 0, 3) is adjacent to (0, 1, 2). The truth of this for our (d'_i) implies the same relationship for the (d_i). We want to see that two positions in the Tower of Hanoi differ by just one move if and only if the corresponding points are adjacent in our triangular graph.

To see this, we consider the possible moves from a position p. From any position, there are at most three moves, of two types.

1) The smallest disc can be moved from its peg to either of the other pegs. These moves can be made from any position.
2) The second smallest of the uppermost discs, i.e. the smaller of the uppermost discs on the pegs which the smallest disc is not on, can be moved from its peg to the other such peg. An empty peg acts as if it had an infinitely large and immovable disc on it, but when there are two empty pegs, there is no move of this type. That is, there are only two moves, of type 1, from the perfect positions.

Consider a move of the first type from a position p to a position q and its effect on the distances d_i. The calculations will be identical until the final stage when one will be added to one distance and subtracted from another. The same holds for the coordinates d_i' and so p and q will be adjacent in our triangular graph.

Now what happens if we make a move of the second type? If the second smallest uppermost disc is the k-th, then the k discs $0, 1, \ldots, k-1$ are all on the same peg. From the symmetry of the situation, let us assume position p has discs $0, 1, \ldots, k-1$ on peg 0 and disc k on peg 1 and we want to move it to peg 2 to obtain position q. In computing the distances for p and q, everything is identical for $n-1, n-2, \ldots, k+1$. We can ignore these discs – the only effect is that this permutes the t values in the algorithms, but we only need to examine the set of distances.

Using the known result (2) that it takes $2^k - 1$ moves to move the first k disks from one peg to another, we see that for position p, the distances are these:

for position p:	for position q:
$d_0 = 2^k - 1 + 1 + 2^k - 1 = 2^{k+1} - 1,$	$d_0 = 2^k - 1 + 1 + 2^k - 1 = 2^{k+1} - 1,$
$d_1 = 2^k - 1,$	$d_1 = 1 + 2^k - 1 = 2^k,$
$d_2 = 1 + 2^k - 1 = 2^k,$	$d_2 = 2^k - 1.$

So we see that p and q are adjacent in our triangular graph.

So every pair of positions differing by one move in the Tower of Hanoi corresponds to an adjacent pair of points in our triangular graph. But a point in the triangular graph has: 2 adjacent points if it is at a corner; 4 adjacent points if it is in the interior of an edge; 6 adjacent points otherwise. The three corners are well behaved; both adjacent points are one move from the corner. For all other points I claim that only three of the adjacent points satisfy (4). Renumber the pegs so that $D_0(d'_0) = 1$, i.e. disc 0 is on peg 0. Suppose that the first zero value in the binary representation of d_0 is the k-th digit, i.e. $D_k(d'_0) = 0$, but $D_j(d'_0) = 1$ for $j = 0, 1, \ldots, k-1$. Renumber so that $D_k(d'_1) = 1$. Then the binary representations have the forms

$$d'_0 = \ldots 011 \ldots 1,$$
$$d'_1 = \ldots 100 \ldots 0,$$
$$d'_2 = \ldots 000 \ldots 0.$$

It is clear that we can subtract one from d'_0 and add it to either of the other coordinates while preserving (4). Adding one to d'_0 gives us $\ldots 100 \ldots 0$ and (4) can only hold if we subtract one from d'_1, getting $\ldots 011 \ldots 1$. If d'_0 is not changed, then we have to add one to either of the other two distances and this gives an end digit of one and so (4) does not hold. Hence the only situations where p and q are adjacent points in our triangular graph are those corresponding to moves in the Tower of Hanoi. This completes the proof of the following.

Theorem 7. *The graph of positions in the Tower of Hanoi with n discs and with adjacency between positions one move apart, is isomorphic to the graph of triples of binary n tuples* (d'_0, d'_1, d'_2) *satisfying* (4) *considered as triangular coordinates in a triangle of edge length* 2^{n-1} *and with adjacency being adjacency in the lattice.*

Figure 24 shows a Tower of Hanoi diagram for $n = 3$, with some disc location vectors and our first example plotted. This picture was described by Scorer, Grundy and Smith [Scorer et al. 1944] by a different process, which I illustrate for the passage from 1 to 2 discs. For 1 disc, there are three positions in a triangle; Figure 25 (left). For 2 discs, we get this repeated three times, once for each peg that the second disc is on; Figure 25 (middle). We place these copies at the corners of a triangle and then reflect each small triangle about the axis through the corner of the big triangle; Figure 25 (right). This reflection is the geometric process associated with the $i \circ j$ operation above and corresponds to the fact that

$p = (1, 1, 0)$
$x = (4, 3, 0)$
$d = (3, 7, 4)$
$d' = (4, 0, 3)$

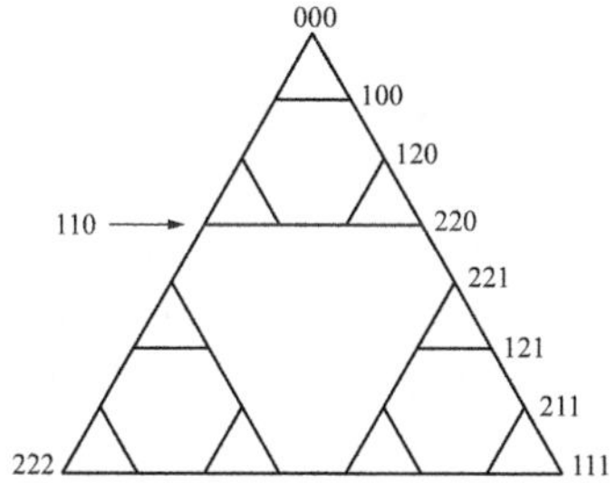

Figure 24. A Tower of Hanoi diagram for $n = 3$.

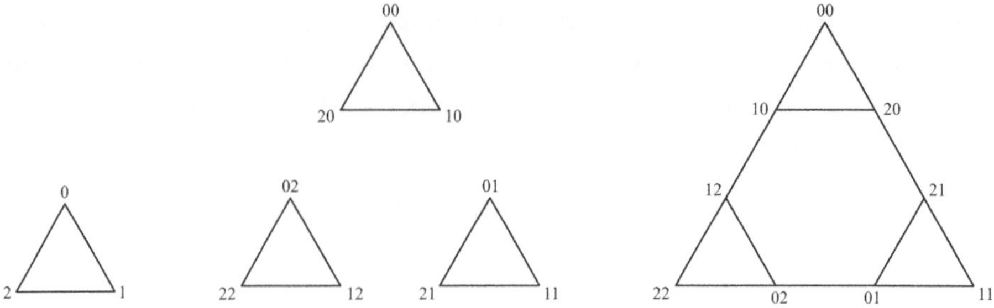

Figure 25. The passage from 1 to 2 discs, from [Scorer et al. 1944].

when we start at i and we want to move a disc from i to j, with $i \neq j$, we have first to move all the previous discs to peg $i \circ j$, so the problem is reduced to a smaller problem, but starting on the different peg $i \circ j$.

Scorer, Grundy and Smith noted that every position, except the perfect ones, has three moves from it, one of each of the kinds discussed above and that the mapping of triples $x = (x_0, x_1, x_2)$ to triples $d' = (d'_0, d'_1, d'_2)$ has order 2. The distances d and the coordinates d'_i really tell us all we could want to know about a position; Figure 24 show the next stage of the graph and the point $p = (1, 1, 0)$ on it). From the symmetry of the triangular pattern, some results can be deduced from properties of the triangle — e.g., δ is the distance of the centroid from a vertex and the average value of d'_i over all p is the distance of the centroid from an edge, i.e., $\delta/2$.

When we think of the Chinese Rings positions as pairs of binary n tuples, each position has either one or two moves and a move consists of shifting a bit from one n-tuple to another according to certain rules. In the Tower of Hanoi, each position is a triple of binary n-tuples and has two moves of the same sort, but generally has a more complex move, though this shifts one from one coordinate to another. The Gray Code permits us to recognise adjacent positions in the Chinese Rings; the triangular coordinates do the same for the Tower of Hanoi. I still feel that the analysis is not quite satisfactory in that there is no formula for the Tower of Hanoi analogous to $G(k) = B(k) \oplus B(\lfloor k/2 \rfloor)$. The standard problem of moving from one perfect position to another, say from 0 to 1, corresponds to moving along the edge $d_2 = 0$ of our triangular graph and the point with coordinates $(2^n - 1 - k, k, 0)$ is the k-th point on the solution path. If we are given the point number, k, we can recreate p by the method after Proposition 6. Early methods of doing this seem rather more complex; see [Hinz 1989a].[1]

[1]Editor's note: for a comprehensive survey of similar topics, see also [Hinz et al. 2013].

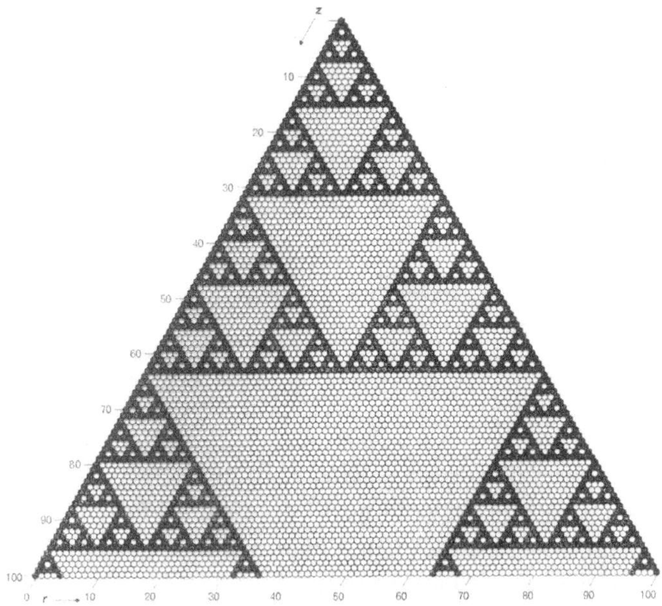

Figure 26. Odd values in Pascal's Triangle. From Siegfried Rösch, *Farbenlehre, auf die Mathematik angewandt Studien am Pascalschen Dreieck*; palette Nr. 15 (Spring 1964), Sandoz AG, Basel.

Related ideas

Ian Stewart, and perhaps others, have pointed out that the triangular pattern arising is the same as the pattern of odd binomial coefficients (BC) in the rows $0, 1, \dots, 2^n - 1$ of Pascal's triangle. This is an easy consequence of the result that $\mathrm{BC}(m, k)$ is odd if and only if $m + (m - k)$ has no carries when done in binary. If we have peg 0 at the top of our triangle, then $\mathrm{BC}(m, k)$ is located at the point $((2^n - 1) - m, k, m - k)$ on the triangular graph and is a legal position if and only if $k \odot m - k = 0$. This also allows us to deduce the number of positions in the m-th row as I did before. This pattern is also the n-th stage in the construction of the fractal called Sierpiński's Gasket (Figure 26). Hinz showed that the average distance between points in these patterns satisfies a fifth-order recurrence and is asymptotic to $466/885$ of the maximal length.

Hinz also considered arrangements of the discs which were not in correct order and asked how many moves were needed to get to a correct order.

Donald Knuth told me about the problem where we imagine the three pegs in a line and one can only move to an adjacent peg. That is, one cannot move from one end to the other. If we want to move the whole pile from one end to the other end, then this variation has a remarkable connection with Hamiltonian

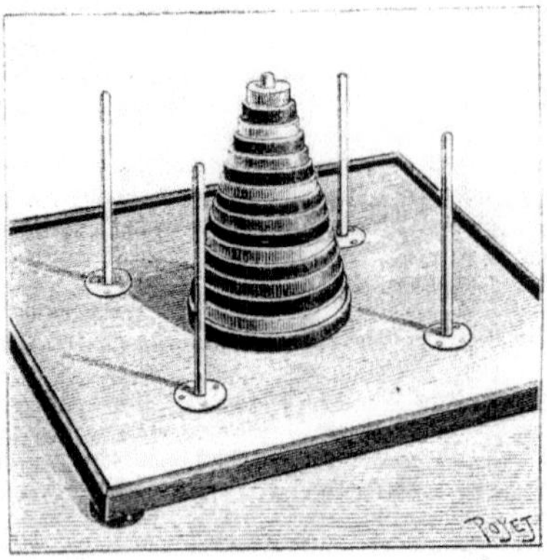

Fig. 5. — Nouvelle Tour d'Hanoï.

No. of Disks	Pegs					
1	1	1	1	1	1	1
2	3	3	3	3	3	3
3	7	5	5	5	5	5
4	15	9	7	7	7	7
5	31	13	11	9	9	9
6	63	17	15	13	11	11
7	127	25	19	17	15	13
8	255	33	23	21	19	17
9	511	41	27	25	23	21
10	1023	49	31	29	27	25

Figure 27. Variant Towers of Hanoi, from Lucas ("Nouveaux jeux scientifiques de M. Édouard Lucas", *La Nature* **17** (1889), 301–303), with a list of best solutions (Joe Celko, Puzzle Column: "Mutants of Hanoi", *Abacus* **1**:3 (1984), 54–57).

circuits which I will leave as an exercise! One can also consider the problem with all movements having to go in the same direction.

There are also unsolved problems. Suppose we have four (or more) pegs (Figures 27, left). Then we can carry out the transfer more easily. Some study reveals a fairly natural method but there are many ways to carry out the transfer in the same number of moves. Hinz has also investigated this and found that this allows one to transfer 64 discs to another peg, in less than six hours using four pegs, compared to some 5×10^9 centuries when using three pegs. No one has yet come up with a proof that this method is really minimal and Knuth suggests that it may be impossible because of the many different ways which give the minimal number of moves. Some best known results are shown on the Figure 27, right.[2]

A number of variations of the Chinese Rings have been devised, several in recent years. In 1891, George E. Everett of Grand Island, Nebraska, obtained a UK patent for the Loony Loop; Figure 28 (left). I have not found a US patent on this. It appeared about 1900 in English puzzle boxes as the Canoe Puzzle. A number of topological variations have appeared more recently. A Hungarian example of the 1980s was called Bogi; Figure 28 (right) is the instructions. Two

[2]We thank one of our reviewers for the following remark: Here the problem on the lower bounds of the number of moves for the k-peg Tower of Hanoi ($k \geq 4$) is unsolved. But in recent years, there are breakthroughs and it seems to be solved completely: namely, for the four-peg case: T. Bousch, "La quatrième tour de Hanoï", *Bull. Belg. Math. Soc. Simon Stevin* **21**:2 (2014), 895–912); for the general case: J. Zerovnik, "Self similarities of the Tower of Hanoi graphs and a proof of the Frame–Stewart conjecture", arXiv 1601.04298v1.

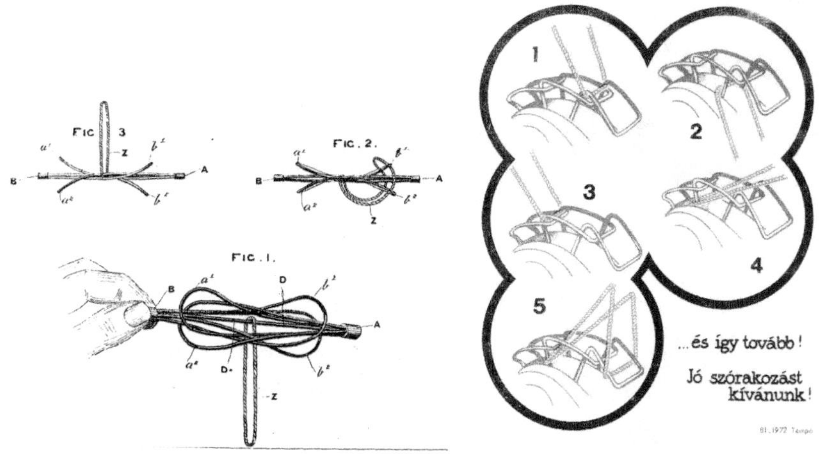

Figure 28. Left: From the patent for the Loony Loop. Right: Bogi instruction sheet.

Figure 29. Brain, bottom of box, and SpinOut, from box.

mechanical versions appeared in the 1980s: The Brain (trademarked by Mag-Nif in 1989); Figure 29, and Spin Out (patented by William Keister in 1972, produced by Binary Arts in 1986); Figure 29 (right).

A similar looking, but quite different, puzzle called Panex, invented by Toshio Akanuma, appeared in Japan in 1983. It looks like two 10-disc piles in two of three channels; see Figure 30 (but the middle channel is not clear). The frame has concealed notches so that a piece cannot move down further than its natural position. With this restriction, one doesn't worry about such trivial matters as putting large discs on smaller. (An essentially identical puzzle was patented in the US

Figure 30. Panex.

by someone else in 1993.) Mark Manasse, Danny Sleator and Victor Wei, at Bell Labs (unpublished study, 1983) have shown that one can move one pile to the centre peg in 4875 moves and one can exchange the piles in a number of moves between 27564 and 31537. They found minimal solutions for up to six levels. At the Fifth Gathering for Gardner in 2002, Nick Baxter gave out a sheet which said that levels seven and eight had been solved, but with current computing power, the next two levels would take 10 and 1200 years. See http://www.baxterweb.com/puzzles/panex.

References

[Anonymous 1879] "Télégraphe multiple imprimeur de M. Baudot", *Ann. Télégraphiques* (3) **6** (1879), 354–389. Says the device was presented at the 1878 Exposition and has been in use on the Paris Bordeaux line for several months. See pp. 361–362 for diagrams and p. 383 for discussion.

[Afriat 1982] Sydney N. Afriat, *The ring of linked rings*, Duckworth, 1982.

[Cardan 1550] J. Cardan, *De subtilitate libri XXI*, J. Petreium, Nuremberg, 1550: , *Liber XV, De incerti generis aut inutilibus subtilitatibus*, pp. 294–295. (= Basel, 1553, pp. 408–409.) French ed. by Richard Leblanc, Paris, 1556, titled *Les Livres d'Hieronymus Cardanus, de la Subtilité et subtiles Inventions, ensemble les causes occultes et les raisons d'icelles*, Book XV, para. 2, p. 291. (= Hieronymi Cardani Opera, Lyon, 1663, vol. 111, p. 587.)

[Crowe 1956] Donald W. Crowe, "The n-dimensional cube and the Tower of Hanoi, *Amer. Math. Monthly* **63** (1956), 29–30.

[Culin 1895] S. Culin, "Games of the Orient", 1895: Section XX: Ryou Kaik Tjyo Delay Guest Instrument (Ring Puzzle), pp. 31–32. Reprinted by Tuttle, Vermont, 1958. Tells the story of Hung Ming (181–234) inventing it.

[Gardner 1957] Martin Gardner, "Mathematical Games", *Sci. Amer.* **196**:5 (May 1957), 150–157, https://www.jstor.org/stable/24940862. Reprinted in his *The Scientific American book of mathematical puzzles and diversions*, Simon & Schuster, 1959, Chap. 6; UK version: *Mathematical Puzzles and Diversions from Scientific American*, Bell, London, 1961; Penguin (without the words "from Scientific American"), 1965.

[Gardner 1972] Martin Gardner, "Mathematical Games", *Sci. Amer.* **227**:2 (Aug 1972), 106–109, https://www.jstor.org/stable/24927411. Expanded version in his *Knotted doughnuts and other mathematical entertainments*, Freeman, 1986, Chap. 2.

[Gray 1947] Frank Gray, US Patent 2,632,058: *Pulse Code Communication.* Applied 13 Nov 1947, patented 17 Mar 1953, 13pp. See especially 1:1–4, 5:73–7:41, 7:67–9:11, Figures 2 and 2A.

[Gilbert 1958] E. N. Gilbert, "Gray codes and paths on the n-cube", *Bell System Technical Journal* **37** (1958), 815–826. Shows there are 9 inequivalent circuits on the 4-cube and 1 on the n-cube for $n = 1, 2, 3$. The latter cases are sufficiently easy that they may have been known before this.

[Gros 1872] Louis A. Gros, *Théorie du baguenodier*. Aimé Vingtrinier, Lyon, 1872. There is supposed to be a photocopy in the library at Oxford, but it couldn't be found when I went some years ago.

[Heath 1972] F. G. Heath, "Origins of the binary code", *Sci. Amer.* **227**:2 (Aug 1972), 76–83, https://www.jstor.org/stable/24927408

[Héraud 1903] E. Héraud, *Jeux et récrátions scientifiques: chimie, histoire naturelle, mathématiques*, 2nd ed., Bailliére et Fils, Paris, 1903. Pages 300–301 shows the same cover as Lucas in 1884. This is not shown in the 1884 ed. of Héraud.

[Hinz 1989a] Andreas M. Hinz, "The Tower of Hanoi", *L'Enseignement Math.* **35** (1989), 289–321. Surveys history and current work. 50 references.

[Hinz 1989b] Andreas Hinz, "An iterative algorithm for the Tower of Hanoi with four pegs", **Computing 42** (1989), 133–140. Studies the problem carefully. 17 references.

[Hinz 1992] Andreas M. Hinz, "Pascal's triangle and the Tower of Hanoi", *Amer. Math. Monthly* **99** (1992), 538–544.

[Hinz et al. 2013] A. M. Hinz, S. Klavžar, U. Milutinović and C. Petr, *The Tower of Hanoi: myths and maths*, Birkhüser/Springer, Basel, 2013.

[Keister 1972] William Keister, US Patent 3,637,215: "Locking Disc Puzzle". Filed 22 Dec 1970; patented 25 Jan 1972. For Spin-Out.

[Legge 1899] *The Yi King*, translated by James Legge, 2nd ed., Clarendon Press, Oxford, 1899 (in the collection *The sacred books of China*). Reprinted as *The I Ching*, Dover, New York, 1963, https://www.biroco.com/yijing/Legge1899.pdf

[Leibniz 1679] G. W. Leibniz, "De progressione dyadica", manuscript of March 1679. His first, unpublished work on the binary system, showing all the arithmetic processes.

[Leibniz 1734] Gottfried Wilhelms Baron von Leibnitz Mathematischer Beweis der Erschaffung und Ordnung der Welt in einem Medaillon an [...] Rudolph August, Weyland Regierenden Herzog zu Braunschw[eig] und Lüneb[urg] entworfen, und an das Licht gestellt von Rud. Aug. Noltenio [= Rudolf August Nolte]. Leipzig, Johann Christian Langenheim, 1734. Scan courtesy of the Bayerische Staatsbibliothek, https://tinyurl.com/leibniz-medallion .

[Longchamps 1883] G. de Longchamps, "Variétés", *J. Mathématiques Spéciales* (2) **2** (1883), 286–287. The article is only signed G. L., but the author is further identified in the index on p. 290.

[Lucas 1884] N. Claus (de Siam) [= Lucas (d'Amiens), i.e., François-Édouard-Anatole Lucas], *La tour d'Hanoï: Jeu de calcul. Science et Nature* **1**:8 (19 Jan 1884), 127–128. His first article on the puzzle. Says it takes $2^n - 1$ moves. Observes that each of the discs always moves in the same cycle of pegs and hence gives the standard rule for doing the solution, which is attributed to the nephew of the inventor, M. Raoul Olive, student at the Lycée Charlemagne. Asks for the minimum number of moves to restore an arbitrary distribution of discs to a start position. Says this is a complex problem in general.

[Murray 1913] Harold James Ruthven Murray, *A history of Chess*, Oxford University Press, 1913, reprinted 1962; reprinted Benjamin Press, Northampton, MA, 1985.

[Ozanam 1723] Jacques Ozanam, *Récréations mathématiques et physiques*, vol. 1, Jombert, Paris, 1723 https://books.google.com/books?id=B61EAAAAcAAJ

[Sato 2013] Ken'ichi Sato, "The Jinkōki of Yoshida Mitsuyoshi", in *Seki, founder of modern mathematics in Japan*, pp. 173–186, edited by E. Knobloch, H. Komatsu and D. Liu, Springer, Tokyo, 2013. Downloadable with a trial subscription at https://www.springerprofessional.de/en/the-jinkoki-of-yoshida-mitsuyoshi/4617862; an abstract can be seen at http://i-wasan.jp/seki/abstract/

Sato_abstract.pdf. Sato writes: "Yoshida Mitsuyoshi (1598–1672) published the Jinkōki first in 1627. This was a problem book of elementary mathematics for everyday use but it also contained many interesting problems which attracted readers. This book became so popular that there have been more than 300 versions published during the Edo era (1603–1868) in Japan."

[Scorer et al. 1944] R. S. Scorer, P. M. Grundy and C. A. B. Smith, "Some binary games", *Math. Gaz.* **28** (no. 280), July 1944, 96–103.

[Slocum and Botermans 1986] Jerry [= Gerald K.] Slocum and Jack Botermans, *Puzzles old & new: how to make and solve them*, Univ. of Washington Press, Seattle, 1986.

[Stibitz 1943] George R. Stibitz, US Patent 2,307,868. Applied 1941, granted 1943. Has a binary counter using the Gray code.

[Takagi 1982] Shigeo Takagi, *Play Puzzle* (in Japanese), part 2, 1982.

[Wallis 1685] J. Wallis, *De algebra tractatus*, 1685. (= *Opera Math.*, Oxford, 1693, vol. II, chap. CXI, *De complicatus annulis*, 472–478.)

[Yü 1958] Ch'ung-En Yü, *Ingenious Ring Puzzle book* (in Chinese), Shanghai Culture Publishing Co., Shanghai, 1958. English translation by Yenna Wu, published by Jerry Slocum, Beverly Hills, CA, 1981. p. 6. States the Chinese Rings were well known in the Sung period (960–1279).

zingmast@gmail.com *London, United Kingdom*

Research Articles

Games of No Chance 5
MSRI Publications
Volume **70**, 2017

A note on polynomial profiles of placement games

JASON I. BROWN, DANIELLE COX,
ANDREW H. HOEFEL, NEIL MCKAY, REBECCA MILLEY,
RICHARD J. NOWAKOWSKI AND ANGELA A. SIEGEL

The polynomial profile of a placement game enumerates the number of different positions. For a subclass of placement games, the polynomial profile is the independence polynomial of a related graph. For several important games, we generate the profiles when the board is a path; in the process, we discover some relationships between them.

1. Introduction

A natural enumeration question for combinatorial games is: "How many legal positions are possible in a game?" Surprisingly, few have actually considered this problem. Farr [7; 8], and Tromp and Farnebäck [20] consider the problem of "counting the number of end positions in GO." Similar enumeration questions are addressed by Hetyei [12], who analyses a game where the number of $\mathcal{P}$-positions (second player win positions) of length n is related to the n-th Bernoulli number of the second kind, and in [17], where it is shown that for the game of TIMBER, on paths, the number of $\mathcal{P}$-positions of length n is related to the Catalan and Fine numbers.

In Section 3, we enumerate the positions of several well-known games. A natural subset of combinatorial games, which we call placement games, are those that consist of placing pieces on a board until the board "fills" and there are no further moves. For each game, except NOGO, we find an auxiliary graph for which a position in the game corresponds to an independent set in the auxiliary graph.

In Section 4, we exhibit bijections between the games with identical generating functions.

All authors' research was supported in part by the Natural Sciences and Engineering Research Council of Canada. This work was part of a workshop and seminar series at Dalhousie University.
MSC2010: 05C31.
Keywords: combinatorial games, independent sets, graphs, independence polynomial, COL, octal games, 012, CIS, SNORT, NOGO.

2. Background

A *placement game* can be abstractly represented as a game on a graph, with the following properties.

- The game begins on a graph that contains no pieces.
- A move is to place a piece on one (or more) vertices subject to the rules of the particular game.
- The rules must imply that if a piece can be placed in a certain position on the board then it was legal to place it in that position at any time earlier in the game.
- Once played, a piece remains on the graph; it is never moved or removed from the graph.

Placement games were first identified during the seminar which led to this paper and have become of interest because of their properties; see [13; 6; 5; 16]. Some known examples of placement games are DOMINEERING [1], COL [2], SNORT [2], NOGO [4] and NODE KAYLES [3; 10]. (Rules for the games considered in this paper are given below.) CHESS and CHECKERS are not in this class of games because pieces are moved and removed and the starting position is not empty. The game of GO is likewise not a placement game since, while pieces are placed and not moved around, they can be removed from the board.

In this paper, we consider several placement games that appear in the literature. In all, a vertex that has not been played on will be called *empty* and at the start all vertices are empty. A move by Left is to place a *blue* piece on one or more (depending upon the rules) uncolored vertices. Similarly, Right places a *red* piece. No two pieces can share the same vertex. The *size* of a piece refers to the number of vertices it occupies when it is placed.

- CIS: Both blue and red pieces have size 1 and no two pieces can be adjacent.
- O12[1]: A blue has size 1 and red piece size 2; pieces are allowed to be adjacent.
- SNORT [2]: Both blue and red pieces have size 1; a blue piece and a red piece cannot be adjacent but two pieces of the same color can be adjacent.
- COL [2]: Both blue and red pieces have size 1; no two blue vertices can be adjacent, neither can two red, but a red can be adjacent to a blue.
- NOGO (also known as *Anti-Atari Go*) [4]: Both blue and red pieces have size 1. Every maximal connected group of blue pieces must include a vertex adjacent to an empty vertex, similarly for any maximal group of red pieces.

[1] O12 on a strip is a partizan octal game (see [15]) and another generalization is NODE KAYLES [3].

game	approx. # positions on P_n	generating function
CIS	2^n	$\frac{3}{1-t-2t^2}$
O12	2.414^n	$\frac{1}{1-2t-t^2}$
COL	2.414^n	$\frac{1+t}{1-2t-t^2}$
SNORT	2.414^n	$\frac{1+t}{1-2t-t^2}$
NOGO	2.769^n	$\frac{(1+t)(1-2t)t}{(1-t+t^2-2t^3)(1-t)}$

Table 1. Rates of growth.

In enumerating positions, we found that there is a bijection between the positions of certain games and the independent sets of an associated graph. These games are examples of *independent placement games*. CIS, COL, SNORT, and O12 are independent placement games, but NOGO is not since it has a "hyperedge" constraint. Enumerating independent sets was considered by earlier researchers. Prodinger and Tichy [19], as well as others, showed that the number of independent sets of a path with n vertices is the $(n+2)$-nd Fibonacci number. They coined the term the *Fibonacci number of a graph G* to mean the number of independent sets of G. For a placement game, placing a piece prevents another piece from occupying the same vertex. For many of the games the number of legal positions is the Fibonacci number of an auxiliary graph. Of interest is not only the Fibonacci number but also the *independence polynomial* of a graph, which is defined as

$$I_G(x) = \sum_{i=0} f_i x^i,$$

where f_i is the number of independent sets of cardinality i (see [14] for example).

Table 1 contains the summary of our findings for the number of legal positions on a path P_n (i.e., n vertices). Note that O12, COL and SNORT have the same generating functions.

Rather than just enumerating all legal positions, a finer measure of a game is to count the number of legal positions with a total of k pieces. Surprisingly, even this is not sufficient to distinguish between the games of COL and SNORT on a bipartite graph (see Theorem 4.2). For full generality, in a hope to distinguish between games, we define a bivariate polynomial. Let P be a placement game played on a board (graph) G. Let n be the number of vertices of G. The *polynomial profile* of P on G is the bivariate polynomial

$$P_{P,G}(x, y) = \sum_{k=0}^{n} \sum_{j=0}^{k} f_{j,k-j} x^j y^{k-j},$$

where $f_{j,k-j}$ is the number of legal positions of P on G which have j Left pieces and $k - j$ Right pieces. Note that $f_{0,0} = 1$ since there is one position with no pieces. Putting $x = y$ gives $P_{P,G}(x, x) = \sum_{i=0}^{n} c_i x^i$, which we will shorten to $P_{P,G}(x)$. This is the polynomial in which c_i is the number of positions with exactly i pieces. Finally, putting $x = y = 1$ counts the total number of positions. We find that these last two objects sometimes give rise to sequences that are listed in the Online Encyclopedia of Integer Sequences (OEIS) [18]. These we note as they occur.

For example, in the game of COL, we have

$$P_{\text{COL},P_3}(x, y) = 1 + 3x + 3y + 6xy + x^2 + y^2 + x^2 y + xy^2;$$
$$P_{\text{COL},P_3}(x) = 1 + 6x + 8x^2 + 2x^3; \quad P_{\text{COL},P_3}(1) = 17.$$

We construct generating functions for the polynomial profiles of these games on a strip. We present only one explicit calculation since the others are similar. (See [9, §1.3 and 1.4] or [11, Chapter 2] for some of the many possible calculation methods.) For the game P, we define

$$GF_P(e, x, y, t) = \sum_{n \geq 0} t^n \sum_{h+i+j=n} f_{h,i,j} e^h x^i y^j,$$

where n is the number of vertices in the path P_n, and $f_{h,i,j}$ is the number of positions with h empty vertices, i Left pieces and j Right pieces. Note, as in O12, i and j may not be the same as the number of colored vertices. In $GF_P(e, x, y, t)$ the coefficient of t^n gives the polynomial that describes all the positions. In practice, we are not interested in h, so we can set $e = 1$ to get

$$GF_P(1, x, y, t) = \sum_{n=0} P_{G,P_n}(x, y) t^n.$$

Several questions suggest themselves.

Question 2.1. For a given independent placement game P, does the closure of the set of roots over all graphs of $P_{P,G}(x)$ cover the complex plane?

Question 2.2. Let P be an independent placement game. Are the coefficients of $P_{P,G}(x)$ unimodal for all graphs G?

Games P and Q are called $\mathcal{G}$-*doppelgänger* if $P_{P,G}(x) = P_{Q,G}(x)$ for all graphs $G \in \mathcal{G}$. We show that COL and SNORT are dopplegänger on bipartite graphs. Under a restricted class of rules, it is shown in [13] that there are no doppelgänger.

Question 2.3. Do there exist placement games P and Q which are doppelgänger for all graphs? Or for any other subclass of graphs other than bipartite? Are there games for which $P_{P,G}(x, y) = P_{Q,G}(x, y)$ for some class of graphs?

3. Profiles of COL, SNORT, CIS, O12, and NOGO on paths

As mentioned in the Introduction, the methods for all the games are similar. We give a proof for COL and omit the others since they similar. We note when the sequences are related to known sequences.

We will be interested in the situation where the board is a strip or path of n vertices which we will denote by P_n. Throughout, we use B to represent a blue (Left) piece and R a red (Right) piece, except in O12 when we'll use RR. In context of the game under consideration, let $f_E(n)$, $f_R(n)$, and $f_B(n)$ be the bivariate polynomials that count the number of positions with, respectively, an uncolored, a red, and a blue rightmost vertex.

3.1. ***The game of* COL.** Given a graph G, with vertices $\{x_1, x_2, \dots, x_n\}$, we define the auxiliary graph G_{COL} with $V(G_{\text{COL}}) = \{x_1, x_2, \dots, x_n\} \times \{1, 2\}$. Vertices (x_i, p) and (x_j, q) are adjacent if $x_i \sim x_j$ and $p = q$ or if $i = j$ and $p \neq q$. That is, G_{COL} is the Cartesian product of G and K_2.

In a position, a blue vertex x_i is identified with $(x_i, 1)$ and a red vertex x_j with $(x_j, 2)$ and the reverse identification for an independent set of G_{COL}. This is a bijection between the positions in COL and independent sets of G_{COL} which forms the proof of the result.

Theorem 3.1. *Let G be a graph then* $P_{\text{COL},G}(x) = I_{G_{\text{COL}}}(x)$ *and thus* COL *is an independent placement game.*

Now we restrict the board to be a path. First we generate the recurrence relations for the positions on P_{n+1}.

Since these are the only three ways a path can end we see that

$$P_{\text{COL},P_{n+1}}(x, y) = f_E(n+1) + f_B(n+1) + f_R(n+1).$$

If a position on P_{n+1} ends with an empty vertex at the right end, the other n vertices can form any legal COL position on P_n; thus, $f_E(n+1) = P_{\text{COL},P_n}(x, y)$. If it ends in a blue (red) vertex then the other n vertices form a legal position that does not end with a blue (red) vertex; therefore,

$$f_B(n+1) = x(f_R(n) + f_E(n)) = xP_{\text{COL},P_n}(x, y) - xf_B(n),$$

n	$P_{\text{COL},P_n}(x,y)$	$P_{\text{COL},P_n}(x)$	$P_{\text{COL},P_n}(1)$
0	1	1	1
1	$1+x+y$	$2x+1+2x$	3
2	$1+2x+2y+2xy$	$2x^2+4x+1$	7
3	$1+3x+3y+6xy+x^2+y^2+x^2y+xy^2$	$1+6x+8x^2+2x^3$	17

Table 2. The first 3 COL polynomials.

likewise,

$$f_R(n+1) = y(f_B(n) + f_E(n)) = yP_{\text{COL},P_n}(x,y) - yf_R(n),$$

and so

$$\begin{aligned} P_{\text{COL},P_{n+1}}(x,y) &= f_B(n+1) + f_R(n+1) + f_E(n+1) \\ &= (1+x+y)P_{\text{COL},P_n}(x,y) - xf_B(n) - yf_R(n). \end{aligned}$$

In the case $x = y$, we have $f_B(n) = f_R(n)$ so that

$$\begin{aligned} P_{\text{COL},P_{n+1}}(x) &= (1+2x)P_{\text{COL},P_n}(x,y) - x(f_B(n) - f_R(n)) \\ &= (1+2x)P_{\text{COL},P_n}(x) - x(P_{\text{COL},P_n}(x,y) - f_E(n)) \\ &= (1+x)P_{\text{COL},P_n}(x) + xP_{\text{COL},P_{n-1}}(x). \end{aligned}$$

Putting $x=1$ gives the number of positions, i.e., $P_{\text{COL},P_{n+1}}(1) = 2P_{\text{COL},P_n}(1) + P_{\text{COL},P_{n-1}}(x,y)$. The first seven coefficients are 1, 3, 7, 17, 41, 99, 239. In [18], this is sequence A001333 *Numerators of continued fraction convergents to sqrt(2).*

Theorem 3.2. *The bivariate generating function for the number of* COL *positions on a path is obtained from*

$$GF_{\text{COL}}(1,x,y,t) = \frac{(1+xt)(1+yt)}{1-(xyt^2+t(1+xt)(1+yt))};$$

the univariate generating function is obtained from

$$GF_{\text{COL}}(1,x,x,t) = \frac{(1+xt)}{1-((1+x)t+xt^2)};$$

and the total number of positions on P_n is $c_n = (1+\sqrt{2})^n + o(1) \simeq 2.414^n$.

Proof. In COL, no adjacent vertices can be colored the same. Therefore, on a path, a position

(1) starts with zero or more empty vertices;

(2) repeated patterns taken from

- B followed by zero or more occurrences of RB followed by at least one E; or
- R followed by zero or more occurrences of BR followed by at least one E; or
- BR followed by zero or more occurrences of BR followed by at least one E; or
- RB followed by zero or more occurrences of RB followed by at least one E;

(3) ends with nothing added;

- B followed by zero or more occurrences of RB; or
- R followed by zero or more occurrences of BR; or
- BR followed by zero or more occurrences of BR; or
- RB followed by zero or more occurrences of RB.

This gives the regular expression

$$E^*((R(BR)^* \mid B(RB)^* \mid RB(RB)^* \mid BR(BR)^*)EE^*)^* \cdot (R(BR)^* \mid B(RB)^* \mid RB(RB)^* \mid BR(BR)^* \mid \epsilon),$$

where ϵ is the empty word. The term $R(BR)^* \mid B(RB)^* \mid RB(RB)^* \mid BR(BR)^*$ occurs twice, and the corresponding expression in the generating function is

$$\frac{xt}{1-xyt^2}+\frac{yt}{1-xyt^2}+\frac{2xyt^2}{1-xyt^2}=\frac{xt+yt+2xyt^2}{1-xyt^2}.$$

The generating function for COL is

$$GF_{\text{COL}}(e,x,y,t)= \left(\frac{1}{1-et}\right)\left(\frac{1}{1-\dfrac{xt+yt+2xyt^2}{1-xyt^2}\dfrac{et}{1-et}}\right)\left(\frac{xt+yt+2xyt^2}{1-xyt^2}+1\right),$$

which gives

$$GF_{\text{COL}}(1,x,y,t)=\frac{(1+xt)(1+yt)}{1-(xyt^2+t(1+xt)(1+yt))};$$

$$GF_{\text{COL}}(1,x,x,t)=\frac{(1+xt)}{1-((1+x)t+xt^2)};$$

and

$$GF_{\text{COL}}(1,1,1,t) = \left(\frac{2+\sqrt{2}}{2\sqrt{2}}\right)\left(\frac{(1)}{1-(1+\sqrt{2})t}\right)+\left(\frac{2-\sqrt{2}}{2\sqrt{2}}\right)\left(\frac{(1)}{1-(1-\sqrt{2})t}\right).$$

From the latter equation we see that the coefficient of t^n in $GF_{\text{COL}}(1, 1, 1, t)$ is

$$c_n = \frac{1-\sqrt{2}}{2}(1-\sqrt{2})^n + \frac{1+\sqrt{2}}{2}(1+\sqrt{2})^n.$$

Since $|1-\sqrt{2}| < 1$ this contribution of this term goes to 0 and so $c_n = (1+\sqrt{2})^n + o(1) \simeq 2.414^n$. □

3.2. *The game of* SNORT. We already know that $P_{\text{SNORT}, P_n}(x) = P_{\text{COL}, P_n}(x)$ but the bivariate polynomials are different.

We define the auxiliary graph G_{SNORT} with $V(G_{\text{SNORT}}) = \{x_1, x_2, \ldots, x_n\} \times \{1, 2\}$. Vertices (x_i, p) and (x_j, q) are adjacent if $i = j$ and $p \neq q$ or if both $x_i \sim x_j$ and $p \neq q$. Another description is that G_{SNORT} is the categorical product of G and K_2 together with the matching edges $((x_i, 1), (x_i, 2))$, $i = 1, 2, \ldots, n$.

In a position, a blue vertex x_i is identified with $(x_i, 1)$ and a red vertex x_j with $(x_j, 2)$ and the reverse identification for an independent set of G_{SNORT}. This is a bijection between the positions in SNORT on G and independent sets of G_{SNORT} which forms the proof of the result.

Theorem 3.3. *Let G be a graph then $P_{\text{SNORT},G}(x) = I_{G_{\text{SNORT}}}(x)$ and thus* SNORT *is an independent placement game.*

In SNORT, we would like to build a position on P_{n+1}. Not surprisingly, we have a similar construction as that for COL:

$$f_B(n+1) = x(f_B(n) + f_E(n)) = x P_{\text{SNORT}, P_n}(x, y) - x f_R(n);$$
$$f_R(n+1) = y(f_R(n) + f_E(n)) = y S_{x,y}(n) - y f_L(n).$$

Since $f_B(1) = x$, $f_R(1) = y$, $f_E(1) = 1$ then

$$P_{\text{SNORT}, P_{n+1}}(x, y) = f_B(n+1) + f_R(n+1) + f_E(n+1) \quad (1)$$
$$= (1 + x + y) P_{\text{SNORT}, P_n}(x, y) - y f_B(n) - x f_R(n). \quad (2)$$

In the next table, the last two columns are the same as Table 2.

Theorem 3.4. *The bivariate generating function for the number of* SNORT *positions on a path is given by*

$$GF_{\text{SNORT}}(1, x, y, t) = \frac{(1 - xyt^2)}{1 - (xt + yt + xyt^2 + t(1 - xyt^2))};$$

the univariate polynomial is given by

$$GF_{\text{SNORT}}(1, x, x, t) = \frac{(1 + xt)}{1 - ((1 + x)t + xt^2)};$$

and the total number of positions on P_n is $c_n = (1 + \sqrt{2})^n + o(1) \simeq 2.414^n$.

3.3. *The game of* CIS. If only one colored piece were being played then every independent set would correspond to a legal position and, as reported in [19] and other papers, the number of independent sets on a path with n vertices is the $(n+2)$-nd Fibonacci number.

Given a graph (board) G with $V(G)=\{a_1, a_2, \ldots, a_n\}$ we construct the auxiliary graph G_{CIS} where $V(G_{\text{CIS}})=\{a_1, a_2, \ldots, a_n\}\times\{1, 2\}$ and $((b, c), (d, e))\in E(G_{\text{CIS}})$ if b is adjacent to d. This is also known as the strong product of G and K_2.

Theorem 3.5. *Let G be a graph then $P_{\text{CIS},G}(x)=I_{G_{\text{CIS}}}(x)$ where $I_{G_{\text{CIS}}}(x)$ is the independence polynomial of G_{CIS}; thus,* CIS *is an independent placement game.*

Proof. We construct a bijection between the legal CIS positions on G and the independent subsets of $V(G_{\text{CIS}})$. A position with i blue pieces on $B=\{a_{b_1}, a_{b_2}, \ldots, a_{b_i}\}$ and $k-i$ red pieces on $R=\{a_{r_1}, a_{r_2}, \ldots, a_{r_{k-i}}\}$ is paired with the set of vertices

$$BR=\{(b, 1) : b\in \mathrm{B}\}\cup\{(r, 2) : r\in \mathrm{R}\}$$

in G_{CIS}. Since $B\cup R$ is independent, so is BR. Any independent set in G_{CIS} can be partitioned into two sets: those with coordinate 1 and those with coordinate 2. The first set is the set of blue pieces and the other is the set of red pieces; the combined set of vertices is an independent set so this is a legal position. □

The recurrence relations are

$$P_{\text{CIS},P_{n+1}}(x, y)=P_{\text{CIS},P_n}(x, y)+(x+y)P_{\text{CIS},P_{n-1}}(x, y).$$

Note that $P_{\text{CIS},P_0}(x, y)=1$ and $P_{\text{CIS},P_1}(x, y)=1+x+y$.

Theorem 3.6. *The bivariate generating function for the number of* CIS *positions is obtained from*

$$GF_{\text{CIS}}(1, x, y, t)=\frac{(1+x+y)}{1-t-xt^2-yt^2};$$

the univariate polynomial is obtained from

$$GF_{\text{CIS}}(1, x, x, t)=\frac{(1+2x)}{1-t-2xt^2};$$

and the number of positions on P_n is $\frac{1}{3}(4\times 2^n+(-1)^{n+1})$.

In particular, the sequence $\{c_n\}=\{1, 3, 5, 11, 21, 43, \ldots\}$ is the Jacobsthal numbers (see A001045 in [18]).

The generating function for the game k-CIS, where there are pieces of k different colors, can be found in a similar fashion.

Corollary 3.7. *In k-*CIS *(that is,* CIS *played with k colors), the generating function is*

$$GF_{k\text{-CIS}}(1, 1, 1, t) = \frac{1+kt}{1-t-kt^2}.$$

We leave the proof to the reader, but note that for $k = 3, 4, 5, 6, 7, 8$ these are the sequences A006130, A006131, A015440, A015441, A015442, and A015443 respectively in [18].

3.4. *The game of* O12. We can define an auxiliary graph G_{O12} for a graph G. Let $V(G_{\text{O12}}) = V(G) \cup E(G)$ and $(a, b) \in E(G_{\text{O12}})$ if one of the following holds:

(i) $a \in V(G)$, $b \in E(G)$, and $a \in b$;

(ii) $a, b \in E(G)$ and $a \cap b \neq \varnothing$.

In other words, G_{O12} is the line graph of G plus the vertices of G where a vertex of G is adjacent to all its incident edges.

Theorem 3.8. *Let G be a graph. Then $P_{\text{O12},G}(x) = I_{G_{\text{O12}}}(x)$, and thus* O12 *is an independent placement game.*

On a general graph, the profile is a symmetric polynomial.

Theorem 3.9. *Let G be a graph on n vertices then $P_{\text{O12},G}(x) = \sum_{i=0}^{n} c_i x^i$ is symmetric, that is, $c_i = c_{n-i}$. If $n = 2m$ and then c_m has the same parity as the number of perfect matchings in G. Moreover, in $P_{\text{O12},G}(x, y) = \sum_{i=0}^{n} \sum_{j=0}^{n} c_{i,j} x^i y^j$ we also have $c_{i,j} = c_{n-i-2j,\, j}$.*

Proof. The proof of all the statements comes from one observation. Let P be a position on G with j Right dominoes and $i - j$ Left pieces and, consequently, $n - (i - j) - 2j = n - i - j$ empty vertices. Interchange empty vertices and Left pieces to get a position with j Right dominoes and $n - i - j$ Left pieces, that is, a position with $n - i$ pieces. Moreover, this is a bijection except, possibly, for the position in which all the vertices of G are occupied by all Right dominoes (the dominoes form a perfect matching and $i = 0$, $j = \frac{1}{2}n$) which is matched to itself. Therefore $c_{i,j} = c_{n-i-2j,j}$. Now

$$c_k = \sum_{j=0}^{k} c_{k-j,j} = \sum_{j=0}^{k} c_{n-k-j,j} = c_{n-k}.$$

If G has $2m$ vertices, then every perfect matching is matched to itself and all the other positions with m pieces are paired off, so the parity of c_m is the same as that of the number of perfect matchings of G. □

Considering just paths, the recurrence relation is

$$P_{\text{O12},P_{n+1}}(x, y) = (1+x) P_{\text{O12},P_n}(x, y) + y P_{\text{O12},P_{n-1}}(x, y).$$

The coefficients of $P_{\text{O}12,P_n}(x)$ (i.e., 1, 1, 1, 1, 3, 1, 1, 5, 5, ...) are the Delannoy numbers; see A008288 [18].

Theorem 3.10. *The bivariate generating function for the number of* O12 *positions on a path is obtained from*

$$GF_{\text{O}12}(1, x, y, t) = \frac{1}{1-((x+1)t+yt^2)};$$

the univariate polynomials are obtained from

$$GF_{\text{O}12}(1, x, x, t) = \frac{1}{1-((x+1)t+xt^2)};$$

and the total number of positions on P_n is $c_n = (\sqrt{2}+1)^{n+1}/(2\sqrt{2})+o(1)$.

The sequence of numbers is 1, 2, 5, 12, 29, 70, 169, etc., which is the sequence of Pell Numbers, A000129 in [18]. When played on K_n the number of positions is the sequence A005425 in [18], which is related to the Hermite polynomials.

3.5. *The game of* NOGO. The empty-vertex-adjacency constraint is a hyperedge condition and so there is no auxiliary graph whose independent sets correspond to the positions in the games. Consequently, NOGO is not an independent placement game.

The recurrence relations for NOGO positions are trickier to generate via considering the last vertex because they do not always arise out of a smaller legal position. These exceptions can be easily identified though.

Consider a position on P_{n+1}. If this position ends with an unoccupied vertex at the right end, the other n vertices can

- form any legal NOGO position on P_n,
- be n blue pieces,
- be n red pieces,
- be i blue vertices which is then followed by legal position on P_{n-i}, $1 \le i \le n-2$ that starts with a red vertex, or
- as in the previous but with interchanging blue and red.

Thus

$$P_{\text{NOGO},P_{n+1}}(x, y) = f_E(n+1) + f_R(n+1) + f_B(n+1).$$

It follows that

$$f_E(n+1) = P_{\text{NOGO},P_n}(x, y) + x^n + y^n + \sum_{i=1}^{n-2}(f_R(n-i)x^i + f_B(n-i)y^i);$$

$$f_B(n+1) = x(f_B(n) + f_E(n)) = x(P_{\text{NOGO},P_n}(x, y) - f_R(n));$$

and likewise

$$f_R(n+1) = y(P_{\text{NOGO},P_n}(x,y) - f_B(n)).$$

Thus

$$P_{\text{NOGO},P_{n+1}(x,y)} = x(P_{\text{NOGO},P_n}(x,y) - f_E(n)) + y(P_{\text{NOGO},P_n}(x,y) - f_E(n)) \\ + P_{\text{NOGO},P_n}(x,y) + x^n + y^n + \sum_{i=1}^{n-2}(f_R(n-i)x^i + f_B(n-i)y^i).$$

Putting $y = x$ gives

$$P_{\text{NOGO},P_{n+1}}(x) \\ = (2x+1)P_{\text{NOGO},P_n}(x) - 2f_E(n) + 2x^n + \sum_{i=1}^{n-2} x^i(P_{\text{NOGO},P_{n-i}}(x) - f_B(n-i)).$$

The total number of positions, 1, 5, 15, 41, 113, 313, 867, 2401, ..., was already known to Tromp and Farnebäck [20] and the sequence is A102620 in [18].

Theorem 3.11. *The bivariate generating function for the number of* NOGO *positions on a path is obtained from*

$$GF_{\text{NOGO}}(1,x,y,t) = \frac{t(1-xyt^2)(1-xt-yt)}{((1-xt)(1-yt)-t-xyt^3)(1-xt)(1-yt)};$$

the univariate polynomial is obtained from

$$GF_{\text{NOGO}}(1,x,x,t) = \frac{(1+xt)(1-2xt)t}{((1-xt)^2-t-2xt^3)(1-xt)};$$

and the total number of positions on P_n is $c_n = 2.769296^n + o(1)$.

4. Relationships between games

As mentioned in the Introduction, if games have the same profile there is the possibility of a bijection between the positions.

4.1. ***Relationship between* COL *and* SNORT.** We now show that the enumeration of positions for COL and SNORT on bipartite graphs are equal.

Lemma 4.1. *Let G be a bipartite graph and let k be a nonnegative integer. The number of legal* COL *positions with k pieces on G is the same as the number of legal* SNORT *positions with k pieces on G.*

Proof. Number the vertices of G with distinct but not necessarily consecutive positive integers such that one color class consists of even numbers and the other odd numbers. We define two transformations.

A : (SNORT $\rightarrow$ COL) Let S_k be a SNORT position with k pieces. Let H be the subgraph of vertices occupied by a piece, and let H' be a connected component of H (necessarily all the vertices of H' are occupied by blue pieces or all by red pieces). Let $x \in V(H')$ be the least numbered vertex in H'. If x is even then in H' interchange red and blue pieces on all the odd numbered vertices. This component forms a legal COL position since no two adjacent vertices are occupied by the same colored piece. Do this for each component and we have a legal COL position of k pieces in G.

B : (SNORT $\rightarrow$ COL) Let C_k be a COL position with k pieces. Let H be the subgraph of vertices occupied by a piece, and let H' be a connected component of H. In H' all vertices occupied by B will be in one color class (i.e., odd or even) and the vertices in the other will be occupied by R. Let $x \in V(H')$ be the least numbered vertex in H'. If x is even then change the pieces in all the odd numbered vertices to the same as that occupying x, and leave the others as they are. If x is odd then change nothing. This component forms a legal SNORT position since no two adjacent vertices have different colored pieces. Do this for each component and we have a legal COL position of k pieces in G.

Let S_k be a SNORT position with k pieces then $B(A(S_k)) = S_k$. Let C_k be a COL position of k pieces then $A(B(C_k)) = S_k$. Therefore we have a bijection between the positions of k pieces and the lemma is proved. □

This result gives the following.

Theorem 4.2. *If G is a bipartite graph then $P_{\text{COL},G}(x) = P_{\text{SNORT},G}(x)$. In particular, $P_{\text{COL},G}(1) = P_{\text{SNORT},G}(1)$, i.e., the number of positions on G is the same for* COL *and* SNORT.

4.2. *The relationship between* COL *and* O12.

Theorem 4.3. *Let n be a positive integer; then*

$$P_{\text{COL},P_{n+1}}(x) = 2x\,P_{\text{O12},P_n}(x) + P_{\text{COL},P_n}(x),$$

$$P_{\text{O12},P_{n+1}}(x) = x\,P_{\text{O12},P_n}(x) + P_{\text{COL},P_n}(x).$$

Proof. For each equality, we give a bijection between the positions.

First, we prove $P_{\text{COL},P_{n+1}}(x) = 2x\,P_{\text{O12},P_n}(x) + P_{\text{COL},P_n}(x)$.

The COL positions on P_{n+1} that start with an empty vertex are paired with the COL positions on P_n (i.e., P_{n+1} minus the first vertex).

Consider the COL positions with k pieces on P_{n+1} that start with a blue piece. We will transform this in to an O12 position with $k-1$ pieces on P_n by starting just after the beginning blue piece and converting the pieces as we progress to the other end of the path by the following rules. For ease of translation, we will change the O12 colors: *aqua* replaces blue and *crimson* replaces red.

(1) A blue piece becomes aqua.

(2) A red piece preceded by a blue piece becomes aqua.

(3) An empty vertex followed by a red piece are both replaced by a crimson domino.

(4) An empty vertex not followed by a red piece is left empty.

For the reverse, starting from a O12 position with $k-1$ pieces on P_n:

(1) Start with a new blue piece.

(2) An aqua piece which now is preceded by an empty or a red piece becomes blue, otherwise it is replaced by a red piece.

(3) A crimson domino is replaced by an empty vertex followed by a red piece.

(4) An empty vertex is left empty.

It is clear that two "blue-start" COL positions map to different O12 positions and that two different O12 positions map to different "blue-start" COL positions. Also, one piece in one game is mapped to one piece in the other. So the two sets have the same cardinality and the COL position has one extra piece.

This leaves the COL positions that start with a red piece. For these, in the previous transformation rules interchange "blue" and "red".

Now we prove $P_{O12,P_{n+1}}(x) = xP_{O12,P_n}(x) + P_{COL,P_n}(x)$.

The O12 positions on P_{n+1} that start with an aqua piece correspond to all the positions of O12 positions on P_n. Using the transformations, the O12 positions with k pieces on P_{n+1} that start with

(a) a crimson piece correspond to all the COL positions with k pieces on P_n that start with a red piece;

(b) those that start with an empty-aqua pair of vertices correspond to a COL positions that starts with a blue piece; and

(c) those that start with an empty-empty pair of vertices correspond to a COL positions that starts with an empty vertex. □

In [18], it is mentioned that Clark Kimberling in (Mar 09 2012) showed that A008288 is jointly generated with A035607 via an array of coefficients of polynomials $u(n,x)$. Initially, $u(1,x) = v(1,x) = 1$, for $n > 1$, $u(n,x) = xu(n-1,x) + v(n-1)$ and $v(n,x) = 2xu(n-1,x) + v(n-1,x)$. These are the same recursions as in the theorem. No proof is referenced in [18] but our proof gives a combinatorial game theory explanation.

References

[1] E. R. Berlekamp. Blockbusting and Domineering, *J. Combin. Theory Ser. A*, **49**:67–116, 1988.

[2] E. R. Berlekamp, J. H. Conway, and R. K. Guy, *Winning ways for your mathematical plays*, A K Peters, 2001.

[3] H. L. Bodlaender and D. Kratsch, Kayles and nimbers, *J. Algorithms*, **43**:106–119, 2002.

[4] C. Chou, O. Teytaud, and S. Yen, Revisiting Monte-Carlo tree search on a normal form game: Nogo, pp. 73–82 in *Applications of evolutionary computation*, Lecture Notes in Computer Science **6624**, Springer, 2011.

[5] Sara Faridi, Svenja Huntemann and Richard J. Nowakowski, Games and complexes, I: Transformation via ideals, in this volume.

[6] Sara Faridi, Svenja Huntemann and Richard J. Nowakowski, Games and complexes, II: Weight games and Kruskal–Katona type bounds, in this volume.

[7] G. E. Farr, The Go polynomials of a graph, *Theoret. Comp. Sci.*, **306**:1–18, 2003.

[8] Graham Farr and Johannes Schmidt, On the number of Go positions on lattice graphs, *Inform. Process. Lett.*, **105**:124–130, 2008.

[9] Philippe Flajolet and Robert Sedgewick, *Analytic combinatorics*, Cambridge University Press, 2009.

[10] Adrien Guignard and Eric Sopena, Compound Node-Kayles on paths, *Theoret. Comp. Sci.*, **410**:2033–2044, 2009.

[11] I. P. Goulden and D. M. Jackson, *Enumerative combinatorics*, Wiley-Interscience, New York, 1983.

[12] Gábor Hetyei, Enumeration by kernel positions, *Adv. in Appl. Math.*, **42**(4):445–470, 2009.

[13] Svenja Huntemann, Richard J. Nowakowski, Doppelgänger games, *Recreat. Math. Mag.* **1**, 2014.

[14] Vadim E. Levit, Eugen Mandrescu, The independence polynomial of a graph: a survey, *Proceedings of the 1st International Conference on Algebraic Informatics*, Aristotle Univ. Thessaloniki, 2005, pp. 233–254.

[15] G. A. Mesdal, Partizan splittles, pp. 447–461 in *Games of No Chance* 3, Cambridge Univ. Press, 2009.

[16] Rebecca Milley and Gabriel Renault, Dead ends in misère play: the misère monoid of canonical numbers, *Discrete Math.* (2013) **313**:2223-223.

[17] Richard J. Nowakowski, Gabriel Renault, Emily Lamoureux, Stephanie Mellon, and Timothy Miller, The game of Timber, *J. Combin. Math. Combin. Comput.*, (2013) **85**: 213–225.

[18] *The on-line encyclopedia of integer sequences*, published electronically at http://oeis.org, 2010. Last visited June 2013.

[19] H. Prodinger and R. F. Tichy, Fibonacci numbers of graphs, *Fibonacci Quart.*, (1982) **20**(1):16–21.

[20] J. Tromp and G. Farnebäck, Combinatorics of Go, *Proceedings of 5th International Conference on Computer and Games*, Torino, Italy, May 2006.

jason.brown@dal.ca — *Mathematics and Statistics, Dalhousie University, Halifax, NS, Canada*

danielle.cox@msvu.ca — *Mount St. Vincent University, Halifax, NS, Canada*

andrew.hoefel@gmail.com — *Queen's University, Kingston, ON, Canada*

neilmckay@dal.ca — *Mathematics and Statistics, Dalhousie University, Halifax, NS, Canada*

rmilley@grenfell.mun.ca — *Computational Mathematics, Grenfell Campus, Memorial University of Newfoundland, Corner Brook, NL, Canada*

r.nowakowski@dal.ca — *Mathematics and Statistics, Dalhousie University, Halifax, NS, Canada*

siegel@mathstat.dal.ca — *Mathematics and Statistics, Dalhousie University, Halifax, NS, Canada*

Games of No Chance 5
MSRI Publications
Volume **70**, 2017

A PSPACE-complete graph nim

KYLE BURKE AND OLIVIA C. GEORGE

We build off G. Stockman's game NIMG to create a version named NEIGHBORING NIM. By reducing from GEOGRAPHY, we show that this game is PSPACE-hard. The games created by the reduction share strong similarities with UNDIRECTED (VERTEX) GEOGRAPHY and regular NIM, though these are both solvable in polynomial-time. This application of graphs can be used as a form of game sum with any rulesets, not only NIM.

1. Background

1.1. ***Algorithmic combinatorial game theory.*** Most of the results here revolve around the computational complexity of determining which player has a winning strategy from a given game position. There exist faster algorithms to solve this problem for some rulesets than for others. For each ruleset, we consider the computational problem that could be solved by such an algorithm. We will refer to both the ruleset and problem by the same name.

We strongly encourage readers unfamiliar with these topics to refer to [1].

1.2. ***Terminology.*** A small amount of nonstandard terminology is used:

- We use the word *sticks* to refer to the objects in nim heaps. Thus, a nim heap of size six contains six sticks.
- An *optimal sequence set* is a set of sequences of plays for both players such that any move deviating from all of the sequences results in an $\mathcal{N}$-position (meaning, the $\mathcal{N}$ext player has a winning move). No move in that sequence should be nonoptimal for either player. Thus, if a player does not know whether they have a winning strategy, adhering to an optimal sequence is at least as good as any other move.

1.3. NIM. NIM is an impartial game played on a collection of heaps, each with a nonnegative number of sticks. On a player's turn, they choose a nonempty pile

MSC2010: primary 91A46; secondary 68Q17.
Keywords: nim, graph nim, neighboring nim, combinatorial game theory, impartial games.

and remove as many sticks as desired (at least one) from that pile. A player loses when they cannot remove sticks (all piles are empty).

NIM is a classic impartial game, being the basis of Nimbers and Sprague–Grundy theory [8; 5]. NIM has lots of nice properties, from easy evaluation of games to obvious composition of two NIM games (the sum is just a new NIM game).

1.4. NIMG. NIM has been extended to incorporate graphs so that nim heaps are assigned to either edges or vertices. There are three different versions of the game named NIMG. In all three versions, a turn consists of both traversing an edge of the graph and removing sticks from a visited element.

1.4.1. *Edge-heap NimG.* Fukuyama describes NIMG where nim heaps are embedded into the edges of the graph [4]. On each turn, the current player chooses an edge to traverse (which has at least 1 stick on it) and removes any number of sticks from that edge. The next player then starts on the vertex on the other end of that edge and must choose an adjacent edge for their move. When there are no more edges with sticks adjacent to the current vertex, the current player loses. Many results for this game are known on complete graphs [2].

1.4.2. *Vertex-heap NimG.* In VERTEX NIMG, players similarly move from one vertex to another, but heaps are connected to the vertices instead of edges [9]. The two variants can be easily described here as: "remove sticks, then move" and "move, then remove sticks". In both cases, a player loses if they cannot complete their turn. The main topic of this paper is a variant of "move, then remove".

1.5. *Geography.* We will use GEOGRAPHY to show the PSPACE-hardness of NEIGHBORING NIM. There are many flavors of GEOGRAPHY; we use the term to refer to DIRECTED VERTEX GEOGRAPHY. This impartial game is played on a directed graph; each turn begins with a vertex already chosen. The current player's turn consists of selecting an arc leaving the chosen vertex that leads to a vertex that hasn't yet been visited during the game. The next player then starts their turn with the resulting vertex selected. We formally describe the ruleset as follows:

Definition 1.1 (GEOGRAPHY — DIRECTED VERTEX). Geography positions are described by $G = (V, E)$ and $v \in V$. Move options for (G, v) are all (G', v') where

- $(v, v') \in E$,
- $V' = V \setminus \{v\}$,
- E' is the subset of E induced by V', and
- $G' = (V', E')$.

GEOGRAPHY is known to be PSPACE-complete [6; 7].

2. NEIGHBORING NIM

We define the ruleset NEIGHBORING NIM to be similar to the "move, then remove" version of NIMG, but also allow players to choose to play on the same vertex as the last move as though each vertex has a self-loop. Note that standard NIM is equivalent to a game of NEIGHBORING NIM on a complete graph with each heap on a separate vertex. A more formal definition follows.

Definition 2.1 (NEIGHBORING NIM). NEIGHBORING NIM positions are described by $G = (V, E)$, $w : V \to \mathbb{N}$, and $x \in V$. The options for (G, w, x) are all (G, w', x') where $w' : V \to \mathbb{N}$ and

- $x' = x$ or $\{x, x'\} \in E$,
- $w'(x') < w(x')$, and
- $\forall v \in V \setminus \{x'\} : w'(v) = w(v)$.

Our main result for this paper is that NEIGHBORING NIM is PSPACE-hard. Since our analysis uses graphs with a small number of sticks on each vertex, we define a version of the game with a bounded number of sticks per vertex.

Definition 2.2 (k-NEIGHBORING NIM). k-NEIGHBORING NIM is the same ruleset as NEIGHBORING NIM, except that the weight function w has bounded range $[0, k]$.

We are able to show that 2-NEIGHBORING NIM is PSPACE-complete, and thus c-NEIGHBORING NIM is also PSPACE-complete for any constant $c \geq 2$. The case for 1-NEIGHBORING NIM is solvable in polynomial time, since this game is equivalent to UNDIRECTED (VERTEX) GEOGRAPHY [3]. Thus, if P $\neq$ PSPACE, allowing a second stick on vertex-heaps is enough to increase the computational hardness of determining the winning player!

3. Computational complexity of NEIGHBORING NIM

3.1. ***PSPACE-hardness.*** The following is the main result of this paper:

Theorem 3.1 (hardness). NEIGHBORING NIM *is PSPACE-hard.*

We will show the hardness of this problem by reducing from the game GEOGRAPHY, which is PSPACE-hard [6].

Proof. Given any GEOGRAPHY position, we will give an algorithm to construct an equivalent NEIGHBORING NIM state, meaning that there is a win in the GEOGRAPHY position exactly when there is a win in corresponding NEIGHBORING NIM position. First we will describe the method for generating these positions, then prove their equivalence.

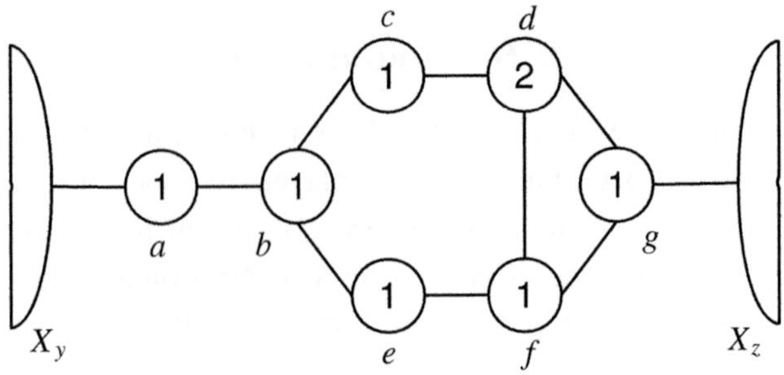

Figure 1. Our main gadget: reduce each directed edge from y to z to the undirected weighted graph shown here.

Let GG be a GEOGRAPHY position on the directed and unweighted graph $G = (V, E)$. We define a new undirected graph, $G' = (V', E')$ with weights on the vertices $w : V' \to \mathbb{N}$ in the following way: $\forall v \in V$: let $X_v \in V'$ and set $w(X) = 1$. Also, $\forall (y, z) \in E$: (edge directed from y to z) let $a_{y,z}$, $b_{y,z}$, $c_{y,z}$, $d_{y,z}$, $e_{y,z}$, $f_{y,z}$, $g_{y,z} \in V'$ where, ignoring the (y, z)-subscripts,

- $w(a) = w(b) = w(c) = w(e) = w(f) = w(g) = 1$,
- $w(d) = 2$, and
- $\{(X_y, a), (a, b), (b, c), (c, d), (b, e), (e, f), (d, f), (d, g), (f, g), (g, X_z)\} \subset E'$.

See Figure 1 for a visual description of this gadget.

We soon show that the resulting G' is the graph for our NEIGHBORING NIM position equivalent to GG. The only final step in the reduction is to declare that if GG has a starting vertex v then $X_v \in V'$ is the starting vertex (where the previous play had been made) in our game and $w(X_v)$ is set to 0 instead of 1.

To complete the reduction, we must show that the structure in Figure 1 "acts" like a directed edge in GEOGRAPHY. Thus, we must prove:

- Moving "backwards" is a losing play. If the previous play was at X_z, then a backwards play would be to remove the only stick at $g_{(y,z)}$. A backwards play results in an $\mathcal{N}$-position.
- The same player moving into the gadget should also move out. If a player moves from X_y to $a_{(y,z)}$, then in an optimal sequence of plays, the same player will move from $g_{(y,z)}$ to X_z.

We prove the former in Lemma 3.2 and the latter in Lemma 3.3. The result is that each of these gadgets (as in Figure 1) in the NEIGHBORING NIM position works just like a (directed) edge in GEOGRAPHY. Trying to go backwards will result in losing and, if players play optimally, they both might as well continue normally through each gadget.

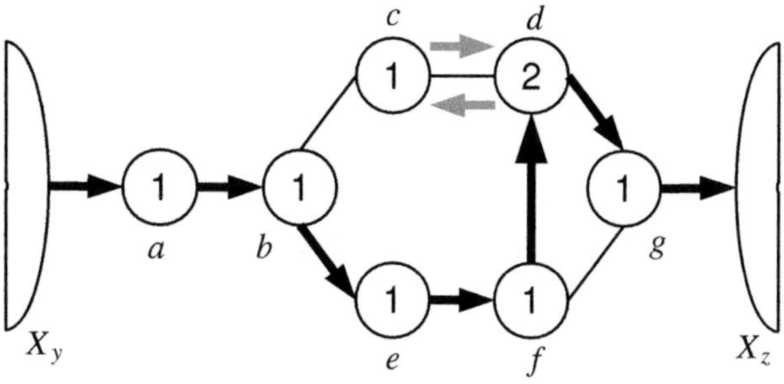

Figure 2. This sequence of moves is "safe" for both players to traverse the gadget. The two gray arrows indicate the potential additional moves in the second sequence. Each move assumes that exactly one object is taken from a vertex.

Lemma 3.2 (don't go backwards). *Any play from* (X_z) *to* $g_{(y,z)}$ *(for all* y*) results in an* $\mathcal{N}$*-position.*

(See Appendix A for a proof of this claim.) This implies that our gadgets are directed: if a player tries to go "backwards" from an X-vertex to an i-vertex, the opponent will have a winning strategy.

To finish showing that our gadget acts like a directed edge, we must prove that "nothing can go wrong" during a regular forward traversal of the structure. To this end, we find two sequences that constitute an optimal sequence set through the gadget, thus showing that neither player benefits from deviating from the sequence. In order to get from one end of the gadget (as in Figure 1) to the other, the following sequence of moves suffices (let Alice and Bob be our two players; we will again ignore subscripts): Alice "takes" a, Bob takes b, Alice takes e, Bob takes f, Alice decrements d by 1, Bob takes g, Alice takes X_z. Note that the same player (in this example, Alice) who chooses to take a also moves to X_z. The other sequence is where Bob takes c instead of g; here Alice will take the remaining object at d and Bob will be forced to take g, rejoining with the first sequence. See Figure 2 for a visual description of the safe sequences. We must prove that neither player benefits from deviating from these sequences. To do this, we show that any deviation is a losing move.

Lemma 3.3 (stick to the script). *Let the notation* $k(p)$ *denote taking* p *objects from vertex* k *in a turn. Then, after the plays* $(\ldots, X_y(1), a(1))$, *any play deviating from the following sequences is a losing move*:

$$(b(1), e(1), f(1), d(1), g(1), X_z(1)),$$
$$(b(1), e(1), f(1), d(1), c(1), d(1), g(1), X_z(1)).$$

(See Appendix B for the proof of this claim.) This implies that once a player makes an appropriate move onto the gadget (playing on an a-node) any "safe" sequence of moves in the gadget results in that same player making the play at the opposite X node. The two above claims combined show that our gadget correctly models a directed edge in a graph just between the X nodes.

Thus, for any edge (y, z) in our GEOGRAPHY position GG, the move to $a_{(y,z)}$ will result in the same player moving to X_z as desired. Also, since we proved players shouldn't go backwards, this game is equivalent to GG; the first player has a winning strategy in GG exactly when the first player has a winning strategy in this NEIGHBORING NIM position.

Thus, NEIGHBORING NIM is PSPACE-hard. □

The hardness of VERTEX NIMG follows directly.

Corollary 3.4 (VERTEX NIMG hardness). VERTEX NIMG *is PSPACE-hard.*

Proof. Neighboring Nim is a special case of VERTEX NIMG where all vertices have self-loops. Thus, VERTEX NIMG is also PSPACE-hard. □

3.2. *Speculation on completeness.* Unfortunately, NEIGHBORING NIM is not automatically PSPACE-complete as games could take a number of moves exponential in the size of the description of the game. For example, a vertex can have a number of sticks exponential in the amount of bits needed to express that number and the rest of the graph. We leave this unsolved as Open Problem 6.1. There are good arguments to conjecture either way.

On one hand, it seems to not be inside PSPACE. Games can last an exponential number of turns, so the game trees are extremely tall. A straight-forward brute-force traversal can't be performed in polynomial space.

On the other hand, it might be inside PSPACE. Although there are many EXPTIME-hard rulesets, the authors know only of loopy examples. This means they can have positions that repeat during the course of a game, which cannot occur in NEIGHBORING NIM. Additionally, Nim heaps are well-understood; perhaps increasing the size of the heaps doesn't greatly increase the difficulty of finding strategies. It may also be that if there are only m different heap sizes on vertices, you can substitute them with a set of m significantly smaller sizes.

3.3. *PSPACE-complete versions.* We can sidestep this problem a bit by using our bounded-heap-size version of the game.

Corollary 3.5 (2-NEIGHBORING NIM completeness). k-NEIGHBORING NIM *is PSPACE-complete for any $k \geq 2$.*

Proof. The result of the reduction from Theorem 3.1 is always a 2-NEIGHBORING-NIM position. Thus, the PSPACE-hardness holds for this subset of positions as well. The positions are in PSPACE because k bounds the maximum number of moves per vertex. □

4. Generalization

This graph-embedding technique works with games other than NIM. Given a graph, assign different game states to the vertices, and use similar rules: players may make one move legal in the game in any vertex neighboring the last play. We define this formally.

Definition 4.1 (NEIGHBORING-R). Given any ruleset R, NEIGHBORING-R has positions of the form $G = (V, E)$, $w : V \to \text{positions}(R)$, and $x \in V$. The left options for (G, w, x) are (G, w', x') where $w' : V \to \text{positions}(R)$ and

- $x' = x$ or $\{x, x'\} \in E$,
- $w'(x')$ is a left option of $w(x')$, and
- $\forall v \in V \setminus \{x'\} : w'(v) = w(v)$.

The right options are defined analogously: $w'(x')$ must be a right option of $w(x')$.

4.1. ***Inequivalent positions.*** This new definition allows a NEIGHBORING NIM vertex to contain multiple heaps instead of only a single heap. Although each nim position is equivalent to a single heap, that equivalence doesn't carry over in the neighboring situation. Consider the two NEIGHBORING NIM games in Figure 3. (The previous move was made on the leftmost vertex in both cases.) The values of the games embedded in the left vertices are both 0, the values on the middle vertices are both 0, and the values of the rightmost games are both $*$. However, the overall value of the positions are not equivalent.

In the top game, there are no move options, so the value is 0. In the bottom position, the next player can move to the middle vertex, even though the value of the nim game there is also zero. After that move there are exactly two moves remaining. Thus, the initial game has exactly three moves remaining and has a value of $*$.

4.2. ***Generalized hardness.*** The next result allows us to say something about the hardness of graph-embedded versions of many impartial games.

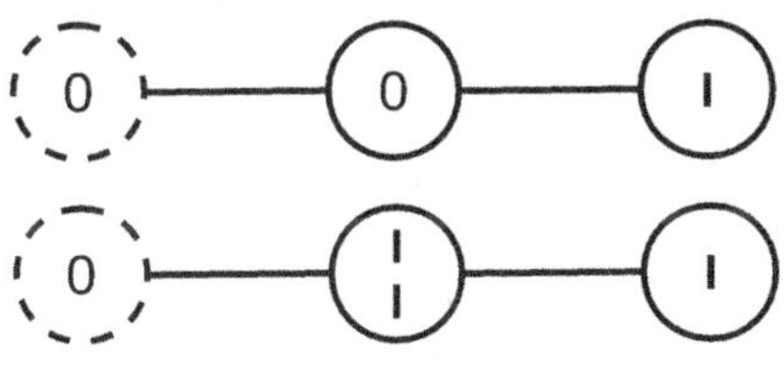

Figure 3. Two NEIGHBORING NIM positions. In both, the last move was made on the dashed vertex. The value of the top game is zero, the value of the bottom game is $*$.

Theorem 4.2 (neighboring-R hardness). *For any ruleset R which has positions identical to* 0, $*$, *and* $*2$, NEIGHBORING-R *is PSPACE-hard.*

Proof. Two positions are *identical* if they have isomorphic game trees. Replacing the nim heaps of size 1 and 2 with $*$ and $*2$, respectively, in the reduction of Theorem 3.1 doesn't change the winnability of the resulting games. Thus, the reduction applies to R. □

5. Conclusions

Building on algorithmic work analyzing different versions on NIMG, we present NEIGHBORING NIM, a new PSPACE-hard game.

An interesting aspect of the hardness of NEIGHBORING NIM is the juxtaposition with VERTEX GEOGRAPHY. 1-NEIGHBORING NIM is the same ruleset as UNDIRECTED VERTEX GEOGRAPHY, which is solvable efficiently [3]. 2-NEIGHBORING NIM, however, is PSPACE-hard.

Furthermore, we can replace NIM and apply the graph-embedding concept to any other ruleset R to create NEIGHBORING-R.

6. Future work

There are many extensions to the work described here. The most prominent is certainly the unknown completeness of NEIGHBORING NIM with any number of sticks.

Open Problem 6.1. NEIGHBORING NIM $\in$ PSPACE?

Additionally, the computational hardness of other flavors of NIMG remains unsolved.

Open Problem 6.2. What is the computational complexity of EDGE NIMG?

Open Problem 6.3. What is the computational complexity of VERTEX NIMG on graphs without self-loops?

Other explorable problems include the hardness of other versions of NEIGHBORING-R.

Open Problem 6.4. Is NEIGHBORING-R PSPACE-hard if R includes any positions *equivalent* to $*$ and $*2$?

(Note that Open Problem 6.4 is a stronger statement than shown here because equivalent does not necessarily mean identical.)

Open Problem 6.5. For which other computationally easy rulesets R is NEIGHBORING-R hard?

Open Problem 6.6. Are there strictly partisan positions of a ruleset R that can be used to show NEIGHBORING-R is hard? How small can the game trees be to get a hard game?

7. Acknowledgments

The authors would like to thank those who played NEIGHBORING NIM with them, including three excellent Wittenberg University students: Deanna Fink, Dang Mai and Ernie Heyder, as well as Professor Doug Andrews who defeated the first author over and over again. We would also like to thank Professor Adam Parker for listening to initial versions of the hardness proof and proofreading this paper.

Appendix A: Proof of Lemma 3.2

Lemma 3.2 (don't go backwards). *Any play from* (X_z) *to* $g_{(y,z)}$ *(for all* y*) is suboptimal.*

We will refer to the player who moves from X_z to g (we will leave out the subscript for the internal vertices) as the "foe" while the other player is the "hero". We will show that the hero has a winning strategy after a backwards move. We can now look at two cases, each depending on the state of the game outside the gadget.

The first is the case where the move from a to X_y would be a winning play. In this case, the hero can next move from g to d and take both of the objects there. The foe has two options, both of which, we show, allow the hero to win:

(1) *The foe moves to* c. In this case the hero must choose to go to b. The foe can now either choose to move to a — in which case the hero will gladly move to X_y and win as we assumed — or to e. Then the hero simply takes the object at f and, as there are no more moves, the hero has won.

(2) *The foe moves to* f. The hero must then take e and the foe must take b. The hero can then move to c and win the game.

The second major case assumes that the move from a to X_y is a losing play. Here, the hero can still move to d (from g) but will take only one of the objects. Now the foe has three options: taking the other object at d, moving to c or moving to f. We show all to be losses:

(1) *Foe moves to* c. Now the hero should take the remaining object at d. The following sequence must occur: foe must take f, hero at e, foe at b, hero at a, followed by the foe at X_y, a losing move by our assumption.

(2) *Foe takes the remaining object at* d. The hero will choose to take c, so the foe must take b. The hero can then take a, forcing the foe to take X_y, a losing move by our assumption.

(3) *Foe takes f.* The hero should then take e so the foe must take b. Again, the hero can take a, so the foe must move to X_y, a losing move by our assumption.

Thus, it is a losing play to move from an X-vertex to a g-vertex.

Appendix B: Proof of Lemma 3.3

Lemma 3.3 (stick to the script). *Let the notation $k(p)$ denote taking p objects from vertex k in a turn. Then, after the plays $(\ldots, X_y(1), a(1))$ any play deviating from the following sequences is a losing move*:

$$(b(1), e(1), f(1), d(1), g(1), X_z(1)),$$
$$(b(1), e(1), f(1), d(1), c(1), d(1), g(1), X_z(1)).$$

We continue by analyzing all possible deviations from these sequences and show that they result in a loss. In this claim, we will refer to the deviating player as the foe and the other player as the hero. We will show that the foe loses in each case. It may be helpful to refer to Figure 2 during these case descriptions.

(1) $c(1)$ *instead of* $e(1)$. Here we have two subcases: either moving from g to X_z is a winning (result is a $\mathcal{P}$-position) or losing (an $\mathcal{N}$-position) move. If it's in $\mathcal{N}$, then the hero can respond to $c(1)$ with $d(1)$. If the foe then chooses $g(1)$, the hero can take the remaining stick in d with $d(1)$; $f(1)$ and $e(1)$ must follow with the hero winning. If the foe instead chooses $f(1)$, the hero can win instantly by choosing $e(1)$. For the foe's last chance, they could select $d(1)$, removing the other stick from d. The hero should respond with $g(1)$. The foe will lose by selecting $f(1)$, because the hero will win at $e(1)$, but the foe will also lose with $X_z(1)$, an $\mathcal{N}$-position, as assumed.

   If $X_z(1)$ is instead leaves the board in $\mathcal{P}$, the hero should respond to $c(1)$ with $d(2)$. The foe could choose $f(1)$, but the hero can then win with $e(1)$. Instead, the foe can choose $g(1)$ in which case the hero can choose $X_z(1)$ and win, as assumed.

(2) $d(2)$ *instead of (the first)* $d(1)$. Here the hero has a simple move to win. By taking $c(1)$ there are no further moves and the foe has lost.

(3) $g(1)$ *instead of (the first)* $d(1)$. The hero can respond with $d(1)$. This leaves two different adjacent vertices with 1 object apiece and no other adjacent nonempty vertices. Either move by the foe results in one remaining move and a win for the hero.

(4) $d(1)$ *instead of* $c(1)$. The hero can respond with $c(1)$ and win.

(5) $d(1)$ *instead of* $X_z(1)$. This cannot happen in the second sequence, but if it happens in the first, the hero can respond with $c(1)$ and win.

Thus, any deviation from the two sequences specified in the claim puts the game in an $\mathcal{N}$-position.

References

[1] Erik D. Demaine and Robert A. Hearn. Playing games with algorithms: Algorithmic combinatorial game theory. In Michael H. Albert and Richard J. Nowakowski, editors, *Games of No Chance* 3, volume 56 of *Mathematical Sciences Research Institute Publications*, pages 3–56. Cambridge University Press, 2009.

[2] Lindsay Erickson. Nim on complete graphs. *Ars Combin.*, 119:117–127, 2015.

[3] Aviezri S. Fraenkel, Edward R. Scheinerman, and Daniel Ullman. Undirected edge geography. *Theor. Comput. Sci.*, 112(2):371–381, 1993.

[4] Masahiko Fukuyama. A nim game played on graphs. *Theor. Comput. Sci.*, 1-3(304):387–399, 2003.

[5] P. M. Grundy. Mathematics and games. *Eureka*, 2:198–211, 1939.

[6] David Lichtenstein and Michael Sipser. Go is polynomial-space hard. *J. ACM*, 27(2):393–401, 1980.

[7] C. H. Papadimitriou. *Computational Complexity*. Addison Wesley, Reading, Massachusetts, 1994.

[8] R. P. Sprague. Über mathematische Kampfspiele. *Tôhoku Math. J.*, 41:438–444, 1935.

[9] Gwendolyn Stockman. Presentation: The game of nim on graphs: NimG, 2004. Available at http://www.aladdin.cs.cmu.edu/reu/mini_probes/papers/final_stockman.ppt.

kwburke@plymouth.edu

Department of Computer Science and Technology, Plymouth State University, Plymouth, NH, United States

Games of No Chance 5
MSRI Publications
Volume **70**, 2017

A nontrivial surjective map onto the short Conway group

ALDA CARVALHO AND CARLOS PEREIRA DOS SANTOS

This paper explores the general question "Is there a natural habitat for the short Conway group?" by looking for a ruleset with a legal position for each short game value. Surprisingly, a ruleset with this property exists in combinatorial game theory literature and it is implemented in Siegel's CGSuite software. A proof that KONANE is an affirmative answer to the question is presented, making it the first known universal ruleset.

1. Introduction

The main result of this paper states that all the short combinatorial games are game values of particular positions of KONANE (Theorem 14 in Section 4). To prove this result constructively, two instrumental lemmas giving the needed building "pieces" are used (Lemmas 12 and 13 in Section 3). Some fundamental results of combinatorial game theory, like reduction concepts and the largest game value of the day n, are also used. Readers can find these results in Section 3. The next paragraphs of this introduction and Section 2 have some basics: the rules of KONANE, the state of the art and motivation for this research, and the definitions of *habitat* and *universality of a ruleset* (Definitions 3 and 7). Readers who know the rules of KONANE and are fluent in combinatorial game theory may wish to proceed to Lemma 12, Lemma 13 and Theorem 14 (Sections 3 and 4).

One of the principal goals of combinatorial game theory (CGT) is the study of combinatorial rulesets with the following properties ([1; 2; 5; 15] are fundamental references, [7] is a complete survey):

- There are two players who take turns moving alternately.
- No chance devices such as dice, spinners, or card deals are involved, and each player is aware of all the details at all times.

This work was partially funded by Fundação para a Ciência e a Tecnologia through projects UID/MAT/04721/2013 (Pereira dos Santos) and UID/Multi/00491/2013 (Carvalho).
MSC2010: 13F55, 91A46.
Keywords: combinatorial game theory, universality, habitat.

- Even ignoring the alternating condition, the play must end in a finite number of moves, and the winner is often determined on the basis of who made the last move. Under normal play, the last player wins, while in misère play, the last player loses.

We distinguish between multiple meanings of the word *game* by using the words *ruleset* and *game*. The word *ruleset* has a concrete meaning related to some particular set of rules (KONANE, AMAZONS, NIM are examples of rulesets). The word *game*, by contrast, has the abstract mathematical meaning defined by Conway [2; 5]. When we speak of the *game value* of a game, we are emphasizing that it is being considered in this latter sense, as an algebraic object which can be compared for equality with, or added to, other games.

The *options* of a game are all those positions which can be reached in one move. In CGT, games can be expressed recursively as $G = \{\mathcal{G}^L \mid \mathcal{G}^R\}$ where $\mathcal{G}^L$ are the Left options and $\mathcal{G}^R$ are the Right options of G. The *followers* of G are all the games that can be reached by all the possible sequences of moves from G.

Conway made a recursive construction based in a transfinite sequence of *days* [5]. His inductive definition constructs the proper class of combinatorial games (we will only consider the normal play). The games with finite sets of options $\mathcal{G}^L$ and $\mathcal{G}^R$ are called *short games* and are born before the day ω. The games born on day ω or after are called *long games*.

Founding fathers of CGT observed that independent components naturally arise in several rulesets. The analysis of these decompositions led to the definition of a *disjunctive sum*.

Definition 1 (disjunctive sum). Let G and H be games. Then,

$$G + H = \{\mathcal{G}^L + H, G + \mathcal{H}^L \mid \mathcal{G}^R + H, G + \mathcal{H}^R\}$$

(note that if G is a game and $\mathcal{S}$ a set of games, $A + \mathcal{S} = \{G + H : H \in \mathcal{S}\}$).

The proper class of combinatorial games (short and long) with the disjunctive sum is an abelian group. In fact, with a suited order relation (motivated by the game practice), it is an abelian group with a partial order.

Definition 2 (short Conway group). The *short Conway group* is the subgroup of the proper class of combinatorial games containing the short Normal-play games.

A classical example of a combinatorial ruleset is NIM, first studied by C. Bouton [3]. NIM is played with piles of stones. On his turn, each player can remove any number of stones from any pile. The winner is the player who takes the last stone. NIM is an example of an impartial ruleset: Left options and Right options are the same for the game and all its followers. The values involved in NIM are called nimbers (stars):

$$*k = \{0, *, \dots, *(k-1) \mid 0, *, \dots, *(k-1)\}.$$

It is a surprising fact that all impartial rulesets take only nimbers as values (Sprague–Grundy theorem; see [8; 16]).

2. NIM dimension and the concept of habitat

The Sprague–Grundy theorem states that for every impartial G there is a nonnegative integer n such that $G = *n$. It is also well known that in partizan rulesets we still can construct nimbers (see [1; 2; 5; 15]). This motivates some very natural questions.

Definition 3. Let $\mathcal{S}$ be a set of combinatorial game values. A ruleset A is a *habitat* of $\mathcal{S}$ if for every $G \in \mathcal{S}$ there is a position of A with game value equal to G.

Berlekamp asked the question "What is the habitat of $*2$?" [9]. It was possible to generalize the question: "For a given ruleset, what is the largest n such that $*2^n$ is the game value of a legal position?" [10; 11; 12; 13]. This led to the definition of nim dimension and to the proposal of some processes to analyze the problem.

Definition 4. A combinatorial ruleset has *nim dimension n* if it contains a position with game value $*2^n$ but not $*2^{n+1}$. A ruleset has *infinite* nim dimension if all the nimbers can be constructed. It has *null*, or $\varnothing$, nim dimension if $*$ cannot be constructed.

Observation 5. A combinatorial ruleset A has nim dimension n if it is a habitat of $\{0, *, *2, *4, \dots, *2^n\}$ and it is not habitat of $\{0, *, *2, *4, \dots, *2^n, *2^{n+1}\}$.

The rulesets KONANE and AMAZONS are "case studies" related to *nim dimension.* KONANE is a classical Hawaiian ruleset, considered very interesting by CGT researchers [4; 6]. In the starting position, the checkered board is filled in such a way that no two stones of the same color occupy adjacent squares. In the opening, two adjacent pieces of the board are removed. After this, a player moves by taking one of his stones and jumping orthogonally over an opposing stone into an empty square. The jumped stone is removed. A player can make multiple jumps on his turn but cannot change direction mid-turn. Multiple jumps are not mandatory. The winner is the player who makes the last move. KONANE is implemented in Siegel's CGSuite, a fundamental computational tool for CGT research [14]. CGSuite, as well as the main result of this paper, regards the generalized version of KONANE.

Convention. In the *generalized* version of KONANE two stones of the same color can occupy adjacent cells.

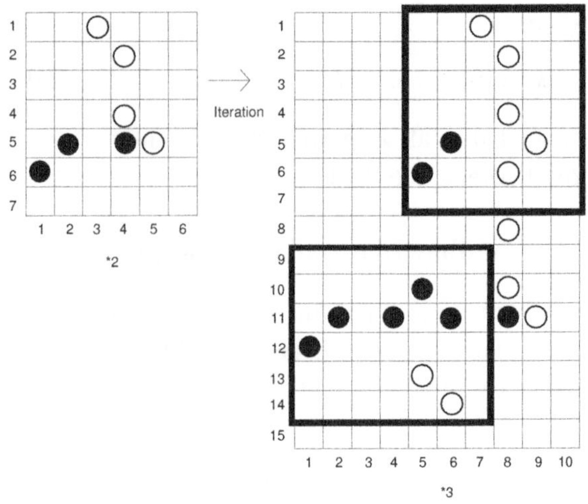

Figure 1. Nimbers in KONANE.

AMAZONS is a combinatorial ruleset invented in 1988 by Walter Zamkauskas. It is played on a checkered board and each move consists of two parts: moving one of one's own amazons one or more empty cells in a straight line (orthogonally or diagonally), exactly as a queen moves in CHESS; it may not cross or enter a cell occupied by an amazon of either color or a stone. After moving, the amazon shoots a stone from its landing cell to another cell, using another queen-like move. This stone may travel in any orthogonal or diagonal direction and, like an amazon, cannot cross or enter a cell where another stone has landed or an amazon of either color stands. The winner is the player who makes the last move. As usual in CGT, Left plays with the black stones and Right with the white ones in both rulesets.

In [12], the *fractal process* was proposed in order to prove that the nim dimension of KONANE is infinite. The basic idea was to construct the $*n$ using the previous $*(n-1)$. Figure 1 shows an iteration to obtain a $*3$ from a previous $*2$.

In [13], the *algebraic process* was proposed. The idea was the implementation of an algebraic table where the entries were the "needed stuff" to construct a $*n$. The amazons moves "created" the entries of the table splitting the initial position into two disjoint components. With this process, it was possible to construct the first known $*4$ in AMAZONS (Figure 2).

Conjecture 6. *The nim dimension of* AMAZONS *is* 2.

Finally, in [10], the *embedding process* was proposed. Considering an initial ruleset A, the idea was to find a construction process in some other ruleset B and embed it in A. Obviously, B should be somehow better understood than A. It was shown that the nim dimension of TRAFFIC LIGHTS is infinite (embedding in

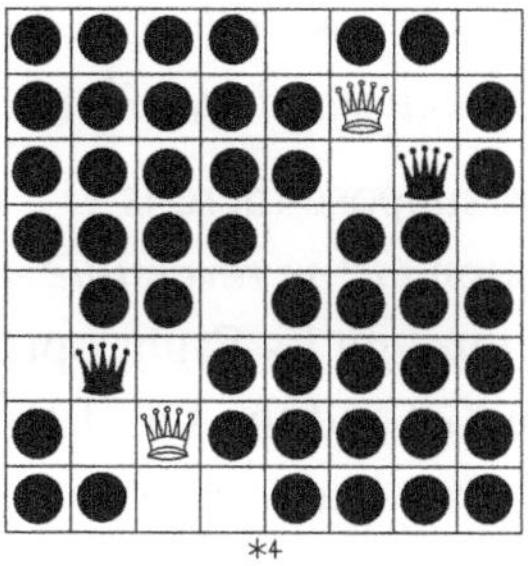

Figure 2. $*4$ in AMAZONS.

it constructions made in REGIO).

It is possible to generalize Berlekamp's original question even further.

Definition 7. A ruleset A is *universal* if it is a habitat of the short Conway group.

Natural is the new question "Is there a natural habitat for the short Conway group?". A classical universal ruleset was not known. In the following sections this state changes. With a constructive idea very similar to the *fractal process*, a universal ruleset is revealed.

3. Preliminary results

The main purpose of this text is to answer the general question "Is there a habitat for the short Conway group?". In Section 4, we prove that all the short combinatorial games are game values of KONANE's positions, and so, this well-known ruleset is a habitat of the short Conway group. In this section, some needed previous lemmas are proved.

KONANE is a very rich combinatorial ruleset with several interesting values; see Figure 3, left and middle. Figure 3, right, shows an example with value 1 in the generalized version.

We now recall some important results of combinatorial game theory.

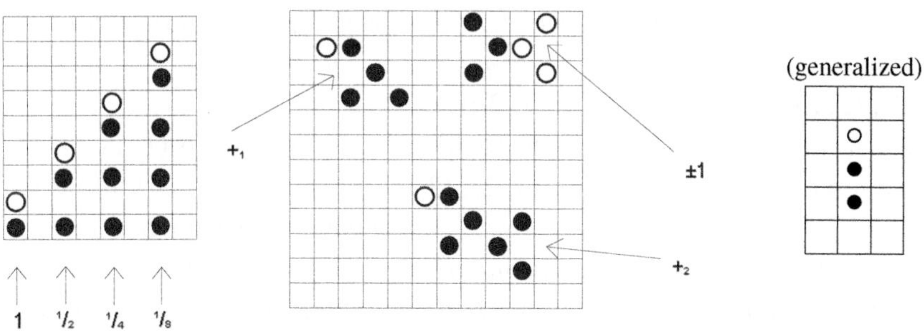

Figure 3. Left and middle: several game values in KONANE. Right: the value 1 in generalized KONANE.

Definition 8. For a game G, suppose $G_1^L, G_2^L \in \mathcal{G}^L$ with $G_2^L \geq G_1^L$. Then the Left option G_1^L is said to be *dominated* by G_2^L (or that G_2^L *dominates* G_1^L).

Definition 9. For a game G, suppose there is $G^L \in \mathcal{G}^L$ and $G^{LR} \in (G^L)^{\mathcal{R}}$ with $G^{LR} \leq G$. Then the Left option G^L is *reversible*, and sometimes, to be specific, G^L is said to be *reversible through* its Right option G^{LR}. In addition, G^{LR} is called a *reversing* option for G^L and the set of Right options of G^{LR}, $(G^{LR})^{\mathcal{L}}$, is a *replacement set* for G^L (eventually empty).

Theorem 10 (reductions of combinatorial games; see [2, pp. 60–63]). (a) *Let G be a game and suppose $G_1^L, G_2^L \in \mathcal{G}^L$ such that G_1^L is dominated by G_2^L. Then $G = \{\mathcal{G}^L \setminus \{G_1^L\} \mid \mathcal{G}^R\}$ in the sense of the equality of games.*

(b) *Let G be a game and suppose that G^L is a Left option of G reversible through G^{LR}. Then $G = \{\mathcal{G}^L \setminus \{G^L\}, (G^{LR})^{\mathcal{L}} \mid \mathcal{G}^L\}$ in the sense of the equality of games.*

Analogous versions of (a) *and* (b) *hold for Right.*

A reversible move for Left is one which Right can promise to respond to in such a way that prospects are at least as good as they were before. In any context, Right promises "if you ever choose option A_1 of G then I will immediately move to A_1^R". So, Left just chooses the option A_1 if he intends to follow up Right's move to A_1^R with an immediate response to one of A_1^R's Left options. If he plans some other move elsewhere, he might just as well start with that. We say that A_1 *reverses through* A_1^R to the Left options of A_1^R. If the Left options of A_1^R are empty then we say that A_1 *reverses out*. G is in *canonical form* if G and all of G's followers have no dominated or reversible options.

The next one is also very well known.

Theorem 11 [1, p. 119]. *The largest game born by day n is n.*

Because the proof for the main result is constructive, we will need some patterns to build the construction. The next two lemmas are useful KONANE patterns.

Lemma 12 (rubber bands). *Consider the* KONANE *pattern shown in Figure 4.*

The position P_n is a $(2n+3)\times 5$-rectangle. The black pieces occupy the cells $(1, 2), (2, 2), (1, 4), (2, 4), \ldots, (1, 2n), (2, 2n)$.

The white pieces occupy the cells $(1, 0), (1, 2n+2)$, *the cells* $(1, 1), (1, 3), \ldots, (1, 2n+1)$, *and the cells* $(3, 2), (4, 2), (3, 4), (4, 4), \ldots, (3, 2n), (4, 2n)$.

Let $P_n \backslash (1, 2k+1)$ be the exposed positions without *the white piece in the position* $(1, 2k+1)$. *Call $a(k)$ the number of black stones above the line $2k+1$ and $b(k)$ the number of black stones bellow the line $2k+1$. The game values of P_n and $P_n \backslash (1, 2k+1)$ are the following*:

(1) *P_n has value* 0 (*there are no moves*).

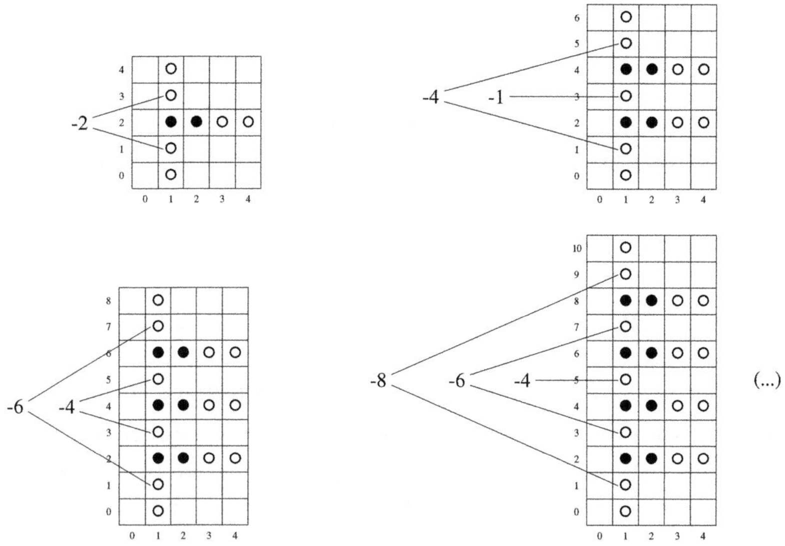

Figure 4. KONANE pattern for rubber bands.

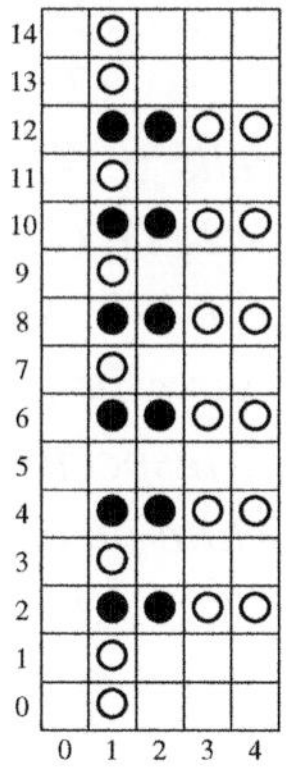

Figure 5. Illustrative case.

(2) $P_2\backslash(1,3)$ *has value* -1.

(3) $P_n\backslash(1,2k+1)$ *different from* $P_2\backslash(1,3)$ *and has value* $-\max(a(k),b(k))$.

Previous observation: For ease, let us consider a very particular case to understand the idea behind the general case; see Figure 5.

If Right plays $(1,7)\mapsto(1,5)$ the value becomes $\{-8 \mid -6\} = -7$ (simplicity rule). If Right plays $(1,3)\mapsto(1,5)$ the value becomes $\{-4 \mid -2\} = -3$ (simplicity rule).

So, the original position is $\{ \mid -7, -3\}$ which, by domination, is equal to $\{ \mid -7\} = -8$. Right's move must be made in the opposite direction of the largest number of black stones.

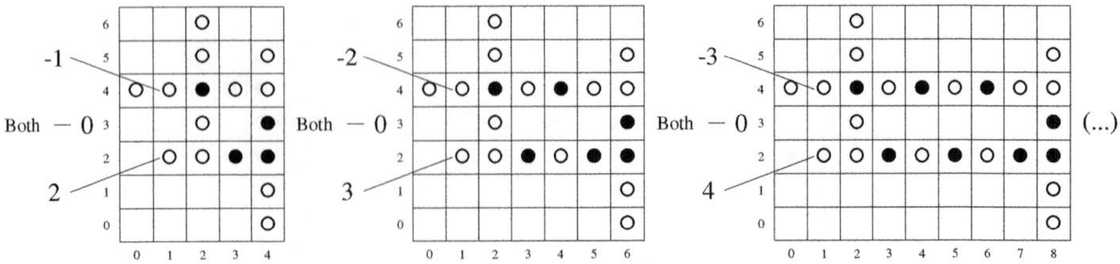

Figure 6. KONANE pattern for taps.

Proof. The cases P_1 and P_2 can be analyzed with pure calculations. Consider P_n with $n > 2$.

The idea for the general case P_n with $n > 2$ follows as in the previous observation: Right's move must be made in the opposite direction of the largest number of black stones. After removing the piece in the cell $(1, 2k+1)$, the game value is $\{ \mid -\max(a(k), b(k)) + 1\}$ which is equal to $-\max(a(k), b(k))$. □

Lemma 13 (taps). *Consider the* KONANE *pattern shown in Figure 6.*

The position P_n is a $7 \times (2n+3)$-rectangle.

The black pieces occupy the cells $(2n+2, 2)$, $(2n+2, 3)$ and the cells $(2, 4), (3, 2), (4, 4), (5, 2), \ldots, (2n, 4), (2n+1, 2)$.

The white pieces occupy the cells $(0, 4), (1, 2), (1, 4), (2, 2), (2, 3), (2, 5), (2, 6)$, the cells $(2n+2, 0), (2n+2, 1), (2n+2, 4), (2n+2, 5)$, and the cells $(3, 4), (4, 2), (5, 4), \ldots, (2n, 2), (2n+1, 4)$.

Let $P_n \backslash (1, 2)$ and $P_n \backslash (1, 4)$ be the exposed positions without *the white pieces in the positions $(1, 2)$ and $(1, 4)$, respectively. Let $P_n \backslash \{(1, 2), (1, 4)\}$ be the exposed positions* without both *white pieces in the positions $(1, 2)$ and $(1, 4)$. The game values of P_n, $P_n \backslash (1, 2)$, $P_n \backslash (1, 4)$ and $P_n \backslash \{(1, 2), (1, 4)\}$ are the following*:

(1) *P_n has value 0 (there are no moves).*

(2) *$P_n \backslash (1, 2)$ has value $n+1$.*

(3) *$P_n \backslash (1, 4)$ has value $-n$.*

(4) *$P_n \backslash \{(1, 2), (1, 4)\}$ has value 0.*

Proof. In $P_n \backslash (1, 4)$, Right can take, one by one, all the n black pieces of the line 4. So, the game value of $P_n \backslash (1, 4)$ is $-n$.

In $P_n \backslash (1, 2)$, Left can take, one by one, all the n white pieces of the line 2 plus the white piece of the cell $(2, 3)$, leaving the value 0 in the line 4. The game value of $P_n \backslash (1, 2)$ is $n+1$.

In $P_n \backslash \{(1, 2), (1, 4)\}$, there are exactly n moves for both players (the capture of the white stones of the line 2 and the capture of the black stones of the line 4). So, the first player loses and the game value of $P_n \backslash \{(1, 2), (1, 4)\}$ is 0. □

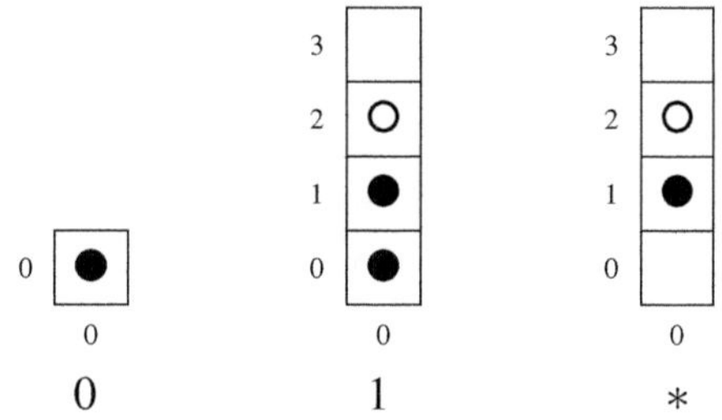

Figure 7. Initial games 0, 1, and $*$ (day 1).

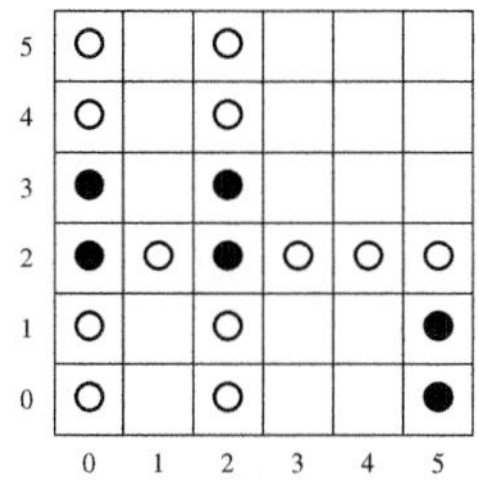

Figure 8. Joining of rubber bands.

4. KONANE is a universal ruleset

The proof for the main result is constructive. The games of day n are constructed with games of the previous days, generating a recursive process Γ such that

$$G = \Gamma(\mathcal{G}^L, \mathcal{G}^R).$$

The "design type" of $\Gamma(\mathcal{G}^L, \mathcal{G}^R)$ must be the same as the "design types" of the games in $\mathcal{G}^L$ and $\mathcal{G}^R$. In our proof, the "design type" is basically the shape. All the games are constructed in rectangular areas. The rectangular areas of the set of the options generate a new game in a larger rectangular area which will be useful to build new rectangles.

Theorem 14. *All the short combinatorial games are game values of particular positions of* KONANE.

Proof. First, we present the initial games 0 (day 0) and 1 and $*$ (day 1) in Figure 7. It is possible to join rubber bands without changing the values. For example, consider the position for the value 1 shown in Figure 8.

Second, we present the *connecting scheme* in Figure 9. If, at some point, a black stone moves to the cell (5, 1) removing the stones in (0, 1), (2, 1) and (4, 1), the game value becomes 1. Moreover, *and this is one of the principal ideas of the proof*, it is possible to choose rubber band sizes such that the incomplete captures reverse out (Theorem 11 and Lemma 12).

Third, we present the general idea of the recursion in Figure 10. In the initial

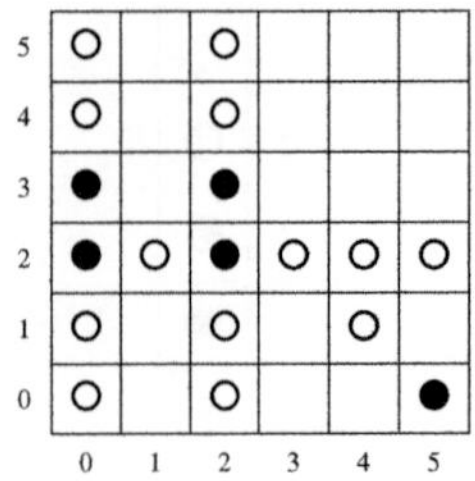

Figure 9. Connecting scheme example.

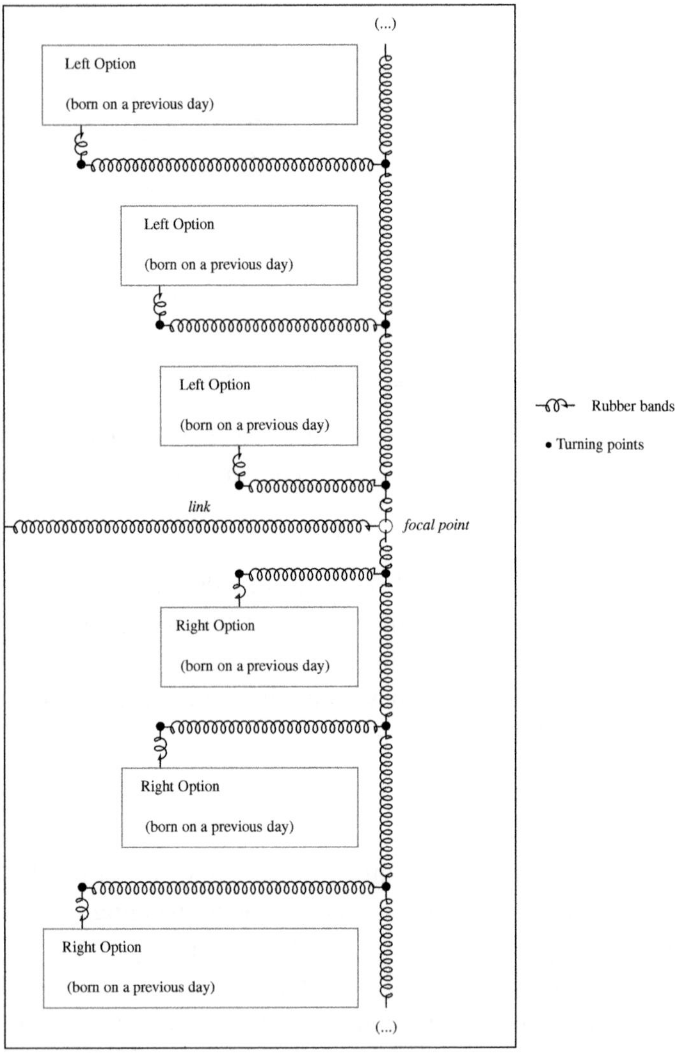

Figure 10. Recursion example.

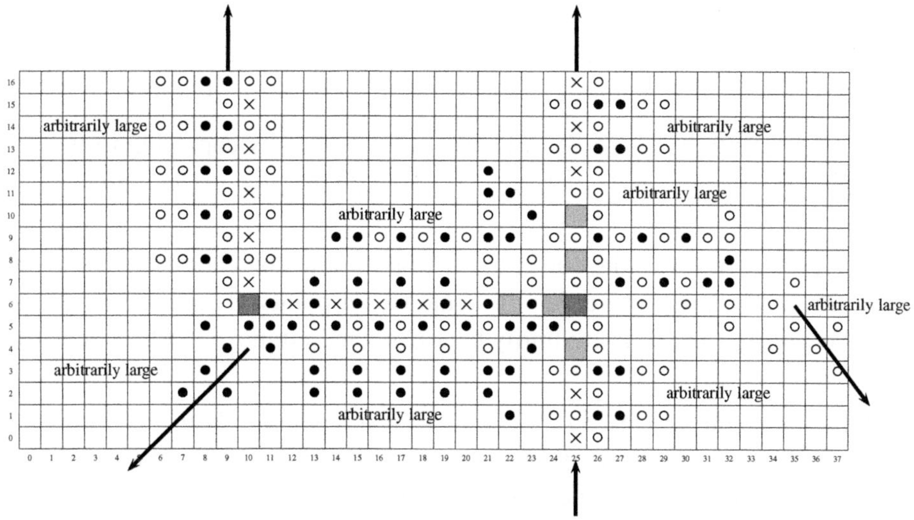

Figure 11. Turning point example.

position, the *focal point* is the pair of cells occupied by the only pair of stones that can move.

The construction of each rectangular position is made with *rubber bands*, *turning points* based in *taps*, *one link* (itself a rubber band) and *rectangular positions* of the previous days. It is important to observe that there are no moves in these rectangular positions. The focal point is occupied by opposing stones only after a complete capture through the link; after that, a game value of the previous days is obtained.

The rubber bands are arbitrarily large. So, when we construct a game of the day n with options of the previous days, we choose the rubber band sizes in such a way that for every incomplete Left capture (incomplete Right capture) exists an integer $G^{LR} \leq -n$ ($G^{RL} \geq n$). Therefore, these options reverse out. This is possible due to Lemma 12.

The turning points are constructed with the following idea: when Left (Right) captures to a turning point he makes a threat larger than or equal to n (smaller than or equal to $-n$). To defend the threat, Right (Left) has only one good option, creating again a threat smaller than or equal to $-n$ (larger than or equal to n). After these two forced moves, the only way for Left (Right) to defend the second threat is to choose an option of the previous days. The idea is based again in reversibility: instead of constructing directly an option G^L (G^R), we construct an option $\{T \geq n, \ldots \mid \{G^L, \ldots \mid T' \leq -n, \ldots\}\}$ ($\{\{T \geq n, \ldots \mid G^R, \ldots\} \mid \ldots, T' \leq -n\}$).

Fourth, we present the details of the turning points. Consider the picture showing a turning point in Figure 11. The crosses represent options adjacent to rubber bands that reverse out. So, when we have a sequence of captures made by a dark piece in the column 25, we only need to analyze the moves to the gray and dark gray cells. The turning point is the cell (25, 6) (dark gray).

(1) If Left captures to the cell (25, 4), Right replies by capturing to the cell (22, 4) obtaining more than or equal to $-n$ points (Lemma 12). So, the Left move to (25, 4) reverses out.

(2) If Left captures to the cell (25, 8), Right replies by capturing to the cell (25, 7) obtaining more than or equal to $-n$ points. This happens due to Lemma 13 (the tap is arbitrarily large) So, the Left move to (25, 8) reverses out.

(3) If Left captures to the cell (25, 10), she opens completely the tap, turning it equal to zero (Lemma 13). Right replies capturing to the cell (22, 10) and, afterwards, Right's stone in (22, 10) will capture to (22, 8), obtaining at least $-n$ points.

After the Left move to (25, 6), a threat in (35, 6) is created (at least n points).

(1) If Right captures to the cell (24, 6), Left replies by capturing to the cell (24, 7). Afterwards, the move from (24, 7) to (20, 7) will get at least n points. So, the Right move to (24, 6) reverses out.

(2) If Right captures to the cell (22, 6), Left replies by capturing to the cell (22, 7), obtaining at least n points. So, the Right move to (22, 6) reverses out.

The good Right move is to (10, 4), creating a threat in (10, 4) (at least $-n$ points). Left has to go up to an option that is a game of the previous days. □

The proof that KONANE is an universal ruleset was constructive. Using the process illustrated in the proof, as an example, we finish the paper presenting the impressive construction of the game $\{0, \uparrow *, \pm 1 \,||\, -1, \{* | -1\}\}$ (game of the day 3). Figure 12 illustrates a terminal position useful for future constructions (it is a rectangle). If we remove the white stone of the cell (37, 29) while putting a black stone in the cell (38, 29), a situation that occurs after a capture of the white pieces of the row 29, we get the game $\{0, \uparrow *, \pm 1 \,||\, -1, \{* | -1\}\}$. Observe that the colored rectangles are the options of the previous days.

References

[1] M. H. Albert, R. J. Nowakowski, D. Wolfe. *Lessons in play: an introduction to combinatorial game theory*, A K Peters, 2007.

[2] E. R. Berlekamp, J. Conway, R. Guy. *Winning ways*, Academic Press, London, 1982.

[3] C. L. Bouton. "Nim, a game with a complete mathematical theory", *Ann. Math.* 3 (2), 35–39, 1902.

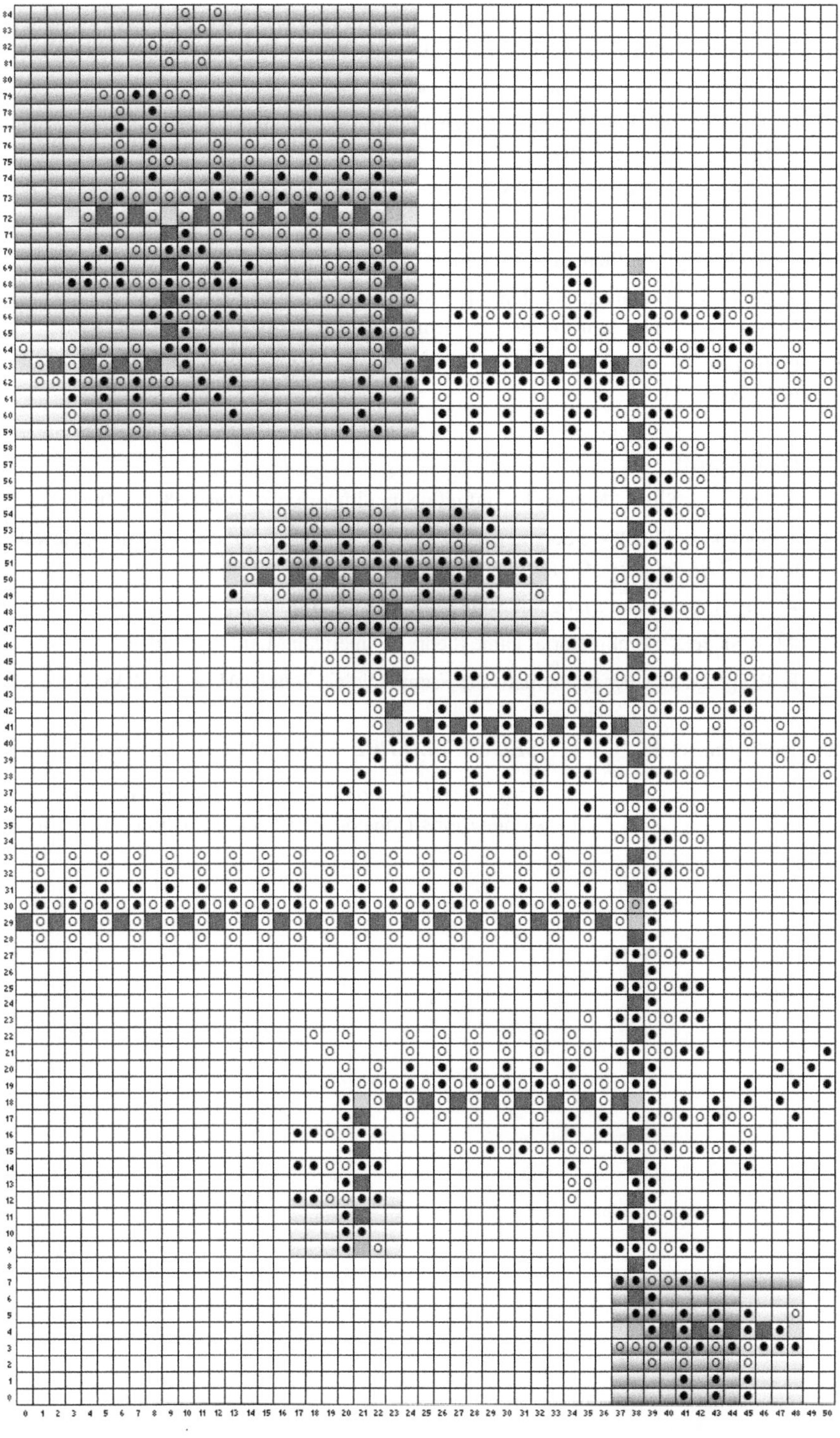

Figure 12. Terminal position for future constructions.

[4] A. Chan, A. Tsai. "$1 \times n$ Konane: a summary of results", *More games of no chance*, R. J. Nowakowski (ed.), *MSRI Publ.*, 42, Cambridge University Press, 331–339, 2002.

[5] J. Conway. *On numbers and games*, Academic Press, 1976.

[6] M. D. Ernst. "Playing Konane mathematically: a combinatorial game-theoretic analysis", *UMAP J.*, 16, 95–121, 1995.

[7] A. S. Fraenkel. "Combinatorial games: selected bibliography with a succinct gourmet introduction", *Electr. J. Combin.*, 2012.

[8] P. M. Grundy. "Mathematics and games", *Eureka*, 1939.

[9] R. K. Guy. "Unsolved problems in combinatorial games", *Games of no chance*, R. J. Nowakowski (ed.), *MSRI Publ.*, 29, Cambridge University Press, 475–49, 1996.

[10] C. Santos. "Embedding processes in combinatorial game theory", *Discrete Math.*, 159, 675–682, 2011.

[11] C. Santos. *Some notes on impartial games and Nim dimension*, PhD Thesis, University of Lisbon, 2010.

[12] C. Santos, J. Silva. "Konane has infinite Nim dimension", *Integers* **8**(1), 2008.

[13] C. Santos, J. Silva."Nimbers in partizan games", *Games of no chance* 4, R. J. Nowakowski (ed.), *MSRI Publ.*, 63, Cambridge University Press, (2015).

[14] A. Siegel. Combinatorial game suite, *http://www.cgsuite.org/*

[15] A. N. Siegel. "Combinatorial game theory", American Math. Soc., 2013.

[16] R. P. Sprague. "Über mathematische Kampfspiele", *Tohoku Math. J.*, 1935.

acarvalho@adm.isel.pt *ISEL & CEMAPRE, University of Lisbon, Lisboa, Portugal*

cmfsantos@fc.ul.pt *Center for Functional Analysis, Linear Structures and Applications University of Lisbon, Lisboa, Portugal*

Games of No Chance 5
MSRI Publications
Volume **70**, 2017

Games and complexes I: Transformation via ideals

SARA FARIDI, SVENJA HUNTEMANN
AND RICHARD J. NOWAKOWSKI

Placement games are a subclass of combinatorial games which are played on graphs. We will demonstrate that one can construct simplicial complexes corresponding to a placement game, and this game could be considered as a game played on these simplicial complexes. These complexes are constructed using square-free monomials.

1. Introduction

We will demonstrate a relationship between a subclass of combinatorial games, such as DOMINEERING and COL, and algebraic structures defined on simplicial complexes. There are two relationships, one via the maximal legal positions and the other through the minimal illegal positions. We will begin by giving the necessary background, first from combinatorial game theory, then from combinatorial commutative algebra.

For a game, *perfect information* means that both players know which game they are playing, on which board, and the current position. No *chance* means that no dice can be rolled or cards can be dealt, or any other item involving probability can be used.

Definition 1.1. A *combinatorial game* is a 2-player game with perfect information and no chance, where the two players are *Left* and *Right* (denoted by L and R respectively) and they do not move simultaneously. Then a game is a set P of *positions* with a specified starting position. *Rules* determine from which position to which position the players are allowed to move. A *legal position* is a position that can be reached by playing the game from the starting position (which is legal) according to the rules. Moving from position P to position Q is

The first and third authors' research was supported in part by the Natural Sciences and Engineering Research Council of Canada.
MSC2010: 13F55, 91A46.
Keywords: combinatorial games, simplicial complexes, square-free monomial ideal, placement game.

called a *legal move* if both P and Q are legal positions and the move is allowed according to the rules. Q is usually called an option of P.

In this paper, a combinatorial game will be denoted by its name in SMALL CAPS. Well-known examples of combinatorial games are CHESS, CHECKERS, TIC-TAC-TOE, GO, and CONNECT FOUR. Examples of games that are not combinatorial games include bridge, backgammon, poker, and snakes and ladders.

Although games usually have a "winning condition" associated to them, i.e., rules as to which player wins, for the purposes of this paper games do not need to have a notion of winning identified.

We will assume that the board on which games are played is a graph (or can be represented as a graph). A space on a board is then equivalent to a vertex and we use the two terms interchangeably.

Definition 1.2. A *strong placement game* is a combinatorial game which satisfies the following:

(i) The starting position is the empty board.

(ii) Players place pieces on empty spaces of the board according to the rules.

(iii) Pieces are not moved or removed once placed.

(iv) The rules are such that if it is possible to reach a position through a sequence of legal moves, then any sequence of moves leading to this position consists only of legal moves.

The TRIVIAL placement game on a board is the strong placement game that has no additional rules.

A *basic position* is a board with only one piece placed. Any position, whether legal or illegal, in a strong placement game can be decomposed into basic positions.

The concept of a placement game originates in Brown et al [2] where condition (iv) is replaced by the condition that if it is legal to place a piece at one point, it must have been legal at any point before. We call this type of game a "medium placement game". A "weak placement game" is a combinatorial game that satisfies the above conditions (i) through (iii).

Note that (iv) implies that every subposition of a legal position is also legal.

Placement games were only recently defined formally by Brown et al. in [2], even though several placement games, for example TIC-TAC-TOE or DOMINEERING, have been known and studied for a long time. In this work, we will consider strong placement games exclusively.

Throughout this paper, "placement game" refers to a strong placement game. Here are three more we will use as examples.

R		R	L	R		L	$\cong$	R		R	$+$	R		L		

Figure 1. A COL position that is the disjunctive sum of two COL positions.

Definition 1.3. In SNORT, players may not place pieces on a vertex adjacent to a vertex containing a piece from their opponent.

Definition 1.4. In COL, players may not place pieces on a vertex adjacent to a vertex containing one of their own pieces.

Definition 1.5. In NOGO, at every point in the game, for each maximal group of connected vertices of the board that contain pieces placed by the same player, one of these needs to be adjacent to an empty vertex.

In these games, the pieces only occupy one vertex each, which is in fact not necessary. For example in CROSSCRAM [8] and DOMINEERING [1] the players' pieces occupy two adjacent vertices.

Definition 1.6. The *disjunctive sum* between two positions of combinatorial games G and H is the position in which a player can play in one of G and H but not both simultaneously.

Assuming implicitly that placement games are part of a disjunctive sum implies that a board might be filled with more pieces of one player than of the other. Making this assumption is very useful since in many placement games the board might "break up" into the disjunctive sum of smaller boards.

Example 1.7. For an example, consider COL played on the path P_7. Then the position on the left of Figure 1 is equivalent to the one in which the middle space is "deleted" (on the right), i.e., it is equivalent to the disjunctive sum of the two COL positions on the right, one of which has two Right pieces but no Left pieces.

For a placement game G and a board B, let

$$f_i(G, B)$$

denote the number of positions with i pieces played, regardless of which player the pieces belong to. If the game and board are clear from context, we shorten the notation to f_i.

Definition 1.8 (Brown et al. [2]). For a game G played on a board B, the *game polynomial* is defined to be

$$P_{G,B}(x) = \sum_{i=0}^{k} f_i(G, B)x^i.$$

$P_{G,B}(1)$ is then the total number of legal positions of the game.

The motivation for game polynomials came from Farr [6] in 2003 where the number of end positions and some polynomials of the game GO were considered, and work in this area was continued by Tromp and Farnebäck [10] in 2007 and by Farr and Schmidt [7] in 2008. Even though GO is not a placement game since pieces are removed, it shares many properties with this class of games. Thus it was natural for the authors of [2] to consider the concept of game polynomials for placement games.

We will now introduce concepts from combinatorial commutative algebra that we will need to construct simplicial complexes equivalent to placement games.

Definition 1.9. A *simplicial complex* Δ on a finite vertex set V is a set of subsets (called *faces*) of V with the conditions that if $A \in \Delta$ and $B \subseteq A$, then $B \in \Delta$. The *facets* of a simplicial complex Δ are the maximal faces of Δ with respect to inclusion. A *nonface* of a simplicial complex Δ is a subset of its vertices that is not a face. The *f-vector* $(f_0, f_1, \dots, f_k)$ of a simplicial complex Δ enumerates the number of faces f_i with i vertices. Note that if $\Delta \neq \varnothing$, then $f_0 = 1$.

In the algebraic literature, the f-vector of a complex is usually indexed from -1 to $k-1$ as this is the "dimension" of the face (the number of vertices minus 1). Due to the connection between placement games and simplicial complexes, we have chosen the combinatorial indexing.

Recall that an *ideal* I of a ring $R = R(+, \cdot)$ is a subset of R such that $(I, +)$ is a subgroup of R and $rI \subseteq I$ for all $r \in R$.

Let k be a field and $R = k[x_1, \dots, x_n]$ a polynomial ring. Given a simplicial complex Δ on n vertices, we can label each vertex with an integer from 1 to n. Each face F (resp. nonface N) of Δ can then be represented by a square-free monomial of R by including x_i in the monomial representing the face F (resp. the nonface N) if and only if the vertex i belongs to F (resp. N). We then have the following (see [3] and [4] for more information).

Definition 1.10. The *facet ideal* of a simplicial complex Δ, denoted by $\mathcal{F}(\Delta)$, is the ideal generated by the monomials representing the facets of Δ. The *Stanley–Reisner ideal* of a simplicial complex Δ, denoted by $\mathcal{N}(\Delta)$, is the ideal generated by the monomials representing the minimal nonfaces of Δ.

Definition 1.11. The *facet complex* of a square-free monomial ideal I, denoted by $\mathcal{F}(I)$, is the simplicial complex whose facets are represented by the square-free monomials generating I. The *Stanley–Reisner complex* of a square-free monomial ideal I, denoted by $\mathcal{N}(I)$, is the simplicial complex whose faces are represented by the square-free monomials not in I.

To clarify these concepts, we will give two examples.

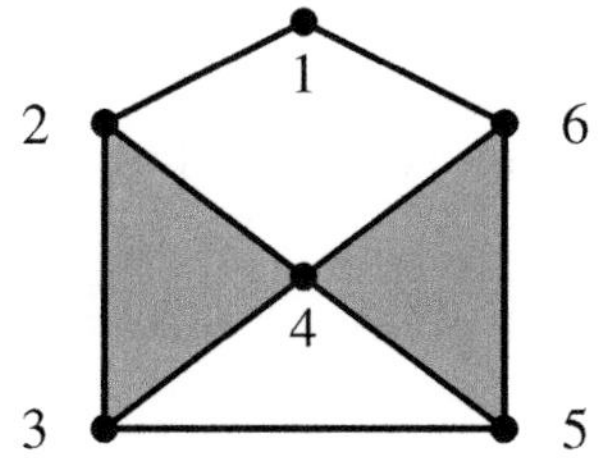

Figure 2. An example of a simplicial complex.

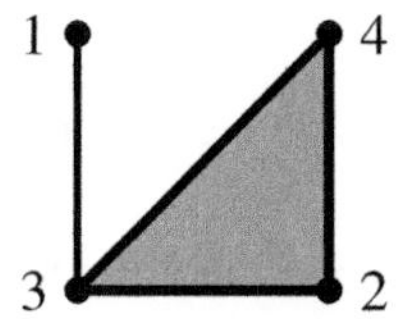

Figure 3. Facet complex of $I = \langle x_1x_3, x_2x_3x_4\rangle$.

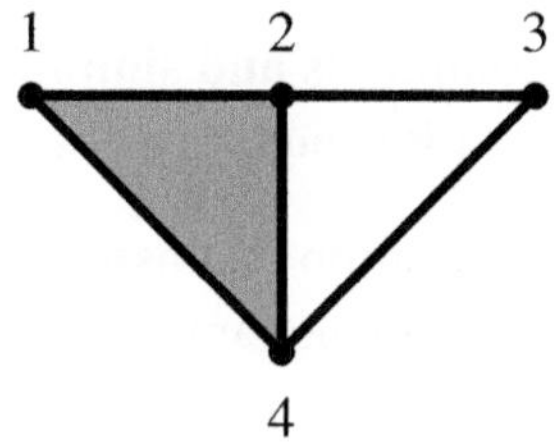

Figure 4. Stanley–Reisner complex of $I = \langle x_1x_3, x_2x_3x_4\rangle$.

Example 1.12. Consider the simplicial complex Δ in Figure 2 with the labeling of the vertices as given.

The facet ideal of Δ then is

$$\mathcal{F}(\Delta) = \langle x_1x_2, x_1x_6, x_2x_3x_4, x_3x_5, x_4x_5x_6\rangle,$$

and the Stanley–Reisner ideal of Δ is

$$\mathcal{N}(\Delta) = \langle x_1x_3, x_1x_4, x_1x_5, x_2x_5, x_2x_6, x_3x_4x_5, x_3x_6\rangle.$$

Example 1.13. Consider the square-free monomial ideal $I = \langle x_1x_3, x_2x_3x_4\rangle$. The facet complex $\mathcal{F}(I)$ is given in Figure 3 and the Stanley–Reisner complex $\mathcal{N}(I)$ is given in Figure 4.

It is clear that the facet operators are inverses of each other, i.e., $\mathcal{F}(\mathcal{F}(\Delta)) = \Delta$ and $\mathcal{F}(\mathcal{F}(I)) = I$, from their definitions. This is also true of the Stanley–Reisner operators: A minimal nonface of $\mathcal{N}(I)$ is a minimal monomial generator of I, thus a generator of I, showing $\mathcal{N}(\mathcal{N}(I)) = I$. Similarly, since $\mathcal{N}(\Delta)$ contains

1	2	3

Figure 5. Labeling P_3.

L	R	L	$x_1y_2x_3$	L		R	x_1y_3
R	L	R	$y_1x_2y_3$	R		L	y_1x_3

Figure 6. Maximum legal positions for COL on P_3.

all monomials representing nonfaces, a square-free monomial not in $\mathcal{N}(\Delta)$ has to be a face of Δ, thus $\mathcal{N}(\mathcal{N}(\Delta)) = \Delta$.

This shows that both the facet and the Stanley–Reisner operators give a bijection between the set of all square-free monomial ideals in n variables and the set of all simplicial complexes on n vertices.

2. Constructing monomials and simplicial complexes from placement games

We will now introduce a construction that associates a set of monomials and a simplicial complex to each placement game.

Given a placement game G on a board B, we can construct a set of square-free monomials in the following way: First, label the basic positions by $1, 2, \ldots, n$. For each legal position we then create a square-free monomial by including x_i if Left has played in position i and y_j if Right has placed in position j. The empty position (before anyone has started playing) is represented by 1.

Example 2.1. Consider COL played on the path P_3. We label the basic positions, in this case the spaces of the board, as given in Figure 5.

The maximum legal positions and their corresponding monomials are given in Figure 6.

Using these monomials, we can build a simplicial complex $\Delta_{G,B}$ on the vertex set $V = \{x_1, \ldots, x_n, y_1, \ldots, y_o\}$ by letting a subset F of V be a face if and only if there exists a square-free monomial m representing a legal position such that each element of F divides m.

Definition 2.2. A simplicial complex that can be constructed from a placement game G on a board B in this way is called a *legal complex* and is denoted by $\Delta_{G,B}$.

Example 2.3. Consider COL played on the path P_3. Using the notation from Example 2.1, we get the legal complex Δ_{COL,P_3} as given in Figure 7.

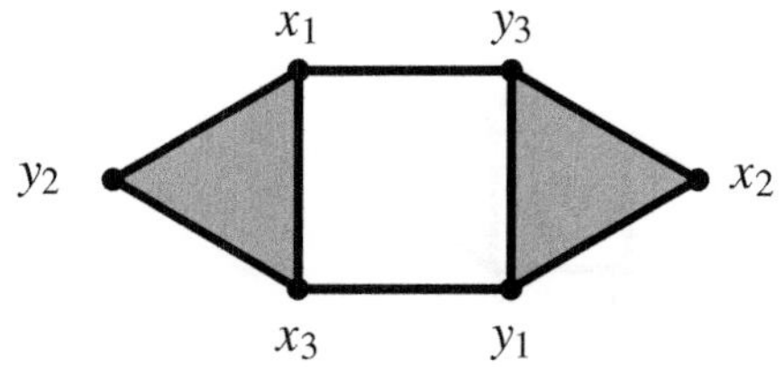

Figure 7. The legal complex Δ_{COL,P_3}.

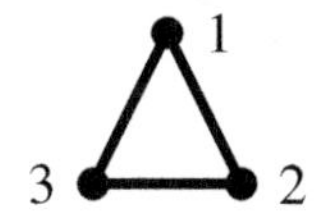

Figure 8. Labeling C_3.

Observe that the maximum legal positions of a game, i.e., the positions in which no piece can be placed by either Left or Right (so the game ends), correspond to the facets of $\Delta_{G,B}$ and thus uniquely determine $\Delta_{G,B}$.

In game theoretic terms, the f-vector of a legal complex $\Delta_{G,B}$ indicates that there are f_i legal positions with i pieces in the game G, regardless if pieces belong to Left or to Right. Thus for placement games the entries of the f-vector of the legal complex $\Delta_{G,B}$ are the coefficients of the game polynomial $P_{G,B}$. Therefore we have the following.

Proposition 2.4.

$$\begin{aligned} f_i(G, B) &= \textit{number of legal positions in } G \textit{ with } i \textit{ pieces played on } B, \\ &= \textit{number of degree } i \textit{ monomials representing legal positions in } G, \\ &= \textit{number of faces with } i \textit{ vertices in } \Delta_{G,B}, \end{aligned}$$

and we can use any of these concepts to find f_i.

This also justifies using the same notation for the coefficients of a game polynomial as for entries of a f-vector.

We now give three more examples for the construction of monomials and simplicial complexes.

Example 2.5. The cycle C_3 is labeled as in Figure 8.

Now consider COL on C_3. The monomials corresponding to the maximum legal positions are

$$\{x_1y_2,\ x_1y_3,\ x_2y_3,\ y_1x_2,\ y_1x_3,\ y_2x_3\}.$$

Also consider SNORT played on P_3 and C_3. The maximum monomials then are

$$\{x_1x_2x_3,\ y_1y_2y_3,\ x_1y_3,\ x_3y_1\}$$

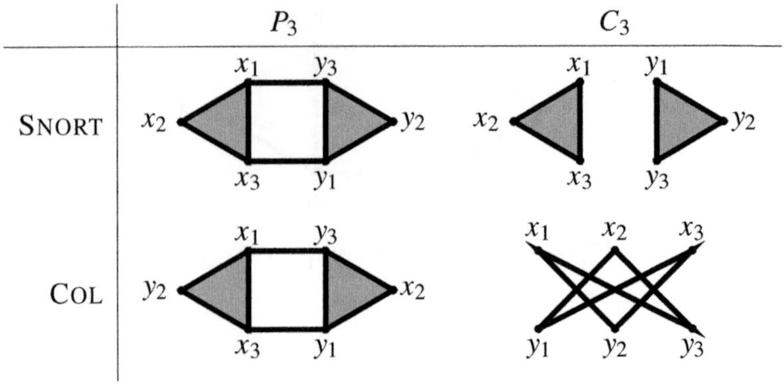

Figure 9. The legal complexes $\Delta_{\text{SNORT},P_3}$, $\Delta_{\text{SNORT},C_3}$, Δ_{COL,P_3}, and Δ_{COL,C_3}.

and

$$\{x_1x_2x_3,\ y_1y_2y_3\},$$

respectively.

The legal complexes of all three games are given in Figure 9.

Note that the legal complexes of COL and SNORT on P_3 are isomorphic. This is true whenever COL and SNORT are played on a bipartite graph; see [9].

3. The ideals of a placement game

Through the monomials that represent legal or illegal positions of a game, we can also associate square-free monomial ideals with a placement game.

Definition 3.1. The *legal ideal*, $\mathcal{L}_{G,B}$, of a placement game G played on the board B is the ideal generated by the monomials representing maximal legal positions of G.

Definition 3.2. The *illegal ideal*, $\mathcal{ILL}_{G,B}$, of a placement game G played on the board B is the ideal generated by the monomials representing minimal illegal positions of G.

Definition 3.3. The *illegal complex*, sometimes called the *auxiliary board* [2], of a placement game G played on the board B, is the simplicial complex whose facets are represented by the monomials of the minimal illegal positions of G. It is denoted by $\Gamma_{G,B}$.

The authors in [2] introduce the auxiliary board for "independence placement games", which is the class of placement games for which the illegal complex is a graph. The term "independence game" was chosen since the independence sets of $\Gamma_{G,B}$ (considered as a graph) correspond to the legal positions of G played on B, i.e., the faces of $\Delta_{G,B}$.

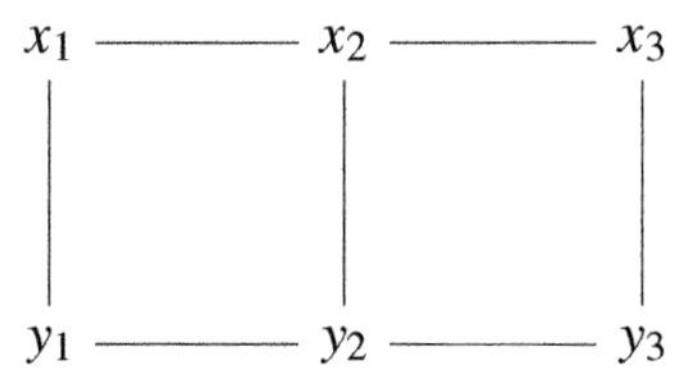

Figure 10. The illegal complex Γ_{COL, P_3}.

Proposition 3.4. *For a placement game G played on a board B we have the following*:

(1) $\mathcal{L}_{G,B} = \mathcal{F}(\Delta_{G,B})$,

(2) $\mathcal{ILL}_{G,B} = \mathcal{F}(\Gamma_{G,B}) = \mathcal{N}(\Delta_{G,B})$.

Proof. (1) The facets of $\Delta_{G,B}$ represent the maximal legal positions of G. Thus $\mathcal{F}(\Delta_{G,B})$ is the ideal generated by the monomials representing the maximal legal positions, which is $\mathcal{L}_{G,B}$ by definition.

(2) The facets of $\Gamma_{G,B}$ are represented by the monomials of the minimal illegal positions of G, which by definition generate $\mathcal{ILL}_{G,B}$, proving the first equality.

Since the faces of $\Delta_{G,B}$ represent the legal positions of G, the minimal nonfaces of $\Delta_{G,B}$ represent the minimal illegal positions, which generate $\mathcal{ILL}_{G,B}$. Thus $\mathcal{ILL}_{G,B} = \mathcal{N}(\Delta_{G,B})$. □

Example 3.5. Consider COL played on the path P_3 with labels as in Example 2.1. We then have the legal ideal

$$\mathcal{L}_{\text{COL}, P_3} = \langle x_1 y_2 x_3, y_1 x_2 y_3, x_1 y_3, y_1 x_3 \rangle$$

and the illegal ideal

$$\mathcal{ILL}_{\text{COL}, P_3} = \langle x_1 x_2, x_2 x_3, y_1 y_2, y_2 y_3 \rangle.$$

The illegal complex Γ_{COL, P_3} is given in Figure 10.

4. Playing games on simplicial complexes

In this section we show that games can be played on the illegal or legal complex rather than the board.

Since the facets of the illegal complex represent the minimal illegal positions, we can play on $\Gamma_{G,B}$, instead of playing G on the board B, according to the following rules.

Illegal Ruleset. (1) Left may only play on vertices labeled x_i, while Right may only play on vertices labeled y_i.

(2) Given a facet, pieces played may not occupy all the vertices of the facet.

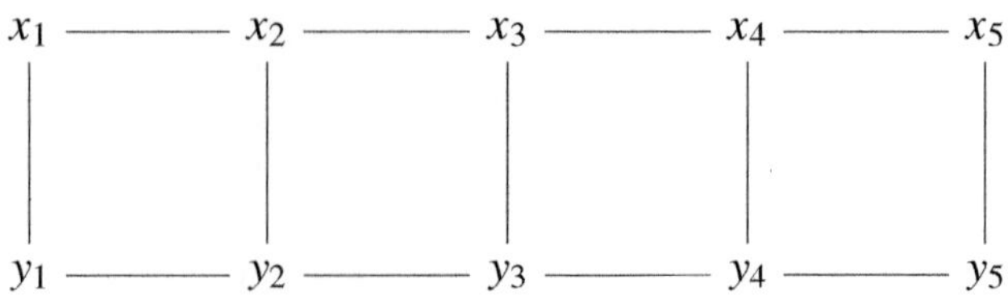

Figure 11. The illegal complex Γ_{COL,P_5}.

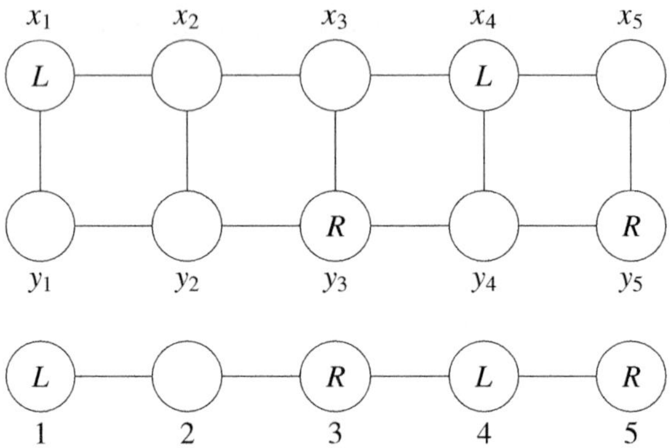

Figure 12. A legal position on Γ_{COL,P_5} and on P_5.

Since the facets of $\Gamma_{G,B}$ are the minimal illegal positions, any vertex set that does not contain all the vertices of any facet is a legal position of G. Thus playing on $\Gamma_{G,B}$ according to the above rules results in legal positions.

Example 4.1. Consider COL played on P_5. Since pieces may not be placed on the same space, or pieces by the same player placed side by side, the facets of Γ_{COL,P_5} then consist of the edges between x_i and y_i, between x_i and x_{i+1}, and between y_i and y_{i+1}. It is given in Figure 11.

Playing on the vertices x_1, y_3, x_4, y_5 is legal since we never have both vertices of an edge. This position is shown on the top of Figure 12, while the bottom shows the corresponding position played on P_5.

The next example of an illegal complex has a facet of cardinality 3.

Example 4.2. Consider NOGO played on the path P_3. The legal ideal is

$$\mathcal{L}_{\text{NOGO},P_3} = \langle x_1x_2, x_1x_3, x_1y_3, x_2x_3, y_1x_3, y_1y_2, y_1y_3, y_2y_3\rangle,$$

while the illegal ideal is

$$\mathcal{ILL}_{\text{NOGO},P_3} = \langle x_1x_2x_3, y_1y_2y_3, x_1y_1, x_1y_2, x_2y_2, x_2y_3, x_3y_3, y_1x_2, y_2x_3\rangle.$$

The illegal complex is given in Figure 13.

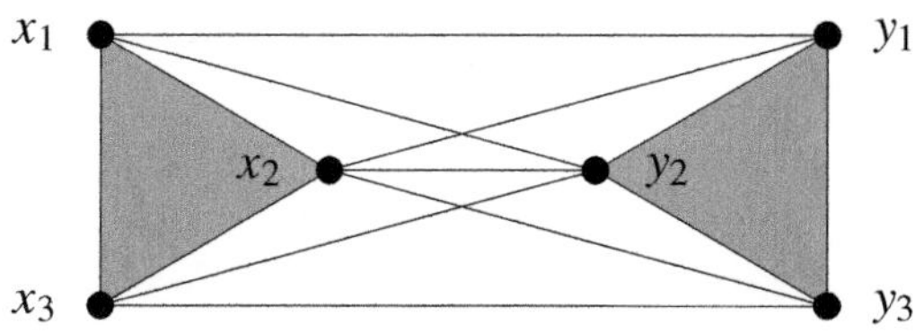

Figure 13. The illegal complex Γ_{NoGo,P_3}.

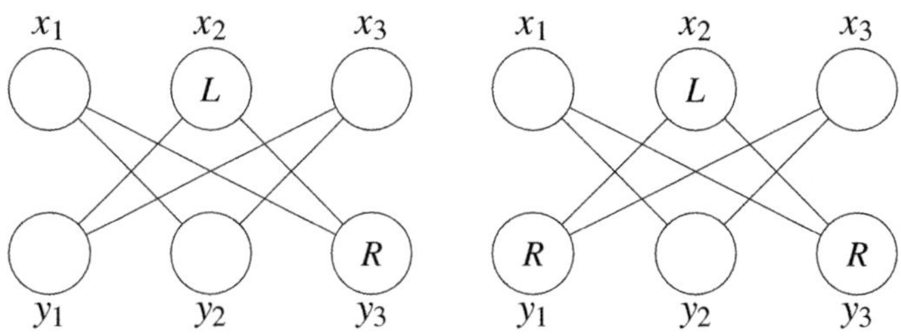

(A) legal position (B) illegal position

Figure 14. A legal and an illegal position when playing on Δ_{COL,C_3}.

Then playing on x_1 and x_2 is legal (they form a face, but not a facet), while playing on x_1, x_2, and x_3 is illegal.

Similarly, playing on the legal complex $\Delta_{G,B}$ according to the following rules is also equivalent to playing G on B.

Legal Ruleset. (1) Left may only play on vertices labeled x_i, while Right may only play on vertices labeled y_i.

(2) The set of occupied vertices needs to be a face of $\Delta_{G,B}$.

Example 4.3. Consider COL played on C_3. The position on the left in Figure 14 is legal, while the one on the right is illegal when playing on the complex.

Notice that both the legal complex and the illegal complex give a representation of the game *and* the board. Thus, we can use the two complexes interchangeably, which is advantageous since sometimes the illegal complex is simpler than the legal complex (for example, the legal complex of COL played on P_5 has facets with 5 vertices, while in the illegal complex the facets have 2 vertices).

The next theorem recapitulates these discussions.

Theorem 4.4. *Given a placement game G played on a board B, there exist simplicial complexes Δ and Γ such that G is equivalent to the game with the Illegal Ruleset played on Γ, and equivalent to the game with the Legal Ruleset played on Δ.*

Proof. As shown above, $\Delta = \Delta_{G,B}$ the legal complex and $\Gamma = \Gamma_{G,B}$ the illegal complex satisfy this. □

5. Discussion

From the construction of legal complexes from placement games, there are several questions that arise naturally.

One question of interest is a possible reverse construction. In other words, we are looking at what conditions a simplicial complex has to satisfy to be a legal complex. In [5] we explore this question further.

Another natural direction to pursue is how the algebra of a square-free monomial ideal I (such as Cohen–Macaulayness, localization/deletion-contraction) affects the rulesets of the games played on the simplicial complexes $\mathcal{F}(I)$ and $\mathcal{N}(I)$.

References

[1] E. R. Berlekamp, J. H. Conway, & R. K. Guy, *Winning ways for your mathematical plays,* A K Peters, Ltd., 2001.

[2] J. I. Brown, D. Cox, A. Hoefel, N. McKay, R. Milley, R. J. Nowakowski, A. A. Siegel. Polynomial profiles of placement games. In this volume.

[3] W. Bruns, J. Herzog. *Cohen–Macaulay rings*, Cambridge Studies in Advanced Mathematics, 39. Cambridge University Press, Cambridge, 1993.

[4] S. Faridi. The facet ideal of a simplicial complex. *Manuscripta Math.* 109(2):159–174, 2002.

[5] S. Faridi, S. Huntemann, R. J. Nowakowski. Games and complexes II: weight games and Kruskal–Katona type bounds. In this volume.

[6] G. E. Farr. The go polynomials of a graph. *Theoret. Comp. Sci.*, 306:1–18, 2003.

[7] G. Farr and J. Schmidt. On the number of go positions on lattice graphs. *Information Processing Letters*, 105:124–130, 2008.

[8] M. Gardner. Mathematical games: cram, crosscram and quadraphage: new games having elusive winning strategies. *Scientific American*, 230(2):106–108, 1974.

[9] S. Huntemann. Simplicial complexes of placement games. Master's Thesis, Dalhousie University, 2013.

[10] J. Tromp and G. Farnebäck. Combinatorics of go. In *Proceedings of 5th International Conference on Computer and Games*, Torino, Italy, May 2006.

faridi@mathstat.dal.ca *Department of Mathematics and Statistics, Dalhousie University, Halifax, Nova Scotia, Canada*

svenja.huntemann@dal.ca *Department of Mathematics and Statistics, Dalhousie University, Halifax, Nova Scotia, Canada*

rjn@mathstat.dal.ca *Department of Mathematics and Statistics, Dalhousie University, Halifax, Nova Scotia, Canada*

Games of No Chance 5
MSRI Publications
Volume **70**, 2017

Games and complexes II: Weight games and Kruskal–Katona type bounds

SARA FARIDI, SVENJA HUNTEMANN
AND RICHARD J. NOWAKOWSKI

A strong placement game G played on a board B is equivalent to a simplicial complex $\Delta_{G,B}$. We look at weight games, a subclass of strong placement games, and introduce upper bounds on the number of positions with i pieces in G, or equivalently the number of faces with i vertices in $\Delta_{G,B}$, which are reminiscent of the Kruskal–Katona bounds.

1. Introduction

Our goal in this paper is to study complexes of placement games (Definition 1.1). In [3] we demonstrated that to a placement game G played on a board B one can associate a simplicial complex $\Delta_{G,B}$, where G can be considered as a game played on $\Delta_{G,B}$.

The main question addressed in this paper is: What complexes can be legal complexes of a placement game?

We give partial answers to this question in specific cases: when the board is a path, a cycle, or a complete graph (also see [5]).

We begin by introducing some of the concepts needed. A complete introduction is given in [3].

Definition 1.1. A *strong placement game* is a combinatorial game played on a graph which satisfies the following:

(i) The starting position is the empty board.

(ii) Players place pieces on empty spaces of the board according to the rules.

(iii) Pieces are not moved or removed once placed.

Faridi's and Nowakowski's research was supported in part by the Natural Sciences and Engineering Research Council of Canada.
MSC2010: primary 91A46; secondary 13F55.
Keywords: combinatorial games, simplicial complexes, placement game.

(iv) If it is possible to reach a position through a sequence of legal moves, then any sequence of moves leading to this position consists only of legal moves.

The TRIVIAL placement game on a board is the placement game that has no additional rules.

Throughout this paper "placement game" refers to a strong placement game. Since placement games are played on a graph, we use the terms board and graph, and space and vertex interchangeably.

A *basic position* is a board with only one piece placed. Any position, whether legal or illegal, in a placement game can be decomposed into basic positions.

Definition 1.2. A *simplicial complex* Δ on a finite vertex set V is a set of subsets (called *faces*) of V with the conditions that if $A \in \Delta$ and $B \subseteq A$, then $B \in \Delta$. The *f-vector* $(f_0, f_1, \ldots, f_k)$ of a simplicial complex Δ enumerates the number of faces f_i with i vertices. Note that if $\Delta \neq \varnothing$, then $f_0 = 1$.

The *legal complex* [3], denoted by $\Delta_{G,B}$, is the simplicial complex whose faces correspond to the legal positions of the placement game G played on the board B.

Question 1.3. Is every simplicial complex the legal complex of a placement game?

In respect to this question, we are interested in the possible f-vectors of legal complexes, thus we will consider the following.

The number of positions in G on B with i pieces played, or equivalently the number of faces with i vertices in the legal complex $\Delta_{G,B}$, is denoted by $f_i(G, B)$, or shortened to f_i if the game and board are clear. In this work, we will be considering upper bounds on $f_i(G, B)$. Specifically, we will be considering Kruskal–Katona type bounds for weight games played on a path, on a cycle, or on a complete graph.

2. The Kruskal–Katona theorem

Kruskal [7] and Katona [6] proved that for each pair of nonnegative integers f and i, f can be written in the form

$$f = \binom{n_i}{i} + \binom{n_{i-1}}{i-1} + \cdots + \binom{n_{i-s}}{i-s},$$

where $n_i > n_{i-1} > \cdots > n_{i-s} \geq i - s \geq 1$ are unique. This sum is called the *i-canonical representation of f*.

We can then define the *j-th pseudopower of f*, for $j \geq 1$, as

$$f_i^{(j)} = \binom{n_i}{j} + \binom{n_{i-1}}{j-1} + \cdots + \binom{n_{i-s}}{j-s}.$$

The Kruskal–Katona theorem gives necessary and sufficient conditions for a vector $(f_0, f_1, \dots, f_k)$ with entries from the nonnegative integers to be the f-vector of a simplicial complex. The following is the version proven by Kruskal.

Theorem 2.1 (Kruskal [7]). *Let $(f_0, f_1, \dots, f_k)$, be a sequence of nonnegative integers. The following statements are equivalent*:

(i) *$(f_0, f_1, \dots, f_k)$ is the f-vector of a nonempty simplicial complex.*

(ii) *$f_0 = 1$ and $f_j \le f_i^{(j)}$ for all $1 \le i \le j$.*

(iii) *$f_0 = 1$ and $f_j \ge f_i^{(j)}$ for all $1 \le j \le i$.*

To show that (ii) holds, it is sufficient to show that $f_0 = 1$ and $f_{i+1} \le f_i^{(i+1)}$ for all $i \ge 1$ since all other cases follow. Similarly, to show (iii), showing $f_0 = 1$ and $f_j \ge f_{j+1}^{(j)}$ for all $j \ge 1$ is sufficient. The Kruskal–Katona theorem is usually stated in terms of either one of these.

If the answer to Question 1.3 is "no", then not every vector that is an f-vector of a simplicial complex is also an f-vector of a legal complex. Thus for the remainder, after introducing weight games, we will give improved upper bounds on the entries of an f-vector of a legal complex.

3. Games with weight

In the remainder, we will consider playing pieces of larger size. Specifically, we call the number of connected vertices a piece covers the *weight* of this piece.

Many placement games have pieces of weight greater than 1. For example, in DOMINEERING [1] and CROSSCRAM [4], Left and Right both play dominoes as their pieces, and so their pieces are of weight 2. Also, as we will mention in Remark 4.4, every placement game with weight on a path is equivalent to a partizan octal game.

Example 3.1. Consider the board given in Figure 1. A piece that has weight 4 could for example be played on the vertex set $\{1, 2, 3, 4\}$, but not on the vertex set $\{1, 3, 5, 6\}$ since these vertices are not connected.

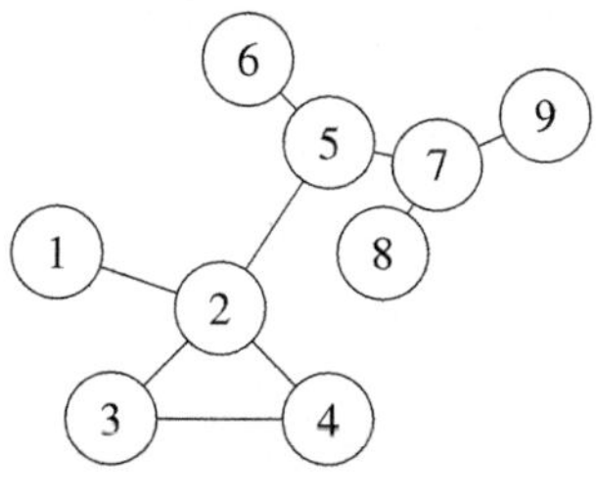

Figure 1. An example board.

We usually assume that every piece of Left has the same weight a, and every piece of Right has the same weight b.

Definition 3.2. A placement game in which the players play pieces of fixed weights is called a *game with weights*. If the game has no rules besides pieces having to be placed on connected sets of empty vertices, we call it a *weight game*. A 2-player weight game will be denoted by $W(a, b)$, where a is the weight that Left plays, while b is the weight that Right plays.

Essentially, the weight game is the TRIVIAL placement game with weights.

In [5], it is shown that the game $W(a, a)$ played on a path or a cycle is equivalent to another placement game in which both Left and Right play pieces of weight 1. This is not necessarily true though if we force every basic position to be legal, as the following discussion shows.

Consider a placement game G in which both Left and Right play pieces of weight 1 and every basic position is legal. Since the basic positions in this case consists of Left or Right occupying a single vertex, we have n Left and Right basic positions each, where n is the number of vertices of the board. Thus we have that the number of legal positions with one piece is the number of basic positions, namely $f_1 = 2n$.

This also implies that a weight game $W(a, b)$ where f_1 is odd is not equivalent to a placement game where both Left and Right play pieces of weight 1 and basic positions are legal. Weight games with f_1 odd indeed exist, as seen in the following example.

Example 3.3. Consider $W(1, 2)$ played on P_2. The basic positions are

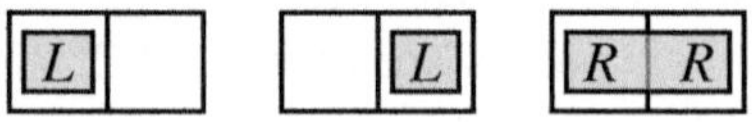

and thus if all basic positions are legal, then $f_1 = 3$.

Suppose the weight of the Left pieces is a and the weight of the Right pieces b and without loss of generality $a \leq b$, then Left would be able to place at most $\lfloor n/a \rfloor$ pieces on a board of n vertices. If we place a mix of Left and Right pieces or just Right pieces, the number of pieces we are able to place will be equal or less. Thus if the f-vector of the legal complex is $(f_0, f_1, \ldots, f_k)$, then

$$k \leq \max\{\lfloor n/a \rfloor, \lfloor n/b \rfloor\}.$$

Proposition 3.4. *For legal complexes corresponding to games on any board of n vertices with pieces of weight* 1, *we have*

$$f_i \leq \binom{n}{i} 2^i.$$

for $i \geq 0$.

Proof. We will consider the number of positions with i pieces of weight 1 in the placement game that has no additional rules, i.e., the TRIVIAL placement game. As we add rules to this game to get other placement games with pieces of weight 1, the number of positions decreases, thus the number of such positions in TRIVIAL gives the maximum. In TRIVIAL, there are $\binom{n}{i}$ ways to choose i spaces to place pieces; for each there are 2 choices: either a Left piece, or a Right piece. Our claim now follows. □

We will now look at how playing pieces of specified weight on different classes of boards influences the f-vector of the corresponding legal complex. The classes of boards we specifically look at are paths, cycles, and complete graphs.

Note that the f-vector of a weight game gives an upper bound on the f-vector of a game with the same weights. Thus the formulae for the weight games in the following sections give bounds for games with weight.

In [5], these formulae are also generalized to t-player weight games.

4. Playing on the path P_n

In this section, we study placement games played on the path P_n, $n \geq 1$, in which Left plays pieces of weight a and Right pieces of weight b.

Proposition 4.1. *If a simplicial complex is the legal complex of a weight game $W(a, b)$ played on P_n then*

$$f_1 = \begin{cases} 0 & \text{if } a, b > n, \\ n-a+1 & \text{if } a \leq n \text{ and } b > n, \\ n-b+1 & \text{if } a > n \text{ and } b \leq n, \\ 2n-a-b+2 & \text{if } a, b \leq n. \end{cases} \tag{1}$$

Proof. We are measuring the number of legal basic positions. If $a, b > n$, then neither Left nor Right can place a piece, thus $f_1 = 0$. If $n \geq a$, then placing one piece of weight a on a strip of length n is equivalent to placing one piece of weight 1 (think of the left-most end of the piece) on a strip of length $n-(a-1) = n-a+1$, so the second and third case follow. Similarly, for the final case

$$f_1 = (n-a+1)+(n-b+1) = 2n-a-b+2. \qquad \square$$

Proposition 4.2. *In a weight game $W(a, b)$ played on P_n, the number of positions with one Left and one Right piece is*

$$N_{LR} = \begin{cases} 0 & \text{if } a+b > n, \\ 2\binom{n-a-b+2}{2} & \text{if } a+b \leq n. \end{cases}$$

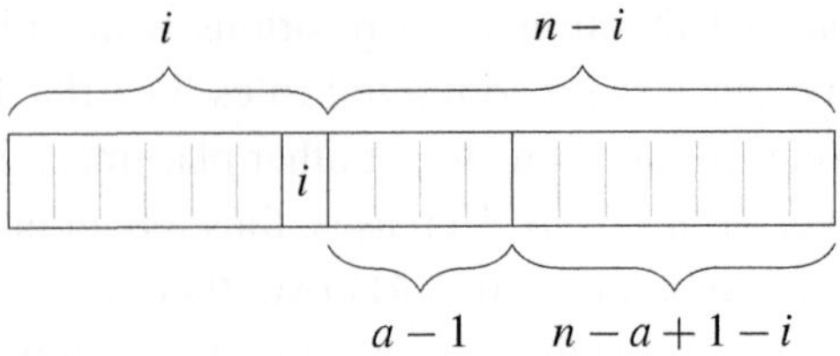

Figure 2. Proof to Proposition 4.2: placing a piece of weight a on a path.

The number of positions with two Left pieces or two Right pieces, respectively, is

$$N_{LL} = \begin{cases} 0 & \text{if } 2a > n, \\ \binom{n-2a+2}{2} & \text{if } 2a \leq n; \end{cases} \qquad N_{RR} = \begin{cases} 0 & \text{if } 2b > n, \\ \binom{n-2b+2}{2} & \text{if } 2b \leq n. \end{cases}$$

For the legal complex of such a game we have

$$f_2 = N_{LR} + N_{LL} + N_{RR}. \tag{2}$$

Proof. To find N_{LR} when $n \geq a + b$, we only consider the case in which the Left piece is the left-most piece. The other case is symmetric. We will first place the Left piece in position i. To be able to fit a Right piece to the right of this, we have $1 \leq i \leq n - a - b + 1$. The strip to the right then has length $n - (i + a - 1) = n - a + 1 - i$ (see Figure 2). Thus we have $n-a+1-i-(b-1) = n-a-b+2-i$ choices to place the Right piece (see Proposition 4.1). Thus the number of position with Left on the left and Right on the right is

$$\begin{aligned} &\sum_{i=1}^{n-a-b+1} (n-a-b+2-i) \\ &\quad = (n-a-b+1)(n-a-b+2) - \sum_{i=1}^{n-a-b+1} i \\ &\quad = (n-a-b+1)(n-a-b+2) - \tfrac{1}{2}(n-a-b+1)(n-a-b+2) \\ &\quad = \tfrac{1}{2}(n-a-b+1)(n-a-b+2) \\ &\quad = \binom{n-a-b+2}{2}. \end{aligned}$$

Then $N_{LR} = 2\binom{n-a-b+2}{2}$.

Similarly, the number of positions with Left on the left and right for $n \geq 2a$ and Right on the left and right for $n \geq 2b$ respectively, then are

$$N_{LL} = \binom{n-2a+2}{2}, \quad N_{RR} = \binom{n-2b+2}{2}.$$

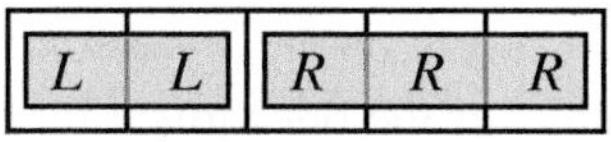

Figure 3. An example position for $W(2, 3)$ on P_5.

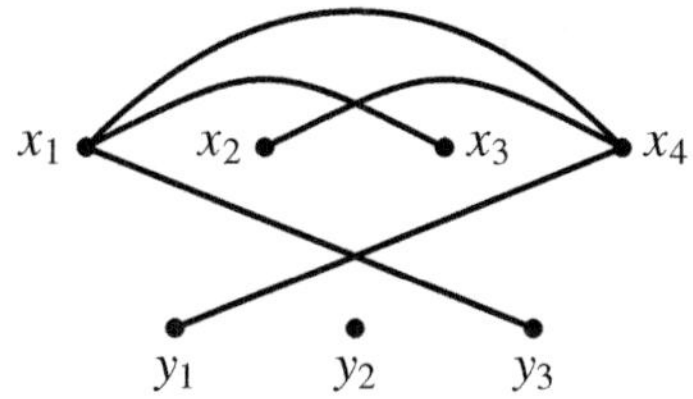

Figure 4. The legal complex $\Delta_{W(2,3),P_5}$.

Since these are the only three possibilities for pairs of pieces, (2) follows immediately. □

It is easy to see that if $a = b = 1$, then the previous two bounds are

$$f_1 = 2n; \quad f_2 = 4\binom{n}{2}.$$

These are the bounds given in Proposition 3.4.

Example 4.3. Consider $W(2, 3)$ on the path P_5. Let x_i represent a Left piece occupying the spaces i and $i + 1$, and similarly for y_i. For example, the position in Figure 3 is represented by $x_1 y_3$. The corresponding simplicial complex is given in Figure 4.

By Propositions 4.1 and 4.2 we have

$$\begin{aligned} f_0 &= 1, \\ f_1 &= 2n - a - b + 2 = 7, \\ f_2 &= \binom{n-2a+2}{2} + 2\binom{n-a-b+2}{2} = 5, \end{aligned}$$

and since $\max\{\lfloor n/a \rfloor, \lfloor n/b \rfloor\} = 2$, we get the f-vector $(1, 7, 5)$, which can be verified from the simplicial complex.

To compare this with the Kruskal–Katona bound, we first need to find the i-canonical representations and calculate the j-th pseudopowers:

$$f_1 = \binom{7}{1}, \quad f_1^{(2)} = \binom{7}{2} = 21,$$

$$f_2 = \binom{3}{2} + \binom{2}{1}, \quad f_2^{(3)} = \binom{3}{3} + \binom{2}{2} = 2, \quad f_2^{(1)} = \binom{3}{1} + \binom{2}{0} = 4.$$

Then $f_2 = 5 < f_1^{(2)} = 21$, $f_3 = 0 < f_2^{(3)} = 2$, and $f_1 = 7 > f_2^{(1)} = 4$, showing that the formulae in Propositions 4.1 and 4.2 give, at least for this example, improved necessary conditions for a vector to be the f-vector of a legal complex of a placement game played on a path over the ones given in the Kruskal–Katona theorem.

We will now show that for fixed a and b and sufficiently large n, then the bound in Proposition 4.2 on f_2 is better than the Kruskal–Katona bound. By the Kruskal–Katona theorem we have

$$f_2 \leq f_1^{(2)} = \binom{2n-a-b+2}{2} = \tfrac{1}{2}[4n^2 + n(6-4a-4b) + g(a,b)],$$

where $g(a,b)$ is a function in a and b, whereas Proposition 4.2 gives

$$\begin{aligned} f_2 &= \binom{n-2a+1}{2} + \binom{n-2b+1}{2} + 2\binom{n-a-b+1}{2} \\ &= \tfrac{1}{2}[4n^2 + 2n(6-4a-4b) + h(a,b)], \end{aligned}$$

where $h(a,b)$ is a function in a and b. Since $a, b \geq 1$, and thus $6-4a-4b < 0$, we have $\frac{1}{2}[4n^2+2n(6-4a-4b)+g(a,b)] < \frac{1}{2}[4n^2+n(6-4a-4b)+h(a,b)]$ for sufficiently large n, showing that as n grows larger our bound becomes increasingly better than the Kruskal–Katona bound.

Remark 4.4. The game O12 is the weight game $W(1,2)$. It is mentioned by Brown et al. in [2] that this game played on a path is equivalent to the partizan Octal game where Left removes one piece and Right two, and both have the possibility to split the heap. It is easy to see that weight games played on a path are all equivalent to a specific partizan Octal game.

5. Playing on the cycle C_n

Consider Left playing pieces of weight a and Right pieces of weight b on a cycle of length $n \geq 3$. For this board, the "left" end of a piece is the end in counter-clockwise direction.

Proposition 5.1. *If a simplicial complex is the legal complex of $W(a,b)$ played on C_n then*

$$f_1 = \begin{cases} 0 & \text{if } a, b > n, \\ n & \text{if either } a \leq n \text{ or } b \leq n \text{ but not both}, \\ 2n & \text{if } a, b \leq n. \end{cases} \tag{3}$$

Proof. The left end of a piece can be placed on any of the n spaces if its weight is less than n, no matter if it is a Right or Left piece. □

Proposition 5.2. *If a simplicial complex is the legal complex of $W(a,b)$ played on C_n then*

$$f_2 = N_{LL} + N_{LR} + N_{RR} \tag{4}$$

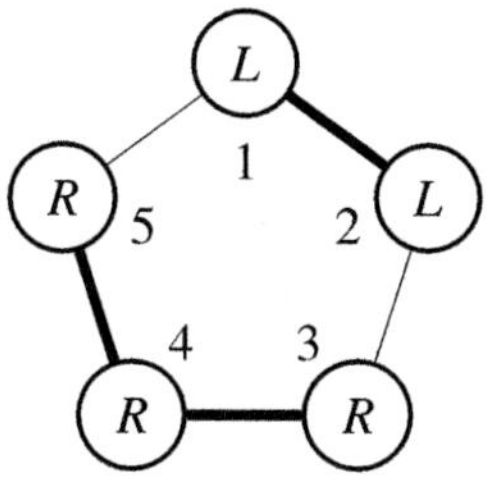

Figure 5. An example position for $W(2, 3)$ on C_5.

where

$$N_{LL} = \begin{cases} 0 & \text{if } 2a > n, \\ \frac{1}{2}n(n-2a+1) & \text{if } 2a \le n, \end{cases} \qquad N_{RR} = \begin{cases} 0 & \text{if } 2b > n, \\ \frac{1}{2}n(n-2b+1) & \text{if } 2b \le n, \end{cases}$$

are the number of positions with two Left pieces, respectively two Right pieces, and

$$N_{LR} = \begin{cases} 0 & \text{if } a+b > n, \\ n(n-a-b+1) & \text{if } a+b \le n, \end{cases}$$

is the number of positions with one Left and one Right piece.

Proof. We will first look at the number of positions with two Left pieces if $n \ge 2a$. There are n choices for placing the first piece. Placing the second piece is equivalent to placing one piece on the path P_{n-a}, i.e., there are $(n-a)-(a-1)$ choices for placing the second piece. Due to symmetry, there are then $\frac{1}{2}n(n-2a+1)$ positions of this form. Similarly, the number of positions with two Right pieces is $\frac{1}{2}n(n-2b+1)$ if $n \ge 2b$.

To count the number of positions with one Left and one Right piece when $n \ge a+b$, we first place the Left, then the Right piece. There are n choices for placing the Left piece. Placing the Right piece is then equivalent to placing a piece of weight b on the path P_{n-a}, i.e., there are $(n-a)-(b-1)$ choices for placing the second piece. Thus, there are $n(n-a-b+1)$ positions of this form. □

If $a = b = 1$, then the previous two bounds are

$$f_1 = 2n, \quad f_2 = 4\binom{n}{2},$$

which are the bounds given in Proposition 3.4.

Example 5.3. Consider $W(2, 3)$ on the cycle C_5. Let x_i represent a Left piece whose left end is on space i, and similarly for y_i; e.g., the position in Figure 5 is represented by $x_1 y_3$. The corresponding legal complex is given in Figure 6.

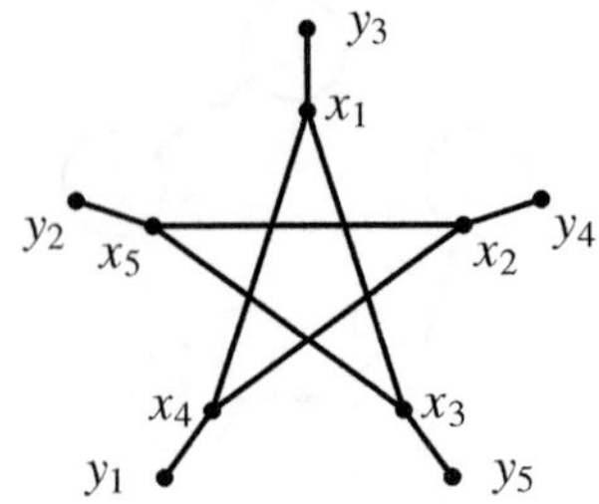

Figure 6. The legal complex $\Delta_{W(2,3),C_5}$.

By Propositions 5.1 and 5.2 we have

$$\begin{aligned} f_0 &= 1, \\ f_1 &= 2n = 10, \\ f_2 &= \tfrac{1}{2}n(n-2a+1)+n(n-a-b+1) = 10, \end{aligned}$$

and since $\max\{\lfloor n/a \rfloor, \lfloor n/b \rfloor\} = 2$, we get the f-vector $(1, 10, 10)$, which can be verified from the simplicial complex.

We will compare these with the Kruskal–Katona bound. The i-canonical representations and j-th pseudopowers are

$$f_1 = \binom{10}{1}, \quad f_1^{(2)} = \binom{10}{2} = 45,$$

$$f_2 = \binom{5}{2}, \quad f_2^{(3)} = \binom{5}{3} = 10, \quad f_2^{(1)} = \binom{5}{1} = 5.$$

Then $f_2 = 10 < f_1^{(2)} = 45$, $f_3 = 0 < f_2^{(3)} = 10$, and $f_1 = 10 > f_2^{(1)} = 5$, showing that for this example Propositions 5.1 and 5.2 give improved necessary conditions for a vector to be the f-vector of a legal complex of a placement game played on a cycle over the ones given in the Kruskal–Katona theorem.

Similar to placement games on a path, we have that for fixed a and b and sufficiently large n the bound in Proposition 5.2 on f_2 is better than the Kruskal–Katona bound. By the Kruskal–Katona theorem we have

$$f_2 \le f_1^{(2)} = \binom{2n}{2} = \tfrac{1}{2}[4n^2 + n(-2)],$$

whereas Proposition 5.2 gives

$$\begin{aligned} f_2 &= \tfrac{1}{2}n(n-2a+1) + \tfrac{1}{2}n(n-2b+1) + n(n-a-b+1) \\ &= \tfrac{1}{2}[4n^2 + n(4-4a-4b)] < \tfrac{1}{2}[4n^2 + n(-2)], \end{aligned}$$

since $a, b \ge 1$ implies $4 - 4a - 4b \le -4$, showing that as n grows larger our bound becomes increasingly better than the Kruskal–Katona bound.

6. Playing on the complete graph K_n

Finally, we will consider placement games played on a complete graph of n vertices in which Left places pieces of weight a and Right pieces of weight b.

Proposition 6.1. *If a simplicial complex is the legal complex of $W(a, b)$ played on K_n then*

$$f_k = \sum_{l=0}^{k} \left(\frac{\prod_{i=0}^{k-l-1} \binom{n-ia}{a}}{(k-l)!} \right) \left(\frac{\prod_{j=0}^{l-1} \binom{n-(k-l)a-jb}{b}}{l!} \right) \tag{5}$$

for $k \geq 0$.

Proof. Playing a piece of weight a on the complete graph with n vertices is equivalent to deleting a vertices from the graph. Thus placing a second piece on the graph is equivalent to placing a piece on the complete graph on $n - a$ vertices.

Also, since every pair of vertices is connected, playing a piece of weight a is equivalent to playing a pieces of weight 1, thus there are $\binom{n}{a}$ choices for placing the piece.

Thus playing s pieces of weight a we have

$$\frac{\prod_{i=0}^{s-1} \binom{n-ia}{a}}{s!}$$

choices. Then playing $k-l$ pieces of weight a and l pieces of weight b (assuming without loss of generality we place the pieces of weight a first) we have

$$\frac{\prod_{i=0}^{k-l-1} \binom{n-ia}{a}}{(k-l)!} \frac{\prod_{j=0}^{l-1} \binom{n-(k-l)a-jb}{b}}{l!}$$

different positions.

To get the total number of positions with k pieces played, we let l range from 0 to k and add the terms, giving (5). □

If $a = b$, then the previous bound becomes

$$\begin{aligned} f_k &= \sum_{l=0}^{k} \frac{n(n-1)\cdots(n-(k-l)a+1)(n-(k-l)a)\cdots(n-ka+1)}{(k-l)!l!(a!)^k} \\ &= \frac{n!}{(n-ka)!(a!)^k} \sum_{l=0}^{k} \frac{1}{k!} \binom{k}{l} = \frac{n!}{(n-ka)!k!(a!)^k} \sum_{l=0}^{k} \binom{k}{l} \\ &= \frac{n!}{(n-ka)!k!(a!)^k} 2^k. \end{aligned}$$

If $a = b = 1$, then this becomes

$$f_k = \frac{n!}{(n-k)!k!}2^k = \binom{n}{k}2^k,$$

which is the bound given in Proposition 3.4.

If we assume without loss of generality that $a \leq b$, then we have

$$\begin{aligned} f_k &= \sum_{l=0}^{k} \frac{n(n-1)\cdots(n-(k-l)a-lb+1)}{(k-l)!l!(a!)^{k-l}(b!)^l} \\ &= \frac{n!}{k!}\sum_{l=0}^{k} \frac{\binom{k}{l}}{(a!)^{k-l}(b!)^l(n-(k-l)a-lb)!} \\ &\leq \frac{n!}{k!}\sum_{l=0}^{k} \frac{\binom{k}{l}}{(a!)^k(n-kb)!} = \frac{n!}{(n-kb)!k!(a!)^k}2^k. \end{aligned}$$

We can similarly find a lower bound. Thus

$$\frac{n!}{(n-ka)!k!(b!)^k}2^k \leq f_k \leq \frac{n!}{(n-kb)!k!(a!)^k}2^k.$$

For fixed a, b, and k, we then have

$$n(n-1)\cdots(n-ka+1)\frac{2^k}{k!(b!)^k} \leq f_k \leq n(n-1)\cdots(n-kb+1)\frac{2^k}{k!(a!)^k},$$

and since

$$n(n-1)\cdots(n-ka+1) \geq (n-ka+1)^{ka} \text{ and } n(n-1)\cdots(n-kb+1) \leq n^{kb},$$

this implies

$$C'(n-ka+1)^{ka} = C'n^{ka} + O(n^{ka-1}) \leq f_k \leq Cn^{kb},$$

where C and C' are constants depending on a and k, respectively b and k.

Also note that $W(a, b)$ played on the complete graph K_n is the least restrictive game on the most connected board. Thus the formula in Proposition 6.1 gives upper bounds for any placement game with weights on any board.

Example 6.2. Consider $W(2, 2)$ and let the board be the complete graph K_4. Let $x_{i,j}$ represent a Left piece occupying the vertices i and j, and similarly for $y_{i,j}$. For example the position in Figure 7 is represented by $x_{1,4}y_{2,3}$. The corresponding simplicial complex is given in Figure 8.

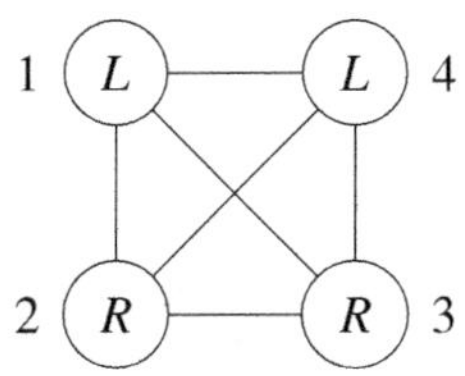

Figure 7. An example position for $W(2,2)$ on K_4.

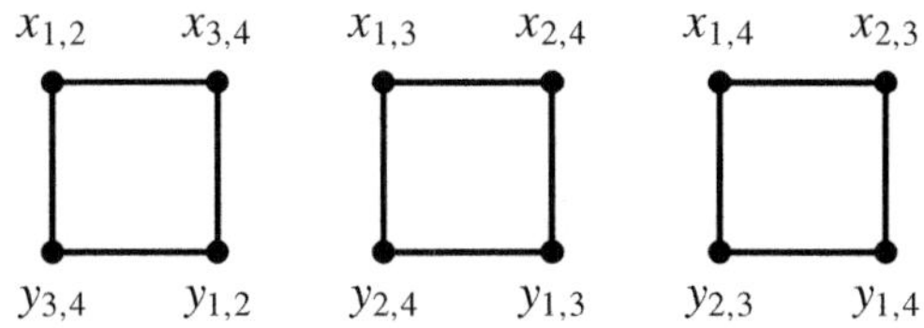

Figure 8. The legal complex $\Delta_{W(2,2),K_4}$.

By Proposition 6.1 we have

$$f_0 = 1,$$
$$f_1 = \binom{n}{a} + \binom{n}{b} = 12,$$
$$f_2 = \frac{\binom{n}{a}\binom{n-a}{a}}{2} + \binom{n}{a}\binom{n-a}{b} + \frac{\binom{n}{b}\binom{n-b}{b}}{2} = 12,$$

and since $\max\{\lfloor n/a \rfloor, \lfloor n/b \rfloor\} = 2$, we get the f-vector $(1, 12, 12)$, which can be verified from the simplicial complex.

The i-canonical representations and the j-th pseudopowers are

$$f_1 = \binom{12}{1}, \qquad f_1^{(2)} = \binom{12}{2} = 66,$$
$$f_2 = \binom{5}{2} + \binom{2}{1}, \quad f_2^{(3)} = \binom{5}{3} + \binom{2}{2} = 11,$$
$$f_2^{(1)} = \binom{5}{1} + \binom{2}{0} = 6.$$

Then $f_2 = 12 < f_1^{(2)} = 66$, $f_3 = 0 < f_2^{(3)} = 11$, and $f_1 = 12 > f_2^{(1)} = 6$, showing that for this example the formula in Proposition 6.1 gives improved necessary conditions for a vector to be the f-vector of a legal complex.

We will now show that for fixed a and b and sufficiently large n, the bound in Proposition 6.1 for f_2 is better than the Kruskal—Katona bound. By the Kruskal-Katona theorem we have

$$f_2 \leq f_1^{(2)} = \binom{\binom{n}{a} + \binom{n}{b}}{2} = \frac{1}{2}\Big[\binom{n}{a}\Big(\binom{n}{a} + 2\binom{n}{b} - 1\Big) + \binom{n}{b}\Big(\binom{n}{b} - 1\Big)\Big],$$

whereas Proposition 6.1 gives

$$\begin{aligned} f_2 &= \frac{1}{2}\binom{n}{a}\binom{n-a}{a} + \frac{1}{2}\binom{n}{b}\binom{n-b}{b} + \binom{n}{a}\binom{n-a}{b} \\ &= \frac{1}{2}\left[\binom{n}{a}\left(\binom{n-a}{a} + 2\binom{n-a}{b}\right) + \binom{n}{b}\binom{n-b}{b}\right]. \end{aligned}$$

Recall that $f(n) = O(g(n))$ means that $f(n) \le Cg(n)$ for some positive constant C. Then $f(n) = O(n^k)$ means that $f(n)$ is bounded by a polynomial of degree at most k. Also recall that $f(n) = g(n) + O(n^k)$ means $f(n) - g(n) = O(n^k)$.

Since

$$\begin{aligned} \binom{n}{i} &= \frac{1}{i!}\left(n^i - n^{i-1}\tfrac{1}{2}i(i-1) + O(n^{i-2})\right) \text{ for } i \ge 2, \\ \binom{n-i}{j} &= \frac{1}{j!}\left(n^j - n^{j-1}\tfrac{1}{2}j(j+2i-1) + O(n^{j-2})\right) \text{ for } j \ge 2, \end{aligned}$$

it easily follows that $\binom{n-a}{a} + 2\binom{n-a}{b} \le \binom{n}{a} + 2\binom{n}{b} - 1$ and $\binom{n-b}{b} \le \binom{n}{b} - 1$. Thus

$$\begin{aligned} &\frac{1}{2}\left[\binom{n}{a}\left(\binom{n-a}{a} + 2\binom{n-a}{b}\right) + \binom{n}{b}\binom{n-b}{b}\right] \\ &\qquad < \frac{1}{2}\left[\binom{n}{a}\left(\binom{n}{a} + 2\binom{n}{b} - 1\right) + \binom{n}{b}\left(\binom{n}{b} - 1\right)\right], \end{aligned}$$

showing that the new bound is better than the Kruskal–Katona bound as n grows larger.

We have not compared the bounds for f_k with $k > 2$ since it is difficult to find the i-canonical representation of f_{k-1} in this case.

7. Discussion

A general question is to find sufficient conditions for a simplicial complex to be a legal complex. Since it is already not easy to find necessary conditions for a vector to be the f-vector of a legal complex, this seems to be very hard and much further work is needed.

References

[1] E. R. Berlekamp, J. H. Conway, R. K. Guy, *Winning ways for your mathematical plays*, A K Peters, Ltd., 2001.

[2] J. I. Brown, D. Cox, A. Hoefel, N. McKay, R. Milley, R. J. Nowakowski, A. A. Siegel, Polynomial profiles of placement games. In this volume.

[3] S. Faridi, S. Huntemann, R. J. Nowakowski. Games and complexes I: transformation via ideals. In this volume.

[4] M. Gardner. Mathematical games: cram, crosscram and quadraphage: new games having elusive winning strategies. *Scientific American*, 230(2):106–108, 1974.

[5] S. Huntemann. Simplicial complexes of placement games. Master's thesis, Dalhousie University, 2013.

[6] G. Katona. A theorem of finite sets. *Theory of graphs* (Tihany, 1966), pp. 187–207. Academic Press, New York, 1968.

[7] J. B. Kruskal. The number of simplices in a complex. In *Mathematical optimization techniques*, pp. 251–278. Univ. of California Press, Berkeley, CA, 1963.

faridi@mathstat.dal.ca *Department of Mathematics and Statistics, Dalhousie University, Halifax, Nova Scotia, Canada*

svenja.huntemann@dal.ca *Department of Mathematics and Statistics, Dalhousie University, Halifax, Nova Scotia, Canada*

rjn@mathstat.dal.ca *Department of Mathematics and Statistics, Dalhousie University, Halifax, Nova Scotia, Canada*

Games of No Chance 5
MSRI Publications
Volume **70**, 2017

Chromatic nim finds a game for your solution

MICHAEL J. FISHER AND URBAN LARSSON

We play a variation of nim on stacks of tokens. Take your favorite increasing sequence of positive integers and color the tokens according to the following rule. Each token on a level that corresponds to a number in the sequence is colored red; if the level does not correspond to a number in the sequence, color it green. Now play nim on an arbitrary number of stacks with the extra rule: if all top tokens are green, then you can make *any* move you like. On two stacks, we give explicit characterizations for winning the normal play version for some popular sequences, such as Beatty sequences and the evil numbers corresponding to the 0's in the famous Thue–Morse sequence. We also propose a more general solution which depends only on which of the colors "dominates" the sequence. Our construction resolves a problem posed by Fraenkel at the BIRS 2011 workshop in combinatorial games.

1. Introduction

At the workshop in combinatorial games in BIRS 2011, Aviezri Fraenkel posed the following intriguing problem: find nice (short/simple) rules for a 2-player combinatorial game for which the $\mathcal{P}$-positions are obtained from a pair of complementary Beatty sequences [Be]. We begin by solving this problem, by defining a class of heap games, dubbed bichromatic nim, or just chromatic nim, and then later in Section 4, we explain some background to the problem. In Section 3, we solve a similar game on arithmetic progressions. In Section 5, we discuss the general environment for chromatic nim on two heaps. At last, in Section 6, we study the famous evil numbers, also known as the indexes of the 0's in the Thue–Morse sequence.

2. Bichromatic nim finds a game for your complementary Beatty solution

Let S denote a subset of the positive integers. We let the i-th token in a stack be *red* if $i \in S$, and otherwise the i-th token is *green*. We play a take-away game on $k \geq 0$ copies of such stacks of various finite sizes. Classical nim rules are always

MSC2010: 11B05, 91A46.
Keywords: combinatorial games, density, gaps, topology.

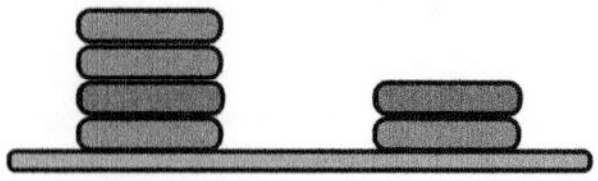

Figure 1. The 2-stack ϕ^2-chromatic nim position (4, 2), where $\phi = \frac{1}{2}(1+\sqrt{5})$.

allowed; any number of tokens can be removed from precisely one of the stacks. In addition, if no heap size belongs to S, then *the position is green* and any move is legal; in particular, it is now allowed to lower all stacks to 0. Another way to identify a green position is to look at the stacks from above. If you see only green tokens, then the position is green. Two players alternate moving and a player who cannot move loses. Note that if S is the set of positive integers, then the game is k-pile nim (because all tokens are red). If S is the empty set, then the game is 1-pile nim, independently of k, because all tokens are green (for a reader who likes to compute so-called Grundy values of impartial games, in this special case it obviously means that the Grundy value is the total number of tokens). We call this game S-chromatic nim.

Let $\beta > 2$ be irrational and let $S = \{\lfloor \beta n \rfloor\}$, for n running over the positive integers. Then the red tokens are determined by: the i-th token is red if and only if there is an n such that $i = \lfloor \beta n \rfloor$. We play on two stacks, and, since the sequences are regular (as opposed to random), determined by a number β, we call the game (2-stack) β-chromatic nim.

For an example, view Figure 1. Since the position is not green, then only nim type moves are possible. The unique winning move is to remove three tokens from the left-most heap. The easiest way to identify a winning move is to make sure precisely one heap is green and the other one is red. Now you also need to count the number of tokens in the respective stack, colored in the same color as the top token. If these two numbers are identical then you have found your winning move. This idea generalizes as we show in Section 5.

Theorem 1. *Let $\beta > 2$ be irrational. Then a position (x, y) of 2-stack β-chromatic nim is a winning position for the previous player if and only if $(x, y) = (\lfloor \alpha n \rfloor, \lfloor \beta n \rfloor)$ or $(\lfloor \beta n \rfloor, \lfloor \alpha n \rfloor)$, for some $n \in \mathbb{N}$, and where*

$$\frac{1}{\alpha} + \frac{1}{\beta} = 1. \tag{1}$$

Proof. Recall Beatty's theorem [Be]: if (1) is satisfied, then the sequences $(\lfloor \alpha n \rfloor)$ and $(\lfloor \beta n \rfloor)$ are complementary, for $n > 0$; each positive integer occurs in precisely one of the sequences and only once in this sequence. Suppose first that $x = \lfloor \alpha m \rfloor$ and $y = \lfloor \beta m \rfloor$, for some $m \in \mathbb{N}$. If $m = 0$, we are done, so suppose that $m > 0$. Observe that no position of this form is green, since one of the coordinates is $\lfloor \beta m \rfloor$, a red stack height. Hence it suffices to show that there is no nim option of the same form. Note that (1) together with $\beta > 2$ implies that $\alpha > 1$. Hence decreasing just one of the heaps cannot give a position of the same form; it follows since, by Beatty's theorem, the sequences $(\lfloor \alpha n \rfloor)$ and $(\lfloor \beta n \rfloor)$ are complementary.

Suppose next that the pair (x, y) is not of the given form. If the smaller heap is empty then the current player removes all tokens in the higher stack as well, which solves this case. Otherwise there are positive integers $m \geq n$ falling into one of these four cases:

Case 1: $x = \lfloor \alpha m \rfloor$, $y = \lfloor \beta n \rfloor$, $m > n$.

Case 2: $x = \lfloor \alpha m \rfloor$, $y = \lfloor \alpha n \rfloor$, $n > 0$.

Case 3: $x = \lfloor \alpha n \rfloor$, $y = \lfloor \beta m \rfloor$, $m > n$.

Case 4: $x = \lfloor \beta m \rfloor$, $y = \lfloor \beta n \rfloor$, $n > 0$.

Notice that none of the positions represents a position of the form in the theorem; the first and third since the sequences are strictly increasing and the second and fourth by complementarity. Hence, our task is to find a legal move to a position of the form in the theorem for each case.

The position (x, y) given by the second case is green and so it is an $\mathcal{N}$-position. This follows from complementarity of the sequences $(\lfloor \alpha i \rfloor)$ and $(\lfloor \beta i \rfloor)$, namely since $x = \lfloor \alpha m \rfloor$ there is no integer i such that $\lfloor \beta i \rfloor = x$ and similarly for y. For Case 1, it is clear that the current player can lower the x stack to $x = \lfloor \alpha n \rfloor$.

For the third case, by $m > n$ since $\beta > 2$ we get $\lfloor \beta m \rfloor > \lfloor \beta n \rfloor$, so that the desired nim move on the y-stack is to lower it to the position $(\lfloor \alpha n \rfloor, \lfloor \beta n \rfloor)$. The fourth case is similar, but the lowering is on the x-stack, motivated by $\lfloor \beta m \rfloor \geq \lfloor \beta n \rfloor > \lfloor \alpha n \rfloor$, which follows since (1) gives $1 < \alpha < 2 < \beta$ and by $n > 0$. (The latter inequality excludes the terminal position $(x, y) = (0, 0)$ which of course is also of the form $(x, y) = (\lfloor \alpha n \rfloor, \lfloor \beta n \rfloor)$, for some $n \in \mathbb{N}$). □

3. Games with arithmetic progression solutions

Next, let us study a generalization of the game rules of β-chromatic nim to β an integer; that is, let $\beta \geq 2$ be an integer and let $S = \{\beta n \mid n \in \mathbb{Z}^+\}$. We have the following perhaps not so surprising result in view of Theorem 1. The solution will still consist of complementary sequences, and indeed we have now shifted focus around to the more standard one in CGT (finding a solution for your game).

Theorem 2. *Let $\beta \geq 2$ be an integer. Then a position (x, y) of 2-pile β-chromatic nim, with $x \leq y$, is a previous player winning position if and only if $(x, y) = (0, 0)$*

or

$$(x, y) = (\beta n + t, (\beta - 1)(\beta n + t) + t) \quad (2)$$
$$= (\beta n + t, \beta((\beta - 1)n + t)), \quad (3)$$

for some $n \in \mathbb{N}$ *and some* $t \in \{1, \ldots, \beta - 1\}$.

Proof. Note first that, by the definition of t, $(\beta - 1)n + t$ takes on all the positive integers. Therefore the y-coordinates will take on precisely all multiples of β. For the same reason, the x-coordinates will take on precisely the complement of this set. Let $\mathcal{P}'$ denote all positions (x, y) and (y, x) where x and y are defined by (2). We begin by showing that no option of $(x, y) \in \mathcal{P}'$ is in $\mathcal{P}'$. Since the sets of all x's and y's are complementary, it suffices to prove that only nim type moves are possible. But, by definition, with notation as in (3), each x is green and each y is red.

Let us next prove that from each position $(x, y) \notin \mathcal{P}'$, there is a move to a position in $\mathcal{P}'$.

Here there are positive integers $m \geq n$ falling into one of these cases:

Case 1: $x = \beta m + t$, $y = \beta((\beta - 1)n + t)$, $m > n$;

Case 2: $x = \beta m + t$, $y = \beta n + t$, $n > 0$;

Case 3: $x = \beta n + t$, $y = \beta((\beta - 1)m + t)$, $m > n$;

Case 4: $x = \beta((\beta - 1)m + t)$, $y = \beta((\beta - 1)n + t)$, $n > 0$.

For Case 1, we can reduce the x-stack to $\beta n + t$. For Case 2, both stacks are green, so a move to $(0, 0)$ is possible. For Case 3, we can reduce y to $\beta((\beta - 1)n + t)$. Finally, for Case 4, since $\beta \geq 2$,

$$\beta((\beta - 1)m + t) \geq \beta((\beta - 1)n + t) \geq \beta(n + t) > \beta n + t,$$

so the desired move to $(\beta n + t, \beta((\beta - 1)n + t))$ is possible. □

4. Discussion of the origin to the problematic solution

Typically, game rules of combinatorial games are short and easy to learn, but not always. One should be able to learn the rules of a game without a degree in mathematics. The distinction we are speaking of is play-games versus math-games. We contribute play-game rules to an original math-game problem. The new element is the coloring of the tokens. This takes care of uncountably many problems (disguised as one problem). We already know that there is a countably infinite family of play-game rules for $\mathcal{P}$-positions of distinct complementary Beatty sequences [F82] and that family has been expanded via continued fractions in [DR; LW]. But Wythoff nim [W] is the origin of these type of questions. nim on two heaps provides a nearly trivial mimicking winning strategy and the $\mathcal{P}$-positions are all positions with equal heap sizes. By adjoining these nim $\mathcal{P}$-positions as moves in a new game, Wythoff discovered that the new $\mathcal{P}$-positions will be described by half lines of slope the golden ratio and its inverse. In fact, the game in Figure 1 is $\mathcal{P}$-equivalent to Wythoff nim. Now, the challenge of finding game rules for *any* complementary pair of Beatty sequences was posed in [DR] and resolved in [LHF]. There was a proviso to the solution in [DR]; the game rules must be *invariant*, and this is a new notion to an old description of vector subtraction games from [G]. Many game rules are noninvariant (perhaps the most famous of them all is Fibonacci nim), but the classical ones (subtraction games, nim, Wythoff nim) are invariant in the sense that a rule does not depend on which position it was moved from (apart from the empty-heap condition). The $\star$-operator defined in [LHF] produces invariant games, but only with exponential complexity in log of heap sizes. So the proviso game rules being invariant is nice in one way, but on the other hand, the obtained games cannot easily be played by human beings. The exponential complexity is decreased to polynomial ditto in [FL], but where invariance was relaxed to 2-invariance, a special restricted family of variant games; but although the game rules are polynomial in succinct input size, they remain math-games. Here we study simple game rules: true play-games. They have a somewhat surprising solution (although not as surprising as the solution Wythoff originally discovered in his variation of the game of nim).

5. Other chromatic nim games

Since chromatic nim was capable enough to solve the problem posed at the BIRS 2011 workshop in combinatorial games, we got interested in what properties the game might have in a somewhat more general setting. Let us discuss a natural generalization of the games and sequences from Section 2, still just on 2 stacks. Given a set S, let us call our game S-chromatic nim. The following lemma will allow us to resolve any 2-stack game for sequences with a surplus of green tokens, without further knowledge of the sequence. Let us regard the set S as an increasing sequence of integers $S = \{s_i\}_{i>0}$ and let $\bar{S} = \{\bar{s}_i\}_{i>0}$ denote the unique increasing sequence *complementary* to S on the positive integers (that is if n is a positive integer, then n is in precisely one of the sequences). Then S is *green-dominated* if, for all i, $\bar{s}_i < s_i$, and S is *red-dominated* if, for all i, $\bar{s}_i > s_i$. See also paper [L2014] for a similar construction. Hence it is clear that a sequence cannot be both red-dominated and green-dominated, and of course, "most" increasing sequences are neither. Note that for example β-chromatic nim from Section 2 is green-dominated, since $\beta > 2$. We have the following lemma for any green-dominated game S-chromatic nim.

Lemma 3. *Suppose that S-chromatic nim is green-dominated. Let*

$$\{(a_i, b_i), (b_i, a_i) \mid i \in \mathbb{N}\}$$

denote its set of $\mathcal{P}$-positions, where for all $i \geq 0$, $a_i \leq b_i$. Then, for all $i > 0$, the i-th green token from below is the a_i-th token and the i-th red token from below is the b_i-th token. Therefore, for all i, $a_i < b_i$, which implies that all monochromatic positions are $\mathcal{N}$-positions.

Proof. First of all it is clear that if both heaps are green, then there is a move to $(0, 0)$, so we assume first that one of the heaps is red and the other green. If in addition, the red heap contains the same number of red tokens as the number of green tokens in the green heap, then there is no nim-type move to a position of the same form. But these are the only legal type of moves, so the "$\mathcal{P}$ cannot go to $\mathcal{P}$ property" is satisfied. Now, if the number of red tokens in the red heap is different from the number of green tokens in the green heap, then we have to find a candidate $\mathcal{P}$-position to move to. If there are more red tokens in the red heap than there are green tokens in the green heap, then there is a nim-type move that equalizes the numbers. Hence, only the case for two red heaps remains to be considered. One of the heaps contains no more red tokens than the other. Then, because of the green-dominated property, the other heap contains at least as many green tokens as the number of red ones in the former. Hence a nim-type move suffices to reduce the taller red heap to a green heap with as many green tokens as the number of red ones in the previously smaller red heap. The base case is that since the game is green-dominated, when the first red token appears, then there is a green below, so the above proof applies. □

How do you play to win a green-dominated game? Let us summarize the proof of Lemma 3.

Proposition 4. *If the heaps have different colors and the red heap has r red tokens and the green heap has g green tokens, with $r = g$, then there is no winning move for the current player. Otherwise, the first player should remove $r - g$ red tokens from the red heap if $r > g$ and $g - r$ green tokens from the green heap if $g > r$, in either case keeping the color of the changed heap the same. If both heaps are green you move to $(0, 0)$. If both heaps are red and r is the number of red tokens in the smaller heap, then play in the larger (or equally sized) heap so that it becomes green with r green tokens.*

Proof. This is a direct consequence of Lemma 3. Notice again, for the last sentence, this is always possible, because of the green-dominated property. □

For β-chromatic nim with a rational $\beta > 2$, we know a winning strategy via Proposition 4, but do not yet have a complete characterization extending Theorems 1 and 2.

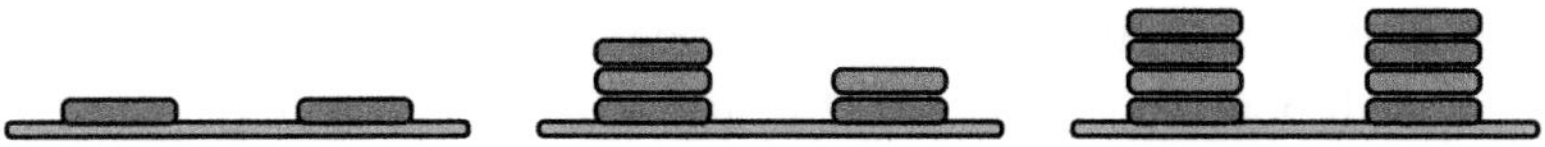

Figure 2. The first three $\mathcal{P}$-positions of $\frac{3}{2}$-chromatic nim.

When S is red-dominated, then Lemma 3 and Proposition 4 are no longer true, and it is easy to see because the base case fails. Let $\beta = \frac{3}{2}$. Then, for β-chromatic nim, $S = \{1, 3, 4, 6, 7, 9, \ldots\}$. This sequence is not green-dominated since, for example, the stack with one token has one red token and no green one. It is red-dominated because $\bar{S} = \{2, 5, 8, 11, \ldots\}$, and so $s_i < \bar{s}_i$ for all i. In fact, the smallest nonterminal $\mathcal{P}$-position is the red position $(1, 1)$, so the conclusions of Lemma 3 and Proposition 4 are false. Let us list the first few $\mathcal{P}$-positions of this game, to obtain intuition for the next result on red-dominated sequences. In Figure 2, we see the three first P-positions of $\frac{3}{2}$-chromatic nim, and the pictures illustrate how colors can be either both red or mixed. But there is a simple explanation to this behavior. We use the following definition.

Definition 5. The game S-chromatic nim is *locally green-dominated* (lgd) at level $d \in S$, if there is some positive integer $k \in \bar{S}$, such that the k-shifted heaps (with the lower $k-1$ tokens removed) induces a local green-dominated game on d tokens, colored according to (a reindexed) $S \setminus \{0, \ldots, k-1\}$. The local green-domination is maximal if the game is not lgd at level $d+1$.

In a sense, the red-dominated games behave like nim on the red tokens, but any local green-dominance has to be compensated for; we can use a local variation of Lemma 3 to compute the $\mathcal{P}$-positions, until the position is not lgd any longer, at which point the old nim-strategy will reappear (see the third picture, Figure 2).

Proposition 6. *Consider a red-dominated game S and d a positive integer. Then (d, d) is a $\mathcal{P}$-position if and only if the game is not locally green-dominated at level d. Otherwise, we consider the locally green-dominated game beginning at some minimal level k; that is, we apply Lemma 3 to compute the $\mathcal{P}$-positions with level k exchanged for level* 1 *(a green token). This computation stops at a level where the lgd is maximal.*

Remark 7. In Figure 2, we obtain the first nonzero $\mathcal{P}$-position as $(1, 1)$. (Only red tokens behave as nim.) The second $\mathcal{P}$-position is of the second type (with $k = 2$), because colors of heaps are mixed. The way to compute the $\mathcal{P}$-positions in a lgd game is to apply the algorithm for green-dominated games, but here starting with the 2nd green layer of tokens rather than the first red layer. Now, already at level 3 ($d = 3$) the lgd becomes maximal, so in fact the green-dominated algorithm terminates in just one step. (Notice here that Definition 5 is satisfied with $k = 2$ and $d = 3$.)

Proof. By Definition 5, the stack-sizes can be partitioned into two classes. Class 1 is the set of all non-lgd's and Class 2 is the set of all (maximal) lgd's. The latter contains the discrete intervals of the form $\{k, \dots, d\}$ as in Definition 5. Since we are discussing red-dominated games, the base case is as the left-most picture in Figure 2 (level 1 is red). Now, there is a smallest green token, at say level k. It will be the first Class 2 token. Since the first red token above this green one exists (by red-dominating), say at level $d > k$, the first Class 2 $\mathcal{P}$-position will be (k, d). The Class 2 $\mathcal{P}$-positions will have to continue as outlined in Lemma 3 until perhaps the level-k adjusted green-dominating property fails. (It does not have to fail, because the set S can still be red-dominated, because the initial layer(s) of red tokens could compensate for example a local green dominated periodic behavior.) If it fails, then there will be a maximal (red) token, say at level $d > k$ for which lgd holds. Then $(d+1, d+1)$ is a $\mathcal{P}$-position. It follows from the fact that the red token at level $d+1$ cannot be paired up with a green token below, because, by Lemma 3, they have already been matched up with lower red tokens. Now the proof follows by induction, since the continuation from level $d+1$ onwards is the same as restarting from level 1, with $d+1$ exchanged for 1. (Because only nim-type moves are allowed, by induction, no lower $\mathcal{P}$-positions can be reached in a single move, and the special move rule for green positions can just as well be used to move to $(d+1, d+1)$, which is $\mathcal{P}$ by induction. □

Now it is easy to combine Proposition 4 with the proof of Proposition 6 to find your winning move. The clue is to identify the non-lgd sequences to know the excess of red tokens that need to be subtracted in order to use the green-dominated result correctly. But, this is easy by the recursive argument. We note that this general winning strategy requires a bottom-up approach and is therefore slow compared to the results in Sections 2 and 3.

6. More fractal rules and strategies

We say that a nonnegative integer is *evil* if it has an even number of ones in its binary expansion; otherwise it is called *odious*. In this section, we let S be the set of evil integers and call the resulting game *evil-chromatic nim*. Thus, in this game, the evil integers are red and the odious integers are green.

The following observation is not strictly needed in the proofs to come, but we prove it anyway.

Observation 8. There cannot be more than two consecutive odious numbers (evil numbers) in any sequence of consecutive integers.

Proof. Suppose $n, n+1, n+2$ are three consecutive odious numbers. Note that n has to be odd or the parity of the number of 1's for $n+1$ will be incorrect. Because of the constraints on n, n has to look like

(i) $n = \underbrace{11\cdots1}_{2i+1\text{ 1's}}$, or

(ii) $n = \underbrace{1*\cdots*}_{2i\text{ 1's}}\textcircled{0}\underbrace{11\cdots1}_{2j+1\text{ 1's}}$,

where (i) consists of an odd number of consecutive 1's and (ii) begins from the left with a 1 followed by 1's and 0's, with an even number of 1's to the left of the circled 0 (a $*$ indicates a 0 or a 1). Then

(1) $n+1 = \mathbf{1}\underbrace{00\cdots0}_{2i+1\text{ 0's}}$, or

(2) $n+1 = \underbrace{1*\cdots*}_{2i\text{ 1's}}\textcircled{1}\underbrace{00\cdots0}_{2j+1\text{ 0's}}$.

But then, in either case, $n+2$ will contain an even number of 1's and thus be evil. The case for evil numbers is similar. □

Definition 9. Let k be a nonnegative integer and let $U \subseteq [k]\cup\{0\} = \{0, 1, \ldots, k\}$ be a subset consisting of consecutive integers. Then the *pseudochromatic number* of U, denoted $\tau(U)$, is defined to be the difference

$$(\#\text{ of green numbers in }U) - (\#\text{ of red numbers in }U).$$

In the special case where $U = [k]\cup\{0\}$, we write $\tau(k)$ for $\tau(U)$.

Lemma 10. *For any nonnegative integer* n, $-1 \le \tau(n) \le 1$.

Proof. The proof will proceed by induction on n. For our base cases, we consider the integers 0 through 7:

integer	binary representation	green/red	τ
0	000	red	−1
1	001	green	0
2	010	green	1
3	011	red	0
4	100	green	1
5	101	red	0
6	110	red	−1
7	111	green	0

Longer lists (of length 2^j) can be built recursively, as shown in Figure 3.

Notice that Figure 3 is constructed by prefixing the block of the binary numbers 00, 01, 10, and 11 with 00, 01, 10, or 11, respectively. Further, observe that blocks A, B, C, and D each have pseudochromatic number equal to 0. Hence, $\tau(15) = 0$.

		τ
	0000	−1
A	**00**01	0
	0010	1
	0011	0
	0100	1
B	**01**01	0
	0110	−1
	0111	0
	1000	1
C	**10**01	0
	1010	−1
	1011	0
	1100	−1
D	**11**01	0
	1110	1
	1111	0

Figure 3. Recursive algorithm for longer lists (of length 2^j).

Moreover, notice the maps illustrated in the figure from A to D and from B to C preserve the evil/odious quality of each integer and its respective pseudochromatic number.

Next assume that our result holds for all j such that $1 \le j < n$. Find $m > 0$ so that $2^{m-1} \le n \le 2^m - 1$. Using the recursive construction described above, we construct the list of length 2^m in Figure 4.

Note that n is either in block C or block D. By induction and the recursive construction of the list, the desired result holds for n. □

Definition 11. Given a positive integer k, we define the *chromatic number* of the set $[k] = \{1, 2, \dots, k\}$, denoted by $\chi(k)$, by $\chi(k) = \tau([k]) + 1$.

The next lemma refines Lemma 10.

Lemma 12. *If k is odious and even, then $\chi(k) = 2$. If k is evil and even, then $\chi(k) = 0$; $\chi(k) = 1$ for every positive odd number k.*

Proof. Suppose that k has binary expansion

$$k = 1\underbrace{00\cdots0}_{z_1 \ge 0\ 0\text{'s}}1\underbrace{00\cdots0}_{z_2 \ge 0\ 0\text{'s}}10\cdots01\underbrace{00\cdots0}_{z_{j-1} \ge 0\ 0\text{'s}}1\underbrace{00\cdots0}_{z_j \ge 0\ 0\text{'s}},$$

where k has j 1's in positions $i_1, i_2, \dots, i_j$ (reading from left-to-right) and the z_ℓ's give the number of 0's after each one. For the purposes of this proof, we let

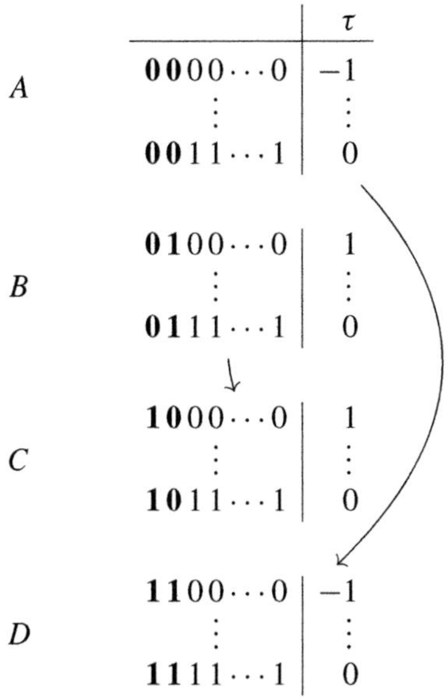

Figure 4. Recursive algorithm for longer lists (of length 2^m).

$k(i_\ell)$ denote the number in binary notation derived from k by changing the i_ℓ-th 1 and any nonzero bit associated to a power of 2 less than it to a 0.

For example, if $k = 10011001$, then $i_1 = 7$, $i_2 = 4$, $i_3 = 3$, $i_4 = 0$, $z_1 = 2$, $z_2 = 0$, $z_3 = 2$, $z_4 = 0$, and $k(i_3) = k(3) = 10010000$ (note that the one counted by i_1 is associated with 2^7).

Next, for a given positive integer k and an associated i_ℓ, we define

$$L(k, i_\ell) = k(i_\ell) + \left\{ \left(\sum_{r=0}^{i_\ell - 1} c_r 2^r \right)_2 : c_r = 0 \quad \text{or} \quad c_r = 1 \right\}$$

if $\ell > 1$ and

$$L(k, i_1) = \left\{ \left(\sum_{r=0}^{i_1 - 1} c_r 2^r \right)_2 : c_r = 0 \quad \text{or} \quad c_r = 1 \right\} \setminus \{0\}.$$

For example, if $k = 10011001$, then

$$L(k, i_3) = 10010000 + \{111, 110, 101, 100, 011, 010, 001, 000\}.$$

We are now ready to proceed with the proof. We consider two cases.

Case 1 (k is odd): If k is odious, then $\chi(L(k, i_j)) = -1$, since $L(k, i_j)$ consists of exactly one evil number. Also observe that $\chi(L(k, i_1)) = 1$, since $L(k, i_1) =$

$[2^{i_1-1}]\setminus\{0\}$. However, $\chi(L(k,i_\ell))=0$ for all $1<\ell<j$ since $\chi([2^q]\cup\{0\})=0$ for all $q>0$. Thus, $\chi(k)=1+1-1=1$, as desired.

If k is evil, then $\chi(L(k,i_j))=1$, since $L(k,i_j)$ consists of exactly one odious number. Also observe that $\chi(L(k,i_1))=1$, since $L(k,i_1)=[2^{i_1-1}]\setminus\{0\}$. However, $\chi(L(k,i_\ell))=0$ for all $1<\ell<j$ since $\chi([2^q]\cup\{0\})=0$ for all $q>0$. Thus, $\chi(k)=-1+1+1=1$, as desired.

Case 2 (k is even): If k is odious, then $\chi(L(k,i_\ell))=0$ for all $1<\ell\leq j$ since $\chi([2^q]\cup\{0\})=0$ for all $q>0$. As in the case above, we have $\chi(L(k,i_1))=1$, since $L(k,i_1)=[2^{i_1-1}]\setminus\{0\}$. Hence, $\chi(k)=1+1=2$, as desired.

If k is evil, then $\chi(L(k,i_\ell))=0$ for all $1<\ell\leq j$ since $\chi([2^q]\cup\{0\})=0$ for all $q>0$. Again, we have $\chi(L(k,i_1))=1$, since $L(k,i_1)=[2^{i_1-1}]\setminus\{0\}$. Hence, $\chi(k)=-1+1=0$, as desired. □

Recall that the "mex" of a nonempty set of nonnegative integers is defined to be the minimum excluded element. For example, if $S=\{0,1,2,7,9,13\}$, then $\operatorname{mex}(S)=3$. Using the mex rule and Lemma 3, we characterize the $\mathcal{P}$-positions of evil-chromatic nim in the next theorem.

Theorem 13. *The set of $\mathcal{P}$-positions of evil-chromatic nim $\{(a_i,b_i)\mid i\geq 0\}$ can be computed recursively as follows. The only terminal $\mathcal{P}$-position is $(a_0,b_0)=(0,0)$. Otherwise, $a_n=\operatorname{mex}\{a_i,b_i\mid i<n\}$ and b_n is the smallest evil number such that $a_n<b_n$.*

Proof. The proof of this result follows from Lemma 3 and Lemma 10 above. □

Given the interesting number theory surrounding evil and odious numbers, we are able to refine the last result quite substantially. We say that a nonnegative integer is *vile* if its binary expansion ends in an even number of zeros; otherwise it is called *dopey*.

Theorem 14. *The set of $\mathcal{P}$-positions of evil-chromatic nim $\{(a_i,b_i)\mid i\geq 0\}$ are given by $(a_0,b_0)=(0,0)$, and, for $n>0$, by*

$$b_n=\begin{cases}2n & \text{if } n \text{ is evil},\\ 2n+1 & \text{if } n \text{ is odious},\end{cases}$$

and

$$a_n=\begin{cases}b_n-1 & \text{if } n \text{ is evil and dopey},\\ b_n-2 & \text{if } n \text{ is vile},\\ b_n-3 & \text{if } n \text{ is odious and dopey}.\end{cases}$$

Proof. The proof will proceed by induction on n.

Case 1 (n is evil and dopey): We claim that $n-1$ must be evil and vile. If not, then $n-1$ is evil and dopey, odious and dopey, or odious and vile. In either of the first two cases, $n-1$ has binary expansion

$$\underbrace{1**\cdots*1}_{k\text{ 1's}}\underbrace{00\cdots0}_{2m+1\text{ 0's}},$$

where k is even if $n-1$ is evil and k is odd if $n-1$ is odious (where $m \geq 0$ and a $*$ denotes a 0 or a 1). But this implies that $n = (n-1)+1$ is either evil and vile or odious and vile, a contradiction.

Next suppose that $n-1$ is odious and vile. Then $n-1$ has binary expansion

(i) $\underbrace{1**\cdots*0}_{2k+1\text{ 1's}}\underbrace{11\cdots1}_{2m\text{ 1's}}$,

(ii) $\underbrace{1**\cdots*0}_{2k\text{ 1's}}\underbrace{11\cdots1}_{2m+1\text{ 1's}}$,

(iii) $\underbrace{11\cdots1}_{2k+1\text{ 1's}}$, or

(iv) $\underbrace{1**\cdots*1}_{2k+1\text{ 1's}}\underbrace{00\cdots0}_{2m>0\text{ 0's}}$.

Since n is dopey, cases (i) and (iv) are not possible, and since n is evil, cases (ii) and (iii) are not possible.

Hence, $n-1$ is evil and vile. Then $n-1$ has binary expansion

(i) $\underbrace{1**\cdots*0}_{2k+1\text{ 1's}}\underbrace{11\cdots1}_{2m+1\text{ 1's}}$,

(ii) $\underbrace{11\cdots1}_{2k\text{ 1's}}$, or

(iii) $\underbrace{1**\cdots*1}_{2k\text{ 1's}}\underbrace{00\cdots0}_{2m>0\text{ 0's}}$.

Since n is dopey, cases (ii) and (iii) are not possible. In case (iv),

$$b_{n-1} = 2(n-1) = (\underbrace{1**\cdots*0}_{2k+1\text{ 1's}}\underbrace{11\cdots1}_{2m+1\text{ 1's}}0)_2,$$

by induction. Adding the next two consecutive terms after b_{n-1} and using Lemma 12 we observe that

	χ	
b_{n-1}	0	$b_{n-1} = 2(n-1)$
a_i	1	
b_n	0	$b_n = b_{n-1}+2 = 2n$

Based on the chromatic numbers shown above, we must have $a_n = a_i = b_n - 1$. Therefore, if n is evil and dopey, then $b_n = 2n$ and $a_n = b_n - 1$.

Case 2 (n is odious and dopey): We will show that $n-1$ must be odious and vile. If $n-1$ is evil and dopey, then the binary expansion of $n-1$ looks like

$$\underbrace{1**\cdots*1}_{2k\text{ 1's}}\underbrace{00\cdots0}_{2m+1\text{ 0's}}.$$

But, if this was true, then n would be vile. Thus $n-1$ is not evil and dopey. Next we consider what happens if $n-1$ was evil and vile. Then $n-1$ would have binary expansion

(i) $\underbrace{1**\cdots*}_{2k+1\text{ 1's}}0\underbrace{11\cdots1}_{2m+1\text{ 1's}}$,

(ii) $\underbrace{11\cdots1}_{2k\text{ 1's}}$, or

(iii) $\underbrace{1**\cdots*1}_{2k\text{ 1's}}\underbrace{00\cdots0}_{2m>0\text{ 0's}}$.

Case (i) is not possible since n is odious. Further, cases (ii) and (iii) are not possible because n is dopey. Now, if $n-1$ is odious and dopey, then $n-1$ would have binary expansion

$$\underbrace{1**\cdots*1}_{2k+1\text{ 1's}}\underbrace{00\cdots0}_{2m+1\text{ 0's}}.$$

If this was so, then n would be evil and vile, a contradiction. By process of elimination, $n-1$ must be odious and vile. Then $n-1$ has binary expansion

(i) $\underbrace{1**\cdots*}_{2k+1\text{ 1's}}0\underbrace{11\cdots1}_{2m\text{ 1's}}$,

(ii) $\underbrace{1**\cdots*}_{2k\text{ 1's}}0\underbrace{11\cdots1}_{2m+1\text{ 1's}}$,

(iii) $\underbrace{11\cdots1}_{2k+1\text{ 1's}}$, or

(iv) $\underbrace{1**\cdots*1}_{2k+1\text{ 1's}}\underbrace{00\cdots0}_{2m>0\text{ 0's}}$.

Since n is dopey, cases (i) and (iv) are not possible. In case (ii),

$$b_{n-1}=2(n-1)+1=(\underbrace{1**\cdots*}_{2k\text{ 1's}}0\underbrace{11\cdots1}_{2m+1\text{ 1's}}1)_2,$$

and in case (iii),

$$b_{n-1}=2(n-1)+1=(\underbrace{11\cdots1}_{2k+1\text{ 1's}}1)_2,$$

both by induction. Adding the term before b_{n-1} and the two after it, we see by Lemma 12 the results at the top of the next page.

	χ	
a_{i-1}	2	
b_{n-1}	1	$b_{n-1} = 2(n-1)+1$
a_i	2	
b_n	1	$b_n = b_{n-1}+2 = 2n+1$

Based on the chromatic numbers shown above, we must have $a_n = a_{i-1} = b_n - 3$. Therefore, if n is odious and dopey, then $b_n = 2n+1$ and $a_n = b_n - 3$.

Case 3 (n is evil and vile): We will show that $n-1$ is either odious and dopey or odious and vile. To this end, if $n-1$ was evil and dopey, then its binary expansion would look like

$$\underbrace{1**\cdots*1}_{2k\text{ 1's}}\underbrace{00\cdots0}_{2m+1\text{ 0's}}.$$

Since n is evil, this is not possible. If $n-1$ was evil and vile then its binary expansion would look like

(i) $\underbrace{1**\cdots*0}_{2k\text{ 1's}}\underbrace{11\cdots1}_{2m\text{ 1's}}$,

(ii) $\underbrace{1**\cdots*0}_{2k+1\text{ 1's}}\underbrace{11\cdots1}_{2m+1\text{ 1's}}$,

(iii) $\underbrace{11\cdots1}_{2k\text{ 1's}}$, or

(iv) $\underbrace{1**\cdots*1}_{2k\text{ 1's}}\underbrace{00\cdots0}_{2m>0\text{ 0's}}$.

Cases (i), (iii), and (iv) are not possible since n is evil and case (ii) is not possible as n is vile. Thus, $n-1$ is either odious and dopey or odious and vile. If $n-1$ is odious and dopey, then its binary expansion looks like

$$\underbrace{1**\cdots*1}_{2k+1\text{ 1's}}\underbrace{00\cdots0}_{2m+1\text{ 0's}}.$$

Then, by induction,

$$b_{n-1} = 2(n-1)+1 = (\underbrace{1**\cdots*1}_{2k+1\text{ 1's}}\underbrace{00\cdots0}_{2m+1\text{ 0's}}1)_2.$$

Adding the term before b_{n-1} and the one after it, we see by Lemma 12 that

	χ	
a_i	2	
b_{n-1}	1	$b_{n-1} = 2(n-1)+1$
b_n	0	$b_n = b_{n-1}+1 = 2n$

Based on the chromatic numbers shown above, we must have $a_n = a_i = b_n - 2$. Now if $n-1$ is odious and vile, then its binary expansion looks like

(i) $\underbrace{1**\cdots*}_{2k+1\text{ 1's}}0\underbrace{11\cdots1}_{2m\text{ 1's}}$,

(ii) $\underbrace{1**\cdots*}_{2k\text{ 1's}}0\underbrace{11\cdots1}_{2m+1\text{ 1's}}$,

(iii) $\underbrace{11\cdots1}_{2k+1\text{ 1's}}$, or

(iv) $\underbrace{1**\cdots*1}_{2k+1\text{ 1's}}\underbrace{00\cdots0}_{2m>0\text{ 0's}}$.

Because n is vile, cases (ii) and (iii) are not possible. By induction

$$b_{n-1} = 2(n-1)+1 = (\underbrace{1**\cdots*}_{2k+1\text{ 1's}}0\underbrace{11\cdots1}_{2m\text{ 1's}}1)_2$$

in case (i) and

$$b_{n-1} = 2(n-1)+1 = (\underbrace{1**\cdots*1}_{2k+1\text{ 1's}}\underbrace{00\cdots0}_{2m>0\text{ 0's}}1)_2$$

in case (iv). In either case we again have

	χ	
a_i	2	
b_{n-1}	1	$b_{n-1} = 2(n-1)+1$
b_n	0	$b_n = b_{n-1}+1 = 2n$

Thus, if n is evil and vile, then $b_n = 2n$ and $a_n = b_n - 2$.

Case 4 (n is odious and vile): We will show that $n-1$ is either evil and dopey or evil and vile. If $n-1$ was odious and dopey, then its binary expansion would look like

$$\underbrace{1**\cdots*1}_{2k+1\text{ 1's}}\underbrace{00\cdots0}_{2m+1\text{ 0's}}.$$

Since n is odious, this is not possible. If $n-1$ was odious and vile, then it binary expansion would look like

(i) $\underbrace{1**\cdots*}_{2k+1\text{ 1's}}0\underbrace{11\cdots1}_{2m\text{ 1's}}$,

(ii) $\underbrace{1**\cdots*}_{2k\text{ 1's}}0\underbrace{11\cdots1}_{2m+1\text{ 1's}}$,

(iii) $\underbrace{11\cdots1}_{2k+1\text{ 1's}}$, or

(iv) $\underbrace{1**\cdots*1}_{2k+1\text{ 1's}}\underbrace{00\cdots0}_{2m>0\text{ 0's}}$.

Because n is odious, cases (i) and (iv) are not possible and since n is vile, cases (ii) and (iii) are not possible. Thus, $n-1$ is either evil and dopey or evil and vile. If $n-1$ is evil and dopey, then it has binary expansion

$$\underbrace{1**\cdots*1}_{2k\text{ 1's}}\underbrace{00\cdots0}_{2m+1\text{ 0's}}.$$

Thus, by induction,

$$b_{n-1}=2(n-1)=(\underbrace{1**\cdots*1}_{2k\text{ 1's}}\underbrace{00\cdots0}_{2m+1\text{ 0's}}0)_2.$$

Adding the three consecutive terms after b_{n-1} and using Lemma 12 we have

	χ	
b_{n-1}	0	$b_{n-1}=2(n-1)$
a_{i-1}	1	
a_i	2	
b_n	1	$b_n=b_{n-1}+3=2n+1$

Based on the chromatic numbers shown, we must have $a_n=a_{i-1}=b_n-2$. If $n-1$ is evil and vile, then it has binary expansion

(i) $\underbrace{1**\cdots*}_{2k\text{ 1's}}0\underbrace{11\cdots1}_{2m\text{ 1's}}$,

(ii) $\underbrace{1**\cdots*}_{2k+1\text{ 1's}}0\underbrace{11\cdots1}_{2m+1\text{ 1's}}$,

(iii) $\underbrace{11\cdots1}_{2k\text{ 1's}}$, or

(iv) $\underbrace{1**\cdots*1}_{2k\text{ 1's}}\underbrace{00\cdots0}_{2m>0\text{ 0's}}$.

Since n is odious, (ii) is not possible. By induction

$$b_{n-1}=2(n-1)=(\underbrace{1**\cdots*}_{2k\text{ 1's}}0\underbrace{11\cdots1}_{2m\text{ 1's}}0)_2$$

in case (i),

$$b_{n-1}=2(n-1)=(\underbrace{11\cdots1}_{2k\text{ 1's}}0)_2$$

in case (iii), and

$$b_{n-1}=2(n-1)=(\underbrace{1**\cdots*1}_{2k\text{ 1's}}\underbrace{00\cdots0}_{2m>0\text{ 0's}}0)_2$$

in case (iv). No matter what the case, we again have the results at the top of the next page.

	χ	
b_{n-1}	0	$b_{n-1} = 2(n-1)$
a_{i-1}	1	
a_i	2	
b_n	1	$b_n = b_{n-1} + 3 = 2n + 1$

Thus $a_n = a_{i-1} = b_n - 2$. Hence, if n is odious and vile, then $b_n = 2n+1$ and $a_n = b_n - 2$. □

Example. Find the 17509^{17509}th $\mathcal{P}$-position of evil-chromatic nim (17509 is the 2015th prime).

With the help of the computer algebra system Mathematica we know that $q = 17509^{17509}$ is evil and vile. Hence, Theorem 14 tells us that

$$(a_q, b_q) = (2q-2, 2q).$$

References

[BCG] E. R. Berlekamp, J. H. Conway, R. K. Guy, *Winning ways*, **1–2**, Academic Press, London (1982). Second edition, **1–4**, A K Peters, Wellesley/MA (2001/03/03/04).

[Be] S. Beatty, Problem 3173, *Amer. Math. Monthly*, **33** (1926), 159.

[Bo] C. L. Bouton, Nim, A game with a complete mathematical theory, *Ann. Math.* (2), **3** (1901), 35–39.

[DR] E. Duchêne and M. Rigo, Invariant games, *Theoret. Comp. Sci.*, **411** 34–36 (2010), 3169–3180.

[F82] A. S. Fraenkel, How to beat your Wythoff games' opponent on three fronts, *Amer. Math. Monthly* **89** (1982), 353–361.

[F] A. S. Fraenkel, The Rat Game and the Mouse Game, *Games of no chance* 4, MSRI Publications **63**, Cambridge Univ. Press (2015), 95–113.

[FL] A. S. Fraenkel, U. Larsson, Take-away games on Beatty's theorem and the notion of k-invariance, in this volume.

[G] S. W. Golomb, A mathematical investigation of games of "take-away". *J. Combin. Theory* **1** (1966), 443–458.

[L1] U. Larsson, Restrictions of m-Wythoff Nim and p-complementary Beatty sequences, *Games of no chance* 4, MSRI Publications **63**, Cambridge Univ. Press (2015), 127–160.

[L2] U. Larsson, 2-pile Nim with a Restricted number of move-size imitations, *Integers* **9** (2009), Paper G4, 671–690.

[L3] U. Larsson, Blocking Wythoff Nim, *Electr. J. Combin.*, P120 of Volume 18(1) (2011).

[L2012] U. Larsson, The $\star$-operator and invariant subtraction games, *Theoret. Comput. Sci.*, **422** (2012), 52–58.

[L2014] U. Larsson, Wythoff Nim extensions and splitting sequences, *J. Integer Sequences*, **17** (2014), article id. 14.5.7.

[LHF] U. Larsson, P. Hegarty, A. S. Fraenkel, Invariant and dual subtraction games resolving the Duchêne–Rigo conjecture, *Theoret. Comp. Sci.* **412**, 8–10 (2011), 729–735.

[LW] U. Larsson, M. Weimerskirch, Impartial games, whose rule sets produce given continued fractions, preprint.

[R] J. W. Rayleigh. The theory of sound, *Macmillan*, London (1894), 122–123.

[S] A. J. Schwenk, "Take-away games", *Fibonacci Quart.* **8** (1970), 225–234.

[Z] M. Zieve, Take-away games, *Games of No chance*, MSRI Publications **29**, Cambridge Univ. Press (1996), 351–361.

[W] W.A. Wythoff, A modification of the game of Nim, *Nieuw Arch. Wisk.* **7** (1907), 199–202.

mfisher@wcupa.edu *Department of Mathematics, West Chester University of Pennsylvania, West Chester, PA, United States*

urban031@gmail.com *Faculty of Industrial Engineering and Management, Technion - Israel Institute of Technology, Haifa, Israel*

Games of No Chance 5
MSRI Publications
Volume **70**, 2017

Take-away games on Beatty's theorem and the notion of k-invariance

AVIEZRI S. FRAENKEL AND URBAN LARSSON

We formulate three reasonably short game rules for three two-pile take-away games, which share one and the same set of P-positions. This set is comprised of a pair of complementary homogeneous Beatty sequences together with $(0, 0)$. We relate the succinctness of the game rules with the complexity of the P-positions by means of a notion dubbed k-invariance, and introduce the *game-invariance number* for a set of P-positions.

1. Introduction

The first author posed an intriguing problem at the GONC-workshop: "Describe nice/short rulesets for games with so-called complementary Beatty sequences as sets of P-positions." The problem is the opposite of the main field of research in this area, which is to, given a game, search for its set of P-positions. Here we describe three such rule sets, which resolves the question for any pair of complementary homogenous Beatty sequences.

Let us recall the rules of d-Wythoff [10], d a fixed positive integer. The available positions are (x, y), x and y nonnegative integers. The legal moves are

(I) Nim-type: $(x, y) \to (x - t, y)$, if $x - t \geq 0$ and $(x, y) \to (x, y - t)$, if $y - t \geq 0$; $t > 0$.

(II) Extended diagonal type: $(x, y) \to (x - s, y - t)$ if $|t - s| < d$ and $x - s \geq 0$, $y - t \geq 0$; $s > 0, t > 0$.

This game is a so-called impartial take-away game [2], vol. 1. We restrict attention to *normal* play; that is, the player first unable to move loses. For our games it means that the player called upon to move from $(0, 0)$ loses.

Rules (I) and (II) imply that d-Wythoff is a so-called *invariant* [5; 16] (take-away) game; that is, each available move is legal from any position as long as the resulting position has nonnegative coordinates. Every move in any invariant

MSC2010: 91A46.
Keywords: combinatorial game, complementary Beatty sequences, impartial game, k-invariant game, game invariance number.

game is an *invariant move*. In this note we study another type of take-away game, where certain positions have some local restrictions on the set of otherwise invariant moves. Such games are called *variant* [5; 16]. We define these notions in Section 4.

Central to our investigation is Beatty's theorem [1] (predated by Lord Rayleigh [19]): Let $\beta > 2$ be an irrational number and define its *complement*, $\hat{\beta}$, by the equality $\hat{\beta}^{-1}+\beta^{-1}=1$, so that $\hat{\beta}=\beta/(\beta-1)$. This clearly implies $1<\hat{\beta}<2<\beta$. Let $A_n = \lfloor n\hat{\beta}\rfloor$, $B_n = \lfloor n\beta\rfloor$, $A = \cup_{n\geq 1}\{A_n\}$, $B = \cup_{n\geq 1}\{B_n\}$. Beatty's theorem then asserts that A and B are *complementary* sets, that is, $A\cup B=\mathbb{Z}_{\geq 1}$, $A\cap B=\varnothing$. Since $\beta > \hat{\beta} > 1$, the (homogeneous) *Beatty sequences* (A_n) and (B_n) are strictly increasing.

1.1. *Three games.* We formulate three game rules. Let $\beta > 2$ be a fixed irrational and let $d = \lfloor\beta\rfloor$. Fix a pair of nonnegative integers (x, y). Recall that $B_n = \lfloor n\beta\rfloor$ for all n:

(G1) The moves are as in nim on two piles (I), except that, if $B\cap\{x, y\}=\varnothing$, then in addition to the nim-type move a player may also take away $s\in\{0, \dots, d\}$ from the other pile in the same move. This game is denoted by *β-nim.*

(G2) The moves are as in d-Wythoff, subject to (I) and (II), except that if $B\cap\{x, y\}\neq\varnothing$, then only nim-type moves (I) are permitted. This game is denoted by *β-Wynim.*

(G3) The moves are as in d-Wythoff, subject to (I) and (II), except that if $B\cap\{x, y\}\neq\varnothing$, then the pair (s, t), with s and t as in (II), cannot belong to the pair of *β-triangles* defined by

$$\big\{(x, y), (y, x) \mid (x, y) \in \{(1, \lfloor\beta\rfloor), (2, \lfloor\beta\rfloor), (2, \lfloor\beta\rfloor + 1)\}\big\}.$$

This game is denoted by *β^{T}-Wynim.*

The name Wynim derives from Wythoff-nim; in (G3) the T in β^{T} stands for triangles. The main result of this note is as follows.

Theorem 1. *The set of P-positions of β-nim, β-Wynim and β^{T}-Wynim is the same. It is*

$$\mathcal{P} := \bigcup_{n\geq 0}\{(A_n, B_n)\} \cup \bigcup_{n\geq 0}\{(B_n, A_n)\},$$

where $A_n = \lfloor n\hat{\beta}\rfloor$, $B_n = \lfloor n\beta\rfloor$.

We prove this result in Section 3. In Section 4 we develop the distinction between invariant and variant games and relate our findings to certain complexity issues. In the section to come we give some examples.

2. Examples and tables of P-positions

In the proof of the main result and in the examples of sets of P-positions to come, we use the following illustrative notation.

Notation 2. For every $n \geq 0$:

(1) $\Delta A_n := A_{n+1} - A_n$, $\Delta B_n := B_{n+1} - B_n$ are the *gaps*.

(2) $\Delta_n := B_n - A_n$.

(3) $\Delta_n^2 := \Delta_{n+1} - \Delta_n$.

For some (invariant) take-away games on two heaps where short formulas for both the rules and the P-positions are known, such as [10; 12; 14], the coordinates of the P-positions are defined via certain algebraic numbers together with the floor function. Our first example rather uses a well-known transcendental number.

Example 3. In the game of π-Wynim, a player may move as in nim on two piles (I), or, if the position does not contain a coordinate of the form $\lfloor \pi n \rfloor$, deviate at most $\lfloor \pi \rfloor - 1 = 2$ positions from the "main diagonal" as given by the game d-Wythoff; that is use (II) with $d = 3$. The result of this note implies that the P-positions of this game are the set

$$\cup_{n \geq 0} \{ (\lfloor \hat{\pi} n \rfloor, \lfloor \pi n \rfloor), (\lfloor \pi n \rfloor, \lfloor \hat{\pi} n \rfloor) \},$$

the first few of which are displayed in Table 1.

Example 4. Example 3 illustrates Theorem 1 for a member of our second game family, β-Wynim. A further example: Let $d = 2$ in the formula $\beta =$

n	0	1	2	3	4	5	6	7	8	9	10	11	12	13	14
A_n	0	1	2	4	5	7	8	10	11	13	14	16	17	19	20
B_n	0	3	6	9	12	15	18	21	25	28	31	34	37	40	43
Δ_n	0	2	4	5	7	8	10	11	14	15	17	18	20	21	23
Δ_n^2	2	2	1	2	1	2	1	3	1	2	1	2	1	2	2

n	15	16	17	18	19	20	21	22	23	24	25	26	27	28	29
A_n	22	23	24	26	27	29	30	32	33	35	36	38	39	41	42
B_n	47	50	53	56	59	62	65	69	72	75	78	81	84	87	91
Δ_n	25	27	29	30	32	33	35	37	39	40	42	43	45	46	49
Δ_n^2	2	2	1	2	1	2	2	2	1	2	1	2	1	3	1

Table 1. The first few P-positions (A_n, B_n) for β-nim, β-Wynim and β^{T}-Wynim, $\beta = \pi = 3.14159\ldots$

n	0	1	2	3	4	5	6	7	8	9	10	11	12	13
A_n	0	1	2	4	5	7	8	9	11	12	14	15	16	18
B_n	0	3	6	10	13	17	20	23	27	30	34	37	40	44

n	14	15	16	17	18	19	20	21	22	23	24	25	26	27
A_n	19	21	22	24	25	26	28	29	31	32	33	35	36	38
B_n	47	51	54	58	61	64	68	71	75	78	81	85	88	92

Table 2. The first few P-positions (A_n, B_n) for 2-Wythoff, β-nim β-Wynim and β^{T}-Wynim; β as in Example 4.

$\frac{1}{2}(2+d+\sqrt{d^2+4})$ (see d-Wythoff and paper [10]) and with $\hat{\beta}=\beta-d$. Then $\beta=\sqrt{2}+2$, $\hat{\beta}=\beta-2$; note that $\lfloor\beta\rfloor=3$ as in Example 3. The first few P-positions are shown in Table 2. Since $\beta-\hat{\beta}=d=2$, we have $\Delta_n=dn=2n$, so $\Delta_n^2=d=2$ for all $n\geq 0$, and the β-triangles, as in (G3), for both these games, will be $\{(1,3),(2,3),(2,4)\}\cup\{(3,1),(3,2),(4,2)\}$.

Remark 5. It is remarkable that, for $\beta=\frac{1}{2}(2+d+\sqrt{d^2+4})$, d-Wythoff has the same set of P-positions as our three games. In particular, for $d=1$, 1-Wythoff is the classical Wythoff game [2]. For $d=2$, the first few P-positions of the games are displayed in Table 2. In [4] it was shown that from the classical Wythoff game no move can be deleted while preserving the set of P-positions of the classical Wythoff game. In the present note, Wythoff moves were deleted, and the P-positions are still preserved. The difference is that in [4] only *invariant* moves were permitted. See Section 4 for more on the latter topic.

3. Proof of the main result

We preface the proof of Theorem 1 by collecting some facts on the sets $\{A_n\}$ and $\{B_n\}$.

Proposition 6. *For every* $n\geq 0$:

(i) *The only possible gap pairs are*

$$(\Delta A_n, \Delta B_n)\in\big\{(1,\lfloor\beta\rfloor),\ (1,\lfloor\beta\rfloor+1),\ (2,\lfloor\beta\rfloor),\ (2,\lfloor\beta\rfloor+1)\big\}.$$

(ii) $\Delta_n^2=\Delta B_n-\Delta A_n$.

(iii) $\Delta_n^2\in\{\lfloor\beta\rfloor-2,\lfloor\beta\rfloor-1,\lfloor\beta\rfloor\}$.

Proof. (i) This is a well-known result.

(ii) $\Delta_n^2=(B_{n+1}-A_{n+1})-(B_n-A_n)=(B_{n+1}-B_n)-(A_{n+1}-A_n)=\Delta B_n-\Delta A_n$.

(iii) Follows directly from (i) and (ii). □

Example 7. Notice that in Example 3, Table 1, Δ_n^2 assumes all three possible values $\{1, 2, 3\} = \{\lfloor\beta\rfloor - 2, \lfloor\beta\rfloor - 1, \lfloor\beta\rfloor\}$. In Example 4, Δ_n^2 assumes only the value $2 = \lfloor\beta\rfloor - 1$.

Proof of Theorem 1. Since our games are acyclic, it suffices to demonstrate the following two properties for each game:

P $\rightarrow$ N: Every move from any position of the form

$$(A_n, B_n) \quad \text{or} \quad (B_n, A_n) \tag{1}$$

results in a position outside (1).

N $\rightarrow$ P: Given any position outside (1), there exists a move into (1).

For the direction P $\rightarrow$ N we use the same argument for the games (G1) β-nim and (G2) β-Wynim, namely: Suppose that we play from a position of the form (1). The game rules imply that only nim-type moves (I) are permitted so that by complementarity, there is no move to a position of the same form.

For the game (G3) β^{T}-Wynim, we have to show that both

(i) $(A_n, B_n) \rightarrow (A_m, B_m)$ and

(ii) *cross moves* $(A_n, B_n) \rightarrow (B_m, A_m)$ are blocked for every $0 \le m < n$.

(i) By Proposition 6, $(B_n - B_m) - (A_n - A_m) = \Delta_n - \Delta_m \ge \Delta_n - \Delta_{n-1} = \Delta_{n-1}^2 \in \{\lfloor\beta\rfloor - 2, \lfloor\beta\rfloor - 1, \lfloor\beta\rfloor\}$, where the $\ge$ follows from the fact that $\beta > \hat{\beta}$, which implies that Δ_i is a nondecreasing function of i. Therefore the move $(A_n, B_n) \rightarrow (A_m, B_m)$ is either blocked by the triangle move restriction of β^{T}-Wynim (if $\Delta_{n-1}^2 \le \lfloor\beta\rfloor$), or by the $\lfloor\beta\rfloor$-Wythoff constraint (if $\Delta_{n-1}^2 \ge \lfloor\beta\rfloor$).

(ii) Notice that this move is possible only if $A_n > B_m$. Now $(B_n - A_m) - (A_n - B_m) = \Delta_n + \Delta_m$. Similarly to (i), if $\Delta_n + \Delta_m \in \{\lfloor\beta\rfloor - 2, \lfloor\beta\rfloor - 1, \lfloor\beta\rfloor\}$, this forces $m = 0$ and $n = 1$ so that the move is blocked by the β-triangle move restriction; otherwise by the $\lfloor\beta\rfloor$-Wythoff constraint.

For the direction N $\rightarrow$ P, let (x, y), $0 \le x \le y$ be a position not of the form (1). We assume first that this position has a coordinate of the form B_n, so for each game it suffices to show that a nim-type (I) move suffices for moving into (1). If $x = B_n$ then move $y \rightarrow A_n$. If $y = B_n$ and $x > A_n$ then move $x \rightarrow A_n$. If $y = B_n$ and $x < A_n$, complementarity implies that there exists $m < n$ such that either $x = A_m$ so the move $y \rightarrow B_m$ restores (1); or else $x = B_m$, so the move $y \rightarrow A_m$ does it.

Hence we may assume that both x and y are in A, say $x = A_m \le A_n = y$. If $y > B_m$, then the nim-type (I) move $y \rightarrow B_m$ suffices for each game. We may therefore assume that

$$x = A_m \le A_n = y < B_m. \tag{2}$$

Since each of (A_i) and (B_i) is strictly increasing, a nim-type move to a position (1) does not exist, so we have to find a (II) extended diagonal type move for the games β-Wynim and β^{T}-Wynim. Observe that for both these games, this type of move is now unrestricted with $k = \lfloor \beta \rfloor$.

Let $d := y - x$. Then $d = A_n - A_m < B_m - A_m = \Delta_m$. By Proposition 6, Δ_i grows from 0 to Δ_m as i grows from 0 to m, in steps $\Delta_i^2 = \Delta B_i - \Delta A_i \in \{\lfloor \beta \rfloor - 2, \lfloor \beta \rfloor - 1, \lfloor \beta \rfloor\}$, bounded above by $\lfloor \beta \rfloor$. Hence there exists j such that $0 \le d - \Delta_j < \lfloor \beta \rfloor$. Then move $(x, y) \to (A_j, B_j)$. We need to show three things:

(i) $j < m$,

(ii) $y > B_j$,

(iii) $|(y - B_j) - (A_m - A_j)| < \lfloor \beta \rfloor$.

(i) $\Delta_j \le d = y - A_m < B_m - A_m = \Delta_m$. Since Δ_i is an increasing function of i, we have $j < m$.

(ii) $y = A_m + d > A_j + d \ge A_j + \Delta_j = B_j$.

(iii) $|(y - B_j) - (A_m - A_j)| = |(y - A_m) - (B_j - A_j)| = |d - \Delta_j| < \lfloor \beta \rfloor$.

On the other hand, for the game β-nim and a position of the form in (2), by Proposition 6(i) a nearest lower P-position is attainable by an extended "horizontal" nim-type move. Precisely, since $\Delta B_n \in \{\lfloor \beta \rfloor, \lfloor \beta \rfloor + 1\}$ we can lower $y = A_n$ to B_i, where $i \ge 0$ is such that $B_i < A_n < B_{i+1}$, and $x = A_m$ to A_i, that is move $(A_m, A_n) \to (A_i, B_i)$. By the (I) nim-type move we have to show that $A_i < A_m$. But the definition of i together with (2) give $B_i < A_n < B_m$ which, by $i < m$, implies $A_i < A_m$.

Thus the set $\mathcal{P}$ is indeed the set of P-positions for our three games. □

4. The notion of k-invariance and game complexity

Let us continue our discussion of variant versus invariant games from the introduction and Remark 5, and relate it to the complexity of numbers and games. We think of an integer as the simplest number, followed by the rationals, algebraic numbers and transcendental numbers, the most complex numbers.

Our three games are, in fact, "minimally variant" in the sense that all their positions can be partitioned into precisely two sets, namely,

$$\{(A_i, A_j) \mid i, j \in \mathbb{Z}_{>0}\} \quad \text{and} \quad \{(B_i, A_j), (A_i, B_j), (B_i, B_j) \mid i, j \in \mathbb{Z}_{\ge 0}\},$$

such that, for each game, for each set, the possible moves are invariant. This observation motivates a weakening of the notion of invariance to *k-invariance*, $k \in \mathbb{Z}_{>0}$.

Definition 8. Let X be a subset of the set of positions (j-tuples of nonnegative integers) of a game G on j heaps. Then m (also a j-tuple of nonnegative integers, but not all 0) is an *invariant move* in X, if for all $x \in X$, $x - m$ is an option, provided $x - m$ is a position in G.

Definition 9. Let G be a game and X a subset of all positions in G. Then m is a *variant move* in X if there exist $x, y \in X$ such that both $x - m$ and $y - m$ are positions in G, and $x - m$ is an option but $y - m$ is not.

Definition 10. Let $k \in \mathbb{Z}_{>0}$. A game G is k-*invariant* if

- its set of positions can be partitioned into k subsets, such that, within each subset X, each allowed move is invariant in X;
- for any partition $\sqcup X_i$ of G's positions into $< k$ subsets, there is an i and an m such that m is a variant move in X_i.

If a game G is not k-invariant for any $k \geq 1$, then it is ∞-*invariant*. If $k = 1$, then G is *invariant* (the second item does not apply). If $k \neq 1$, then G is *variant*.

The games in this paper are all 2-invariant. The "mouse game" in [6] is 2-invariant. On the other hand, the $\star$-operator for invariant subtraction games [16; 15] produces an invariant game, the "mouse trap" [13], with the same sets of P-positions as the mouse game. The game Mark [7; 8] is ∞-invariant (of course, any ∞-invariant game is variant). It is played on the nonnegative integers: from position n, either subtract one, or move to $\lfloor \frac{1}{2} n \rfloor$.

Let S be a set of d-tuples of nonnegative integers. Suppose that $P(G) = S$ where G is a k-invariant game, and for all $l < k$ there is no l-invariant game H such that $P(H) = S$, then we say that the set of P-positions S is k-*game-invariant*. Note that, for any set S there is a *trivial* $|S|$-*invariant game* (perhaps $|S| = \infty$): no move is possible from a position in S and each position not in S has a move to 0. Hence, we may ask, for any given set S of P-positions (at least there is the trivial game), what its game-invariance number is (a positive integer or infinity).

For example, we wonder if the game-invariance number for the P-positions of Mark is finite.

Let us return to the setting in this paper, and let $\gamma = \beta - \hat{\beta}$. It appears that the complexity of γ, the simplicity of the game rules and the game-invariance number are related. If γ is an integer, the game-invariance number is one, and moreover, by (I) and (II) the game rules are "short"; see also Example 4. For our three games, γ is not necessarily an integer, the game rules are longer and the 1-invariance is replaced by 2-invariance. To shed more light on these suggested relationships, it might be well to investigate whether γ rational, algebraic [5; 17], or transcendental has any effect on the length of the game rules and the game-invariance number.

We close this section with a problem which requires a little background. The succinct input size of a given ordered pair of integers (x, y) is $\log(xy)$. The time complexity of deciding whether a given ordered pair (x, y) is of the form (A_n, B_n) is polynomial in $\log(xy)$; see [10, §3]. In [16, Main Theorem] it is demonstrated that, given the set $\mathcal{P}$ in Theorem 1, there is an *invariant* game for which the time complexity of determining whether a given ordered pair (s, t) is a legal move, but it is exponential in $\log(st)$. The game invariance number is one, but in general short game rules are not known. In [5; 17] polynomial time and invariant game rules are determined for the set $\mathcal{P}$ when γ is restricted to some specific algebraic numbers of degree 2.

We state a problem related to the games in this note, concerning the relation between game-invariance number and short game rules.

Problem. Let $\beta > 2$ be irrational, and let

$$\mathcal{Q} = \bigcup_{n\geq 0}\{(A_n, B_n)\} \cup \bigcup_{n\geq 0}\{(B_n, A_n)\},$$

where $(A_n)_{n\geq 1}$ and $(B_n)_{n\geq 1}$ are complementary Beatty sequences defined by β, and $A_0 = B_0 = 0$. Suppose that $\gamma = \beta - \hat{\beta}$ is transcendental, and consider a game G with $\mathcal{Q}$ as its set of P-positions. Can G have game-invariance number 1, if there is a polynomial time algorithm, in $\log(x, y)$, for finding a winning move from any given N-position (x, y)?

By the results of this paper, we know that game invariance number 2 is attainable. Perhaps this problem exists even if γ is algebraic, or even if γ is a noninteger rational number. Perhaps it holds even if the Beatty sequences A and B are not complementary [9].

The notion of k-invariance is also interesting in a somewhat different context. In [18] certain k-invariant 2-heap subtraction games with a finite number of (variant) moves are studied and it was shown that they embrace computational universality.

Many heap games in the literature have move-size dynamic rules (e.g., Fibonacci nim), blocking maneuvers (e.g., blocking Wythoff nim), depend on positions moved to, rather than position moved from, and so on, and new variations yet to come. The notion of k-invariance in this paper is only intended as a small guide for a larger classification in the future.

5. Discussion

We have formulated three reasonably short game rules for three 2-invariant games, which have identical sets of P-positions. Observe however that the rules contain a partial information about the set of P-positions (but neither $\hat{\beta}$ nor the density 1 is disclosed). Is it possible to find short 2-invariant game rules, without disclosing any part of the P-positions?

Suppose that we fix irrational $\beta > 2$ and then increase the density of the pairs of sequences from 1 to say an arbitrary number $\zeta > 1$ (or decreases to a density < 1) where α is defined via $1/\alpha + 1/\beta = \zeta$, and for all n, $A_n = \lfloor \alpha n \rfloor$. (That is, for all β, $\alpha \neq \hat{\beta}$.) Given the new sets of P-positions, is there still a short/succinct 2-invariant description for game rules without disclosing irrationals or/and the joint density of the sequences? It is unknown to us whether there exist invariant rules for such games (see also [16], [9] for similar problems). What are the game-invariance numbers for such sets of P-positions?

Acknowledgment

We thank Eric Duchêne for some informative discussion on invariance vs. k-invariance of games.

References

[1] S. Beatty, Problem 3173, *Amer. Math. Monthly* **33** (1926) 159; Solution [to Problem 3177, numbering in **33** was erroneous] in **34** (1927) 159–160.

[2] E. R. Berlekamp, J. H. Conway and R. K. Guy [2001-2004], *Winning Ways for Your Mathematical Plays*, Vol. 1-4, A K Peters, Wellesley, MA, 2nd edition: vol. 1 (2001), vols. 2, 3 (2003), vol. 4 (2004).

[3] E. Duchêne, A. S. Fraenkel, S. Gravier and R. J. Nowakowski, Another bridge between NIM and WYTHOFF, *Australas. J. Combin.* **44** (2009) 43–56.

[4] E. Duchêne, A. S. Fraenkel, R. J. Nowakowski and M. Rigo [2010], Extensions and restrictions of Wythoff's game preserving its P positions, *J. Combin. Theory Ser. A* **117**, 545–567.

[5] E. Duchêne and M. Rigo, Invariant games, *Theoret. Comput. Sci.* **411**(34-36) (2010) 3169–3180.

[6] A. S. Fraenkel, The Rat game and the Mouse game, *Games of No Chance 4*.

[7] A. S. Fraenkel, Aperiodic subtraction games, *Electr. J. Combin.* (The Zeilberger Festschrift volume) Vol. 18(2) (2011) P19 12pp.

[8] A. S. Fraenkel, The vile, dopey, evil and odious game players, *Discrete Math.* 312 (2012) 42–46, special volume in honor of the 80th birthday of Gert Sabidussi.

[9] A. S. Fraenkel, Beating your fractional Beatty game opponent and: What's the Question to Your Answer?, Advances in Combinatorics, in Memory of Herbert S. Wilf, Proc. Waterloo Workshop on Computer Algebra, I. S. Kotsireas and E. V. Zima, eds., May 2011, Springer (2013), 175–186.

[10] A. S. Fraenkel, How to beat your Wythoff games' opponent on three fronts, *Amer. Math. Monthly* **89** (1982) 353–361.

[11] A. S. Fraenkel and M. Lorberbom, Nimhoff games, *J. Combin. Theory* (Ser. A) **58** (1991) 1–25.

[12] U. Larsson, Restrictions of m-Wythoff Nim and p-complementary sequences, *Games of No Chance 4*.

[13] U. Larsson, "The $\star$-operator and invariant subtraction games", preprint (to be renamed to: Permutation games and the $\star$-operator) http://arxiv.org/1009.4220

[14] U. Larsson, Blocking Wythoff Nim, *Electr. J. Combin.*, 18 (2011) P120.

[15] U. Larsson, The $\star$-operator and invariant subtraction games, *Theoret. Comput. Sci.* Vol. 422, (2012) 52–58.

[16] U. Larsson, P. Hegarty and A. S. Fraenkel, Invariant and dual subtraction games resolving the Duchêne-Rigo conjecture, *Theoret. Comput. Sci.* **412** (2011) 729–735.

[17] U. Larsson, M. Weimerskirch, Impartial games whose rulesets produce given continued fractions, in this volume.

[18] U. Larsson, J. Wästlund, From heaps of matches to the limits of computability, *Electr. J. Combin.*, 20 (2013) P41.

[19] J. W. S. Rayleigh [1937], *The Theory of Sound*, MacMillan, London, 1st ed. 1877, 2nd ed. 1894, reprinted 1928, 1929, 1937. There is also a Dover 1945 publication, which is a reprint of the 2nd. The "sound" equivalent of Beatty's theorem, is stated, without proof, in sect. 92a, pp. 122-123, in the 1937 reprint of the 2nd ed., the only one we saw. Kevin O'Bryant told us that he checked the 1st edition and the statement is not there.

fraenkel@wisdom.weizmann.ac.il *Department of Computer Science and Applied Mathematics, Weizmann Institute of Science, Rehovot, Israel*

urban031@gmail.com *Department of Industrial Engineering and Management, Technion - Israel Institute of Technology, Haifa, Israel*

Games of No Chance 5
MSRI Publications
Volume **70**, 2017

Geometric analysis of a generalized Wythoff game

ERIC FRIEDMAN, SCOTT M. GARRABRANT,
ILONA K. PHIPPS-MORGAN,
A. S. LANDSBERG AND URBAN LARSSON

Combinatorial 2-player games can be studied from different perspectives. Traditionally the goal has been to acquire a perfect strategy, and to this purpose an efficient procedure (polynomial in succinct input size) is required. However, most combinatorial games are intrinsically hard to analyze; success is limited to a small number of games with predominant "mathematical structure". The classical games of Nim (Bouton 1901) and Wythoff Nim (Wythoff 1907) are easy to analyze rigorously, but already seemingly modest variants, like (p, q)-GDWN (Larsson 2012, 2014), appear to withstand log-polynomial descriptions. Therefore, development of new methods is highly desirable.

Here, we use methods from physics, such as *renormalization*, in an attempt to understand the larger geometry of a game's P-positions (safe positions for the Previous player), rather than their exact configurations (Friedman et al. 2007, 2009). By studying evolution diagrams of a general class of *linear extensions* of Nim, Wythoff Nim and GDWN, we observe that P-positions often distribute uniformly along lines (a.k.a. P-beams), visually separated from the move lines. Given a fundamental hypothesis, a *filling property* which generalizes directly from Wythoff Nim, we formulate natural equations on the slopes and densities of P-positions along these lines; here, a key innovation, a *reorganization* model, guides us in selecting the relevant rules (move lines). The exceptional case of the symmetric (p, q)-GDWN is interesting, because of observed quasi-log repetitive fluctuations, and these games have defied all previous analysis.

1. Introduction to the class Linear Nimhoff

In this paper we study a class of combinatorial games using renormalization-based techniques from physics in combination with computer simulations. This approach leads to a probabilistic geometric analysis of the underlying structure

MSC2010: 91A46.
Keywords: combinatorial game, GDWN, Nim, Wythoff Nim, renormalization, reorganization.

and behavior of a game. A number of interesting features are revealed, including observations of quasi log-periodic fluctuations. Our class of games, dubbed *Linear Nimhoff*, is a generalization of the classical Wythoff's game [Wythoff 1907] and the more recent GDWN [Larsson 2012a].

The *renormalization* approach to games involves:

1) Identifying broad, overall patterns in games (on a course grained level) by focusing on their scaling, asymptotic, and/or global, probabilistic behavior;
2) Analyzing these patterns using scaling/course-graining techniques derived from self-consistency conditions;
3) A novel method for this paper, a *reorganization* model, which filters out game rules that do not contribute to the broad overall patterns.

Two of the best-known impartial combinatorial games are Nim and Wythoff's modification of Nim, a.k.a Wythoff's game. In two-pile Nim, players alternate in removing tokens from a pile of their choosing, with the player who removes the last token declared the winner. Wythoff's game is an extension of two-pile Nim in which players, in addition to being able to remove tokens from a selected pile (as in Nim), also have the option of removing the same number of tokens from both piles simultaneously. Both games can be trivially but conveniently recast in terms of a marker moving on a semi-infinite, two-dimensional integer grid. The marker's (x, y) positional coordinates indicate the current number of tokens in each pile; the lower left corner $(0, 0)$ of the grid represents the game's terminal position. In Nim, the marker can slide leftwards or downwards; in Wythoff's game, a diagonal move (down and to the left) is also allowed. Both of these games have been well studied and are considered completely "solved" in the sense that a complete specification of the P- and N-positions[1] in these games is possible (see, e.g., [Bouton 1901; Wythoff 1907; Berlekamp et al. 1982]).

In *Linear Nimhoff*, the game marker can move not only leftwards, downwards, and diagonally, but along other prescribed directions as well. As we shall see, this seemingly simple extension leads to nontrivial *geometric* structures associated with the P-positions of the game. Mostly, we find that the P-positions in Linear Nimhoff lie along certain diffuse *lines*, and otherwise, in some specific cases exhibiting various quasi-log-periodic behavior within diverging *P-beams* [Larsson 2012a]. We show how the slopes and densities of these lines (or in the case of beams, the mean values of slopes and densities) can be computed via a semi-heuristic geometrical technique (adapted from renormalization models in physics) first described in [Friedman et al. 2007; Friedman et al. 2009].

[1]In keeping with standard terminology, a P-position in a combinatorial game is a winning position for the **P**revious player (i.e, the player who just moved to that position); an N-position is a winning position for the **N**ext player to move.

The results rely on a probabilistic description of the games' underlying structure, and in view of recent results on a symmetric restriction GDWN [Larsson 2012a; Larsson 2014], we will see that this assumption should be relaxed in some cases. Therefore we define classes of games, guided by the observed behaviour of their outcomes.

(i) The *strict class* of Linear Nimhoff will consist of games for which the P-positions satisfy a probabilistic description along lines (each probabilistic line is described unambiguously by a slope and a density).

(ii) The *nonprobabilistic class* will contain any other Linear Nimhoff game.

The main purpose of this paper is to explain the probabilistic geometry of games in the strict class. For games in the nonprobabilistic class, we argue that some general behavior will carry over, in spite of apparent fluctuations gradually transforming the probabilistic geometry. Experimental data show that, in most cases, if the probabilistic geometry breaks up into something else, then the new patterns will satisfy some *quasi log-periodic fluctuations* centred in the values obtained by the strict class computations.

(iii) The *QLPF class* (pronounced culpif) is a subset of the nonprobabilistic class (ii). It contains games for which the P-positions (P-beams) follow some quasi log-periodic fluctuations. An approximate scaling factor can be computed, by experimental data, often indicating log-periodic regions void of P-positions, which heuristically shows why the intersection of this class with the strict class (i) is empty.

(iv) The class of *relaxed Linear Nimhoff* is as QLPF, but where a relaxed geometric assumption (allowing for forbidden regions to separate fluctuated P-beams instead of probabilistic P-lines) includes games with log-periodic fluctuations to the class in (i). In particular, the mean values of the densities and slopes of the P-beams will satisfy the equations as obtained in (i).

The type of visual fluctuations in relaxed Linear Nimhoff can be very hard to describe; we rather suggest an empirical classification of those games. By generalizing the outcomes of games, by counting the number of P-positions as options (a method adapted from Blocking the Queen games [Cook et al. 2015]), we obtain more accurate predictions. It remains an open problem to determine if the classes (i) and (iv) are nonempty. Through many experiments, we believe that classes (iii) and (iv) are the same, so we often identify QLPF with relaxed Linear Nimhoff, although we are not yet aware of any method to prove this.

1.1. ***Game rules.*** Linear Nimhoff is an impartial combinatorial game played by moving a marker along positions on a semi-infinite, two-dimensional integer grid along given half-lines. A *position* $X = (x, y)$ of a game is a vector in $\Re^2$

whose coordinates are nonnegative integers; the set of all positions constitutes the *position space* of the game.

A *rule* $r = (a, b)$ is a vector in $\Re^2$ other than $(0, 0)$, whose coordinates are nonnegative integers. The *ruleset* is a set $R = \{r_i = (a_i, b_i)$, for $i = 1 \dots n)\}$ of rules, for some $n \geqslant 2$; by assumption, the ruleset of Linear Nimhoff always includes the Nim-rules $(a_1, b_1) = (1, 0)$ and $(a_n, b_n) = (0, 1)$. The ruleset designates the legal moves in the game: from a position X, one can select any rule $r \in R$ and move along this vector to any *valid* position $X - kr$, for k a positive integer. That is, from a general position $X = (x, y)$, a player can move to any position in the set $\{(x - ka, y - kb) : k \in \mathbb{Z}_{>0}, (a, b) \in R, x - ka \geq 0, y - kb \geq 0\}$. The class GDWN [Larsson 2012a] is a restriction of the general class Linear Nimhoff in that $(a, b) \in R$ implies also $(b, a) \in R$. The class (p, q)-GDWN consists of all games of the form $\{(1, 0), (q, p), (1, 1), (p, q), (0, 1)\}$, $0 < p < q$ integers.

A player loses the game when no legal moves remain available, as occurs when the position $(0, 0)$ is reached.

Since the rules imply that there are no infinite sequences of moves, every position in Linear Nimhoff can be uniquely characterized as being a P-position or an N-position. The set of all P-positions in the game is denoted $\mathcal{P}$, and the set of all N-positions is $\mathcal{N}$.

Figure 1 illustrates the allowed moves for one instantiation of the game (under ruleset $R = \{(1, 0), (2, 1), (3, 3), (0, 1)\}$). Observe that two-pile Nim and Wythoff Nim constitute special cases of Linear Nimhoff, with rulesets $\{(1, 0), (0, 1)\}$ and $\{(1, 0), (1, 1), (0, 1)\}$, respectively.

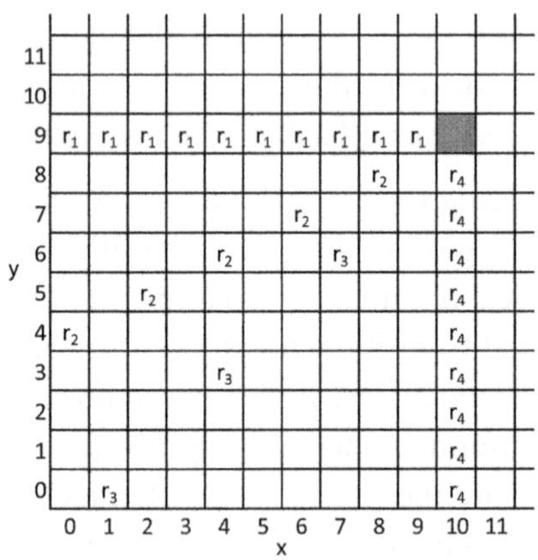

Figure 1. Legal moves in Linear Nimhoff. The figure illustrates the positions available to a player from the starting position shown (filled square) under rules $\{r_1 = (1, 0), r_2 = (2, 1), r_3 = (3, 3), r_4 = (0, 1)\}$.

2. Observations on the P-Positions of Linear Nimhoff

Figure 2 depicts the locations of the P-positions in Linear Nimhoff for a variety of different rulesets that we believe belong to the strict class.

For reference, Figures 2a,b illustrate the special cases of two-pile Nim and Wythoff's game, respectively. In Nim, the P-positions lie along the main diagonal, while in Wythoff's game the P-positions lie near two lines passing through the origin with slopes $\phi^{-1} = \frac{-1+\sqrt{5}}{2}$ and $\phi = \frac{1+\sqrt{5}}{2}$. The remainder of Figure 2 illustrates what occurs for more general cases in the strict class. In Figure 3, we show an instance of the class (iv) — the game (3, 5)-GDWN; that is the ruleset $R = \{(1, 0), (5, 3), (1, 1), (3, 5), (0, 1)\}$, indicating a quasi-log-periodic behavior.

The computer simulations reveal the following:

(1) In a typical game of Linear Nimhoff in the strict class, the P-positions approximate certain lines passing through the origin (see Figures 2c–f). By experimental data [Larsson 2012a], some of these "lines" will tend to diverge, but apparently never beyond certain bounds; see also Section 5, where the subclass GDWN of Linear Nimhoff is discussed. Therefore, and similar to this

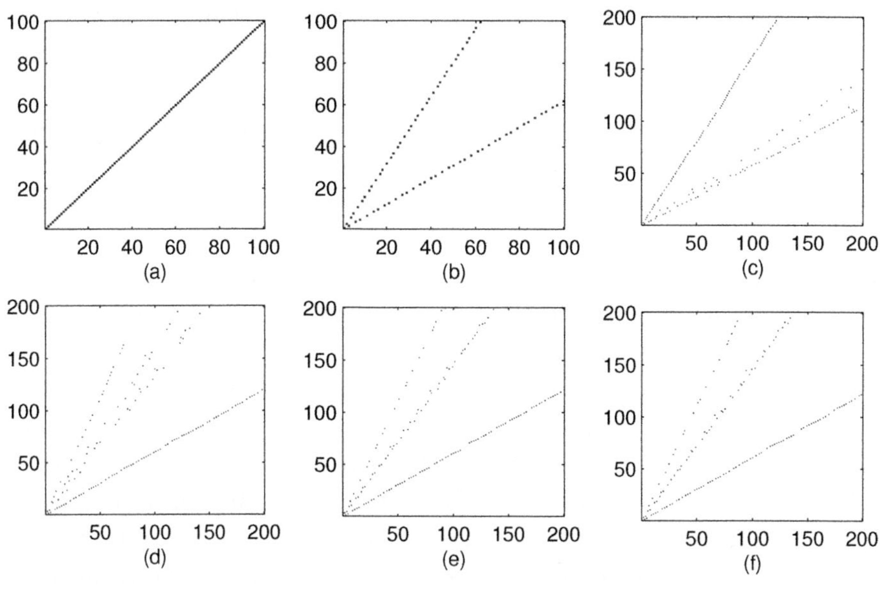

Figure 2. The locations of the P-positions (in black) in the x-y plane for various rule sets in Linear Nimhoff. (a) $R = \{(1, 0), (0, 1)\}$ (i.e., two-pile Nim); (b) $R = \{(1, 0), (1, 1), (0, 1)\}$ (i.e., Wythoff's game); (c) $R = \{(1, 0), (3, 2), (1, 1), (0, 1)\}$; (d) $R = \{(1, 0), (1, 1), (2, 3), (1, 2), (0, 1)\}$; (e) $R=\{(1,0),(1,1),(1,2),(0,1)\}$; (f) $R=\{(1,0),(1,1),(1,2),(1,8),(0,1)\}$.

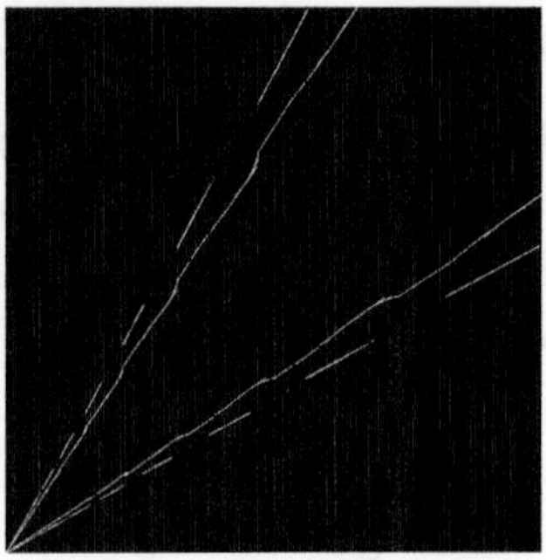

Figure 3. The location of P-beams (in white in this picture) of (3, 5)-GDWN for $x \leqslant 32600$, $y \leqslant 32600$; pixels have been coarsened for better visibility.

reference, we refer to these lines as either *P-lines* or *P-beams*, depending on conjectured behavior as in the strict class (i) or the QLPF class (iii), respectively.

In many cases, though, the P-positions exhibit very modest scatter about the P-lines; when computation is extended beyond a few hundred, the width of the scatter remains small in relation to the scale of the P-line. This includes some GDWN games, such as (1, 2)-GDWN and (2, 3)-GDWN [Larsson 2014]. Certain (p, q)-GDWN games appear to belong to the "diverging beams" class rather than the "small scatter along lines" class (we define this class in Section 5.1). In the case of the strict class, we assume that, along any given line, the proportion of the number of P-positions (below given x-coordinates) does not vary beyond certain bounds; the distribution is more or less uniform on large spatial scales (i.e., density variations are purely local). In QLPF it turns out that we instead compute mean values of the density and slope of a P-beam, see also Section 5, which appears to coincide with the slope and density of an imagined P-line, but instead representing a "centre" of a P-beam.

In most observed examples, if the rules of a Linear Nimhoff game are not symmetric then the P-positions are "uniformly distributed", and the suggested notion would be "P-line", but we have found an exception: for the ruleset $R = \{(1, 0), (5, 3), (8, 5), (1, 1), (3, 5), (0, 1)\}$, the fluctuations from the game $R = \{(1, 0), (5, 3), (1, 1), (3, 5), (0, 1)\}$ survive the introduction of the new rule, $r' = (8, 5)$. On the other hand, there are many symmetric rulesets where "P-line" appears to be the correct notion; see Section 6.1.

(2) The number of P-lines which appear in the figures (for games in the strict class) as well as their slopes and densities, depends nontrivially on the particular ruleset under consideration. For games such as Nim, Wythoff's Nim, (1, 2)-GDWN and some more GDWN games [Larsson 2012a], the number of P-lines

present equals the *number-of-rules-minus-one*: there are n P-lines and there are $n+1$ rules in the ruleset. Suppose that R is in the strict class, and let $\Delta(R) = \#R - \#$(P-lines associated with R). A comparison of Figures 2e and f illustrates that the addition of a new rule to a ruleset does not always create a new P-line. For (p, q)-GDWN games it is conjectured [Larsson 2012a] that the geometric behavior is as for Wythoff's game, unless (p, q) is of a very special form, as will be defined in Section 5.1: we get $\Delta = 1$ in the latter case, but $\Delta = 3$ in the former case. As we will see, the estimates of slopes and densities of P-lines in the strict class requires $\Delta = 1$. If this is not the case, then we provide an algorithm to reduce the number of rules, described as a novel *reorganization* model in Section 4.

3. Analysis of Linear Nimhoff

The intention of this section is to characterize the overall geometric structure of the P-positions in the strict class of Linear Nimhoff. Towards this end, we will forego determining the precise locations of P-positions in favour of a more global geometrical description that quantifies the number, slopes, and densities of the P-lines. The motivation behind this approach is as follows: recent work [Zeilberger 2001; Zeilberger 2004; Friedman et al. 2007; Friedman et al. 2009] suggests that some combinatorial games — including some that are presumed to be computationally "hard" — may display certain regularities which are manifest in the underlying geometric structure of the game's P-positions.

In particular, such games may simultaneously display both order (in the sense of a regular underlying geometry) and disorder (in the form of scatter about this regular structure). While the disordered component (i.e., scatter) is believed to be associated with a game's complexity and may resist analytical treatment, the ordered component may be tractable to analysis and yield critical new insights into the game [Friedman et al. 2007; Friedman et al. 2009]. For this reason, in this section, the overall goal is to calculate the number, slopes, and densities of the P-lines in the strict class of Linear Nimhoff, where scatter along lines is the most prominent feature.

A position X in Linear Nimhoff is called a *parent* of position Y, if a player can move from X to Y in one turn, and Y a *child* of X — in standard terminology the children are the "options". Given a rule $r \in R$, we define X to be a *parent under* r of Y, and Y to be a *child under* r of X, if one can move from X to Y using r. The *slope* $s(r)$ of a rule $r = (a, b)$ in Linear Nimhoff is denoted by $s(r) = b/a$.

3.1. ***Forbidden regions.*** In this section, we build a rigorous model for the analysis of Linear Nimhoff. We locally extend the ruleset R, to allow for any nonempty set of rules, so that for example R can be void of nim-type rules, or it may contain instead of (0, 1) a rule such as $(0, k)$, any $k > 0$. The reason for this relaxation of R is that the scatter-along-lines type of geometry also appear frequently in this bigger class.

For $\alpha, \beta \in \Re \cup \{\infty\}$, with $0 \leqslant \alpha < \beta$, then $F_{\alpha,\beta} = \{(x, y) : \alpha \leqslant y/x \leqslant \beta\} \subset \Re^2$ is a *forbidden region* if $F \cap P$ is finite, and the region is *sharp* if, for any $\epsilon > 0$, there are infinitely many P-positions (x, y) with ratios y/x in each of the intervals $(\alpha - \epsilon, \beta)$ and $(\alpha, \beta + \epsilon)$, or $\alpha = 0$, $\beta = \infty$ respectively.

Let $B \subset P(R)$. Sometimes this set is a P-beam, or perhaps even a P-line. Let $\alpha = \liminf y/x \leqslant \limsup y/x = \beta$, for $(x, y) \in B$. Then B is a *P-line* if $\alpha = \beta$. In general, if $\alpha \leqslant \beta$, then B is a *P-beam* if it contains no forbidden region. Thus we generalize notation and let $\Delta(R) = \#R - \#$P-beams.

Hypothesis 1. For each forbidden region F, there exists a single rule $r \in R$ by which it is possible to move from almost all positions in F to a P-position; in other words, almost all N-positions within a given forbidden region are parents of P-positions under the *same* rule r.

For example in the game of Nim, there are precisely two sharp forbidden regions, $F_{0,1}$ and $F_{1,\infty}$ respectively. Note that, given Hypothesis 1, the positions in a forbidden region may have moves to P-positions via other rules as well. This is exemplified in the middle forbidden region of Wythoff Nim $F_{\phi^{-1},\phi}$, where each position of the form (x, b_n), $|b_n - x| < n$ or (b_n, y), $|b_n - y| < n$ has not only a diagonal move, but also a Nim-type move to a P-position.[2]

Consider $r \in R$. We write $F' = F'(r)$ to denote almost all positions in a forbidden region F, satisfying Hypothesis 1; if $\forall x \in F' \, \exists y \in P : y + kr = x$, for some positive integer k, we say that the rule *r fills the forbidden region F* (with N-positions).

Lemma 2. *Consider a ruleset R and suppose that r fills the forbidden region* $F_{\alpha,\beta}(r)$. *If F is sharp and Hypothesis 1 holds, then* $\alpha < s(r) < \beta$.

Proof. Suppose that $s(r) = \beta + \delta$, for some $\delta > 0$. By Hypothesis 1, $\exists y \in P : y + kr = x, \forall x \in F \cap N$. Then, $\forall (u, v) : \beta < v/u < \beta + \delta, \exists k' \in \mathbb{Z}_{>0}, x \in F \cap N : x + k'r = (u, v)$, which gives $y + (k + k')r = (u, v)$, and so there is a move from (u, v) to the P-position y. But, if $F_{\alpha,\beta}(r)$ is sharp, there are infinitely many P-positions of the form (u, v), so $\delta \leqslant 0$. The lower bound is analogous. □

[2]The ideas presented here are even more general. We may, for example, exclude one or both Nim-type moves; in the game $R = \{(1, 1)\}$, the set of P-positions is $\{(0, x), (x, 0) : x \geqslant 0\}$. There are more simple examples of R, for which we can justify the value of the definition of forbidden regions. Take for example the ruleset $R = \{(x, 2x), (2x, x)\}$. The reader may check that the P-positions are precisely the positions of the forms $(0, x)$, $(x, 0)$ or $(2x, 2x)$, so that we have two sharp forbidden regions, again $F_{0,1}$ and $F_{1,\infty}$, but for a different reason than that of Nim. Note that, in each example mentioned in this paragraph, the various sets $F \cap P$ are empty.

See also Figure 8 (the picture to the left) in Section 5. From Lemma 2 it follows that no two sharp forbidden regions can be filled by the same game rule.

Observation 3. Let $F(r)$ and $G(r')$ be two distinct sharp forbidden regions with fill rules r and r' respectively. Then $r \neq r'$.

Lemma 4. *Given Hypothesis 1, then* $\Delta(R) \geqslant 1$.[3]

Proof. By Observation 3, there can be at most one sharp forbidden region per rule. Recall $\Delta(R) = \#R - \#(\text{P-beams associated with } R)$. Each sharp forbidden region is bounded by a P-beam on each side, except for the case of nim-type moves in R, in which case there is only one P-beam on one of the sides. □

3.2. *Equivalence classes.* In Wythoff's game, the observation that there is a P-position in every row, column, and diagonal proves crucial to its analysis. A similar situation exists in Linear Nimhoff except we require more general terms than rows, columns, and diagonals. For this purpose, we will use equivalence classes to define sets with similar properties.

Consider some rule r. We define an equivalence relation $\sim_r$ as follows: for positions X and Y, $X \sim_r Y$ if either Y is a child under r of X, Y is a parent under r of X, or Y equals X. The equivalence classes under $\sim_r$ are sets of colinear points lying along lines with the same slope as r. Let C_r be the set of all equivalence classes under $\sim_r$. For example, $C_{(1,0)}$ is the set of all rows in the game's two-dimensional position space, $C_{(0,1)}$ is the set of all columns, and $C_{(1,1)}$ is the set of all diagonals with slope 1.

In Wythoff's game, the statement that there is exactly one P-position in every diagonal is equivalent to the statement that every set in $C_{(1,1)}$ contains exactly one P-position. We generalize this idea, which suffices to prove a strengthening of Lemma 2.

Theorem 5. *Let $r \in R$ and suppose $F = F_{\alpha,\beta}$ is a forbidden region such that $\alpha < s(r) < \beta$. Given Hypothesis 1, the rule r fills F' if and only if almost all sets in C_r contain exactly one P-position.*

Proof. Suppose that the rule r fills F'. Then each position $X \in F' \cap N$ has a P-position as a child under r. Since $r \in R$, by definition of a P-position, no equivalence class in C_r contains more than one P-position. Since $F \cap P$ is finite, the forward implication follows.

Conversely, assume there exists exactly one P-position in almost all sets in C_r. Then almost all positions are equivalent under $\sim_r$ to a P-position. All but finitely many positions in F are N-positions. Let f be an arbitrary N-position in F. By Lemma 2, all of its parents under r are also in F, so all of them are also

[3] If R does not contain both nim-type move vectors, then $\Delta(R) \geqslant 0$ and $\Delta(R) \geqslant -1$ respectively one or zero nim-type moves.

N-positions. Thus, except for finitely many positions in F, f has a child under r which is a P-position. This is true for all f in F', so the rule r fills F'. □

As a consequence of Lemma 2 we noted that no single game rule r can fill two distinct forbidden regions. We strengthen this in the following assumption:

Hypothesis 6. Let R be a ruleset for which $|R| = n + 1$. Then there are n P-beams and $n + 1$ forbidden regions associated with R. That is, $\Delta(R) = 1$.

3.3. *The slopes and densities of P-lines in the strict class.*

For the strict class of Linear Nimhoff, we impose the following restriction to the family of rulesets:

Hypothesis 7. Let $B \subset P(R)$. If B is a P-beam, then B is a P-line.

The key to understanding the properties of the P-lines in the strict class of Linear Nimhoff lies within an analysis of the forbidden regions. Associated with each forbidden region are various constraints, which collectively can be solved to yield quantitative predictions for the slopes and densities of the game's P-lines. Here we describe how these constraints are constructed. Consider a game of Linear Nimhoff with ruleset R whose P-positions lie within n P-lines. Label these n P-lines $l_1, l_2, \ldots, l_n$ in order of increasing slope, where

$$m_i = \lim_{(x,y)\in l_i} \frac{y}{x}$$

denotes the *slope* of line l_i. Thus, designating the set of all n P-lines as $L = \{l_1, l_2, \cdots, l_n\}$, we have $\forall\, l_i, l_j \in L, i < j \Rightarrow m_i < m_j$. These n P-lines divide the plane into $n+1$ forbidden regions, which are filled under $n+1$ distinct rules. Let $R = \{r_1 = (1, 0), r_2 = (a_2, b_2), \cdots, r_{n+1} = (a_{n+1}, b_{n+1}), r_{n+1} = (0, 1)\}$ be a set of $n+1$ fill rules, labeled by increasing slope, i.e. $\forall\, r_i, r_j \in R, i < j \Rightarrow s(r_i) < s(r_j)$. Recall that R always contains rules $r_1 = (1, 0)$ and $r_{n+1} = (0, 1)$, since these two rules are responsible for filling the forbidden regions bordering the x and y axes, respectively.

Hypothesis 8. For any P-line $\ell_i = \{(x_n, y_n)\}$, the projected density along the x-axis $\lambda_i = \lim n/x_n$ exists, for (x_n) increasing.

Denote by c_r^+ and c_r^- those classes in C_r, for which membership of $(0, y)$ implies $y > 0$ and $y < 0$ respectively (y rational).

Lemma 9. *If ℓ is a P-line with slope greater than $s(r)$, then $c_r^- \cap \ell = \varnothing$. If ℓ is a P-line with slope smaller than $s(r)$, then $c_r^+ \cap \ell = \varnothing$.*

Proof. The proof is a geometric argument displayed in Figure 4. □

Therefore, P-lines with slopes greater than $s(r)$ only contribute P-positions to sets in C_r that lie along lines with positive y-intercepts, while P-lines with

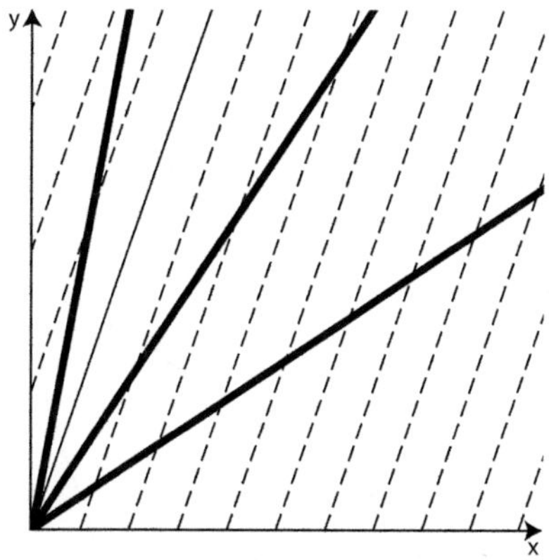

Figure 4. In this depiction, dashed lines represent the equivalence classes comprising C_r; the heavy solid lines are P-lines; the thin solid line is the line of slope $s(r)$ passing through the origin.

slopes less than $s(r)$ only contribute P-positions to sets in C_r that lie along lines with positive x-intercepts.

As we describe next, by dividing up C_r into two parts in this manner, each rule in R (save for rules $(1, 0)$ and $(0, 1)$) will give rise to two geometric constraints in the form of algebraic equations. Solving these equations yields analytical values for the slopes and densities of the P-lines. Numerical simulations of Linear Nimhoff under different rulesets show full agreement with these predicted values (to within numerical uncertainty). We prove the following theorem.

Theorem 10. *Suppose that a ruleset R, with $n+1$ rules, satisfies Hypotheses 1-8. Then the following system of equations holds*:

$$\sum_{j=1}^{i-1} \frac{\lambda_j}{b_i - m_j a_i} = 1 \ \ \forall\, i \in \{2, 3, \dots, n+1\}, \tag{1}$$

$$\sum_{j=i}^{n} \frac{\lambda_j}{m_j a_i - b_i} = 1 \ \ \forall\, i \in \{1, 2, \dots, n\}. \tag{2}$$

Proof. To find the first half of the constraints, let $r_i = (a_i, b_i)$ be an arbitrary element in R other than $r_1 = (1, 0)$. Combining all the P-positions within P-lines with slope less than $s(r_i)$, by Theorem 5, we get exactly one P-position in (almost) every set in C_{r_i} that lies along a line with positive x-intercept.

The P-line l_j with slope $m_j < s(r_i)$ and density (per unit x) λ_j will on average contribute P-positions to a fraction

$$\frac{\lambda_j}{b_i - m_j a_i}$$

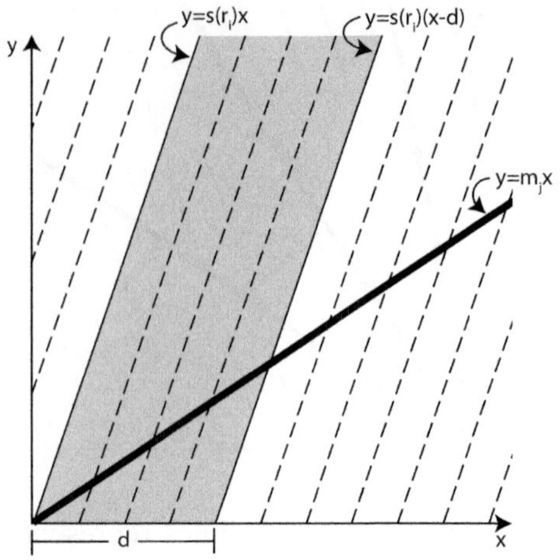

Figure 5. Fractional contribution of P-positions by a P-line.

of the sets in C_{r_i} that lie along lines with positive x-intercepts. Namely, the heavy solid line in Figure 5 represents a P-line with slope m_j and density (per unit x) λ_j; the dashed lines represent the equivalence classes associated with rule $r_i = (a_i, b_i)$, with slope $s(r_i) = b_i/a_i$. Two of these equivalence classes (depicted as thin solid lines in the figure) have been distinguished, and the area between them shaded. Observe that the section of the P-line intersecting the shaded region has horizontal extent $s(r_i)d/(s(r_i) - m_j)$, and hence the expected number of P-positions along this segment of the P-line is $\lambda_j s(r_i)d/(s(r_i)-m_j)$. Note also that the total number of equivalence classes which intersect this segment of the P-line is db_i, since the horizontal spacing between adjacent equivalence classes is $1/b_i$. Thus, the P-line contributes P-positions to a fraction $\lambda_j s(r_i)/b_i/(s(r_i) - m_j)$ of equivalence classes. Substituting $s(r_i) = b_i/a_i$ yields the desired ratio $\lambda_j/(b_i - m_j a_i)$.

Because there is exactly one P-position in almost every set in C_{r_i}, the sum of these fractions over all P-lines with slope less than $s(r_i)$ must equal 1. As a result of their ordered labeling, the P-lines with slope less than $s(r_i)$ are exactly l_1 through l_{i-1}. This yields the equation

$$\sum_{j=1}^{i-1} \frac{\lambda_j}{b_i - m_j a_i} = 1.$$

This equality holds for all r_i in R except for $r_1 = (1, 0)$. The rule $(1, 0)$ is excluded because there are no P-lines with slopes less than $s(1, 0)$, and there are no sets in $C_{(1,0)}$ that lie along lines with positive x-intercepts.

Similarly, to find the second half of the constraints, let $r_i = (a_i, b_i)$ be some arbitrary element in R' other than $r_{n+1} = (0, 1)$. Combining all the P-lines with slope greater than $s(r_i)$, we get exactly one P-position in almost every set in C_{r_i} that lies along a line with positive y-intercept. Geometry shows that the P-line l_j with slope $m_j > s(r_i)$ and density (per unit x) λ_j will contribute P-positions to a fraction

$$\frac{\lambda_j}{m_j a_i - b_i}$$

of the sets in C_{r_i} that lie along lines with positive y-intercepts. Because there is exactly one P-position in every set in C_{r_i}, the sum of these fractions over all P-lines with slope greater than $s(r_i)$ must equal 1. As a result of their ordered labeling, the P-lines with slope greater than $s(r_i)$ are exactly l_i through l_n. This yields the equation

$$\sum_{j=i}^{n} \frac{\lambda_j}{m_j a_i - b_i} = 1.$$

This equality holds for all r_i in R except for $r_{n+1} = (0, 1)$. The rule $(0, 1)$ is excluded because there are no P-lines with slopes greater than $s(0, 1)$, and there are no sets in $C_{(0,1)}$ that lie along lines with positive y-intercepts. □

4. Finding P-lines when $\Delta > 1$: a reorganization model

We wish to refine the renormalization approach in the previous sections, and to this purpose we introduce a more dynamic *reorganization* model, which allows us to temporarily relax Hypothesis 6. It describes a reorganization of P-positions within given P-lines if and only if we "adjoin" a nonfill rule (defined in the next paragraph) to a given ruleset. This model is consistent with computer simulations, as well as with results and conjectures in previous work [Larsson 2012a] (the previous results concern a sometimes trivial reorganization where locations of P-positions stay fixed).

In the preceding section, it was shown how to compute the slopes and densities of the P-lines in the strict class of Linear Nimhoff assuming the game's *fill rules* are known, that is whenever $\Delta = 1$, and in this case we make an assumption by saying $\text{fill}(R) = R$. Since this work does not concern the precise location of P-positions, but rather the precise asymptotics of density and slopes of lines, we ignore local influence of any subset of rules, as long as the overall geometry is preserved, and so in the computation of densities and slopes, the rules that are not fill rules must be ignored (to assure $\Delta = 1$). Here we suggest a recursive algorithm for determining which rules in a ruleset R are fill rules.

The simplest case, say R^0, has two rules, namely $(1, 0)$ and $(0, 1)$. Here, the game is Nim, and the P-positions lie along a single P-line of slope and density both equal to 1. Hence in this case, $\text{fill}(R^0) = R^0$. Now, consider $R^0 \cup \{(2, 1)\}$.

It is easy to prove (by induction) that the P-positions are the same as Nim, so the adjoined rule (2, 1) is redundant to the geometry of the game; it is clearly not a fill rule, because $\text{fill}(R^0 \cup \{(2,1)\}) = R^0$. If we on the other hand adjoin the diagonal Wythoff type move, we know that $\text{fill}(R^0 \cup \{(1,1)\}) = R^0 \cup \{(1,1)\}$. Next consider the game $R = R_0 \cup \{(2,1),(1,1)\}$. Here we have again $\Delta = 1$ (see Section 5.1), so no rule is redundant to the geometry of the game. If we want to compute the number of P-lines in R recursively, the order of adjoining moves is clearly important; we must begin by adjoining the rule (1, 1) to R_0. If we start with (2, 1), the procedure would instead point towards Wythoff's game, which is wrong. Yet, another example is the game $R = R_0 \cup \{(2,2),(1,1)\}$. This game is clearly Wythoff Nim, but if we were to at first adjoin (2, 2) to the rules of Nim, then this move should be included, since it splits Nim's P-positions. However the game $R_0 \cup \{(2,2)\}$ is not Wythoff Nim, and $\Delta(R) = 2$, and the rule (2, 2) is irrelevant (since it is included in (1, 1)). We therefore suggest the rule (a, b) be tested for inclusion before the rule (a', b') if $a + b \leqslant a' + b'$ (in case of multiples we may instead assume they have been removed before starting the algorithm). We next describe the iteration step.

Let $r = (a, b)$ be an element of R other than $(1, 0)$ and $(0, 1)$, and let $R \setminus \{r\}$ be a set containing $n - 1$ rules (a_i, b_i) for which $a + b \geqslant c + d$ for all $(c, d) \in R$. Assume we know the slopes and densities of the P-lines for the game with ruleset $R \setminus \{r\}$. Let L be the set of P-lines for the game $R \setminus \{r\}$, labeled l_1 to l_n in order of increasing slope, and let i be the number of P-lines with slope less than $s(r)$. Consider the inequalities

$$\sum_{j=1}^{i} \frac{\lambda_j}{b - m_j a} \leq 1, \quad \sum_{j=i+1}^{n} \frac{\lambda_j}{m_j a - b} \leq 1. \tag{3}$$

The first (second) inequality holds if there is, on average, at most one P-position per set in C_r lying along a line with positive x-intercept (y-intercept). If one of the P-lines has slope equal to $s(r)$, the second inequality will contain a division by zero, and should be treated as not holding.

If both inequalities are satisfied, in the strict case, when r is adjoined to the ruleset $R \setminus \{r\}$, the P-positions can reposition themselves within the same P-lines such that each set in C_r contains at most one P-position. Therefore, in this model, the slopes and densities of the P-lines for the game with ruleset R will be the *same* as those for the game with the simpler ruleset $R \setminus \{r\}$ (which, by this recursive argument, was presumed to have been previously studied).

In contrast, if one or both of the inequalities is not satisfied, then if r is added to the ruleset, the overall geometry of the P-lines must change in order for there to be at most one P-position in each set in C_r.

By this reorganization model, the elements of R can therefore be assumed in bijective correspondence to the forbidden regions, which allows us to calculate the slopes and densities of the P-lines using Theorem 10.

4.1. ***The general class $R = \{(1,0), r, (0,1)\}$.*** As an illustration, we now use the above methods to solve the general class of Wythoff-like games with rulesets of the form $R = \{(1,0), (a,b), (0,1)\}$. The first step is to determine which rules have an effect on the geometry. Here, we need only concern ourselves with rule $r = (a,b)$, since $(1,0)$ and $(0,1)$ are always fill rules. To test whether r affects the geometry, we must consider the simpler game with ruleset $R \setminus \{r\}$. For this game, the ruleset is $\{(1,0), (0,1)\}$, which is equivalent to Nim and is known to have a single P-line with slope and density (per unit x) both equal to 1. Next, we must count the number, q, of P-lines with slope less than $s(r)$. If $a < b$, then $s(r) = b/a > 1$, which implies that the one P-line has slope less than $s(r)$ and thus $q = 1$. Otherwise, $s(r) = b/a \leq 1$, which implies that the one P-line has slope greater than or equal to $s(r)$ and $q = 0$. In the first case where $a < b$, the two inequalities (3) from Section 4 are

$$\sum_{j=1}^{1} \frac{\lambda_j}{b - m_j a} = \frac{1}{b-a} \leqslant 1, \tag{4}$$

and

$$\sum_{j=2}^{1} \frac{\lambda_j}{m_j a - b} = 0 \leqslant 1. \tag{5}$$

Because $a < b$ and both a and b are integers, $b - a \geq 1$, which implies that $1/(b-a) \leq 1$ and hence both inequalities are satisfied. Therefore, $r = (a,b)$ will have no effect on the overall geometry, and the game with ruleset $R = \{(1,0), (a,b), (0,1)\}$ with $a < b$ will still have a single P-line with slope and density (per unit x) both equal to 1, just as in two-pile Nim. Similarly, if $b < a$, then $q = 0$ and the two inequalities are

$$\sum_{j=1}^{0} \frac{\lambda_j}{b - m_j a} = 0 \leq 1, \tag{6}$$

and

$$\sum_{j=1}^{1} \frac{\lambda_j}{m_j a - b} = \frac{1}{a-b} \leq 1. \tag{7}$$

Since $b < a$ and both a and b are integers, $a - b \geq 1$, which implies that $1/(a-b) \leq 1$, and thus both inequalities are satisfied. Therefore, we conclude that $r = (a,b)$ will have no effect on the overall geometry, and the game with ruleset $R = \{(0, 1), (a, b), (1, 0)\}$ with $b < a$ will still have a single P-line with slope and density (per unit x) both equal to 1.

All that remains is the case in which $a = b$. In this case, $s(r) = 1$, which is the same as the slope of the single P-line in the game with ruleset $\{(0, 1), (1, 0)\}$. As a result, the second inequality contains a division by zero and is treated as not holding. Therefore, the rule (a, a) *will* have an effect on the overall geometry of the P-lines and is in R'. This means that $R' = \{(0, 1), (a, a), (1, 0)\}$. Applying the methods of Section 3.3 — see (1) and (2) — we find that this game will have two P-lines satisfying

$$\frac{\lambda_1}{a - m_1 a} = 1, \quad \lambda_1 + \lambda_2 = 1, \quad \frac{\lambda_2}{m_2 a - a} = 1, \quad \frac{\lambda_1}{m_1} + \frac{\lambda_2}{m_2} = 1.$$

Solving this system of four equations yields predictions for the slopes and densities of the P-lines:

$$m_1 = \frac{-1 + \sqrt{1 + 4a^2}}{2a}, \quad \lambda_1 = \frac{1 + 2a - \sqrt{1 + 4a^2}}{2},$$
$$m_2 = \frac{1 + \sqrt{1 + 4a^2}}{2a}, \quad \lambda_2 = \frac{1 - 2a + \sqrt{1 + 4a^2}}{2}.$$

For the special case of Wythoff's Game where $r = (1, 1)$, these predictions yield the standard result. The prediction for the case $r = (a, a)$ for other values of a is consistent with [Connell 1959].

5. The class (p, q)-GDWN

Through several experiments, the QLPF class has been observed in Linear Nimhoff; early fluctuations appear to stabilize to a quasi-log-periodic behavior within each P-beam. So far, almost every such observation is contained in a proper subclass of Generalized Diagonal Wythoff Nim (GDWN) [Larsson 2012a; Larsson 2014]. The class GDWN simplifies the equations, because the rules are symmetric $(p, q) \in R$ if and only if $(q, p) \in R$, and so the P-positions are also symmetric, (x, y) is a P-position if and only if (y, x) is also. The games with only one additional symmetric rule,

$$R = \{(1, 0), (q, p), (1, 1), (p, q), (0, 1)\},$$

where $p < q$ are positive integers, have attained most attention so far, and this general class is also dubbed (p, q)-GDWN. This is where we mostly observed QLPF games; more specifically they appear when $(p, q) \notin \{(1, 2), (2, 3)\}$ is either a Wythoff pair or a dual Wythoff pair. The first few such pairs are displayed in Tables 1 and 2, respectively.

n	1	2	3	4	5	6	7	8	9	10	11	12	13	14
A_n	1	3	4	6	8	9	11	12	14	16	17	19	21	22
B_n	2	5	7	10	13	15	18	20	23	26	28	31	34	36

Table 1. The first few Wythoff pairs (A_n, B_n).

n	1	2	3	4	5	6	7	8	9	10	11	12	13	14
A_n+1	2	4	5	7	9	10	12	13	15	17	18	20	22	23
B_n+1	3	6	8	11	14	16	19	21	24	27	29	32	35	37

Table 2. The first few dual Wythoff pairs (A_n+1, B_n+1).

The Wythoff pairs are of the form $(A_n, B_n) = (\lfloor \phi n \rfloor, \lfloor \phi^2 n \rfloor)$ whereas the dual Wythoff pairs are of the form $(A_n+1, B_n+1) = (\lfloor \phi n + 1 \rfloor, \lfloor \phi^2 n + 1 \rfloor)$, for some $n > 0$. Here, we collect these pairs as the *WdW-pairs*, and we denote the set $\Omega = \{(p, q) : (p, q) \text{ is a WdW-pair}\}$. Note that, viewed as an increasing sequence of pairs of integers, Ω is the total order

$$\Omega = \{(1, 2), (2, 3), (3, 5), (4, 6), (4, 7), (5, 8), (6, 10), (7, 11), \dots\},$$

with alternating entries from the two sequences.

The first case where "P-lines" gets distorted, creating fluctuated P-beams, is the game $(3, 5)$-GDWN; $(3, 5)$ is the third WdW-pair, and it appears that fluctuations are not possible for the pairs $(1, 2)$ and $(2, 3)$, because the outer P-beams are too stable — they are generated greedily [Larsson 2012a; Larsson 2014]. The QLPF behavior is discussed in Figures 6, 7, 8 and 9 for the game of $(3, 5)$-GDWN. Previous work focused on "the split", the existence of a forbidden region between the beams (rather than the mean slopes of the P-beams) and obtained two decimal conjectures for the bounds of the forbidden region, namely ≈ 1.74 and ≈ 1.57, as can also be extrapolated from Figure 3. Our new computations give rather the mean of the slopes (of course we obtain a much higher precision in the new estimates), and they confirm the previous observations: ≈ 1.760145300 and ≈ 1.537962520; see also Figure 9.

In general for (p, q)-GDWN, it is convenient to note that $m_1 = 1/m_4$, $m_2 = 1/m_3$, $m_1 = \lambda_1/\lambda_4$ and $m_2 = \lambda_2/\lambda_3$. Thus, to compute the conjectured mean slopes of P-beams, it suffices to solve the system of densities

$$1 = \lambda_1 + \lambda_2 + \lambda_3 + \lambda_4,$$

$$\lambda_1 = p - \frac{q\lambda_1}{\lambda_4}, \quad 1 = \frac{\lambda_1\lambda_4}{\lambda_4 - \lambda_1} + \frac{\lambda_2\lambda_3}{\lambda_3 - \lambda_2}, \quad 1 = \frac{\lambda_2\lambda_3}{q\lambda_2 - p\lambda_3} + \frac{\lambda_2\lambda_3}{q\lambda_3 - p\lambda_2} + \frac{\lambda_1\lambda_4}{q\lambda_4 - p\lambda_1}.$$

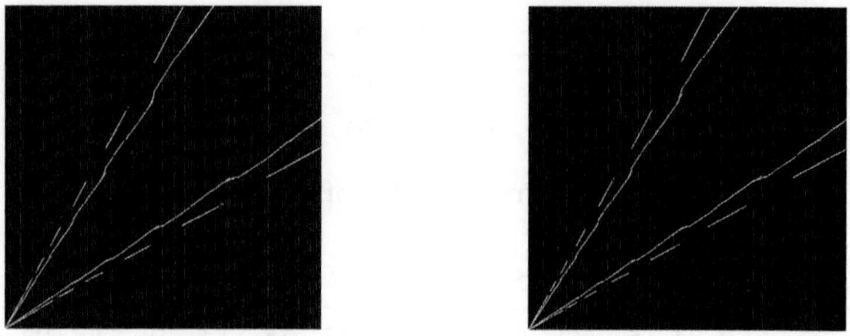

Figure 6. Left: P-positions of (3, 5)-GDWN for $x \leqslant 32600$, $y \leqslant 32600$. Right: P-positions of (3, 5)-GDWN for $x \leqslant 47800$, $y \leqslant 47800$; by an experimental ≈ 1.478 scaling, one may conjecture a geometric "log-invariance" of P-positions.

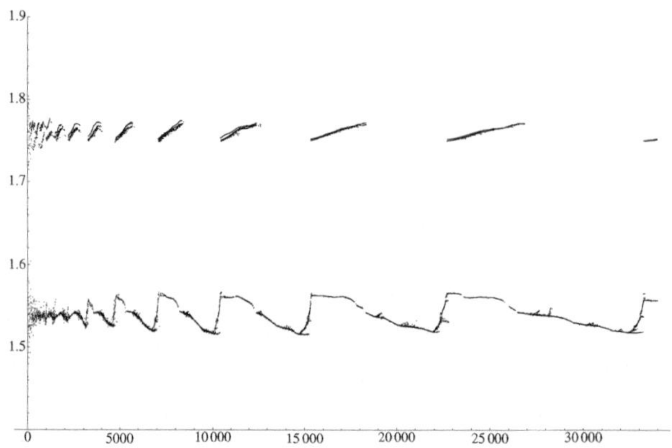

Figure 7. The game (3, 5)-GDWN: the ratios $y/x > 1$ whenever (x, y) is a P-position, for $x \leqslant 35000$.

A reasonable conjecture is that, if a quasi log-periodic behavior has not started to appear within a few thousand multiples of the move rule, then the P-beam will be a P-line with bounded, perhaps $o(\log x)$, scatter; see figures in [Larsson 2012a] for the games (1,2)-GDWN and (2,3)-GDWN.[4] [5]

For several cases, the distribution along the x-axis appears to be nonuniform, when computations are extended beyond a few hundreds; see below figures

[4]By rigorous methods a split of P-beams is demonstrated for the two cases $(p, q) = (1, 2)$ and $(2, 3)$ [Larsson 2014], but the method did not suffice to obtain any precision to the extent of that split. This work is the first significant progress on that subject.

[5]For Maharaja Nim [Larsson et al. 2013], the type of scatter has a structure which is possible to express via a *Dictionary* on algebraic words. The scatter in Linear Nimhoff appears more random, and indeed the methods in Maharaja Nim depend on the big restriction of (1, 2)-GDWN where only Knight type moves are adjoined to Wythoff Nim.

Figure 8. The fill-rule properties of the rules (0, 1), left, (1, 1), middle, and (3, 5), right, respectively visualized for the game (3, 5)-GDWN, parents of the P-positions under the respective fill rule are displayed in black for $x \leqslant 10000$, $y \leqslant 10000$.

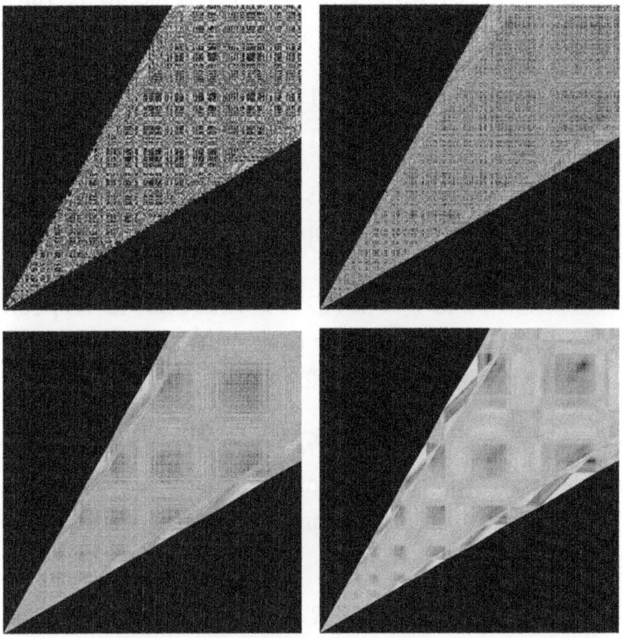

Figure 9. Initial fluctuations appear to stabilize to a quasi-log-periodic geometry in the game (3, 5)-GDWN. The first picture consists of a few hundred pixels (on each axis), and the last picture displays $\sim$ 20000×20000 pixels. The coloring scheme in the picture is that white pixels are P-positions, and the remaining colors are N-positions. Except for possibly the first picture, only the patterns of N-positions are visible (unless zooming in). Black means here that the pixel sees only one P-position (viewing along the rules of the game), yellow means the pixels detects two P-positions, and for red it detects three (or more) P-positions.

concerning the game (3, 5)-GDWN, which is the "first" game observed — the only (p, q)-GDWN games in the strict class, with $(p, q) \in \Omega$, appear for $(p, q) = (1, 2)$ and $(p, q) = (2, 3)$, where P-positions distribute uniformly along the P-lines. Nevertheless, by adapting the mean values to the slopes and densities, again the experimental data agree with the values computed in [Larsson 2012a], up to three-digit precision.

The properties of forbidden regions and fill rules appear to still govern the overall behavior, but the assumption of a uniform distribution along the x-axis does not hold any longer. A reasonable guess is that there is uniform behavior from the point of view of each filling rule r (rather than along the x-axis), which then stabilizes the patterns to a sufficient degree (but we do not develop this idea further here).

The slopes of the upper P-lines of (1,2)-GDWN satisfy the pair of fourth degree equations

$$(z-1)\left(\frac{1}{2z-1}-\frac{1}{z-2}\right)=\frac{w^2-1}{2w-1}, \qquad \frac{z^2-1}{z}=(w-1)\frac{-w^2+2w+2}{w},$$

which show that the hypothesis of this work is consistent with previous work, and moreover they provide a huge improvement in numerical precision of the previously conjectured upper slopes; previously four digits, and now $w \approx 2.24772558355773557$ and $z \approx 1.47779977527220012$ respectively (and the densities of the four P-lines, in order of increasing slope, are ≈ 0.11021167, 0.25912616, 0.382936586 and 0.24772558).

5.1. ***A linear Nimhoff relaxation of the Wythoff/dual Wythoff pair conjecture.*** In this section, we show that the inequalities (3) from Section 4 are consistent with an asymetric relaxation of the conjecture that (p, q)-GDWN splits if and only if $(p, q) \in \Omega$ [Larsson 2012a].

Theorem 11. *Consider the ruleset R = Wythoff Nim and a pair of positive integers (p, q). The proposition $(p, q) \in \Omega$ if and only if* $\operatorname{fill}(R \cup (p, q)) = R \cup (p, q)$ *is consistent with the reorganization model using the inequalities* (3).

Proof. By adjoining the vector (p, q) to Wythoff Nim, we get two cases for the inequalities:

$$\sum_{j=1}^{1}\frac{\lambda_j}{q-m_j p}\le 1, \qquad \sum_{j=2}^{2}\frac{\lambda_j}{m_j p-q}\le 1, \tag{8}$$

or

$$\sum_{j=1}^{2}\frac{\lambda_j}{q-m_j p}\le 1, \qquad \sum_{j=3}^{2}\frac{\lambda_j}{m_j p-q}\le 1, \tag{9}$$

because $m_1 = \phi^{-1}$ and $m_2 = \phi$. Also $\lambda_1 = \phi^{-2}$ and $\lambda_2 = \phi^{-1}$, and where the cases are $1 < q/p < \phi$ and $\phi < q/p$ respectively.

Case 1. The inequalities are $\dfrac{\phi^{-2}}{q-\phi^{-1}p} \leqslant 1$ and $\dfrac{\phi^{-1}}{\phi p-q} \leqslant 1$, by (8). Hence

$$\phi^{-1} \leqslant \phi q - p \quad \text{and} \quad \phi^{-1} \leqslant \phi p - q \tag{10}$$

if and only if there is no new split of P-lines when adjoining the move vector (p, q), with $1 < q/p < \phi$, to Wythoff Nim. The first inequality in (10) is trivially true, so it suffices to verify the second. The dual Wythoff pairs are of the form $(p, q) = (2, 3), (4, 6), (5, 8), \ldots$ and the ratios are smaller than ϕ. In this case, the conjecture is that there is a split (an inequality does not hold). But, for example $(p, q) = (4, 5)$ is not a dual Wythoff pair, so it should not split (inequalities hold). Obviously the first inequality holds since $q > p$. For the second inequality we compute

$$\phi^{-1} \leqslant \phi\lfloor\phi n + 1\rfloor - \lfloor\phi^2 n + 1\rfloor$$

if and only if

$$\begin{aligned}\phi &\leqslant \phi\lfloor\phi n + 1\rfloor - \lfloor\phi^2 n\rfloor,\\ 0 &\leqslant \phi\lfloor\phi n\rfloor - \lfloor\phi n\rfloor - n,\\ 0 &\leqslant \phi^{-1}\lfloor\phi n\rfloor - n,\\ \phi n &\leqslant \lfloor\phi n\rfloor,\end{aligned}$$

which is false, because ϕ is irrational.

For the other direction, suppose that $q/p < \phi$, but (p, q) is not a dual Wythoff pair. Then, by complementarity, either

$$1 < p = \lfloor\phi n + 1\rfloor$$

with $2 < q < \lfloor\phi^2 n + 1\rfloor$, or

$$1 < p = \lfloor\phi^2 n + 1\rfloor$$

with $2 < q \leqslant \lfloor\phi^2 n + 1\rfloor + \lfloor\phi n\rfloor$.

In either case, we must prove that the second inequality in (10) holds. In the first case, it suffices to justify

$$\begin{aligned}\phi^{-1} &\leqslant \phi\lfloor\phi n + 1\rfloor - \lfloor\phi^2 n\rfloor,\\ -1 &\leqslant \phi\lfloor\phi n\rfloor - \lfloor\phi n\rfloor - n,\\ n - 1 &\leqslant \phi\lfloor\phi n\rfloor - \lfloor\phi n\rfloor,\\ n - 1 &\leqslant \phi^{-1}\lfloor\phi n\rfloor,\\ \phi(n-1) &\leqslant \lfloor\phi n\rfloor.\end{aligned}$$

In the second case, we justify

$$\begin{aligned}
\phi^{-1} &\leqslant \phi\lfloor\phi^2 n+1\rfloor - \lfloor\phi^2 n+1\rfloor - \lfloor\phi n\rfloor,\\
\phi^{-1} &\leqslant \phi^{-1}\lfloor\phi^2 n+1\rfloor - \lfloor\phi n\rfloor,\\
0 &\leqslant \lfloor\phi^2 n\rfloor - \phi\lfloor\phi n\rfloor,\\
0 &\leqslant n - \phi^{-1}\lfloor\phi n\rfloor,\\
\phi n &\geqslant \lfloor\phi n\rfloor.
\end{aligned}$$

Case 2. The inequalities are

$$\frac{\phi^{-2}}{q-\phi^{-1}p} + \frac{\phi^{-1}}{q-\phi p} \leqslant 1, \tag{11}$$

by the first inequality in (9) and $0 \leqslant 1$ by the second. We wish to prove that (p, q) is a Wythoff pair if and only if there is no new split of P-lines when adjoining the move vector (p, q) with $\phi < q/p$ to Wythoff Nim.

The inequality (11) is equivalent with

$$1 \leqslant q - p - \frac{qp}{q-p}\phi^{-3}. \tag{12}$$

By letting $(p, q) = (\phi n, \phi^2 n)$ be a Wythoff pair, we get

$$1 \leqslant n - \frac{\lfloor\phi^2 n\rfloor\lfloor\phi n\rfloor}{n}\phi^{-3},$$

which simplifies to

$$(n^2-n)\phi^3 \geqslant \lfloor\phi^2 n\rfloor\lfloor\phi n\rfloor \geqslant (\phi n-1)(\phi^2 n-1) = \phi^3 n^2 - \phi^2 n - \phi n + 1.$$

This holds if and only if $0 \geqslant 1$. Hence, the hypothesis of a new P-line is consistent with the conjecture for (p, q)-GDWN in this case. Next, we must check that if (p, q) with $q/p > \phi$ is not a Wythoff pair, then inequality (12) holds. In case $p = \lfloor\phi n\rfloor$, then the problem reduces to justifying the inequality

$$(m^2-m)\phi^3 \geqslant \lfloor\phi n+m\rfloor\lfloor\phi n\rfloor$$

for any $m > n$. We get

$$\begin{aligned}
(m^2-m)\phi^3 &\geqslant (\phi n+m)\phi n,\\
m^2\phi^3 &\geqslant \phi^2 n^2 + \phi m n + m\phi^3,\\
m^2\phi^2 + m^2\phi &\geqslant \phi^2(m^2-(m+n)(m-n)) + \phi(m^2-(m-n)m) + m\phi^3,\\
m(\phi^2+\phi) &\leqslant \phi^2(m+n)(m-n) + \phi(m-n)m,
\end{aligned}$$

which holds because $m > n$. Hence, no new P-line is introduced if $q/p > \phi$ and (p, q) is not a Wythoff pair. □

The games in GDWN have symmetric rulesets, and so, to justify that conjectures in previous work is consistent with the reorganization model, it suffices to prove that introducing the move vector (q, p), to the Linear Nimhoff game $R = \{(1, 0), (1, 1), (p, q), (0, 1)\}$, $(p, q) \in Q$, introduces a fourth P-line (in the case $(p, q) \leqslant (2, 3)$) or otherwise P-beam. Because computational experiments have shown that P-beams are not always P-lines, a general proof could be a bit more demanding. Perhaps the model will even be refuted in some cases of fluctuating P-beams, although experimental results point towards that the model holds also in this interesting QLPF-case.

In the other cases, it is known from [Larsson 2012a] that if $(p, q) \notin Q$, then $\text{fill}(R) = R \setminus \{(p, q), (q, p)\}$. This means that the slopes and densities of any such game are identical to Wythoff's game. The individual locations of those P-positions can be completely different from those of Wythoff Nim, if $q/p > \phi$, as is illustrated in [loc. cit.] for (2,4)-GDWN and (7,12)-GDWN. If $q/p < \phi$, then in fact the P-positions are identical [loc. cit.] (which holds also for Linear Nimhoff games of the form $\{(1, 0), (1, 1), (p, q), (0, 1)\}$). The "scatter along a P-line" can vary hugely; for example for (7, 12)-GDWN, around x-coordinate 40000, only the first digit in the ratio y/x for an "upper" P-position ≈ 1.6 has been experimentally confirmed.

Using terminology in this study, Theorem 11 can be restated as follows.

Corollary 12. *If $R = \{(1, 0), (1, 1), (p, q), (0, 1)\}$ is in the strict class, then there are three P-lines for the rule set R if and only if $(p, q) \in Q$.*

In this case, we believe that Linear Nimhoff is in the strict class, but in going from one adjoined move of Wythoff Nim to two adjoined moves, we restate a conjecture from [Larsson 2012a]. Given the three P-lines from Corollary 12, we can justify numerically for the first few games that our reorganization model corresponds to the conjectures, even though several of these games are believed to be QLPF-games, with an even more interesting behavior.

Conjecture 13 [Larsson 2012a]. *Consider $R = \{(1, 0), (q, p), (1, 1), (p, q), (0, 1)\}$. There are four P-beams (P-lines in case of $(p, q) = (1, 2)$ or $(2, 3)$) for the move set R if and only if $(p, q) \in Q$. Otherwise there are two P-lines of slopes and densities as for Wythoff's game. In case of P-beams, this class of games belongs to the QLPF class of relaxed Linear Nimhoff.*

By Corollary 12, we know that $\text{fill}(R) = R = \{(1, 0), (1, 1), (p, q), (0, 1)\}$ whenever $(p, q) \in Q$, and so there are three P-lines. Experimentally it seems that $m_1 \approx \phi^{-1}$ and $\lambda_1 \approx \phi^{-2}$ as for Wythoff Nim. Thus it is a delicate matter to introduce the new rule (q, p). If the approximate values are in fact equalities, then for the Wythoff pairs, we get one set of inequalities, and for the dual Wythoff pairs, we get another set. We display first the equations for the three conjectured

P-lines for the game $\{(1, 0), (1, 1), (p, q), (0, 1)\}$. For the P-lines with positive x-intercept, we get exactly one P-position in every set in C_{r_i} that lies along a line with positive x-intercept:

$$\sum_{j=1}^{i-1} \frac{\lambda_j}{b_i - m_j a_i} = 1 \quad \forall i \in \{2, 3, 4\}, \tag{13}$$

For those with positive y-intercept,

$$\sum_{j=i}^{n} \frac{\lambda_j}{m_j a_i - b_i} = 1 \quad \forall i \in \{1, 2, 3\}.$$

Altogether,

$$\frac{\lambda_1}{1-m_1} = 1, \qquad \frac{\lambda_1}{q-m_1 p} + \frac{\lambda_2}{q-m_2 p} = 1, \qquad \lambda_1 + \lambda_2 + \lambda_3 = 1,$$

and

$$\frac{\lambda_1}{m_1} + \frac{\lambda_2}{m_2} + \frac{\lambda_3}{m_3} = 1, \qquad \frac{\lambda_2}{m_2 - 1} + \frac{\lambda_3}{m_3 - 1} = 1, \qquad \frac{\lambda_3}{m_3 p - q} = 1.$$

Computer explorations give that the slope m_3 is greater (smaller) than p/q if (p, q) is a Wythoff pair (dual Wythoff pair). (Note that the slope $0 < p/q < 1$.)

Thus, in the case of a *Wythoff pair*, to justify that the new rule (q, p) is part of the rule set, it suffices to show that the second inequality does not hold, i.e., that

$$\frac{\lambda_1}{m_1 p - q} + \frac{\lambda_2}{m_2 p - q} + \frac{\lambda_3}{m_3 p - q} > 1. \tag{14}$$

(Since there is no P-line below the rule (p, q), the first inequality is trivially satisfied.)

In the case of a *dual Wythoff pair*, to justify that the new rule (q, p) is part of the rule set, and since there is exactly one P-line below the rule (p, q), we must show that one of the inequalities does not hold, i.e., that

$$\frac{\lambda_1}{q - m_1 p} > 1 \quad \text{or} \quad \frac{\lambda_2}{m_2 p - q} + \frac{\lambda_3}{m_3 p - q} > 1. \tag{15}$$

We have verified the inequalities (14) and (15) numerically, for a few initial (p, q) pairs, but we omit further details. Instead we propose the following problem.

Problem 14. Prove the analogue of Theorem 11 in this setting; in particular, the occurrence of a new P-line for the cases $(p, q) = (1, 2)$ and $(2, 3)$, where it has been conjectured that the games are in the strict class.

6. A scheme built on observed reflections in fluctuation

A subclass of the (p, q)-GDWN games exhibits quasi-log-periodic fluctuations, summarized in Tables 3 and 4, where "?" means that it is uncertain whether a visual inspection indicates an integer number of "half" log-periods.

Let the numbers $a^{-1} < 1$ and $a > 1$ denote the mean slopes of the outer P-beams, whereas $b^{-1} < 1$ and $b > 1$ are the mean slopes of the inner P-beams, respectively. If a is the slope of the 1st line (the mean slope of the top P-beam), the log-period of bouncing from the the fourth line along $(0, 1)$ to the 1st line and back along $(1, 0)$ is $\log a - \log a^{-1} = 2\log a$. The log-period of bouncing from the fourth line to the third line along $(1, 1)$, then back to the fourth line along $(1, 0)$ is $\log \frac{a-1}{b-1}$.

Analogously, we study

$$\xi_{\mathrm{a}}(a, b) = \frac{2\log a}{\log(a-1) - \log(b-1)}, \quad \xi_{\mathrm{b}}(a, b) = \frac{2\log b}{\log(a-1) - \log(b-1)},$$

$$\xi_{\mathrm{ab}}(a, b) = \frac{\log b - \log a^{-1}}{\log(a-1) - \log(b-1)}, \quad \xi_{\mathrm{ba}}(a, b) = \frac{\log a - \log b^{-1}}{\log(a-1) - \log(b-1)},$$

in Tables 3 and 4, looking for integers whenever possible; the entry (a^{-1}, a) corresponds to the visible number of log-periods for ξ_{a}, (b^{-1}, b) to ξ_{b}, (a^{-1}, b) to ξ_{ab}, and (b^{-1}, a) to ξ_{ba}. The type of bouncing is displayed in Figure 10, where the white dotted box corresponds to the $2\log a$ bounce, and the red-black zig-zag pattern in the lower $(7, 4)$-forbidden sector corresponds to the $\log \frac{a-1}{b-1}$ bounce. The blue box is analogous, but here the relation of bounce is measured between the first and third P-beam. In both these cases, we visually detect four periods. Sometimes the visual inspection indicates that we should rather count the number

a^{-1}	–	$a, 3$ or $b, 2.5$	$b, 4$	$a, 6.5$	$a, 7?$	$b, 8.5$	$a, 11$	$b, 10$
b^{-1}	–	$a, 2.5$	$a, 4$	$a, 6 \approx b, 6$	?	$b, 8$	$b, 10$	$a, 10$
p	1	3	4	6	8	9	11	12
q	2	5	7	10	13	15	18	20

Table 3. Geometric behavior for the Wythoff pairs.

a^{-1}	–	$a, 3$	$a, 5$	$b, 6.5$	?	$a, 9$	$a, 11$	$a, 13$
b^{-1}	–	$b, 2.5$	$b, 4$	$b, 6$	$b, 6?$	$b, 8$	$b, 10$	$b, 12$
p	2	4	5	7	9	10	12	13
q	3	6	8	11	14	16	19	21

Table 4. Geometric behavior for the dual Wythoff pairs.

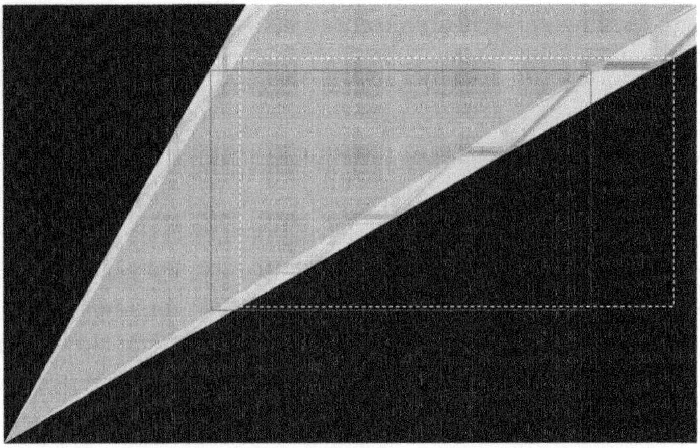

Figure 10. Visual interpretation of $\xi_{\mathrm{ba}}(a, b)$ (blue) and $\xi_{\mathrm{ab}}(a, b)$ (white dashed) for the game (4, 7)-GDWN.

of "half" log-periods (here "half" means either a bounce with the inner or outer P-beam). We leave it as an open problem to justify these integer of half integer approximations in terms of the game rules.

Note that the log-ratio for $\xi_{\mathrm{a}}(a, b)$ appears to approximate the lower sequence of the Wythoff pairs. Tables 5 and 6 seem to indicate that each one of the four log ratios could contribute to explain the "bouncing" between P-beams, and how

(1, 2)	[2.247725584, 1.477799775, 1.687531557]
(3, 5)	[1.760145300, 1.537962520, 3.270827298]
(4, 7)	[1.768998972, 1.601914235, 4.656921040]
(6, 10)	[1.697534360, 1.589805226, 6.308758458]
(8, 13)	[1.662070300, 1.582782860, 7.966064987]
(9, 15)	[1.678278787, 1.607189817, 9.353017596]
(11, 18)	[1.656365788, 1.598892313, 11.01365251]
(12, 20)	[1.668937168, 1.615891660, 12.39876038]
(14, 23)	[1.653187304, 1.608112999, 14.06105137]

Table 5. Wythoff pairs (p, q), [slope a, slope b, $\xi_{\mathrm{a}}(a, b)$] (initial 9 digits).

(2, 3)	[1.739269208, 1.408430574, 1.865590884]
(4, 6)	[1.638930839, 1.482391155, 3.515816330]
(5, 8)	[1.675293133, 1.547469192, 4.917904008]
(7, 11)	[1.640358498, 1.550853934, 6.574385950]
(9, 14)	[1.621326573, 1.552514257, 8.234019700]
(10, 16)	[1.640881924, 1.578215903, 9.625732990]

Table 6. Some dual Wythoff pairs (p, q), [slope a, slope b, $\xi_{\mathrm{a}}(a, b)$].

Figure 11. Initial P-positions for two one-rule-extensions of (3, 5)-GDWN. To the left, the adjoined rule is $r = (4, 7)$, and to the right the adjoined rule is $r = (5, 8)$.

the fluctuations become stable.

The log-ratio $\xi(a, b) > 2$ gives fluctuations. Is this a requirement for fluctuations to become permanent?

Tables 5 and 6 display bounces between the P-beams, which indicates mostly integer log-ratio, or otherwise half of integer log-ratio.

In Figure 11, we display a variation in the behavior of adjoining a new rule to the game of (3, 5)-GDWN. In the left-most picture, the rule (4, 7) has been adjoined. The behavior satisfies the reorganization model and a new P-line has occured. To the right, it appears that the new rule, which is (5, 8), is not quite able to change the quasi-log periodic behavior of (3, 5)-GDWN; neither is it clear whether a new P-line (or P-beam) appears (according to the reorganization model, a new P-beam should appear).

6.1. ***Questions.*** Experimental data [Larsson 2010] give uniformly distributed P-positions along P-lines also for the games

$$\{(1, 2), (2, 3)\}\text{-GDWN} = \{(1, 0), (2, 1), (3, 2), (1, 1), (2, 3), (1, 2), (0, 1)\},$$

$\{(1, 2), (2, 3), (3, 5)\}$-GDWN and $\{(1, 2), (2, 3), (3, 5), (5, 8)\}$-GDWN. For each of these games a new Fibonacci-type pair has been adjoined, and data shows that this gives birth to two new symmetric P-lines.

For the game $\{(1, 2), (2, 3), (3, 5), (5, 8), (8, 13)\}$-GDWN, however, computations do not seem to indicate a new split; only the same number of P-beams as for the game $\{(1, 2), (2, 3), (3, 5), (5, 8)\}$-GDWN can be distinguished (although the P-positions are clearly different). These types of questions relate to the inequalities (3), and in this spirit one would also like to justify the hypothesis, from an arXiv preprint version of [Larsson 2012a], that the game $\{(x, y) : x, y \leqslant 5\}$ has exactly five (uniform) upper P-lines, and similar problems.

7. Discussion

We remark here that the game of Linear Nimhoff is itself a special case of some other general games that have been studied in the literature — "vector subtraction games" [Golomb 1966] a.k.a. "invariant games" [Duchêne et al. 2010; Larsson et al. 2011; Larsson 2012b], "vector addition games" [Larsson et al. 2013], and the "n-vectors game" [Duchêne et al. 2009]. The special restrictions on Linear Nimhoff give rise to unique features not apparent in these more general games. Another related class of games is "Nimhoff" [Fraenkel et al. 1991]; they focus on games between Nim and Wythoff. Thus, this class differs from ours in that they are not concerned with linear rules (our class is the most general on what we regard as linear extensions of 2-pile Nim). Another difference is that the paper [Fraenkel et al. 1991] concerns structures of Grundy-values of games (bridging the complexity class between Nim and Wythoff's game). In this paper we restrict attention to the patterns of P-positions (corresponding to Grundy-value 0).

In this work we have characterized the overall geometric structure of the P-positions in the game of Linear Nimhoff. More specifically, our analysis has produced highly accurate quantitative predictions about the number, slopes, and densities of the P-lines observed in the game, predictions which have been subsequently borne out of numerical simulations. Unlike standard game-theoretic techniques commonly used to analyze combinatorial games, the methodology employed here offers a probabilistic/geometric description, rather than an exact, deterministic specification, of the locations of the P-positions. The virtue of this approach is that it has broad explanatory powers and allows one to tackle more complex games for which standard deterministic methods have failed. It is usually believed that such methods need be nonrigorous, but here we build a rigorous model, and instead leave the questions of existence for future study.

Generalized classes of games that include Linear Nimhoff have been defined previously in the literature, but Linear Nimhoff itself has not been extensively analyzed, and its restricted structure gives rise to certain features not readily apparent in these more general games.

Linear Nimhoff can also be seen as a specific form of the more general "n-vectors game", introduced in [Duchêne et al. 2009]. The n-vectors game is defined the same way as Linear Nimhoff, with three distinctions: (i) the vectors in the n-vectors game can exist in any vector space of the form $\Re^p$, (ii) the coordinates of the vectors need not be integers, and (iii) a player can only move to a position that can be expressed as a sum of nonnegative multiples of the vectors in the ruleset. Because in Linear Nimhoff $(1, 0)$ and $(0, 1)$ are always included in the ruleset (by definition), it follows that every position in Linear Nimhoff can be trivially expressed as a sum of nonnegative integer multiples of these two rules, and hence the third distinctive feature of the n-vectors game becomes irrelevant. So Linear Nimhoff with ruleset R is equivalent to the n-vectors game played with the vectors in R which includes rules $(1, 0)$ and $(0, 1)$. We note also that through appropriate changes in bases, it is possible to recast some other n-vectors games into Linear Nimhoff form.

In other games, log-periodicity in P-position density first conjectured using heuristic analysis was later formally proven [Garrabrant et al. 2013].

Since Wythoff's game provides a model for self-organization according to Phyllotaxis [Kappraff et al. 1998], one might want to consider the apparent self-organization in these new games as forms of generalized Phyllotaxis, where the fill rule property plays an analogous significant rule as for Wythoff's game.

Acknowledgements

Garrabrant was supported in part by a W. M. Keck Foundation Summer Research Fellowship; Phipps-Morgan was supported in part through an NSF Division of Undergraduate Education STEP 1 grant. Larsson was supported in part by the Killam Trust.

References

[Bouton 1901] C. L. Bouton, Nim, a game with a complete mathematical theory. *Ann. of Math.* **3** (1901–1902), 35–39.

[Berlekamp et al. 1982] Elwyn R. Berlekamp, John H. Conway, and Richard K. Guy, *Winning ways for your mathematical plays*. Academic Press, Inc., 1982.

[Connell 1959] Ian G. Connell, A generalization of Wythoff's game. *Canad. Math. Bull.* **2** (1959), 181–190.

[Cook et al. 2015] M. Cook, U. Larsson, and T. Neary, A cellular automaton for blocking queen games, *Cellular Automata and Discrete Complex Systems*, 21st IFIP WG 1.5 International Workshop, Automata 2015, Turku, Finland, June 8–10, Proceedings, J. Kari (ed.), LNCS 9099 (2015), 71–84.

[Duchêne et al. 2009] E. Duchêne and S. Gravier, Geometrical extensions of Wythoff's game. *Discrete Math.* **309**:11 (2009), 3595–3608.

[Duchêne et al. 2010] E. Duchêne and M. Rigo, Invariant games. *Theoret. Comput. Sci.* **411**:34-36 (2010), 3169–3180.

[Friedman et al. 2007] E. J. Friedman and A. S. Landsberg, Nonlinear dynamics in combinatorial games: renormalizing Chomp. *CHAOS* **17** (2007), 023117.

[Friedman et al. 2009] E. J. Friedman and A. S. Landsberg, On the geometry of combinatorial games: a renormalization approach, in: M. H. Albert and R. J. Nowakowski (eds.), *Games of No Chance* 3 (MSRI series), Cambridge University Press (2009).

[Fraenkel et al. 1991] A. S. Fraenkel and M. Lorberbom, Nimhoff games. *J. Combin. Theory Ser. A* **58** (1991), 1–25.

[Garrabrant et al. 2013] S. M. Garrabrant, E. J. Friedman, and A. S. Landsberg, Cofinite induced subgraphs of impartial combinatorial games: an analysis of CIS-Nim. *INTEGERS* **13** (2013), #G2.

[Golomb 1966] S. W. Golomb, A mathematical investigation of games of “take-away”. *J. Combinatorial Theory* **1** (1966), 443–458.

[Kappraff et al. 1998] J. Kappraff, D. Blackmore, G. Adamson, Phyllotaxis as a dynamical system: a study in number (Section 6 Wythoff’s Game). Chapter 17 in the book Symmetry in Plants, R. V. Jean and D. Barabé (eds.), Series in Mathematical Biology and Medicine, Vol. 4 (1998), World Scientific.

[Larsson et al. 2011] U. Larsson, P. Hegarty, and A. S. Fraenkel, Invariant and dual subtraction games resolving the Duchêne–Rigo conjecture. *Theoret. Comput. Sci.* **412** (2011), 729–735.

[Larsson 2010] U. Larsson, preprint of: A generalized diagonal Wythoff Nim. http://arxiv.org/abs/1005.1555.

[Larsson 2012a] U. Larsson, A generalized diagonal Wythoff Nim. *INTEGERS* **12** (2012), #G02.

[Larsson 2012b] U. Larsson, The ⋆-operator and invariant subtraction games, *Theoret. Comput. Sci.* **422** (2012), 52–58.

[Larsson 2014] U. Larsson, Wythoff Nim extensions and splitting sequences. *J. Integer Seq.* **17** (2014), Article 14.5.7.

[Larsson et al. 2013] U. Larsson and J. Wästlund, From heaps of matches to the limits of computability. *Electron. J. Combin.* **20** (2013), P41.

[Wythoff 1907] W. A. Wythoff, A modification of the game of Nim. *Nieuw Arch. Wisk.* **7** (1907), 199–202.

[Zeilberger 2001] D. Zeilberger, Three-rowed Chomp. *Adv. Appl. Math.* **26** (2001), 168.

[Zeilberger 2004] D. Zeilberger, Chomp, recurrences, and chaos. *J. Differ. Equations* **10** (2004), 1281.

ejf@icsi.berkeley.edu — *ICSI and Department of EECS, UC Berkeley, Berkeley, CA, United States*

scott@garrabrant.com — *W. M. Keck Science Department, Claremont McKenna, Pitzer, and Scripps Colleges, Claremont, CA, United States*

ilona.k.pm@gmail.com — *W. M. Keck Science Department, Claremont McKenna, Pitzer, and Scripps Colleges, Claremont, CA, United States*

alandsberg22@gmail.com — *W. M. Keck Science Department, Claremont McKenna, Pitzer, and Scripps Colleges, Claremont, CA, United States*

urban031@gmail.com — *Department of Industrial Engineering and Management, Technion - Israel Institute of Technology, Haifa, Israel*

Games of No Chance 5
MSRI Publications
Volume **70**, 2017

Searching for periodicity in Officers

J.P. GROSSMAN

Officers is a take-and-break game in which a move consists of removing a bean from a heap and leaving the remaining beans from that heap in exactly one or two nonempty heaps. It is an open question as to whether or not the Grundy values of this game are eventually periodic; answering this question in the positive for take-and-break games generally requires computing enough Grundy values to explicitly find the period. We contribute to this search by presenting two novel parallelization strategies that, combined with additional optimizations, accelerate the computation of Grundy values by nearly 60 times compared to the previous state of the art. The resulting implementation computes over 5 million values per second, and has computed a total of more than 140 trillion values over the course of 18 months. To date, no period has been found.

1. Introduction

Officers is an impartial game played with beans arranged into heaps. On each move, a player selects a heap with at least two beans and removes exactly one bean from the heap. If at least two beans remain in the heap, then the player may also optionally split the remaining heap into two. Officers is an example of an *octal game* [Berlekamp, Conway and Guy 2001]: a take-and-break game where each legal move can be described as removing k beans from a heap, then possibly splitting any remaining beans, leaving behind exactly 0, 1 or 2 nonempty heaps. The name "octal game" comes from the fact that such games can conveniently be described as a string of octal digits (traditionally preceded by a decimal point with trailing zeros omitted), where the k-th digit after the decimal point encodes the legal moves involving the removal of k beans from a heap. The binary representation of a digit has three bits, with bit m indicating whether or not it is legal to leave exactly m nonempty heaps. Thus, if the k-th digit is zero then it is never legal to remove exactly k beans. Under this encoding, Officers is the game .6: only the first digit after the decimal point is nonzero (only one bean can be

MSC2010: 91A46.
Keywords: Officers, octal games, periodic, Grundy values.

Take... (# beans)		**1**			**2**			
Leave... (# heaps)	**2**	**1**	**0**	**2**	**1**	**0**	***G*(*n*)**	**Period**
.1			✓				$01\overline{0}$	1
.2 (She-loves-me-she-loves-me-not)		✓					$00\overline{1}$	2
.3 (She-loves-me-she-loves-me-not)		✓	✓				$\overline{01}$	2
.4 (Dawson's chess)	✓						0001120311...	34
.5 (She-loves-me-she-loves-me-not)	✓		✓				$\overline{01}$	2
.6 (Officers)	✓	✓					0012012312...	???
.7 (She-loves-me-she-loves-me-not)	✓	✓	✓				$\overline{01}$	2
.15 (Guiles)			✓	✓		✓	$\overline{1101122122}$	10
.16			✓	✓	✓		1001221401...	149459

Table 1. Simple examples of octal games, along with their periods (where known). The period of .16 was established in [Gangolli and Plambeck 1989]. For each game, checkmarks indicate the legal moves. Officers is the only single-digit octal game that is not known to be periodic.

removed), and this digit has bits 1 and 2 set (exactly 1 or 2 nonempty heaps can remain).

Richard Guy famously asked whether all finite octal games are eventually periodic [Guy 1996]. A finite octal game has a finite number of nonzero digits in its representation (equivalently, a finite number of legal moves), and "eventually periodic" means that the sequence $G(n)$, defined by the Grundy values of single heaps of size n, is eventually periodic. Thus far, no finite octal game has been shown to be aperiodic. Table 1 lists some simple examples of octal games along with their associated periods, if known. Very simple octal games can have surprisingly large periods; the period of the game .106 is 328,226,140,474 [Flammenkamp 2012]! Notably, and somewhat frustratingly, Officers is the only single-digit octal game for which Guy's question remains unanswered.

Determining the period of an octal game (if it exists) typically requires computing enough Grundy values to explicitly observe the period. Given the enormous number of values that may be needed, speed is paramount. Accelerating the computation is nontrivial because it is inherently sequential: the value $G(n)$ generally cannot be determined until $G(n-1)$ is known. Our main contribution is to show how the computation can be parallelized, leveraging both cluster computing and multicore architectures. With the additional use of established techniques for optimizing sequential code, we have been able to accelerate the computation of Grundy values for Officers by nearly 60 times relative to the fastest previously known algorithm.

Our work was specifically motivated by a desire to search for periodicity in Officers, and we focus on this game throughout the paper. However, the

```
// Use a byte array 'mex' to compute the mex function
initialize the mex array with zero
for k = 0 ... (n-1)/2
    mex[G[k] ^ G[n-1-k]] = 1;
G[n] = index of first zero in mex array
```

Figure 1. Brute-force computation of $G(n)$ in Officers.

techniques that we present are general, and could be applied to other unsolved octal games.

2. Computing Grundy values in Officers

We begin with a description of how Grundy values are computed for Officers, and we review the rare-value algorithm [Berlekamp, Conway and Guy 2001]. This algorithm, which can be applied to a large number of finite octal games, is the fastest known sequential technique for computing Grundy values.

2.1. *Brute-force computation.* For heaps with at least two beans, we can slightly restate the rules of Officers as follows: a player removes exactly one bean then splits the heap into exactly two heaps, one of which may be empty. This alternate formulation gives us a succinct recursive definition for $G(n)$:

$$\begin{aligned} G(0) &= G(1) = 0, \\ G(n) &= \operatorname*{mex}_{0 \le k \le (n-1)/2} \{G(k) \oplus G(n-1-k)\} \quad \text{if } n \ge 2. \end{aligned} \tag{1}$$

Here $\oplus$ denotes bitwise exclusive-or and mex is the "minimal excluded" function evaluating to the smallest nonnegative integer that does not appear in the specified set. Figure 1 lists equivalent pseudocode for computing $G(n)$. The code uses a "mex array" to mark all legal moves from a heap of size n, then searches for the first unmarked index in the array. This code is both simple and slow: since $(n-1)/2$ loop iterations are needed to compute $G(n)$, computing all Grundy values up to $G(n)$ requires $O(n^2)$ operations.

2.2. *Rare values.* When many Grundy values are computed for Officers, an interesting pattern emerges: when the values $G(n)$ are expressed in binary, very few of these values have an even number of bits set in positions 1, 2, 3, 5, 6, 7 and 8. Grundy values of this form are the *rare values*. Most of the finite octal games that have been studied have a similar property: for some set of bit positions (that varies depending on the specific game), few Grundy values have an even number of bits set in those positions. Just how rare are these rare values in Officers? Over 140 trillion Grundy values have been computed but only 1584 rare values have been discovered; the last known rare value is $G(20627) = 277 = 100010101_2$.

While the specific set of bits defining rare values is somewhat mysterious, some intuition can be gained as to how the set of rare values, once established, remains rare as additional Grundy values are computed. We refer to nonrare values as *common values*, i.e., values with an odd number of bits set in positions 1, 2, 3, 5, 6, 7 and 8. Consider what happens when two values are exclusive-ored. From a simple parity argument, we have:

$$\begin{aligned} \mathit{common} \oplus \mathit{common} &= \mathit{rare} \\ \mathit{rare} \oplus \mathit{common} &= \mathit{common} \\ \mathit{rare} \oplus \mathit{rare} &= \mathit{rare} \end{aligned} \tag{2}$$

Since most of the Grundy values are common, it follows that most of the elements of the set in (1) are rare, so that the first excluded integer, which is $G(n)$, is very likely to be common.

2.3. *Rare-value algorithm.* Suppose an oracle informs us that we have discovered all the rare values in Officers. In that case, to compute subsequent values of $G(n)$ we only need to consider the common values within the mex array. As such, the loop used to mark the mex array only needs to iterate over the expressions $G(k) \oplus G(n-k-1)$ where $G(k)$ is rare, so instead of $\lfloor (n-1)/2 \rfloor$ loop iterations only a fixed number (1584) are required. This means that computing all Grundy values up to $G(n)$ only requires $O(n)$ operations instead of $O(n^2)$. It also implies that the game is eventually periodic, as $G(n)$ only depends on a finite number of previous values. $G(n)$ would be trivially bounded above by 2^{12} since at most 1584 common values are marked in the mex array, so at some point some sequence of 20628 values must be repeated, after which $G(n)$ is periodic. This gives us a fairly poor bound on the period; all we can say is that it must be less than $2^{12\times 20628}$.

In the absence of an oracle we can't assume that there are no more rare values, however we can still exploit the rare value phenomenon to accelerate the computation. We split the loop used to mark the mex array into two parts:

(1) **Mark common values.** We first iterate over $G(k) \oplus G(n-k-1)$ where $G(k)$ is rare, marking common values in the mex array. We then find the first unmarked common value, which becomes our candidate value for $G(n)$. In all likelihood this is the actual value of $G(n)$, since otherwise $G(n)$ is a rare value. To prove this, we need to be able to mark all rare values smaller than the candidate value.

(2) **Mark rare values.** Next we perform the full iteration shown in Figure 1, keeping track of the number of unmarked rare values smaller than the candidate value. As soon as this number reaches zero we have proven that

```
// 1. Find minimum excluded common value "cval"
foreach k such that G[k] is rare
    mex[G[k] ^ G[n-1-k]] = 1;
cval = first unmarked common value in mex array

// 2. Mark rare values less than cval
numr = number of unmarked rare values less than cval
for k = 0 ... (n-1)/2
    val = G[k] ^ G[n-1-k];
    if val < cval and !mex[val]
        mex[val] = 1;
        numr = numr - 1
        break if numr == 0

// 3. G[n] is either cval or a rare value less than cval
if (numr == 0)
    G[n] = cval
else
    G[n] = index of first zero in mex array
```

Figure 2. Rare-value algorithm for computing $G(n)$ in Officers.

the candidate value is $G(n)$, and we can exit the loop. Alternately, if at least one rare value never gets marked then we have discovered a new rare value.

Figure 2 lists pseudocode for the rare-value algorithm. In theory, this is still an $O(n^2)$ algorithm because the number of iterations required in the second loop is only bounded by $\lfloor (n-1)/2 \rfloor$. In practice, however, less than 3000 iterations are required on average to prove that the candidate value is $G(n)$, so in effect the algorithm is only $O(n)$. On a 2.666 GHz Intel Xeon E5430,[1] the rare-value algorithm computes roughly 88,000 values per second.

3. Accelerating the computation

We now describe a number of parallelization and optimization techniques that can be used to accelerate the computation of Grundy values in Officers far beyond the performance of a baseline implementation of the rare-value algorithm.

3.1. ***Parallelization through speculation.*** The rare-value algorithm would be much faster if the second loop could be eliminated. To this end, suppose we simply assume that there are no more rare values, and speculatively compute Grundy values based on this assumption. Figure 3 lists the (greatly simplified) pseudocode for this speculative computation. However, we still need to verify the assumption which effectively requires the original computation, so at first it seems that we haven't actually gained anything.

[1] All performance measurements in this paper were performed on the same machine. Code was written in C and compiled using the Intel compiler with the -fast option.

```
foreach k such that G[k] is rare
    mex[G[k] ^ G[n-1-k]] = 1;
G[n] = first unmarked common value in mex array
```

Figure 3. Speculative computation of $G(n)$ in Officers, assuming no more rare values.

The key observation is that once a large number of speculative values have been computed, we can perform the verification in parallel on a cluster. Specifically, the sequence of values can be partitioned into intervals $[N_j, N_j+1)$, and for each interval a processor can use the rare-value algorithm to verify that the Grundy values in $[N_j, N_j+1)$ are correct assuming that all previous speculative values are correct. Once all intervals have been verified, an inductive proof is formed that the entire sequence of speculative values is correct.

With a sufficiently large cluster we can verify the speculative values as fast as they can be generated, so our overall performance is only limited by how quickly we can perform the speculative computation. A direct implementation of the pseudocode in Figure 3 computes roughly 196,000 values per second, representing a speedup of 2.2 times over the rare-value algorithm.

3.2. ***Sequential optimizations.*** Now that the bulk of the computation has been reduced to a single tight loop, we can apply two standard optimizations to obtain additional speedups. First, all known Grundy values in Officers are less than 512 and can therefore be represented using 9 bits. Moreover, since we are only computing common values, the value of bit 8 is implied by the values of the other bits, so in fact 8 bits suffices to represent the values and we can store $G(n)$ in a byte array. We never need to recover the ninth bit, and the loop used to mark the mex array is unmodified. The size of the mex array is reduced from 512 entries to 256 entries, since dropping bit 8 creates a one-to-one mapping from common values in [0, 512) to numbers in [0, 256). The only change to the computation is that we need to take this mapping into account when we search the mex array for the first unmarked common value. Additionally, we need to validate the assumption that the Grundy values remain less than 512. Using a byte array for $G(n)$ improves the performance to 321,000 values per second.

The second optimization is to fully unroll the main loop, since we know exactly how many iterations there are as well as the value of $G(k)$ in each iteration. Figure 4 lists the code for the unrolled loop. When compiled, the resulting assembly code contains three instructions per iteration. This optimization further improves performance to 850,000 values per second.

Searching the mex array for the first unmarked common value is also implemented as a loop. This search is much less expensive than the main loop,

```
mex[0 ^ G[n-1]] = 1
mex[0 ^ G[n-2]] = 1
mex[1 ^ G[n-3]] = 1
mex[0 ^ G[n-5]] = 1
mex[1 ^ G[n-6]] = 1
mex[1 ^ G[n-9]] = 1
...
G[n] = first unmarked common value in mex array
```

Figure 4. Speculatively computing $G(n)$ in Officers using an unrolled loop.

```
int compute_mex ()
    if (!mex[2])
        return 2;
    if (!mex[3])
        return 3;
    if (!mex[4])
        return 4;
    if (!mex[5])
        return 5;
    if (!mex[8])
        return 8;
...
```

Figure 5. Evaluating the mex function using an unrolled loop.

representing only 10% of the total compute time even after the main loop is unrolled, but we can still obtain some additional performance benefit from unrolling this loop as well. Figure 5 lists the code for the unrolled mex computation; the compiled code again contains three instructions per iteration. With both loops unrolled, performance increases to 912,000 values per second, more than 10 times faster than the baseline rare-value algorithm.

3.3. ***Multithreading.*** Since the value of $G(n)$ depends on $G(n-1)$, it is not immediately clear how the computation can be further parallelized. Observe, however, that most of the computation of $G(n)$ does not depend on $G(n-1)$, which is only required for a single iteration of the unrolled loop. We can therefore use two threads with a baton-passing approach: a thread marks as much of the mex array as it can without knowing $G(n-1)$, then it receives the value of $G(n-1)$ (the baton) from the other thread, then it finishes computing $G(n)$ and passes this value to its partner. Figure 6 lists pseudocode for this algorithm.

This technique naturally generalizes to m threads. Thread j is responsible for computing $G(n)$ where $n \equiv j \pmod{m}$, and the baton comprises the previous $m-1$ Grundy values. Figure 7 plots the performance of this multithreaded computation as the number of threads is increased from 1 to 8 ("Standard mex" curve). The performance with 8 threads is 3,672,000 values per second.

```
mex[0 ^ G[n-2]] = 1
mex[1 ^ G[n-3]] = 1
mex[0 ^ G[n-5]] = 1
...
receive G[n-1] from the other thread
mex[0 ^ G[n-1]] = 1
G[n] = first common value not in mex
pass G[n] to the other thread
```

Figure 6. Multithreading the computation of $G(n)$ using baton passing.

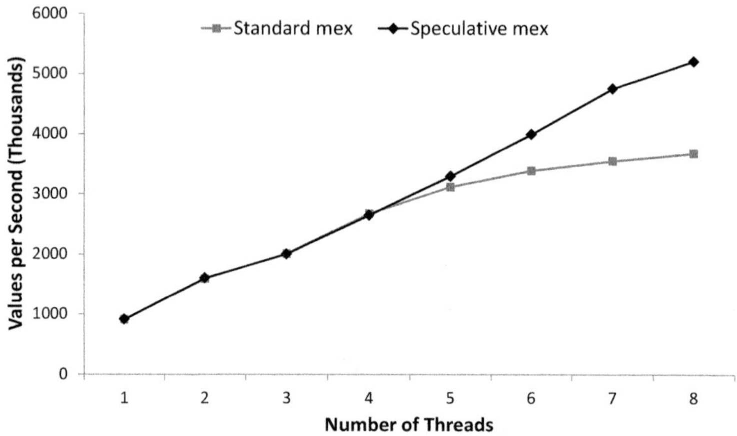

Figure 7. Performance of multithread computation.

In Figure 7 we can see that performance begins to level off beyond 4 or 5 threads. The reason for this is that each thread must wait to receive the previous Grundy values before evaluating the mex function, and must then compute the mex function before forwarding Grundy values to the next thread (Figure 8). This places an upper bound on the performance that can be achieved: the time to compute a value must be at least the time required to evaluate the mex function

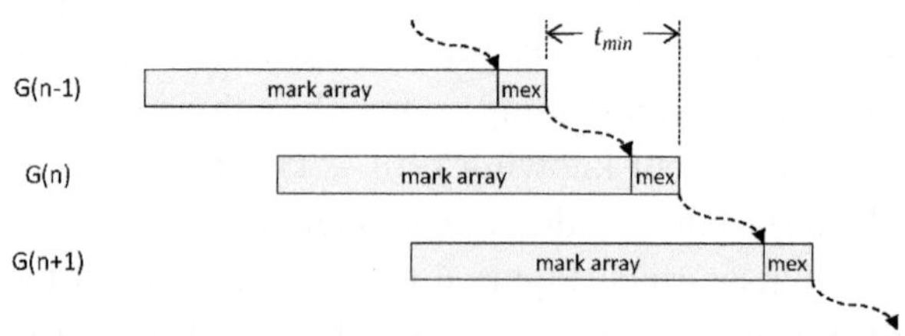

Figure 8. Standard mex computation. A thread waits for the baton, then evaluates the mex function to compute $G(n)$, then passes the baton to the next thread. Performance is limited by the time to compute the mex function plus the time to pass the baton.

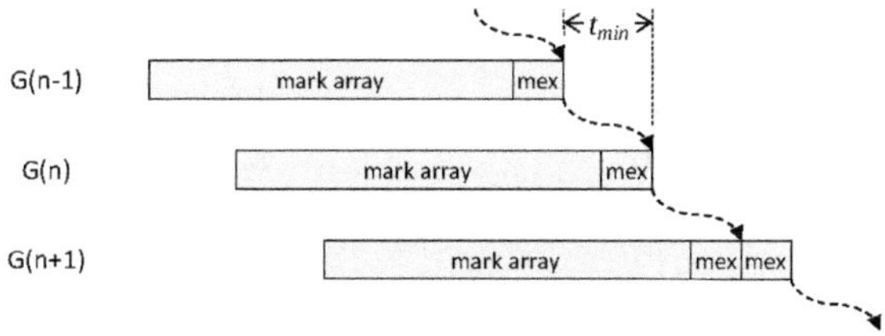

Figure 9. Speculative mex computation. A thread speculatively computes $G(n)$ before receiving the baton, then checks to see if it needs to recompute $G(n)$ after receiving the baton. In the overwhelming majority of cases the speculative computation is correct, so performance is only limited by the time to pass the baton. If the speculative computation is incorrect then the mex computation is repeated, as shown for $G(n+1)$.

plus the inter-thread communication latency. As we approach this upper bound, additional threads offer little incremental benefit.

Suppose that, instead of waiting to receive the previous $m-1$ Grundy values, each thread speculatively evaluates a candidate Grundy value $G'(n)$, even though it cannot finish marking the mex array until it receives data from the previous thread. Under what circumstances will this speculative computation be correct, so that $G(n) = G'(n)$? By definition of the mex function, the value $G'(n)$ will be unmarked, whereas all smaller values will be marked. Once the previous $m-1$ Grundy values are received, a small set of additional entries must be marked (only 5 when $m = 8$). $G'(n)$ is correct if and only if it is not in this set. This determination is computationally inexpensive, requiring only a handful of processor cycles. If it turns out that $G'(n)$ is incorrect then the mex function must be recomputed after the remaining values have been marked, but empirically we have found this case to be extremely rare.

Figure 9 illustrates the speculative mex computation, which has the effect of removing the mex evaluation from the critical path. Now, performance is only limited by the inter-thread communication latency, allowing the computation to scale to a larger number of threads. Figure 7 shows the performance of this modified algorithm ("Speculative mex" curve), which exhibits much better scaling with more than four threads. With 8 threads the performance increases to 5,201,000 values per second — 59 times faster than the rare-value algorithm.

3.4. *Performance summary.* Table 2 summarizes the performance of the algorithms described in the previous sections, showing both incremental speedup as each successive optimization is applied as well as total speedup relative to the baseline rare-value algorithm. The final speculative mex multithreaded algorithm

Algorithm	Values per millisecond	Incremental speedup	Total speedup
Rare values	88	1.0	1.0
Speculative	196	2.2	2.2
Byte array	321	1.6	3.6
Unroll loops	912	2.8	10.4
Multithreaded	3672	4.0	41.7
Speculative mex	5201	1.4	59.0

Table 2. Performance of the various algorithms presented in earlier sections, and for the first row (rare values) in [Berlekamp, Conway and Guy 2001].

takes, on average, 192 ns to compute each Grundy value, or equivalently 512 processor cycles.

4. Searching for periodicity

Ultimately, our goal is to search for periodicity within the sequence of Grundy values. To prove that the sequence is periodic with period p and preperiod d, it suffices to compute values up to $G(2p+2d)$, verifying the periodicity within this range; periodicity for all larger values then follows from a simple induction argument. But how can we efficiently determine candidate values for p and d?

Observe that, if the hypothesis that there are no more rare values beyond $G(20627)$ holds, then each Grundy value only depends on the previous 20628 values. We can therefore search for two identical sequences of 20628 consecutive values. This still sounds like an expensive proposition, but there is an efficient online algorithm for performing this search based on two further simplifications.

The first simplification is to consider the sequence $H(n)$ defined by computing a 64-bit cyclic redundancy check (CRC) of the values $G(n)$, $G(n-1)$, $\ldots$, $G(n-20627)$. We then search for n_0 and n_1 such that $H(n_0) = H(n_1)$, which is much cheaper than comparing two complete sequences of 20628 values. The full CRC computation is expensive, but it does not need to be repeated for each $H(n)$: there is a fast incremental way to compute $H(n)$ from $H(n-1)$, $G(n)$ and $G(n-20628)$ that involves two shifts, two table lookups, and three exclusive-ors. Note that this algorithm may produce false positives so that we still need to do a full sequence comparison for candidate values of n_0 and n_1, but with a 64-bit CRC these false positives are extremely rare.

The second simplification is to restrict n_0 to powers of 2. For each interval $[2k, 2k+1)$ we compute $H(2k)$, then we compare $H(n)$ to $H(2k)$ within the interval. In so doing we almost definitely won't find the smallest values of n_0 and n_1 such that $H(n_0) = H(n_1)$, but we are at least guaranteed to compute

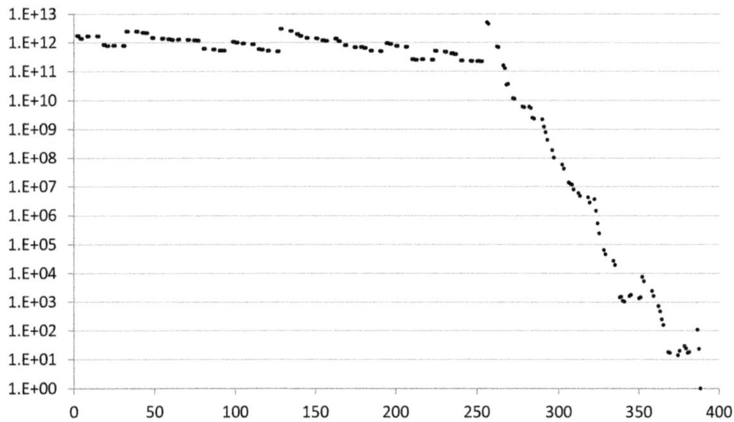

Figure 10. Histogram of common values within the first 140 trillion Grundy values of Officers.

fewer than twice as many values as we need to since there is always a power of 2 between n_0 and $2n_0$.

Augmenting the computation to simultaneously search for periodicity by looking for a repeated CRC value has a small (less than 4%) performance impact, and we are able to process over 5 million Grundy values per second. To date, we have computed over 140 trillion Grundy values without a single candidate CRC match. The largest Grundy value we have observed is 388, whose first (and thus far only) appearance is at $G(7{,}014{,}808{,}364{,}046)$. The last "new" Grundy value we have observed is 380, whose first appearance is at $G(23{,}209{,}561{,}059{,}317)$. Figure 10 shows a histogram of the common Grundy values. A certain amount of structure is evident in the histogram for the values less than 256; the pattern for the remaining values is less clear.

5. Discussion

The hardware used for this work offered a maximum of 8 processor cores, but the performance graph of the speculative mex algorithm in Figure 7 strongly indicates that there is additional performance to be gained by increasing the number of threads. The current trend in commodity processor technology is to provide an increasing number of processor cores per die with little or no increase in clock speed; Intel recently announced their Knight's Landing architecture with over 60 cores [Intel 2014]. This being so, future general-purpose architectures will likely provide additional multithreading speedups for the algorithm described in this paper. It may also be possible to tailor the algorithm for implementation on a graphics processing unit (GPU); recently GPUs have become popular for high-performance computing due to their ability to support a large number of threads performing identical computations.

An alternate approach to improving performance would be to implement the unrolled single-threaded algorithm directly in hardware using a field programmable gate array (FPGA). The algorithm is extremely well suited to such an implementation, and would easily be able to generate a new value on every clock cycle. A modern FPGA can be clocked at around 500 MHz, which would thus generate 500,000,000 values per second — well over 5000 times faster than the baseline rare-value algorithm. At this speed the problem of validating the speculative values becomes quite challenging: a cluster with nearly 6000 processing cores would be required to keep pace with the data produced by a single FPGA!

It is entirely possible that Officers is eventually periodic, but with a period or preperiod so astronomically large that no direct linear search could find the period. If so, all hope is not necessarily lost; analysis of the data produced by the current brute-force search may reveal patterns giving rise to more advanced algorithms that can produce N values per step for some large or increasing N.

The algorithms in this paper are general and could be applied to any octal game that has a set of rare values, as well as certain other take-and-break games such as Grundy's game. Officers happens to be particularly well suited to fast computation on commodity processors for two reasons. First, all known values are less than 512, allowing the computation to be performed using bytes. Second, and more importantly, all the known rare values occur within the first 20628 Grundy values, which means that only the most recent 20628 Grundy values need to be kept in memory for the unrolled speculative computation described in Section 3.1. With such a small memory footprint, the computation data easily fits within the processor cache. Other octal games may not share these properties, but would still benefit from the parallelization techniques that we have presented.

Acknowledgements

The computations described in this paper were performed on the Department of Mathematics and Statistics Compute Cluster at Dalhousie University. We would like to thank Balagopal Pillai for his assistance in using and managing the cluster resources.

References

[Berlekamp, Conway and Guy 2001] E. Berlekamp, J. H. Conway, and R. K. Guy, *Winning ways for your mathematical plays*, vol. 1, A K Peters, Wellesley, MA, 2001.

[Flammenkamp 2012] A. Flammenkamp, "Sprague–Grundy values of octal games", web page, 2012, http://wwwhomes.uni-bielefeld.de/achim/octal.html. Retrieved September 28, 2014.

[Gangolli and Plambeck 1989] A. Gangolli and T. Plambeck, "A note on periodicity in some octal games", *Internat. J. Game Theory* **18**:3 (1989), 311–320.

[Guy 1996] R. K. Guy, "Unsolved problems in combinatorial games", pp. 475–491 in *Games of no chance* (Berkeley, 1994), Math. Sci. Res. Inst. Publ. **29**, Cambridge Univ. Press, 1996.

[Intel 2014] "Intel re-architects the fundamental building block for high-performance computing", press release, Intel Corporation, 2014, http://tinyurl.com/intel-re.

jpg@alum.mit.edu

D. E. Shaw Research, 120 West 45th Street, 39th Floor, New York, New York 10036, United States

Games of No Chance 5
MSRI Publications
Volume **70**, 2017

Good pass moves in no-draw HyperHex: two proverbs

RYAN B. HAYWARD

For a position and a player with a winning move, the *pass-value* is the largest number of free moves the player can allow the opponent so that after these move(s) the player still has a winning move. For a cell-coloring game such as Hex, the pass-value is equivalently the smallest number of cells the opponent needs to color in order to reach a position where the opponent has a winning move. A move is *good* if it increases the pass-value. HyperHex is the hypergraph generalization of Hex: each player has a list of winsets, and wins by coloring all cells of any of her winsets. No-draw HyperHex is the maker-breaker restriction of HyperHex: each player's winset list contains every minimal set that intersects all of the other player's winsets (so draws are not possible). For no-draw HyperHex, we consider two good-move proverbs: your opponent's good move is your good move, and it's never too late for a good move.

1. Introduction

At the 2011 Banff International Research Station Workshop on Combinatorial Game Theory, I asked professional 9-dan Ziang Zhujiu ("Jujo") about the Go proverb *your opponent's good move is your good move* [?]. His instant response was "it's not always true", and of course he is right. For example, in Figure 1 each player's winning move (in the opponent's territory) is the opponent's losing move, so no move is good for both a player and their opponent.

This paper is about good moves in no-draw HyperHex, a game that generalizes Hex. We consider the Go proverb above, and the general proverb *it's never too late*, which when applied to games could be expanded as *it's never too late for a good move*.

Research supported by NSERC, U of A GAMES..
MSC2010: 91A46.
Keywords: game of Hex, maker-breaker game, HyperHex, proverb.

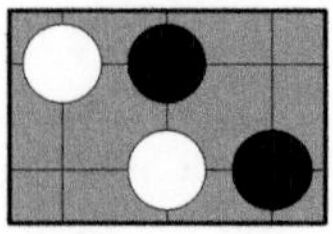

Figure 1. A Go position with no move that wins for both players.

2. No-draw HyperHex

In 1942 Piet Hein invented Polygon, the classic alternate-turn two-player connection game now known as Hex [**?**]. In 1949 John Nash described the game to David Gale, who built a board that was soon in frequent use in Princeton's Fine Hall [**?**; **?**]. Later John Milnor and independently Claude Shannon [**?**] and independently Charles Titus (personal communication with Craige Schensted) [**?**] invented Y, and Shannon created his eponymous (switching) game [**?**]. Y and the Shannon game each generalize Hex. For more on these games, see [**?**; **?**].

HyperHex is a hypergraph generalization of Hex, Y and the Shannon game. The board is a finite set of cells; each player has a collection of winning cell sets, or *winsets*. On each turn, a player colors an uncolored cell; a player wins by coloring all cells of any one of their winsets. Draws are possible; for example, if the board is $\{1, 2, 3, 4\}$ and black and white have respective winsets $\{\{1, 2\}, \{4\}\}$ and $\{\{1, 3\}\}$, then the game move sequence (black: 1, white: 2, black: 3, white: 4) fills the board and yields a draw, since neither player wins.

No-draw HyperHex is the maker-breaker restriction of HyperHex. In a *maker-breaker* game, one player (maker) tries to establish some property — say, form a certain connection — and the other player (breaker) tries to prevent it [**?**]. Maker-breaker games apparently evolved from the Shannon game, in which maker tries to connect two terminal nodes, and breaker tries to thwart this connection.

For a HyperHex game not to end in a draw, it suffices that each possible coloring of all the board cells yields a colored winset for exactly one player, namely that the game is maker-breaker. To satisfy this condition, call the two players M and B, and select any nonempty list of nonempty winsets for M. Remove from M's list any winset that properly contains some other winset (such containing winsets are redundant), and let B's list consist of every minimal cell subset that intersect all of M's winsets. For example, if the board is $\{1, 2, 3, 4\}$ and M's list is $\{\{1, 2\}, \{4\}\}$ then B's list is $\{\{1, 4\}, \{2, 4\}\}$.

A hypergraph is a set of hyperedges (subsets) of a ground set, so in hypergraph terms the winset lists are hypergraphs defined on the same ground set (the set of board cells), M's hypergraph is a *clutter* (no hyperedge properly contains another), and B's hypergraph is the *blocker*, or *transversal* (the set of all hyperedges that

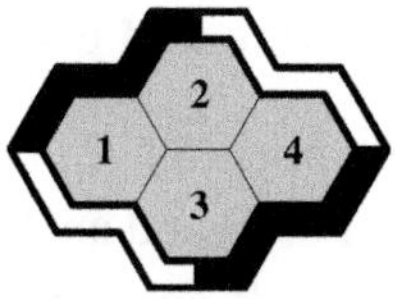

Figure 2. A Hex board.

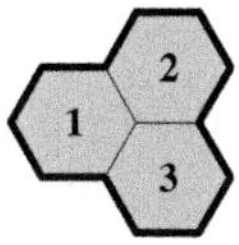

Figure 3. A Y board.

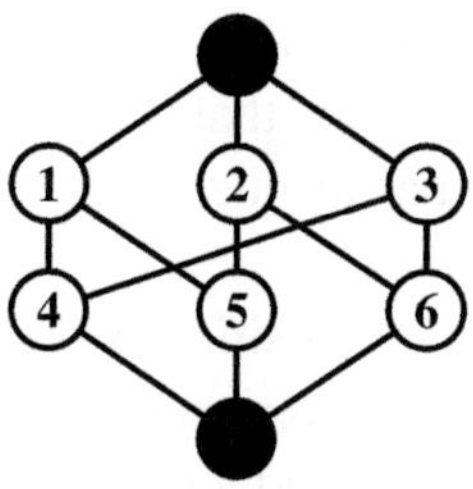

Figure 4. A Shannon board.

intersect all of the other hypergraph's hyperedges) of M's hypergraph. The blocker of a blocker of a clutter is the original clutter (see Corollary 2 of Chapter 2 in *Hypergraphs* by Claude Berge [?]), so in no-draw HyperHex the roles of maker and breaker are interchangeable. So no-draw HyperHex is HyperHex on a clutter and its blocker [?].

Here are some no-draw HyperHex examples. Suppose the cell set is $\{1, 2, 3, 4\}$ and black's winset collection is $\{\{1, 3\}, \{2, 3\}, \{2, 4\}\}$. Then white's winset collection is $\{\{1, 2\}, \{2, 3\}, \{3, 4\}\}$. In Hex, a player wins by connecting their two borders. This HyperHex example is equivalent to Hex on the board in Figure 2.

Or suppose the cell set is $\{1, 2, 3\}$ and black's winset collection is $\{\{1, 2\}, \{1, 3\}, \{2, 3\}\}$. Then white's is the same. In Y, a player wins by connecting all three borders. This HyperHex example is equivalent to Y on the board in Figure 3.

Or suppose the cell set is $\{1, 2, 3, 4, 5, 6\}$ and black's winset collection is $\{\{1, 4\}, \{1, 5\}, \{2, 5\}\{2, 6\}, \{3, 4\}, \{3, 6\}\}$. Then white's is $\{\{1, 2, 3\}, \{4, 5, 6\}, \{1, 2, 4, 6\}, \{1, 3, 5, 6\}, \{2, 3, 4, 5\}\}$. In the Shannon game, black wins by joining two specified terminal nodes, otherwise white wins. This HyperHex example is equivalent to the Shannon game on the network in Figure 4.

3. Pass-value and good moves

For a player P and a HyperHex position H, (H, P) is the game state starting from H in which P moves next. For P and an H with an empty cell c, $H + P(c)$ is the position obtained from H by P-coloring c. $\bar{P}$ is the opponent of P.

In HyperHex, P-coloring a cell — or uncoloring a $\bar{P}$-colored cell — is never disadvantageous for P, since a winning P-strategy can always be modified to accommodate such a change.

Observation 1. For a player P and a HyperHex position H with empty cell c, and for $X = P$ or $\bar{P}$, if P wins (H, X) then P wins $(H + P(c), X)$.

By Observation 1, if P has a second-player win strategy for H then P has a first-player win strategy for H, so each no-draw HyperHex position has one of three outcome-values: *neutral* if each player has a first-player win, *P-win* if P has a second-player win, and *$\bar{P}$-win* if $\bar{P}$ has a second-player win; see Figure 5.

We want a notion of "good move" that is more general than "winning move", so we introduce the notion of pass-value. A similar concept has been studied in Go [**?**].

Informally, the pass-value of a position is the number of times a player can *pass* — make no move but allow the opponent a move — and still leave a winning position. Formally, for a player P and a no-draw HyperHex position H, define the *pass-value* $\nu_P(H)$ as ∞ if a P-winset is already P-colored; otherwise, if P has a winning first-player strategy then $\nu_P(H)$ is the largest number of empty cells t such that P still has a first-player winning strategy after any t cells have been $\bar{P}$-colored; otherwise $\nu_P(H)$ is $-\nu_{\bar{P}}(H)$. Notice that, for all P and H, $\nu_P(H) = -\nu_{\bar{P}}(H)$.

For example, the respective black pass-values of the positions in Figure 5 are (from left) 1, −1, and 0. For a neutral position, each player's pass-value is 0; for a P-win position, P's pass-value is at least 1.

A move is *good* if it increases the player's pass-value, and *wasted* if it does not change it. By Observation 1, every no-draw HyperHex move is good or wasted. A move can increase the pass-value by more than one. For example, see Figure 6. Generalizing this example, it is easy to see that for each $n \geq 4$ there is an $n \times n$ position in which a move changes the pass-value from 0 to $n - 1$.

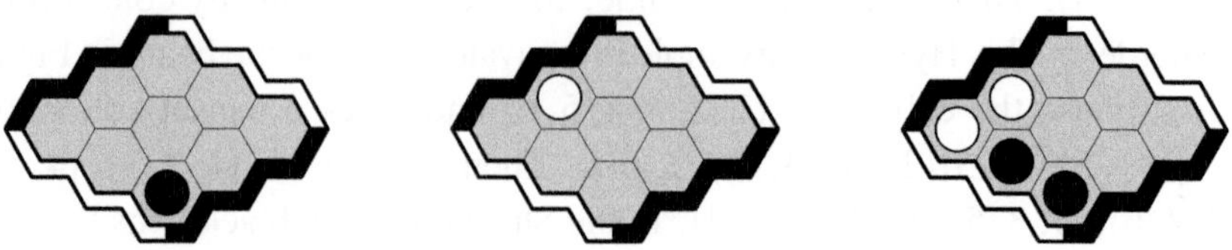

Figure 5. Hex positions with outcome-values black, white, and neutral.

Figure 6. Left: black pass-value 0. Right: black pass-value 2.

4. It's never too late for a good move

In no-draw HyperHex, for a position with finite pass-value, is a good move always available?

If a player is ahead — has positive pass-value, i.e., a 2nd-player win — then sometimes, but not always. For example, in Figure 7 each position has black pass-value 1. In the left position, black has good moves. But consider the right position. If black passes twice, white can color the top two marked cells and then win by playing in either of the other two other marked cells. A similar strategy holds for the four unmarked cells. Thus each black move from this position leaves one of these two white strategies intact, and so leaves the black pass-value at 1. So black has no good move.

But if the player is not ahead, then *it's never too late for a good move*. This follows almost immediately from the definition of pass-value.

Observation 2. For a player P and any no-draw HyperHex position H with $\nu_P(H) = t \leq 0$, P has a good move.

Proof. For $k \geq 1$, a *k-strategy* is a first-player strategy in which k cells are colored on the first move and one cell is colored on each successive move.

Assume P and H are as stated. Thus P has a winning $(1 - t)$-strategy S whose first move is to a set C of $1 - t \geq 1$ cells. Thus, for any cell c in C and the position $H' = H + P(c)$, the $-t$-strategy obtained from S by removing the cell c from the first move is a winning strategy, so $\nu_P(H') \geq t + 1 = \nu_P(H) + 1$, and we are done. □

By Observation 2, white has at least one good move in each position of Figure 7. We leave it as an exercise to the reader to find all such moves.

Figure 7. Left: two black-good moves. Right: no black-good moves.

5. Your opponent's good move is your good move

In Hex, as in Go, it's not always true that *your opponent's good move is your good move.* In Figure 8 both players have five good moves and three moves are good for both players, but in Figure 9 — found by Jonatan Rydh [?] — both players have two good moves but no move is good for both players. The next theorem gives some conditions where some move is good for both players.

Theorem 3. *A neutral no-draw HyperHex position H with t empty cells has a move that is good for both players if*:

- *either player has only one winning move, or*
- $1 \le t \le 5$, *or*
- $1 \le t \le 7$ *and each player has only two winning moves.*

Proof. In a neutral position, a move is good if and only if it is winning, so it suffices to find moves that are winning for both players.

Let $c_1, \ldots, c_t$ be the empty cells of H. Let W_P and $W_{\overline{P}}$ be the respective sets of winning moves for P and $\overline{P}$.

Consider the first part of the theorem. Suppose P has only one winning move, say to cell c_1. Then P has no winning moves in $H + \overline{P}(c_1)$, so c_1 wins for $\overline{P}$, and we are done.

Consider the second part of the theorem. P is neutral so $t \ge 1$. Assume $t \le 5$. We are done by the first part if W_P or $W_{\overline{P}}$ has size one, so assume each set has size at least two, so $t \ge 4$. Since $t \le 5$, one set — say W_P — has size two. Relabel cells if necessary so that $W_P = \{c_1, c_2\}$.

Argue by contradiction: suppose $W_{\overline{P}}$ contains no cell of W_P. Then c_1 does not win for $\overline{P}$, so P has a winning move x in $H + \overline{P}(c_1)$. By Observation 1, x

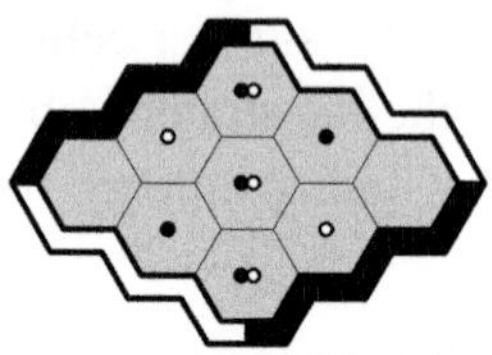

Figure 8. Dots show winning moves. Three moves win for both players.

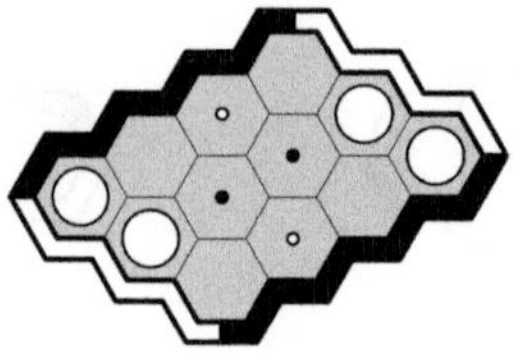

Figure 9. No move wins for both players.

is also P-winning in H, so $x = c_2$. Similarly, c_1 is P's unique winning move in $H + \overline{P}(c_2)$.

First assume $t = 4$. P has a winning reply for any $\overline{P}$-move in $H + \overline{P}(c_1) + P(c_2)$, so $\{c_2, c_3\}$ and $\{c_2, c_4\}$ are P-winsets. Similarly, $\{c_1, c_3\}$ and $\{c_1, c_4\}$ are P-winsets. Thus c_3 wins for P in H (on t;e next move P can color one of c_1, c_2), therefore contradiction ($W_P = \{c_1, c_2\}$).

Next assume $t = 5$. Again, P has a winning reply for any $\overline{P}$-move in $H + \overline{P}(c_1) + P(c_2)$; relabel c_3, c_4, c_5 if necessary so that $\{c_2, c_3\}$ and $\{c_2, c_4\}$ are P-winsets. Similarly, for at least two cells j, k in c_3, c_4, c_5, $\{c_1, j\}$ and $\{c_1, k\}$ are P-winsets. So, for some z in $\{j, k\}$, $\{c_1, z\}$ and $\{c_2, z\}$ are P-winsets, so z wins for P in H; therefore contradiction. Thus the second part of the theorem holds.

Next consider the third part. Again, P is neutral so $t \geq 1$. If $t \leq 5$ we are done, so $t = 6$ or 7. Argue by contradiction: assume that W_P and $W_{\overline{P}}$ have no cell in common, say $W_P = \{c_1, c_2\}$ and $W_{\overline{P}} = \{c_3, c_4\}$.

If P first colors c_1 and then $\overline{P}$ colors c_3 then P has a winning move; similarly, if $\overline{P}$ first colors c_3 and then P colors c_1 then $\overline{P}$ has a winning move. So $H^* = H + P(c_1) + \overline{P}(c_3)$ is neutral and has at most 5 empty cells, so—by the second part—some empty c_w wins H^* for both players.

Notice that $c_w \neq c_2$: otherwise, c_2 wins for $\overline{P}$ in H^*, but also c_1 wins for P in $H' = H + \overline{P}(c_2)$, so c_3 does not win for $\overline{P}$ in $H' + P(c_1)$, so P wins $H^* + \overline{P}(c_2)$, a contradiction. Similarly, $c_w \neq c_4$.

Similarly, some c_x wins for both players in $H + P(c_2) + \overline{P}(c_3)$, some c_y wins for both players in $H + P(c_1) + \overline{P}(c_4)$, some c_z wins for both players in $H + P(c_2) + \overline{P}(c_4)$, and none of c_w, c_x, c_y, c_z are in $\{c_1, c_2, c_3, c_4\}$.

Since $t \leq 7$, at least two of c_w, c_x, c_y, c_z are equal, say $c_w = c_x$. Then P has no winning move in $H + P(c_1) + \overline{P}(c_3) + \overline{P}(c_w)$, and no winning move in $H + P(c_2) + \overline{P}(c_3) + \overline{P}(c_x = c_w)$, so no winning move in $H + \overline{P}(c_w)$ (the only possible winning replies would be c_1 or c_2, but in each case $\overline{P}$ counters with c_3), so c_w wins for $\overline{P}$ in H, so c_w (not in $\{c_1, \ldots, c_4\}$) is in $W_{\overline{P}}$, a contradiction, and we are done. □

Jonatan Rydh's example in Figure 9 shows that the third part of the theorem cannot be strengthened in terms of t. Here is a 6-cell no-draw HyperHex example that shows that the second part of the theorem cannot be strengthened: black's winsets are {1, 3, 4}, {1, 3, 6}, {1, 4, 5}, {1, 5, 6}, {2, 3, 4}, {2, 3, 5}, {2, 4, 6}, {2, 5, 6}; white's winsets are {1, 2}, {1, 3, 6}, {1, 4, 5}, {2, 3, 5}, {2, 4, 6}, {3, 4, 5}, {3, 4, 6}, {3, 5, 6}, {4, 5, 6}. We leave it as an exercise for the reader to verify that the sets of winning opening moves for black and white are {1, 2} and {3, 4, 5, 6}, respectively.

We close with an open problem. Among all neutral no-draw HyperHex positions with t empty cells, w_1 that win for one player, w_2 that win for the other, and no cell that wins for both, what is the smallest possible value of t? We have shown that there is no such t if $w_1 = 1$ or $w_2 = 1$, and that $t = 6$ if $w_1 = w_2 = 2$.

Acknowledgements

I thank Martin Mueller and Ziang Zhujiu for helpful Go comments, Philip Henderson for critiquing a draft and pointing me to Jonatan Rydh's example, Henry Brausen and David Spies for writing supporting software, and the organizers of the BIRS 2011 Workshop for their hospitality. Part of this paper was written while a Visiting Researcher at Université Joseph Fourier in Grenoble; I thank my hosts András Sebö, Frederic Maffray, and Myriam Preissmann for their hospitality.

References

hayward@ualberta.ca *Department of Computing Science, University of Alberta, Edmonton, Canada*

Games of No Chance 5
MSRI Publications
Volume **70**, 2017

Conjoined games: GO-CUT and SNO-GO

MELISSA HUGGAN AND RICHARD J. NOWAKOWSKI

Let $\mathcal{F}$ and $\mathcal{H}$ be two impartial rulesets. We introduce the *conjoined ruleset* $(\mathcal{F} \blacktriangleright \mathcal{H})$ in which the game is played under the $\mathcal{F}$ ruleset and then, when play is no longer possible, to continue under the $\mathcal{H}$ ruleset. The games of GO-CUT and SNO-GO on a path are considered. We give nim-values for positions at the start of Phase 2 for GO-CUT, and for SNO-GO we determine the winner.

1. Introduction

The games STRATEGO, OVID'S GAME, THREE-, SIX-, and NINE-MEN'S MORRIS [3] and also BUILDING NIM [7] are examples of combinatorial games that have two phases. Specifically, in Phase 1 the board is set up and in Phase 2 the game is played. There are many other "math games" (which mathematicians want to analyze regardless of whether people actually want to play them) in which Phase 1 is not even defined, but instances of Phase 2 are analysed. For example, BOXCARS [2], END-NIM [1], PUSH [2], TOPPLING DOMINOES [8] and their variants.

In this paper, we consider playing Phase 1 as a combinatorial game as well as Phase 2 and analyze two specific games. We were introduced to this concept by Kyle Burke and Urban Larsson (personal communication).

To avoid confusion with the multiple meanings of the word "game", we refer to the *ruleset*, which describes the legal moves, and a *position*, which is an instance of the game after several (including zero) legal moves. By necessity, the position also describes the board upon which play takes place.

In an *impartial* game, both players have the same moves available.

Definition 1. Let $\mathcal{F}$ and $\mathcal{H}$ be two impartial rulesets. The *conjoined ruleset* $(\mathcal{F} \blacktriangleright \mathcal{H})$ is to play Phase 1 under the $\mathcal{F}$ ruleset and when play is no longer possible to start Phase 2 which is played under the ruleset of $\mathcal{H}$.

The first author was partially supported by Carleton and Dalhousie Universities and the Canadian Mathematical Society. The second author was partially supported by NSERC.
MSC2010: 91A46.
Keywords: combinatorial game, nim-value, NODE-KAYLES, NOGO, SNORT, CUTTHROAT, GO-CUT, SNO-GO.

Forming a conjoined game allows for an interesting Phase 1 battle before the "real" game begins. Since play in Phase 1 sets up the board, it is convenient to have the corresponding game be a placement game [5], i.e., pieces are placed but not moved or removed. The positions at the beginning of Phase 2 will have structure reflecting the Phase 1 rules, and this allows for some partial analysis. Here we indicate by GAME_I a partizan game converted into an impartial game by allowing both players to place any of the pieces. This paper explores the following two games with two-phase play:

$$\text{GO-CUT} = (\text{NOGO}_I \blacktriangleright \text{CUTTHROAT}_I), \quad \text{and} \quad \text{SNO-GO} = (\text{SNORT}_I \blacktriangleright \text{NOGO}_I).$$

Brief notes about the games: The game NOGO is known as "anti-atari go"[1], but was independently invented by Neil McKay in 2011 (see also [6]). CUTTHROAT was introduced in [9] and the full analysis when played on stars is given in [2]. SNORT is introduced in [3], Vol. 1, and is known as "CATS & DOGS" in Portugal.

Our results cover playing these conjoined games on a path, where the first player who cannot move loses. In Section 2, we obtain the values for all the possible positions of GO-CUT at the start of Phase 2 but we were not able to find the outcomes of an empty path of length n, as a function of n. By contrast, we find the outcome of an empty path for SNO-GO but were not able to find formulas for the corresponding nim-values.

It should be noted that conjoined games and the sequential compound of games [11] have similar definitions. However, the latter is formed from two positions, so that the position for the second phase is known before the game starts and the first phase only determines who plays first in the second phase. In conjoined games, the position for the second phase is determined only after the last play of the first phase.

1.1. ***Basic background.*** The basic Sprague–Grundy theory for impartial games is presented here. Readers should consult any of [3; 2; 10] for further information and proofs:

- The minimum excluded value, *mex*, of a set S is the least nonnegative integer not included in S.
- An *option* of a position H is any position that can be reached in one move.
- Recursively, the *Sprague–Grundy value*, or *nim-value*, of a position H is given by $\mathcal{G}(H) = \text{mex}\{\mathcal{G}(H') \mid H' \text{ is an option of } H\}$. Thus, if a position has no options, it has nim-value 0.
- Let H be a position; the next player to move can win if and only if $\mathcal{G}(H) > 0$.

[1] http://senseis.xmp.net/?AntiAtariGo

- Let p and q be nonnegative integers, then $p \oplus q$ signifies the *nim-sum* or *exclusive or* of p and q. It is obtained by writing p and q in binary and adding without carrying.
- The *disjunctive sum* of positions F and H, written $F + H$, is the game in which a player may move in one component but not both.
- $\mathcal{G}(F + H) = \mathcal{G}(F) \oplus \mathcal{G}(H)$.

2. GO-CUT on a path

Since we are only considering a path we will give only those rules. The generalisations to graphs is left to the reader. We leave the full analysis as an open question for the reader since the authors didn't succeed[2]. In our defence, the game has similarities to BUILDING NIM, which seems to be hard. However, Lemma 3 gives the nim-value (recall, this is denoted by $\mathcal{G}$) of a path at the end of Phase 1.

Definition 2 (GO-CUT). Initial board is a path with n vertices.

Phase 1: On a move a player chooses an uncoloured vertex ($\cdot$) and colours it either red (R) or blue (B) provided that it is contained in a subpath of red (respectively blue) vertices that has at least one end-vertex adjacent to an uncoloured vertex.

When no moves are playable under Phase 1, delete all uncoloured vertices and then delete all maximal paths which have only red vertices or only blue vertices. The game is now a disjunctive sum of components each of which contains both red and blue vertices, that is, nonmonochromatic components.

Phase 2: A player chooses a component from the disjunctive sum, deletes one of the vertices then deletes any resulting monochromatic components.

Thus, at the end of Phase 1, for example, we might have the position $[BB \cdot RBB \cdot RB \cdot R \cdot BB]$ which, after deleting the uncoloured vertices, leaves $[BB] + [RBB] + [RB] + [R] + [BB]$. Now deleting all monochromatic components, gives the starting position for Phase 2 as $[RBB] + [RB]$.

By the rules of NOGO, at the start of Phase 2, a component will consist of i blue vertices followed by j red vertices (or the reverse) for some $i, j > 0$. Call this an (i, j)-component. To extend the notation, we also refer to $(i, 0)$ and $(0, j)$ components but these correspond to empty components.

Lemma 3. *The nim-value of an (i, j)-component is $((i-1) \oplus (j-1)) + 1$.*

[2]The nim-values of GO-CUT starting on a path of n uncoloured vertices, for $n = 1$ through 15 vertices are: 0, 1, 0, 1, 0, 1, 0, 1, 0, 1, 0, 0, 1, 0, 2.

Proof. Clearly, the nim-value of a $(0, j)$- or $(i, 0)$-component is 0. We will refer to an (i, j)-component by (i, j). If i and j are positive then, by induction,

$$\begin{aligned}\mathcal{G}(i, j) &= \operatorname{mex}\{\mathcal{G}(r, j), \mathcal{G}(i, s), 0 \le r \le i-1, 0 \le s \le j-1\} \\ &= \operatorname{mex}\{\{(r-1 \oplus j-1)+1, (i-1 \oplus s-1)+1, \\ &\qquad\qquad 1 \le r \le i-1, 1 \le s \le j-1\} \cup \{0\}\}.\end{aligned}$$

Note that the set $\{(r-1 \oplus j-1), (i-1 \oplus s-1), 1 \le r \le i-1, 1 \le s \le j-1\}$ is the set of nim-values for NIM played with heaps of size $i-1$ and $j-1$ and hence contains $0, 1, \ldots, (i-1 \oplus j-1)-1$ and does not contain $(i-1 \oplus j-1)$. Adding one to each value gives that both 0 and $(i-1 \oplus j-1)+1$ are missing. Since 0 is an option of (i, j), then

$$\begin{aligned}\mathcal{G}(i, j) &= \operatorname{mex}\{\mathcal{G}(r, j), \mathcal{G}(i, s), 0 \le r \le i-1, 0 \le s \le j-1\} \\ &= (i-1 \oplus j-1)+1.\end{aligned}$$ □

3. SNO-GO on a path

We were able to obtain winning strategies for the conjoined games of SNORT_I and NOGO_I on a path. We give the rules for an arbitrary graph so that a useful general tool, Lemma 6, can be introduced.

Definition 4 (SNO-GO). The initial board is a graph with n vertices.

Phase 1: On a move a player chooses an uncoloured vertex $(\cdot)$ and colours it red (R) or blue (B) provided that no red vertex is adjacent to a blue vertex.

Phase 2: When no moves are playable under Phase 1 rules, players can colour an uncoloured vertex red or blue provided that each monochromatic component has at least one vertex adjacent to an uncoloured vertex.

Thus, at the end of Phase 1, for example, we might have the position $[BB \cdot RRR \cdot BB \cdot R \cdot BB]$. At the end of the game, the position may look like $[BB \cdot RRRRBB \cdot R \cdot BB]$.

Definition 5. Let G be a graph, $x, y \in V(G)$ and $z \notin V(G)$. The *merger* of x and y results in the graph G' where $V(G') = (V(G) \setminus \{x, y\}) \cup \{z\}$, and $vw \in E(G')$ if $vw \in E(G)$ or $v = z$ and either $xw \in E(G)$ or $yw \in E(G)$.

The proof of Lemma 6 is clear and left to the reader.

Lemma 6 (reduction). *Let $\mathcal{G}$ be a* SNO-GO *position on a graph G. Suppose x, y are two adjacent vertices that are coloured the same. Let $\mathcal{G}'$ be the position on the board resulting from the merger of x, y where z has the same colour as x and all other vertices retain their colour. The nim-values of $\mathcal{G}$ and $\mathcal{G}'$ are equal.*

As a guide to intuition, consider a path with n vertices where we label the vertices $x_1, \ldots, x_n$. If two adjacent vertices are coloured the same then we can apply Lemma 6 so that each coloured subpath is reduced to size 1 after Phase 1. For example, the position $[BB \cdot RRR \cdot BB \cdot R \cdot BB]$ becomes $[B \cdot R \cdot B \cdot R \cdot B]$ after applying Lemma 6. This notion is summarised in the next Lemma.

Lemma 7. *A* SNO-GO *position on a path of n vertices at the beginning of Phase* 2 *is equal to a path of alternating coloured vertices, each separated by a single uncoloured vertex.*

Proof. At the end of Phase 1, after applying Lemma 6, all consecutive single coloured vertices (red or blue) get amalgamated into a single representative of that colour. If there is a pair of adjacent uncoloured vertices then either of them can be coloured under Phase 1 rules. Also, if x_1 or x_n is uncoloured then Phase 1 play is still possible. Hence, after all reductions, the position will consist of vertices alternating colours with a single uncoloured vertex between them and the end vertices x_1 and x_n are also coloured. □

At the end of Phase 1, we will call any uncoloured vertex a *hole*. Note that a hole will be adjacent to exactly one red and one blue vertex. We relate this game, using Lemma 6, to NODE-KAYLES [4] or equivalently to DAWSON'S CHESS [3].

Definition 8. The impartial game NODE-KAYLES is played on a graph. Players alternately choose a vertex and delete it and all adjacent vertices. The last player to move wins.

Lemma 9. *Given a path on n vertices, let $\mathcal{G}$ be a* SNO-GO *position at the start of Phase* 2 *play. Furthermore, suppose $\mathcal{G}$ has m uncoloured vertices. If $m \geq 2$ then $\mathcal{G}$ is equivalent to* NODE-KAYLES *played on a path with $m-2$ vertices.*

Proof. It is possible that $m = 0$ or $m = 1$. In the first case, all the vertices were coloured the same. In the second, the final position is $B \cdot R$. If $m \geq 2$ playing either of the two holes at the end, without loss of generality $[B \cdot R \ldots] \to [BXR \ldots]$, $X \in \{B, R\}$, leaves an illegal Phase 2 position. Playing an interior hole, e.g., $[B \cdot R \cdot B \cdot R \cdot B \cdot R] \to [B \cdot R \cdot BXR \cdot B \cdot R]$, $X \in \{B, R\}$ eliminates playing in the two adjacent holes as legal moves. This shows that the position is now equivalent to playing NODE-KAYLES on a path of length $m-2$. □

For ease of referencing the players, we assume that Alf plays first on the empty board and Betti plays second.

The outcome class of the sequence of NODE-KAYLES on a path is periodic with period length 34 after a preperiod of 52 and the only $\mathcal{P}$ positions are when n is even. For exact values, see the nim-value sequence for DAWSON'S CHESS in Winning Ways [3], Vol. 1. Our approach is to show that the winning player can ensure to play first at the start at Phase 2, on the equivalent of an odd NODE-KAYLES position, or win if the opponent does not allow 3 or more holes.

Before proving the Main Theorem of this section, we need the following lemmas and conventions.

We partition the path into two pieces: the *outer vertices* consisting of vertices x_1, x_2, x_{n-1}, x_n, and the *interior*, consisting of vertices $x_3, \dots, x_{n-2}$.

Lemma 10. *Let $\mathcal{G}$ be a Phase 1 SNO-GO position on a path of n vertices. If at least one of $\{x_1, x_2\}$ and one of $\{x_{n-1}, x_n\}$ are the same colour then at the end of Phase 1 there will be an even number of holes. If at least one of $\{x_1, x_2\}$ and one of $\{x_{n-1}, x_n\}$ are opposite colours then at the end of Phase 1 there will be an odd number of holes.*

Proof. Let $\mathcal{G}$ be a Phase 1 SNO-GO position where at least one of $\{x_1, x_2\}$ and one of $\{x_{n-1}, x_n\}$ are coloured. At the end of Phase 1, positions will be as in Lemma 7. If at least one of $\{x_1, x_2\}$ and one of $\{x_{n-1}, x_n\}$ are the same colour, then given the alternating pattern of the colours, the number of coloured vertices is odd which implies an even number of uncoloured vertices must be separating them. If at least one of $\{x_1, x_2\}$ and one of $\{x_{n-1}, x_n\}$ of the position are different colours, again given the alternating patterns of colours being separated by single uncoloured vertices, this implies an even number of coloured vertices, separated by an odd number of uncoloured vertices. □

Note that Lemma 11 is referring to the original path before applying Lemma 6.

Lemma 11. *Let $\mathcal{G}$ be a SNO-GO position on a path of $2k+1$ vertices at the end of Phase 1. Let h be the number of holes. If $h = 0, 2$ or $h \geq 3$ and is odd then Alf will win the game.*

Proof. If $h = 0$ then there has been an odd number of moves, the game is over and Alf had the last move.

If $h = 2$ then Phase 2 has no moves but it is Betti's turn to play and so she loses.

If $h \geq 3$ and is odd then there has been an even number of moves $(2k+1-h)$ in Phase 1 and thus Alf moves first in Phase 2. NODE-KAYLES on an odd number of vertices, here $h-2$, is a first player win and so Alf can win the game. □

Theorem 12. *Consider SNO-GO played on a path of n vertices. The initial position is in $\mathcal{P}$ if n is even and in $\mathcal{N}$ if n is odd.*

Proof. First suppose n is odd.

The strategy is for Alf to colour the centre vertex, without loss of generality, blue. Now, until Betti colours an outer vertex (the first outer vertex to be coloured), Alf always plays the same move reflected across the centre vertex. When Betti finally colours an outer vertex there are now several cases to consider. Since

Betti's last move was to colour an outer vertex, without loss of generality, suppose Betti colours x_{n-1} or x_n with X.

If there are two red interior vertices then Alf colours x_1 with the opposite colour from Betti's choice. By Lemma 10, at the end of Phase 1 there will be an odd number of holes, at least 3, and by Lemma 11 Alf will win.

Thus we may suppose that at this point in play, all coloured interior vertices are blue. There are several cases to consider:

(1) Suppose every uncoloured interior vertex is adjacent to at least one blue. That is, it is now illegal to colour an interior vertex red. Alf colours x_1 with X. The number of holes will be 0 if X = blue and 2 if X = red. In both cases, by Lemmas 10 and 11, Alf can force a win.

(2) Suppose there are 4 interior uncoloured vertices, any pair of these are at least distance 3 apart, nor is any adjacent to a blue vertex. Alf colours x_1 with the opposite colour, hence an odd number of holes, and Alf can force at least 3 holes by having two reds separated by a blue vertex.

(3) Similarly, if there is an interior uncoloured vertex, not adjacent to any blue vertex which has an interior blue vertex between it and the closest outer vertex. By symmetry, the reflected vertex is uncoloured and not adjacent to a blue vertex. Again, Alf colours x_1 with the opposite colour, hence an odd number of holes, and Alf can force at least 3 holes by colouring one of the two uncoloured vertices red.

This leaves the situation where the outermost blue interior vertices are followed by at most 5 uncoloured interior vertices and all other uncoloured interior vertices are adjacent to a blue vertex. Since Alf will play symmetry following any Betti move between the two outer blues we can condense the centre to a single blue vertex. The positions that remain to analyze are, where it is Betti to colour one of the, say, left two vertices. The position is one of the following:

(i) $[\cdot\cdot B \cdot\cdot]$,

(ii) $[\cdot\cdot\cdot B \cdot\cdot\cdot]$,

(iii) $[\cdot\cdot\cdot\cdot B \cdot\cdot\cdot\cdot]$,

(iv) $[\cdot\cdot\cdot\cdot\cdot B \cdot\cdot\cdot\cdot\cdot]$,

(v) $[\cdot\cdot\cdot\cdot\cdot\cdot B \cdot\cdot\cdot\cdot\cdot\cdot]$,

(vi) $[\cdot\cdot\cdot\cdot\cdot\cdot\cdot B \cdot\cdot\cdot\cdot\cdot\cdot\cdot]$.

In cases (i) and (ii), Alf forces either 0 or 2 holes by playing symmetrically.

In (iii), if Betti plays x_{n-1}, again Alf forces either 0 or 2 holes by playing symmetrically. Suppose Betti plays to $[B\cdots B\cdots\cdot]$, Alf plays to $[B\cdots B\cdot\cdot B\cdot]$ forcing 0 or 2 holes. If she plays to $[R\cdots B\cdots\cdot]$, then Alf replies $[R\cdots B\cdot\cdot R\cdot]$ forcing 2 holes.

In (iv) and (v) $[\cdot\cdots\cdot B\cdot\cdots\cdot]$, Alf colours the fourth vertex from the end of the side opposite to that of Betti's last move. Alf can now force 2 or 3 holes.

In (vi), from $[\cdot X\cdots\cdot B\cdots\cdots\cdot]$ Alf plays to $[\cdot X\cdot\cdot X\cdot\cdot B\cdots\cdots\cdot]$ and can force 0 or 2 holes. From $[X\cdots\cdots B\cdots\cdots\cdot]$ Alf plays to $[X\cdots\cdots B\cdots\cdots Y]$. Suppose without loss of generality that X is red. Regardless of what Betti plays, Alf can colour a vertex red, on the other side of the centre from X. This generates an odd number (> 1) of holes.

If the board is of even length, Betti plays the reflection symmetry and a similar analysis shows that she can force a win. □

References

[1] M. H. Albert and R. J. Nowakowski. The game of end-nim. *Electr. J. Combin.*, 8(2):#R1, 12pp., 2001. Volume in honour of Aviezri S. Fraenkel.

[2] Michael H. Albert, Richard J. Nowakowski, and David Wolfe. *Lessons in Play*. A K Peters, Ltd., 2007.

[3] E. R. Berlekamp, J. H. Conway, and R. K. Guy. *Winning Ways for your Mathematical Plays*, volume 1-4. A K Peters, Wellesley, MA, 2001-2004. 2nd edition: vol. 1 (2001), vols. 2, 3 (2003), vol. 4 (2004).

[4] H. L. Bodlaender. Kayles on special classes of graphs—an application of Sprague-Grundy theory. In *Graph-Theoretic Concepts in Computer Science* (Wiesbaden–Naurod, 1992), Lecture Notes in Comput. Sci., volume 657, pages 90–102. Springer, Berlin, 1993.

[5] J. I. Brown, D. Cox, A. Hoefel, N. Mckay, R. Milley, R. J. Nowakowski, and A. A. Siegel. Polynomial profiles of placement games. In this volume.

[6] C. Chou, O. Teytaud, and S. Yen. Revisiting monte-carlo tree search on a normal form game: Nogo. In Cecilia Di Chio, Stefano Cagnoni, Carlos Cotta, Marc Ebner, Anikó Ekárt, Anna Esparcia–Alcázar, Juan Merelo, Ferrante Neri, Mike Preuss, Hendrik Richter, Julian Togelius, and Georgios Yannakakis, editors, *Applications of Evolutionary Computation*, volume 6624 of *Lecture Notes in Computer Science*, pages 73–82. Springer, Berlin, 2011.

[7] Eric Duchêne, Matthieu Dufour, Silvia Heubach, and Urban Larsson. Building nim. *Int. J. Game Theory*, 45(4):859–873, 2016.

[8] Alex Fink, Richard J. Nowakowski, Aaron N. Siegel, and David Wolfe. Toppling conjectures. In *Games of No Chance 4*, pages 65–76. Cambridge University Press, 2015.

[9] Sarah K. McCurdy and Richard Nowakowski. Cutthroat, an all-small game on graphs. *INTEGERS*, 5(2):#A13, 13pp., 2005.

[10] Aaron N. Siegel. *Combinatorial Game Theory*. American Math. Society, 2013.

[11] Walter Stromquist and Daniel Ullman. Sequential compounds of combinatorial games. *Theoret. Comput. Sci.*, 119(2):311–321, 1993.

melissa.huggan@dal.ca *Mathematics and Statistics, Dalhousie University, Halifax, Canada*

r.nowakowski@dal.ca *Mathematics and Statistics, Dalhousie University, Halifax, Canada*

Games of No Chance 5
MSRI Publications
Volume **70**, 2017

Impartial games whose rulesets produce given continued fractions

URBAN LARSSON AND MIKE WEIMERSKIRCH

We study 2-player impartial games of the form *take-away* which produce P-positions (second player winning positions) corresponding to *complementary Beatty sequences*, given by the continued fractions $(1; k, 1, k, 1, \dots)$ and $(k+1; k, 1, k, 1, \dots)$. Our problem is the opposite of the main field of research in this area, which is to, given a game, understand its set of P-positions. We are rather given a set of (candidate) P-positions and look for "simple" rules. Our rules satisfy two criteria, they are given by a *closed formula* and they are *invariant*, that is, the available *moves* do not depend on the position played from (for all options with nonnegative coordinates).

1. Introduction

This paper uses ideas from combinatorial game theory, Beatty sequences, and Sturmian words. We have in many cases given the pertinent information in this paper, but have chosen to omit some material on these subjects. The reader who wishes to have more background information on certain topics is directed to the following references.

For standard terminology of impartial removal games on heaps of tokens, see [WW]; for Beatty sequences, see [B]; for k-Wythoff Nim, see [W; F]; for Sturmian words, see [L]; and for continued fractions, see [K].

Our problem is an inverse to that of the main field of research, which for a given an impartial ruleset Γ, (for example, $\Gamma = k$-Wythoff Nim) is to determine the P-positions of Γ (within reasonable time-complexity). Here we rather start with a particular (candidate) set of P-positions and search for "simple" game rules. Let us explain the setting.

Throughout this paper, we will denote the *position* consisting of two heaps of $x \geq 0$ and $y \geq 0$ tokens as (x, y). When the values of x and y are known, we

MSC2010: 11A55, 91A46.
Keywords: combinatorial game, complementary Beatty sequences, continued fraction, impartial game, invariant game, Sturmian word.

adopt the convention that $x \le y$, though in general we regard such a position as an unordered multiset, so we identify (y, x) with (x, y).

Similarly, we let the *move* (u, v) denote a removal of $v > 0$ tokens from one of the heaps and u from the other, where $0 \le u \le v$; thus from the position (x, y), the move (u, v) is ambiguous, being either $(x, y) \to (x-u, y-v)$, provided both $x - u \ge 0$ and $y - v \ge 0$, or $(x, y) \to (x - v, y - u)$, provided both $x - v \ge 0$ and $y - u \ge 0$. Therefore, in general, it is necessary to examine both cases.

Recall that a (homogeneous) *Beatty sequence* is a sequence of integers of the form $(\lfloor n\gamma \rfloor)$, the modulus γ being a positive irrational and n ranging over the non-negative integers, here denoted by $\mathbb{N}$. We are interested in positions of the form

$$(\lfloor n\alpha \rfloor, \lfloor n\beta \rfloor) \tag{1}$$

for $n \in \mathbb{N}$, where $0 < \alpha < \beta$ are irrationals with

$$\alpha^{-1} + \beta^{-1} = 1, \tag{2}$$

that is $1 < \alpha < 2 < \beta$. By (2), the sequences $(\lfloor n\alpha \rfloor)$ and $(\lfloor n\beta \rfloor)$, where n ranges over the positive integers, $\mathbb{Z}^+$, are *complementary*, (see [B]); that is, each positive integer is attained precisely once in precisely one of these sequences.

In k-Wythoff Nim [F], the P-positions correspond to all unordered pairs of the form in (1) with

$$\alpha = [1; k, k, k, \dots] = \frac{2-k+\sqrt{k^2+4}}{2}$$

and

$$\beta = [k+1; k, k, k, \dots] = \frac{2+k+\sqrt{k^2+4}}{2} = \alpha + k,$$

where $x = [x_1; x_2, x_3, x_4, \dots]$ denotes the unique continued fraction expansion, CF, of x:

$$x = x_1 + \cfrac{1}{x_2 + \cfrac{1}{x_3 + \cfrac{1}{x_4 + \cdots}}}.$$

For a variation, in [DR] game rules are examined for P-positions corresponding to the CFs

$$[1; 1, k, 1, k, \dots] \quad \text{and} \quad [k+1; k, 1, k, 1, \dots]. \tag{3}$$

In this paper, we rather study the CF

$$\alpha = \alpha_k = [1; k, 1, k, 1, k, \dots] = \frac{1+\sqrt{1+4/k}}{2},$$

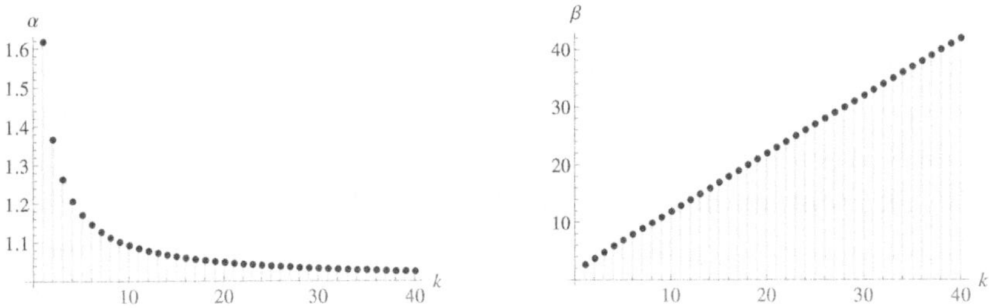

Figure 1. The numbers α_k and β_k for $k \in \{1, 40\}$. See also Figure 5.

with corresponding

$$\beta = \beta_k = [k+1; 1, k, 1, k, 1, \dots] = \frac{k+2+k\sqrt{1+4/k}}{2} = k\alpha + 1 = k\alpha^2.$$

Note that $\alpha_k \in (1, 1+1/k)$ and $\beta_k \in (k+1, k+2)$ (see Figure 1 and Lemma 13).

Notation 1. For each $k \in \mathbb{Z}^+$, for all $n \in \mathbb{N}$, we let $a_n = \lfloor n\alpha \rfloor$, $b_n = \lfloor n\beta \rfloor$, $c_n = a_n - a_{n-1}$ and $d_n = b_n - b_{n-1}$. Moreover we define the following sequences, $A = (a_1, a_2, \dots)$, $B = (b_1, b_2, \dots)$, $C = (c_1, c_2, \dots)$ and $D = (d_1, d_2, \dots)$.

Then, for all $n \in \mathbb{Z}^+$,

$$b_n = \sum_{j=1}^{n} d_j \quad \text{and} \quad a_n = \sum_{j=1}^{n} c_j.$$

Example 2. For $k = 2$,

$$A = (1, 2, 4, 5, 6, 8, 9, 10, 12, 13, 15, 16, 17, 19, 20, 21, 23, 24, \dots),$$
$$C = (1, 1, 2, 1, 1, 2, 1, 1, 2, 1, 2, 1, 1, 2, 1, 1, 2, 1, \dots),$$
$$B = (3, 7, 11, 14, 18, 22, 26, 29, 33, 37, 41, 44, 48, 52, 55, \dots),$$
$$D = (3, 4, 4, 3, 4, 4, 4, 3, 4, 4, 4, 3, 4, 4, 3, \dots).$$

Since $\alpha_k \in (1, 1+1/k)$ and $\beta_k \in (k+1, k+2)$, for each k and all n, we also get that

$$c_n \in \{1, 2\} \quad \text{and} \quad d_n \in \{k+1, k+2\}$$

(see [L]). Moreover, each value is attained infinitely often, a statement which we strengthen in Section 2.

Henceforth, for a fixed $k \in \mathbb{Z}^+$, let

$$S_k = \{(a_n, b_n) \mid n \in \mathbb{N}\}.$$

Note that the special case of S_1 corresponds precisely to the P-positions of Wythoff Nim.

The problem of finding a closed formula ruleset such that the set of all P-positions is identical to S_2 was posed by A. S. Fraenkel at the GONC 2011 workshop at the Banff Centre. Here we resolve the general case for the set of (candidate) P-positions being S_k. Henceforth we omit the word "candidate" and simply talk about sets of P-positions. We have also added the requirement that the ruleset be *invariant* [DR; LHF]; that is, the available moves do not depend on the position (for all options with nonnegative coordinates). This criterion is implicitly fulfilled by many classical removal games, e.g., Nim, k-Wythoff Nim, subtraction games [WW] and S. Golombs take away-games [Go]. Without the requirement of invariance, one may define the most trivial game rules, no move is possible from a position in S, and otherwise each position has a move to $(0, 0)$. On the other hand, the problem of finding invariant (but not necessarily simple) game rules for any set of P-positions, defined by a complementary pair of homogeneous Beatty sequences, was resolved in [LHF]. However those game rules are not simple in the meaning that the only known formula for the invariant moves is exponentially slow in $\log(xy)$. See Figures 2, 3 and 4 for invariant games corresponding to the CFs on page 404, cases $k = 2$.

For many classical games, such as *normal play* Nim and k-Wythoff Nim, the final winning position is unique, namely $(0, 0)$. Given our set of P-positions, S_k, this requirement clearly needs to be satisfied. A convenient way to achieve this is to follow the example of our classical games, to include the Nim rules to our new game. An immediate benefit of doing this is that we automatically satisfy one of the other inherent requirements of the set S_k, namely that there can be at most one P-position in each row and column of $\mathbb{N} \times \mathbb{N}$. Precisely, the desired ruleset Γ_k has the following permitted moves.

Theorem 3. *Let the set S_k be defined by the Beatty sequences where $\alpha = [1; k, 1, k, 1, k, \ldots]$ and $\beta = [k+1; 1, k, 1, k, 1, \ldots]$; that is $S_k = \{(\lfloor n\alpha \rfloor, \lfloor n\beta \rfloor) \mid n \in \mathbb{N}\}$. Then the invariant ruleset $\Gamma = \Gamma_k$ consisting of the following moves has a set of P-positions identical to the set S_k (in all cases, $n, s, t \in \mathbb{Z}^+$):*

Type I (**Nim moves**): $(x, y) \to (x - n, y)$ *or* $(x, y) \to (x, y - n)$.

Type II (**extended diagonal moves**): $(x, y) \to (x - s, y - t)$ *provided that* $|s - t| < k$. *These moves are identical to the moves in k-Wythoff Nim.*

Type III (**extra moves**): *For $i = 1$ to $k - 1$, use the initial value $(f_0^i, g_0^i) = (0, i + 1)$ and define recursively for $n > 0$,*

$$(f_n^i, g_n^i) = (f_{n-1}^i + g_{n-1}^i, kf_{n-1}^i + (k+1)g_{n-1}^i + i).$$

The extra moves for each i are $(f_n^i, g_n^i - 1)$ for $n > 0$.

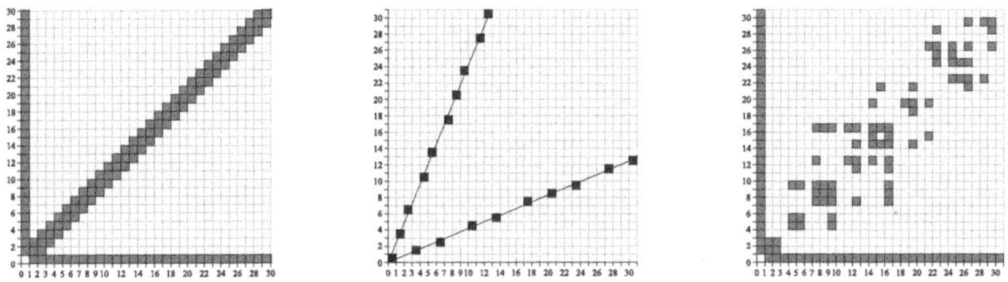

Figure 2. To the left we display the initial moves of the classical game of 2-Wythoff Nim and in the middle its initial P-positions together with the corresponding slopes. To the right we give the initial invariant moves for the game (2-Wythoff Nim)**, with notation as in [LHF], which has P-positions of the same form as those of 2-Wythoff Nim. (The moves are defined via a simple greedy algorithm.)

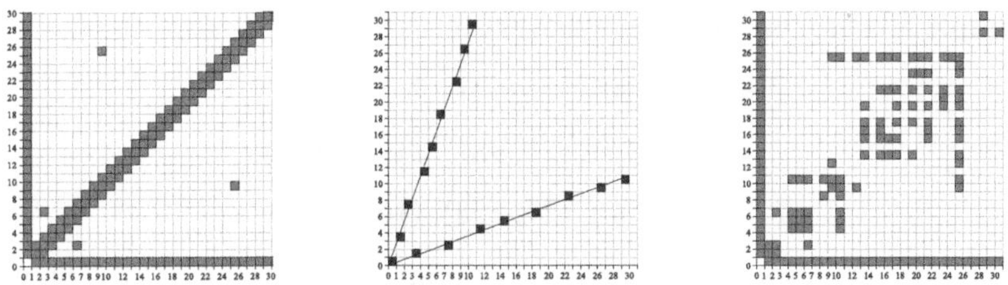

Figure 3. The left-most figure displays the initial moves of our game for $k = 2$ as given in Theorem 3 (see also Example 4 for the extra moves). In the middle we see the P-positions and to the right the initial moves of the invariant game from [LHF] with P-positions identical to the set S_2.

Note that when $n = 0$, the move $(0, i)$ is already in the ruleset as it is a Nim move. In Section 2, we will need to back up the recursion one step and use $(f^i_{-1}, g^i_{-1}) = (-1, 1)$ for all i.

Example 4. For $k = 2$, (and therefore $i = 1$), the extra moves are

$$(2, 6), (9, 25), (35, 96), \ldots$$

An explicit formula for the Type III moves (f^1_n, g^1_n) is given by

$$f^1_n = \frac{(1+\sqrt{3})(2+\sqrt{3})^n + (1-\sqrt{3})(2-\sqrt{3})^n - 2}{4}, \tag{4}$$

$$g^1_n = \frac{(2+\sqrt{3})^{n+1} + (2-\sqrt{3})^{n+1} - 2}{2}. \tag{5}$$

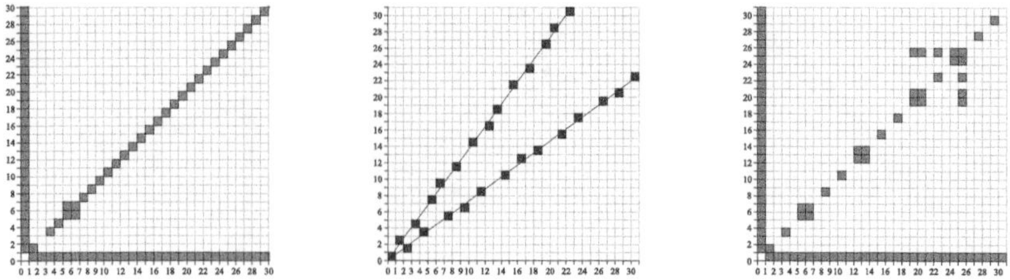

Figure 4. These figures represent the moves (left figure) resolving the P-positions (middle figure) given by the continued fraction (3) with $k = 2$ from [DR]. The right-most figure gives the initial moves of the invariant game from [LHF] with identical P-positions.

Explicit formulas can be found for larger values of k, but are not as succinct, and therefore we have chosen to omit them.

Example 5. For $k = 4$, the extra moves are

$$i = 1 : (2, 10), (13, 63), (77, 372), \dots$$

$$i = 2 : (3, 16), (20, 98), (119, 576), \dots$$

$$i = 3 : (4, 22), (27, 133), (163, 780), \dots$$

The extra moves are necessary for positions of the form $(a_n, b_n - 1)$, where $c_n = 1$ and $d_n = k + 2$. For instance, when $k = 4$, we seek a winning move from the N-position (38, 185) where (38, 186) is a P-position. The previous P-positions are (37, 180), (36, 174), (35, 169) and (33, 163) with differences from one P-position to its predecessor of (1, 6), (1, 6), (1, 5) and (2, 6) respectively. Preceding the nearest lesser difference of (2, 6) are *two* copies of (1, 6) (ignoring the (1, 5)). The winning move uses the largest valid move from the extra move set with $i = 2$, namely the move (20, 98) which moves from (38, 185) to the P-position (18, 87).

The next section develops the machinery to examine these positions and corresponding moves. The final section shows that the rules described in Theorem 3 produce the prescribed set of P-positions.

Remark 6. In the case $k = 1$, the set S_1 is the set of P-positions in 1-Wythoff Nim. The ruleset in Theorem 3 is precisely the ruleset for 1-Wythoff Nim since when $k = 1$, there are no moves of Type III.

Remark 7. In view of Figures 2 and 3, one can see that there is in fact a very succinct description of our games as a modified greedy algorithm. Given an S_k-set of candidate P-positions, the algorithm starts with the moves as in k-Wythoff Nim as a *base set of moves* and then greedily (use for example lexicographic ordering) adjoins an ordered pair of nonnegative integers (x, y), which does not belong to the candidate set of P-positions, as a new move if and only if the move options already defined do not suffice to find a move from (x, y) to any (candidate) P-position. The new move set will be identical to our move set as in Theorem 3.

In this context one might want to explore other complementary Beatty sequences (forming candidate P-positions) and try and describe when similar greedy algorithms define closed formula move sets similar to the ones studied in this paper.

2. The Sturmian word and morphism construction of the Beatty sequence

Here, we lay the groundwork for finding Type III winning moves for positions of the form $(a_n, b_n - 1)$, where $c_n = 1$ and $d_n = k+2$. If the reader wishes to become more familiar with the main structure of the proof of Theorem 3 before reading this section, the details are given in Section 3; all but the one most intricate cases are proved without reference to Section 2. Here we use some terminology from *Sturmian words* and *morphisms* [L]. After some preliminaries, we produce the *characteristic word* which corresponds to the D sequence (this is Lemma 11 which is proved in the Appendix) and thereby gives an alternative description of the B sequence. From it, we find a new characterization of the C and A sequences and note some important properties. Finally, we give an algorithm for finding the desired winning move in Lemma 26.

Lemma 8. *For all $n \in \mathbb{N}$ and $i \in \{1, \dots, k-1\}$, $g^i_{n+1} = (k+2)g^i_n - g^i_{n-1}$.*

Proof. We have that

$$\begin{aligned} g^i_{n+1} &= (k+1)g^i_n + kf^i_n + i \\ &= (k+2)g^i_n - g^i_n + kf^i_n + i \\ &= (k+2)g^i_n - (kf^i_{n-1} + (k+1)g^i_{n-1} + i) + k(f^i_{n-1} + g^i_{n-1}) + i \\ &= (k+2)g^i_n - g^i_{n-1}. \end{aligned}$$

□

2.1. ***The sequence $D = (d_1, d_2, d_3, \dots)$.*** We wish to describe the sequence $D = (d_1, d_2, d_3, \dots)$ via the Sturmian word produced by the morphism

$$\begin{aligned} \varphi(\sigma) &= \sigma\tau^k = \sigma\tau\tau\tau\dots\tau \quad (k \text{ copies of } \tau), \\ \varphi(\tau) &= \sigma\tau^{k+1} = \sigma\tau\tau\tau\dots\tau \quad (k+1 \text{ copies of } \tau), \\ \varphi(uv) &= \varphi(u)\varphi(v). \end{aligned}$$

for any words u, v consisting of the letters σ, τ where the operation is concatenation.

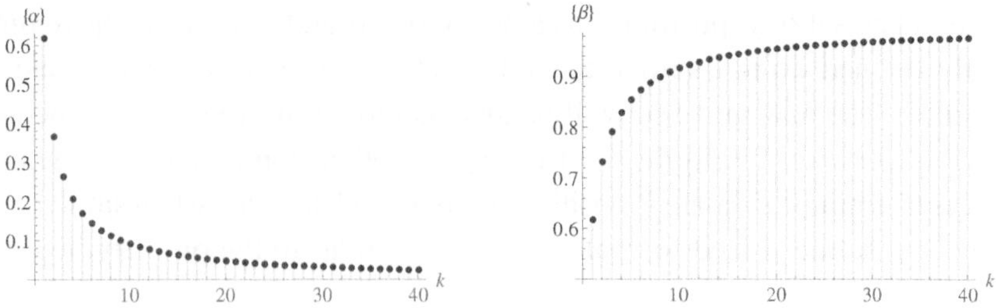

Figure 5. The fractional parts $\{\alpha_k\}$ and $\{\beta_k\}$, for $k \in \{1, 40\}$.

Notation 9. Let w_0 be the word σ, and $w_n = \varphi(w_{n-1})$. Note that w_{n-1} is a prefix of w_n so that

$$W = \lim_{n\to\infty} \varphi^n(w_0)$$

is well-defined.

Example 10. For $k = 2$,

$$W = \sigma\tau\tau\sigma\tau\tau\tau\sigma\tau\tau\tau\sigma\tau\tau\sigma\tau\tau\tau\sigma\tau\tau\tau\sigma\tau\tau\tau\sigma\tau\tau\sigma\tau\tau\tau\ldots$$

Lemma 11. *For all k, $d_j = k+1$ if the j-th letter of W is σ and $d_j = k+2$ if the j-th letter of W is τ.*

We give the proof of Lemma 11 in the Appendix.

2.2. *The sequence* $C = (c_1, c_2, c_3, \ldots)$. Next, we give an equivalent construction of the sequence $C = (c_1, c_2, c_3, \ldots)$ via the D sequence.

Notation 12. The *remainder* or *fractional part* of $x \in \mathbb{R}$ shall be denoted $\{x\} = x - \lfloor x \rfloor$.

Lemma 13. *For all k, $\{\alpha_k\} < 1/k$.*

Proof. Since $1 < \alpha < 2$, $\{\alpha\} = \alpha - 1 = \frac{1}{2}(\sqrt{1+4/k} - 1)$. Then

$$(k+2)^2 > k^2\left(1+\frac{4}{k}\right) \Rightarrow \frac{k+2}{k} > \sqrt{1+\frac{4}{k}} \Rightarrow \frac{1}{k} > \frac{1}{2}\left(\sqrt{1+\frac{4}{k}} - 1\right) = \{\alpha\}. \quad \square$$

Lemma 14. *For all k, $\{\beta_k\} = k\{\alpha_k\}$.*

Proof.

$$\beta = k\alpha + 1 = k + 1 + k\{\alpha\} \Rightarrow \{\beta\} = k\{\alpha\}$$

since $\{\alpha\} < 1/k \Rightarrow k\{\alpha\} < 1$. □

We now compare the sequences C and D, first via an example.

Example 15. For $k=4$,

$$\begin{pmatrix} C \\ D \end{pmatrix} = \begin{pmatrix} 1\,1\,1\,1\,2\,1\,1\,1\,1\,2\,1\,1\,1\,1\,2\,1\,1\,1\,1\,2\,1\,1\,1\,1\,2\,1\,1\,1\,2\,1\,1\,1\,1\,2\,1 & \\ 5\,6\,6\,6\,6\,5\,6\,6\,6\,6\,6\,5\,6\,6\,6\,6\,6\,5\,6\,6\,6\,6\,6\,5\,6\,6\,6\,6\,6\,5\,6\,6\,6\,6\,5 & \cdots \end{pmatrix}$$

If we remove each 5 in the D sequence and the corresponding 1 in the C sequence, what remains in the C sequence is periodic with the value 2 in positions 4, 8, 12, ... and the value 1 otherwise. It turns out that this observation corresponds to an alternative description of the C sequence provided by the D sequence for general k.

Lemma 16. *Suppose that $c_p = c_q = 2$ for $p > 0$ and $q > p$ minimal. Then there are exactly $k-1$ values of i, $p < i < q$ for which $d_i = k+2$.*

Proof. Since $c_p = a_p - a_{p-1} = 2$, Lemma 13 gives $\{p\alpha\} < 1/k$. Let $i \in \{p+1, \dots, q-1\}$ so that $c_i = 1$ and so (the latter inequality is by Lemma 13),

$$0 < \{\alpha\} = \{i\alpha\} - \{(i-1)\alpha\} < 1/k. \tag{6}$$

By Lemma 14, this gives $\{\beta\} = k\{i\alpha\} - k\{(i-1)\alpha\}$. Now, as we have seen, going from b_{i-1} to b_i produces either the difference $d_i = k+1$ or $k+2$. Then, by $\beta \in (k+1, k+2)$, it is clear that the greater value will be attained if and only if there is a $j \in \{1, \dots, k-1\}$ such that $\{(i-1)\alpha\} < j/k < \{i\alpha\}$.

By the last inequality in (6), each j will correspond to a unique i. Hence $d_i = k+2$ occurs exactly $k-1$ times between consecutive occurrences of $c_n = 2$. □

Lemma 17. *Let $k \in \mathbb{Z}^+$. Then $c_n = 2$ if and only if $d_n = k+2$ and $b_n \equiv n \pmod k$.*

Proof. Suppose now that $d_i = k+1$ so that

$$0 < \{\beta\} = \{i\beta\} - \{(i-1)\beta\} < 1. \tag{7}$$

Then $0 < k\{\alpha\} = \{ik\alpha\} - \{(i-1)k\alpha\} < 1$. If in addition $c_i = 2$ we get that $0 < \{ik\alpha\} = k\{i\alpha\} < 1$, so that $0 < \{i\alpha\} - \{\alpha\} = \{(i-1)k\alpha\}/k < 1/k$. This gives that $\{i\alpha\} > 1/k$ so that $c_i = 1$. Thus, we have proved that $c_n = 2$ implies $d_n = k+2$. But then Lemma 16 gives that $b_n - (k+1)n \equiv 0 \pmod k$, for each n such that $c_n = 2$. □

For record keeping purposes, we index the τ in the word W with period k so that

$$W = \sigma\tau_1 \dots \tau_k \sigma\tau_1 \dots \tau_k \tau_1 \sigma\tau_2 \dots$$

Definition 18. A *syllable* of W is a string of letters of the form $\varphi(\sigma)$ or $\varphi(\tau)$; that is, it begins with σ, and ends with the τ which precedes the next σ.

Thus the morphism φ maps letters to syllables. Note that the indexing for the $\varphi(\sigma)$ will depend on the preceding syllable, but each index will appear exactly once. Hence, for all i, we get that

$$\varphi(\tau_i) = \sigma\tau_i\tau_{i+1}\dots\tau_k\tau_1\dots\tau_i.$$

Using this notation, Lemma 17 states that

$$c_n = 2 \text{ if and only if the } n\text{-th letter of } W \text{ is } \tau_k. \tag{8}$$

2.3. *Sums of factors.*

Definition 19. A *factor* of a word is a sequence of consecutive letters. If the factor begins with the first letter of the word, the factor is called a *prefix*. If the factor contains the last letter of a finite word, the factor is called a *suffix*.

Definition 20. For each $i \in \{1, \dots, k-1\}$, let w_0^i be the word $\sigma\tau_1\dots\tau_i$ and $w_n^i = \varphi(w_{n-1}^i)$.

Lemma 21. *For each $i \in \{1, \dots, k-1\}$ and all $n \geq 0$, f_n^i counts the number of copies of τ_k in the word w_n^i and g_n^i counts the number of letters. Note that for $n \geq 1$, g_{n-1}^i counts the number of syllables in the word (which equals the number of copies of σ in w_n^i by construction).*

Proof. Base case: $f_0^i = 0$ and w_0^i contains no τ_k; w_0^i contains $i+1$ letters and $g_0^i = i+1$.

Induction: The morphism φ sends each τ_k to a syllable containing two τ_k and all other letters to a syllable containing a single τ_k, hence the number of copies of τ_k in w_{n+1}^i is $2f_n^i + (g_n^i - f_n^i) = f_{n+1}^i$. The number of letters in the new word is $k+2$ for each letter subtracting one for each σ for a total of $(k+2)g_n^i - g_{n-1}^i = g_{n+1}^i$ by Lemma 8. □

Lemma 22. *A factor of W of length g_n^i contains either g_{n-1}^i or $g_{n-1}^i - 1$ copies of σ. No other number is attainable.*

Proof. By construction, w_n^i has length g_n^i and has g_{n-1}^i copies of σ so g_{n-1}^i is attainable. W is a Sturmian word, and therefore balanced; hence, only one other value is attainable, either $g_{n-1}^i - 1$ or $g_{n-1}^i + 1$. Shift $k+2$ steps to the right in w_n^i. Then we lose two copies of σ and gain one. Hence $g_{n-1}^i - 1$ is the correct value. □

Lemma 23. *For each $i \in \{1, \dots, k-1\}$, and for all $n \geq 1$, (f_n^i, g_n^i) is a P-position.*

Proof. Let $j = g_{n-1}^i$, which is the length of w_{n-1}^i by Lemma 21. By (8), the number of copies of τ_k plus the number of letters in w_{n-1}^i is a_j. By Lemma 21 this equals $f_{n-1}^i + g_{n-1}^i = f_n^i$. By construction and Lemma 8, $b_j = (k+2)j -$ (the number of copies of σ in $w_{n-1}^i) = g_n^i$. □

Definition 24. A factor of W has *index* i if it ends with τ_i for some $i \in \{1, \dots, k-1\}$. A P-position (a_n, b_n) has *index* i if the prefix of W of length n has index i.

Lemma 25. *For a fixed index i, let x be a factor of the word w^i_{n+2}, with the following properties:*

- *x has length g^i_{n+1},*
- *x is not the suffix of w^i_{n+2},*
- *x ends in τ_i.*

Then x contains precisely g^i_n copies of σ. By construction, two equal length factors of W with the same index and the same number of copies of σ will correspond to two equal length factors of C with the same number of copies of 2. Hence the two factor sums in C are equal and the two factor sums in D are equal.

Proof. Note that the statement is vacuous if $n = -2$.

Base case: $n = -1$: If $n = -1$, then x has length $i + 1$; w^i_1 has $i + 1$ syllables, with τ_i in position $i+1$ in the first syllable and in position $i+3-s$ in syllable s for $2 \le s \le i+1$. Hence, the $i+1$ letters ending in τ_i always contain exactly one σ.

Induction: If the terminal τ_i of the factor x is the last letter of a nonterminal syllable of w^i_{n+2}, then the factor contains exactly g^i_n syllables since the terminal τ_i was a result of the output of $\varphi(\tau_i)$, and the previous word w^i_{n+1} has precisely g^i_{n-1} copies of σ in a factor of length g^i_n by induction.

If the terminal τ_i is not the last letter in its syllable, then compare the factor x with the nearest previous factor y for which the terminal τ_i is the last letter in its syllable. If the factors x and y overlap so that there exist nonempty words t, u, v with $y = tu$, $x = uv$, we need to show that the number of copies of σ in t equals the number of copies of σ in v.

Let j be the index of the syllable containing the terminal τ_i of the factor x. If there is no syllable $\varphi(\sigma)$ in v, then the length of v is $(k+1)m$ where $m = j - i$ if $j > i$ and $m = k + j - i$ if $j \le i$; v contains $m - 1$ full syllables plus the terminal partial syllable. Each full and partial syllable contains one σ, so v contains m copies of σ. In other words, the fraction of letters in v which are σ is $1/(k+1)$. If v does contain a syllable $\varphi(\sigma)$, this ratio is unchanged.

For any integer m, the number of copies of σ in any factor of length $(k+1)m$ cannot exceed this ratio since the length of each syllable is $\ge k+1$. Since the number of copies of σ in y is g^i_n, which is maximal by Lemma 22, the number of copies of σ in x, which is at least as many as in y, must also be g^i_n; hence, the number of copies of σ in y equals the number of copies of σ in x.

In the case that x and y do not overlap, note that the maximum distance that x needs to be shifted occurs when the terminal τ_i of x is the leading τ in a syllable ending in τ_i and that this distance is $k-1$ syllables of length $k+2$ plus perhaps a syllable of length $k+1$ plus 2 for a total of $[(k-1)(k+2)]+[k+1]+2=(k+1)^2 < g_2^i$; thus in the induction step, x and y do not overlap only for $n=0$. In this case, y has one syllable of length $k+1$ and i syllables of length $k+2$ yielding $i+1$ copies of σ. We know x has its terminal partial syllable of length $\leq k+1$; thus x has at least as many copies of σ as does y, and since the number of copies of σ in y is maximal by Lemma 22, the number of copies of σ is the same in x and y. □

At the beginning of this section we promised an algorithm for finding a certain winning move. We deliver it here.

Lemma 26. *Let $(x, y) = (a_n, b_n)$ be a P-position with index $i \in \{1, \dots, k-1\}$. From the position $(x, y-1)$, the Type III move (u, v) corresponding to i with $v \leq b_n - 1$ maximal is to a P-position.*

Proof. Find m such that $g_m^i \leq b_n < g_{m+1}^i$. In the first case, if $b_n = g_m^i$, then from $(x, y-1)$, the extra move $(f_m^i, g_m^i - 1)$ is to $(0, 0)$ by Lemma 23. In all other cases, Lemma 25 shows that all factors with index i and length g_{m-1}^i, except the last, in the word w_m^i have the same number of copies of σ; hence the factor sums in C and D have the same sums as in the first case, namely f_m^i and $g_m^i - 1$. Hence the move $(f_m^i, g_m^i - 1)$ is to the P-position (a_{n-j}, b_{n-j}), where $j = g_{m-1}^i$. □

3. The rules are correctly defined

In this section, we prove Theorem 3; that is, we verify that the set S_k is generated as the complete set of P-positions by the ruleset Γ_k.

Definition 27. The *gap* of a P-position (a_n, b_n), denoted δ_n, is

$$\delta_n = b_n - a_n.$$

The *gap difference* between two P-positions (a_m, b_m) and (a_n, b_n) with $m > n$, denoted $\Delta(m, n)$ is

$$\Delta(m, n) = \delta_m - \delta_n.$$

We must check that there is no move connecting any two P-positions (such a "short-circuit" would force at least one of the P-positions to be de facto N and so we had to exclude it from the set S_k) and that every N-position has a move to a position in the set S_k (for otherwise one of the N-positions would be de facto P, and so we had to include it to the set S_k).

Proof, part I—no move connects P to P. By the complementarity of the Beatty sequences, moves of Type I cannot connect any two P-positions.

Note that $\Delta(m, m-1) = k$ or $\Delta(m, m-1) = k+1$. Recall that $a_m - a_{m-1} \leq 2$, so moves of Type II cannot connect (a_m, b_m) and (a_{m-1}, b_{m-1}). If $m-n > 1$, then $\Delta(m, n) \geq 2k$, so moves of Type II cannot connect P-positions in this case either.

It remains to justify that moves of Type III never connect two P-positions. Let (p, q), $p < q$ be an extra move so that $(p, q+1) = (a_i, b_i)$ for some positive integer i, by Lemma 23. From the P-position (a_m, b_m), it is clear that $(a_m - q, b_m - p)$ is not a P-position since $(b_m - p) - (a_m - q) > \delta_m$ and the gap must decrease.

To show that $(a_n - p, b_n - q)$ is not a P-position, assume the contrary, and note that

$$\begin{aligned}(a_n - p, b_n - q) &= (a_n - a_i, b_n - b_i + 1)\\ &= (\lfloor n\alpha\rfloor - \lfloor i\alpha\rfloor, \lfloor n\beta\rfloor - \lfloor i\beta\rfloor + 1). \end{aligned} \tag{9}$$

We have

$$\begin{aligned}\lfloor (n-i)\alpha\rfloor &= (n-i)\alpha - \{(n-i)\alpha\}\\ &= n\alpha - i\alpha - \{(n-i)\alpha\}\\ &= \lfloor n\alpha\rfloor + \{n\alpha\} - \lfloor i\alpha\rfloor - \{i\alpha\} - \{(n-i)\alpha\}.\end{aligned}$$

But then

$$\lfloor (n-i)\alpha\rfloor - \lfloor n\alpha\rfloor + \lfloor i\alpha\rfloor = \{n\alpha\} - \{i\alpha\} - \{(n-i)\alpha\} \tag{10}$$

must be an integer, and is therefore either 0 (if $\{n\alpha\} \ge \{i\alpha\} + \{(n-i)\alpha\}$) or -1 (if $\{n\alpha\} < \{i\alpha\} + \{(n-i)\alpha\}$).

Case 1: $\{n\alpha\} - \{i\alpha\} - \{(n-i)\alpha\} = 0$. Then, (9) and (10) give that

$$a_n - p = \lfloor n\alpha\rfloor - \lfloor i\alpha\rfloor = \lfloor (n-i)\alpha\rfloor = a_{n-i}.$$

Hence, for $(a_n - p, b_n - q)$ to be a P-position, we must have

$$\begin{aligned}b_n - b_i + 1 = b_{n-i} &= \lfloor (n-i)\beta\rfloor\\ &= \lfloor b_n + \{n\beta\} - b_i - \{i\beta\}\rfloor\\ &= b_n - b_i + \lfloor \{n\beta\} - \{i\beta\}\rfloor,\end{aligned}$$

but $\lfloor \{n\beta\} - \{i\beta\}\rfloor$ cannot be 1.

Case 2: $\{n\alpha\} - \{i\alpha\} - \{(n-i)\alpha\} = -1$. Then (10) gives that

$$\lfloor n\alpha\rfloor - \lfloor i\alpha\rfloor = \lfloor (n-i)\alpha\rfloor + 1.$$

By the latter expression, this number, which is strictly greater than zero, can belong either to the set A or B. If

$$\lfloor (n-i)\alpha\rfloor + 1 = b_x \in B,$$

then, for

$$(a_n - p, b_n - q) = (\lfloor n\alpha\rfloor - \lfloor i\alpha\rfloor, \lfloor n\beta\rfloor - \lfloor i\beta\rfloor + 1)$$

to be a nontrivial P-position, we must have that $a_x = b_n - q$ and $b_x = a_n - p$. But, since for all $x > 0$, $b_x > a_x$, this gives

$$a_n - a_i > b_n - b_i + 1,$$

which is false, since $\delta_n > \delta_i$ if $n > i$.

Otherwise,

$$\lfloor n\alpha \rfloor - \lfloor i\alpha \rfloor = a_{n-i+1} \in A$$

so that, by (9) and the definition of a P-position, we must have

$$b_n - b_i + 1 = b_{n-i+1} = \lfloor (n-i+1)\beta \rfloor. \tag{11}$$

However,

$$\begin{aligned} \lfloor (n-i+1)\beta \rfloor &= \lfloor b_n + \{n\beta\} - b_i - \{i\beta\} + b_1 - \{\beta\} \rfloor \\ &= b_n - b_i + b_1 + \lfloor \{n\beta\} - \{b_i\} - \{b_1\} \rfloor. \end{aligned} \tag{12}$$

The last term is either 0 or -1. There are no moves of Type III for the case $k = 1$; thus $k \ge 2$ and $\beta > 3$. Therefore $b_1 \ge 3$, which gives $b_1 + \lfloor \{n\beta\} - \{i\beta\} - \{\beta\} \rfloor \ne 1$, which, by (12), contradicts (11).

Proof, part II — every n has a move to a P. Assume in all cases that $x \le y$.

If (x, y) is an N-position and either $x \in B$ or $y \in B$, then there is a Nim move (Type I) to a P-position. If $x = a_n \in A$, $y \in A$, with $y > b_n$, the Nim move lowering y to b_n is to a P-position.

If $x = a_n$, $y \in A$, $y < b_n - 1$, then $y - x \le \delta_n - 2$. Since the gaps δ_j increase by either k or $k+1$ as j increases by 1, then there is an extended diagonal move (Type II) to a P-position corresponding to δ_j which is nearest $y - x$.

What remains to be shown are winning moves from $x = a_n$, $y = b_n - 1$. If $a_n = a_{n-1} + 2$ or $b_n = b_{n-1} + k + 1$, then the extended diagonal move $(2, k+1)$ or $(1, k)$ moves to the P-position (a_{n-1}, b_{n-1}). Otherwise, Lemma 26 finds the winning Type III move. □

Appendix

All notation in the appendix is local unless stated otherwise. We use theory from [Gl] (further references are given in [Gl]). The words are defined on the alphabet $\{0, 1\}$. For $k \ge 2$ an integer, we are interested in the morphism

$$\theta : 0 \to 1^k 0 \tag{A.1}$$

$$1 \to 1^k 01, \tag{A.2}$$

which we will show corresponds to the positive root

$$\gamma = \frac{\sqrt{k^2+4k}-k}{2} \in \left(\tfrac{1}{2}, 1\right) \tag{A.3}$$

of $x^2 + kx - k = 0$. Namely, by a result in [Gl], we will obtain that

$$\lim_{n\to\infty} \theta^n(1)$$

is the characteristic word c_γ of γ. The density of (the number of 1's in) c_γ is of course γ. Also β as defined in Section 1 equals $\gamma+k+1$ (that is $\gamma = \{\beta\}$). Hence the continued fraction expansion of γ is $\gamma = [0;\, 1, k, \overline{1, k}]$ (where $\bar{x}$ denotes the periodic pattern $x, x, \ldots$).

Let $X = \lim_{n\to\infty} \theta^n(1)$. We show by induction that the Sturmian word $W = \lim_{\to\infty} \varphi(\sigma)$, defined as in Section 2, is identical (via $\sigma \leftrightarrow 0$, $\tau \leftrightarrow 1$) to the word $0X$. That is, we want to show that:

Lemma A.1. *The i-th letter of $0X$ is a 1 if and only if the i-th letter of W is a τ.*

Proof. The first letter in $0X$ and W is 0 and σ respectively; φ acts on its letter, whereas θ does not. Rather, θ acts on the first letter in X. Let us state our induction hypothesis:

Case 1, $x_j = 0$: Then the last letter of the j-th syllable of θ, as in the right-hand side of (A.1), corresponds precisely to the first letter of the $(j+1)$-st syllable of φ.

Case 2, $x_j = 1$: Then the last two letters of the j-th syllable of θ, as in the right-hand side of (A.2), correspond precisely to the first two letters of the $(j+1)$-st syllable of φ.

If these two cases hold for all j, then, by

$$(\text{the length of } \varphi(\sigma)) = (\text{the length of } \theta(0)) = k+1$$

and

$$(\text{the length of } \varphi(\tau)) = (\text{the length of } \theta(1)) = k+2,$$

the infinite words correspond precisely. This follows since then the first k letters in the j-th syllable of X, each a copy of 1, correspond precisely to the last k letters of the j-th syllable of W, each a copy of τ.

Our base case is that the first syllable of X ends with 01 and the second syllable of W begins with $\sigma\tau$, and indeed it holds for the prefixes 01^k01 and $\sigma\tau^k\sigma\tau^{k+1}$ respectively.

But then, comparing the definitions of φ and θ with the paragraph after Case 2, the induction hypothesis gives the claim. □

Define on $\{0, 1\}$ the following three morphisms:

$$E: \begin{matrix} 0 \mapsto 1 \\ 1 \mapsto 0 \end{matrix}, \quad \eta: \begin{matrix} 0 \mapsto 01 \\ 1 \mapsto 0 \end{matrix}, \quad \overline{\eta}: \begin{matrix} 0 \mapsto 10 \\ 1 \mapsto 0 \end{matrix}.$$

A morphism ψ is *Sturmian* if and only if it is a composition of E, η, and $\overline{\eta}$ in any number and order. Furthermore, a morphism ψ is *standard* if and only if it is a composition of E and η in some order. A morphism is *nontrivial* if it is neither E nor the identity morphism.

Suppose $\alpha = [0; 1+d_1, d_2, d_3, \ldots]$, with $d_1 \geq 0$ and all other $d_n > 0$. To the *directive sequence* $(d_1, d_2, d_3, \ldots)$, we associate a sequence $(s_n)_{n\geq -1}$ of words defined by

$$s_{-1} = 1, \; s_0 = 0, \; s_n = s_{n-1}^{d_n} s_{n-2}, \quad n \geq 1.$$

Such a sequence of words is called a *standard sequence*.

For any $n \geq 0$, s_n is a prefix of s_{n+1}, so that $\lim_{n\to\infty} s_n$ is well defined as an infinite word. Moreover, standard sequences are related to characteristic Sturmian words. Each s_n is a prefix of c_α, and we have

$$c_\alpha = \lim_{n\to\infty} s_n.$$

In [Gl], all irrationals $\alpha \in (0, 1)$ such that the characteristic Sturmian word c_α is generated by a morphism are classified. A *Sturm number* is an irrational number $\alpha \in (0, 1)$ that has a continued fraction expansion of one of the following types:

(i) $\alpha = [0; 1+d_1, \overline{d_2, \ldots, d_n}] < \frac{1}{2}$ with $d_n \geq d_1 \geq 1$.

(ii) $\alpha = [0; 1, d_1, \overline{d_2, \ldots, d_n}] > \frac{1}{2}$ with $d_n \geq d_1$.

Observe that if $\alpha = [0; 1+d_1, \overline{d_2, \ldots, d_n}]$ with $d_n \geq d_1 \geq 1$, then

$$1-\alpha = [0; 1, d_1, \overline{d_2, \ldots, d_n}].$$

Hence, α has an expansion of type (i) if and only if $1-\alpha$ has an expansion of type (ii). Accordingly, α is a Sturm number if and only if $1-\alpha$ is a Sturm number and one can show that $c_{1-\alpha}$ is obtained from c_α by exchanging all letters 0 and 1 in c_α, so that

$$c_{1-\alpha} = E(c_\alpha). \tag{A.4}$$

Therefore, we can restrict our attention to characteristic Sturmian words c_α such that α is a Sturm number of type (i).

We say that a morphism ψ *fixes* an infinite word x if $\psi(x) = x$, in which case x is called a *fixed point* of ψ. The following result describes all irrationals $\alpha \in (0, 1)$ such that c_α is a fixed point of a nontrivial morphism.

Theorem A.2 [BS]. *Let $\alpha \in (0, 1)$ be irrational. Then c_α is a fixed point of a nontrivial morphism σ if and only if α is a Sturm number. In particular, if $\alpha = [0; 1+d_1, \overline{d_2, \ldots, d_n}]$ with $d_n \geq d_1 \geq 1$, then c_α is the fixed point of any power of the morphism*

$$\sigma : \begin{array}{l} 0 \mapsto s_{n-1}, \\ 1 \mapsto s_{n-1}^{d_n - d_1} s_{n-2}. \end{array}$$

Now we can prove Lemma 11. Our irrational $\gamma = [0; 1, k, \overline{1, k}]$ (as in (A.3)) is of type (ii) for all k, with $n = 3$, $d_1 = k, d_2 = 1, d_3 = k$, and so we rather apply the theorem to $\alpha = 1 - \gamma = [0; 1+k, \overline{1, k}]$, which is of type (i). For our application, we have that $s_{-1} = 1$, $s_0 = 0$, $s_1 = 0^k 1$ and $s_2 = 0^k 10$, so that the morphism σ in Theorem A.2 corresponds to $0 \to 0^k 10$ and $1 \to 0^k 1$. By (A.4), it is easy to check that the standard morphism $(E\eta)^k \eta$ is identical to $E\sigma = \theta$, so that $E(\lim_{n\to\infty} \sigma^n(0))$ corresponds to the characteristic word c_γ with γ as in (A.3). This concludes the proof. □

References

[B] S. Beatty, Problem 3173, *Amer. Math. Monthly*, **33** (1926) 159.

[BS] J. Berstel and P. Séébold, A characterization of Sturmian morphisms, in: A.M. Borzyszkowski and S. Sokolowski (Eds.), *Mathematical Foundations of Computer Science* 1993, *Lecture Notes in Computer Science*, Vol. 711, *Springer-Verlag*, Berlin, (1993) 281–290.

[DR] E. Duchêne and M. Rigo, Invariant Games, *Theoret. Comp. Sci.*, Vol. 411, 34-36 (2010) 3169–3180.

[F] A. S. Fraenkel, How to beat your Wythoff games' opponent on three fronts, *Amer. Math. Monthly* **89** (1982) 353–361.

[Go] S. W. Golomb, A Mathematical Investigation of Games of Take-Away, *The Journal of Combinatorial Theory*, (1966).

[Gl] A. Glen, Conjugates of characteristic Sturmian words generated by morphisms, *European Journal of Combinatorics archive*, Vol. 25, 7, (2004).

[K] A. Khinchin, *Continued Fractions*. Third edition (English translation). The University of Chicago Press, Chicago (1964).

[L] Lothaire, Algebraic combinatorics on words, Sturmian words, http://www-igm.univ-mlv.fr/berstel/Lothaire/AlgCWContents.html.

[LHF] U. Larsson, P. Hegarty, A. S. Fraenkel, Invariant and dual subtraction games resolving the Duchêne-Rigo Conjecture, *Theoret. Comp. Sci.* Vol. 412, 8-10 (2011) 729–735.

[W] W.A. Wythoff, A modification of the game of Nim, *Nieuw Arch. Wisk.* **7** (1907) 199–202.

[WW] E. R. Berlekamp, J. H. Conway, R.K. Guy, *Winning ways*, **1-2** Academic Press, London (1982). Second edition, **1-4**. A. K. Peters, Wellesley/MA (2001/03/03/04).

urban031@gmail.com — *Department of Industrial Engineering and Management, Technion - Israel Institute of Technology, Haifa, Israel*

weim0024@umn.edu — *School of Mathematics, University of Minnesota, Minneapolis, MN, United States*

Games of No Chance 5
MSRI Publications
Volume **70**, 2017

Endgames in bidding chess

URBAN LARSSON AND JOHAN WÄSTLUND

Bidding chess is a chess variant where instead of alternating play, players bid for the opportunity to move. Generalizing a known result on so-called Richman games, we show that for a natural class of games including bidding chess, each position can be assigned rational upper and lower values corresponding to the limit proportion of money that Black (say) needs in order to force a win and to avoid losing, respectively.

We have computed these values for all three-piece endgames, and in all cases, the upper and lower values coincide. Already with three pieces, the game is quite complex, and the values have denominators of up to 138 digits.

1. Bidding Chess

In chess, positions with only three pieces (the two kings and one more piece) are perfectly understood. The only such endgame requiring some finesse is king and pawn versus king, but even that endgame is played flawlessly by amateur players. In this article we investigate a chess variant where already positions with three pieces exhibit a complexity far beyond what can be embraced by a human.

Bidding Chess is a chess variant where instead of the two players alternating turns, the move order is determined by a bidding process. Each player has a stack of chips and at every turn, the players bid for the right to make the next move. The highest bidding player then pays what they bid to the opponent, and makes a move. The goal is to capture the opponent's king, and therefore there are no concepts of checkmate or stalemate.

As the total number of chips tends to infinity, there is in each position a limit proportion of chips that a player needs in order to force a win. We have computed these limits for all positions with three pieces, and the results (see for example Figure 7) show that already with such limited material, the game displays a remarkable intricacy.

Larsson was partly supported by the Killam Trust (Dalhousie University). Research supported by the Swedish Research Council, the Knut and Alice Wallenberg Foundation, and the Göran Gustafsson Foundation.

MSC2010: 91A46.

Keywords: bidding game, chess, combinatorial game, random turn game, Richman game.

Similar *bidding games* were introduced by David Richman in the 1980s, and presented in [4; 5] only after his tragic death. Bidding chess has been discussed in [1; 2; 3].

There are various reasonable protocols for making bids and handling situations of equal bids [3]. The bids can be *secret*, meaning they are written down on slips of paper and then simultaneously revealed, or *open*, where one player makes a bid and the other chooses between accepting (taking the money) or rejecting (paying the same amount and making a move). The open scheme is in theory equivalent to giving the choosing player a tiebreak advantage.

If the game is played with a small number of chips, the tiebreak scheme and the discreteness of the bidding options can affect the strategy [3]. However, as the number of chips grows, the game approaches a "limit" corresponding to a continuous model where one can bid any real amount. With continuous bidding we may assume that the total amount of money is 1. It turns out then that even a consistent tiebreak advantage is worth less than any positive amount of money.

In the following, we therefore assume that the game is played with continuous money. We will ignore the bidding scheme and the tiebreak rules, since these will affect the outcome under optimal play only when the players' bankrolls are exactly at certain thresholds. Our discussion will focus on analyzing and computing these thresholds.

The bidding player can also make a *negative* bid, meaning that the player who makes the next move will get paid for their trouble. If you have greater bankroll than your opponent, you can therefore either ensure the right to make the next move, or force them to move, by making a sufficiently large negative bid. As we shall see in Section 11, there actually are *zugzwang* positions calling for such bids.

As a "play-game" (rather than "math-game"), we suggest an open scheme where the player who made the last move bids for the next. In the initial position, Black is considered to have made the last move (since it is normally White's turn), and starts the game by bidding for the first move.

Figure 1 shows a position that would be checkmate in ordinary chess. White's

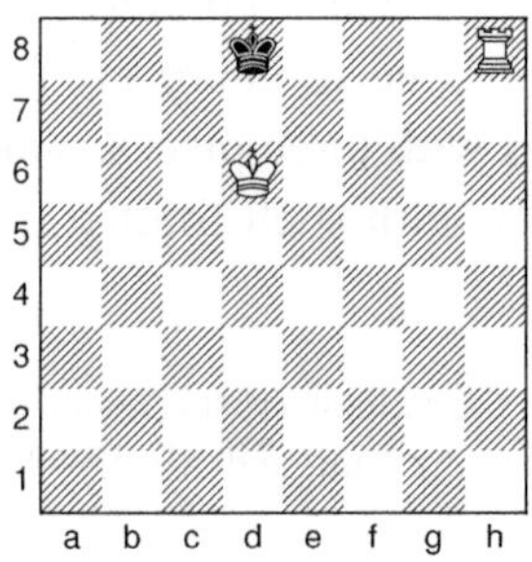

Figure 1. A position with value $\frac{3}{4}$.

rook is threatening the black king, and White will therefore go *all in* (offering all their chips) before the next move. In order for Black to survive, they must win this bid, thereby doubling White's bankroll. Black's best option is now to move their king to one of the squares c7, d7 or e7, adjacent to the white king. At this point, whoever has more money will win the next bid and capture their opponent's king.

The conclusion is that if White's bankroll is larger than $\frac{1}{4}$, they will be able to make one of the next two moves and win, while if it is smaller than $\frac{1}{4}$, Black is able to make two consecutive moves and capture the white king. At the threshold where White has exactly $\frac{1}{4}$ of the money, the outcome depends on the tiebreak scheme, but it still makes sense to say that this position has value $\frac{3}{4}$ for White.

2. Random turn games

One of David Richman's insights was that there is a certain equivalence of bidding games to *random turn* games [4; 5; 6; 7]. In a random turn game, the move order is determined by flipping a coin (just before each move, so you have to make your move before you know who plays next). The position in Figure 1 has value $\frac{3}{4}$ for White also in random turn chess. If White wins the next coin flip, the game is over, while if Black wins it, they will play Kd7 and the next coin flip decides the game.

The equivalence between bidding and random turn games can be understood inductively. If for the moment we disregard the possibility of draws (to which we shall return), we can write $\alpha(P)$ for the probability that White wins the random turn game from a given position P. Let $\alpha(P_w)$ be the probability of White winning from the position P_w obtained after White's best move (that is, conditioning on White winning the next coin flip), and $\alpha(P_b)$ the probability of White winning after Black's best (from their perspective) move. Then provided the coin is fair,

$$\alpha(P) = \frac{\alpha(P_w) + \alpha(P_b)}{2}. \tag{1}$$

There is a bit of circularity in this argument, since the winning probabilities are what defines the "best" moves, so to make the argument rigorous we should consider the probability of white winning in at most n moves, and then take the large n limit (we will return to this issue). But the point is that (1) has an interpretation also for the bidding game. We can think of the values $\alpha(P)$, $\alpha(P_w)$ and $\alpha(P_b)$ as the amounts of money that White can afford Black to have and still win the game. If we know how much money we will need after the next move, both if White and if Black makes that move, then the amount we need in the current position is the average of those two numbers, since we can then bid half their difference and win whether the opponent accepts or rejects. So (1) holds also with that interpretation. By induction, it follows that the amount of money we can allow our opponent to have and still win is the same as our probability of winning the random turn game.

3. Outline

The main results of this study are as follows. Every partizan combinatorial game on a finite number of positions has rational upper and lower values (see Section 6). These values represent, under random turn play, the maximum probabilities that White can obtain of not losing, and of winning, respectively. Under bidding play, the same values represent the amount of money that Black needs in order to force a win and to avoid losing respectively.

For all three-piece chess positions, the upper and lower values coincide. This is proved via computer calculation, see Section 8, and we have no theoretical explanation for why this had to be the case. The calculation revealed that the game is extraordinarily complex. In particular there is a position with king and knight versus king whose (common upper and lower) value has a 138-digit denominator. We also discuss some other positions of special interest; for example the existence of chess positions (with more than three pieces) with distinct upper and lower values (so-called *nontrivial Richman intervals*), and of positions of *zugzwang*, that is, positions requiring negative bids.

4. Examples

In some cases, values of positions in bidding chess can be conveniently calculated by instead analyzing random turn chess. Consider for instance a position with two bare kings; see Figure 2.

In ordinary chess this position is a draw since no king can move to a square adjacent to the other. And if none of the players are willing to take a risk, the random turn game too will be drawn. But a player can guarantee a winning probability of $\frac{1}{2}$ even if the other player is satisfied with a draw. This is a simple

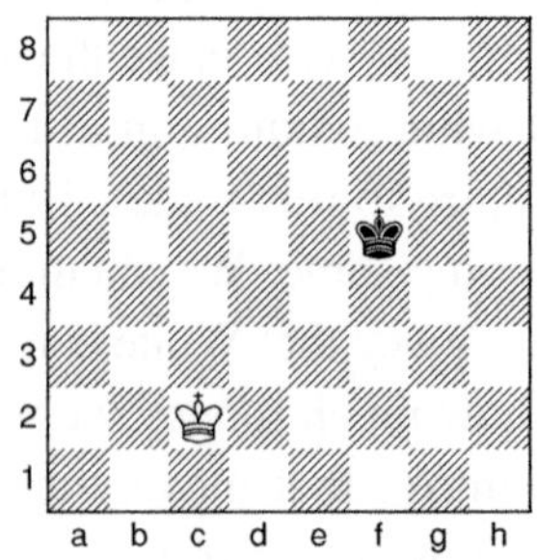

Figure 2. A position with value $\frac{1}{2}$.

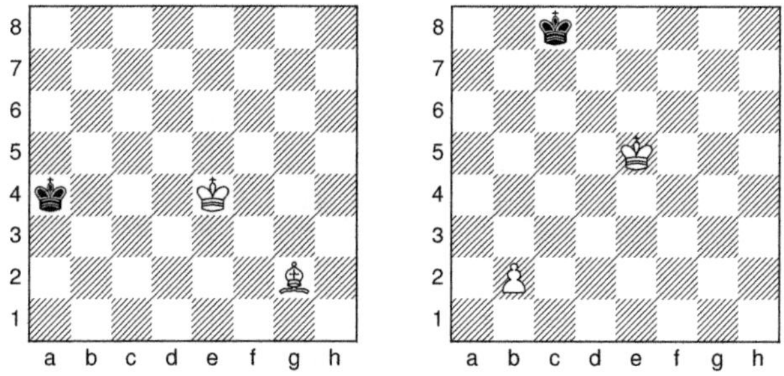

Figure 3. Left: A bishop endgame worth $\frac{9}{16}$. As soon as Black gets to move, the bishop becomes worthless. Right: A pawn endgame worth $\frac{33}{64}$. Whenever Black moves their king to the b-file, the pawn is neutralized.

consequence of the laws of probability: At some point you will get a run of six consecutive moves. Therefore if you consistently move towards your opponent's king every time you get to move, you will at some point be able to get the kings next to each other, giving you a 50% chance of winning on the following turn.

By Richman's equivalence argument it follows that in bidding chess with continuous money, an advantage in bankroll no matter how small will allow you to win the game with two bare kings!

Figuring out how to actually win with a bankroll of $\frac{1}{2} + \epsilon$ is a nice little exercise (the number of moves needed will go to infinity as $\epsilon \to 0$).

Actually that winning strategy can be carried out just as well even if your king is restricted to squares of only one color, say the dark squares. This means that if White has a light-squared bishop (bishop that moves on the light squares) and there are no other pieces except the kings, then as soon as the black king gets to a dark square, the bishop loses its value. It becomes a ghost that can neither attack the black king nor defend the white one.

If we play random turn chess from the position in Figure 3 (left), then in case White wins the first three turns, they can win by playing Bg2–h3–d7xa4 (or any of a number of other ways to capture the black king in three moves). And this is actually the only use White can have of their bishop. If Black wins any of the first three coin flips, they will move their king to a dark square and the chances will be even. White's winning chances are therefore $\frac{1}{8}$ more than Black's, which means that the value is $\frac{9}{16}$.

A similar analysis shows that the position in Figure 3 (right) has value $\frac{33}{64}$. As soon as Black gets to move their king to the b-file, the white pawn will be neutralized, since Black then moves down the b-file and captures the pawn unless White chooses to put the kings next to each other before that. Therefore the only use White can have of their pawn comes from the possibility of capturing the black king in five consecutive moves through b2–b4–b5–b6–b7xc8.

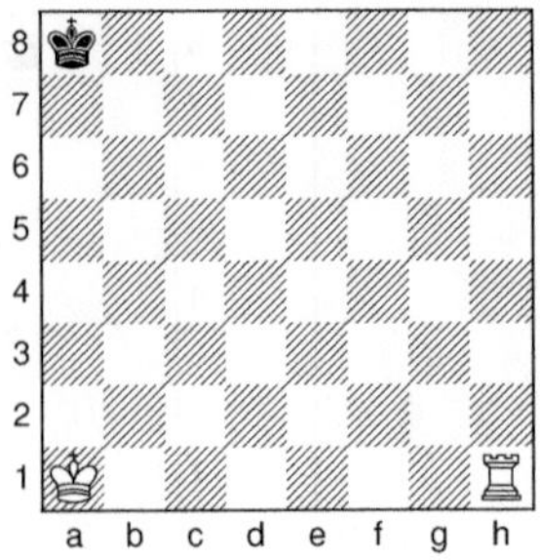

Figure 4. A rook ending worth $1 - 3^6/(2 \cdot 4^6) = \frac{7463}{8192}$.

Other positions can be evaluated by slightly more sophisticated probabilistic arguments. In the position in Figure 4, White's best option is to attack the black king from the side with Rh8. Black on the other hand will play Ka7 (or Kb7), trying to get their king as quickly as possible in close combat with the white king. So the black king will move down the a-file going straight for the white king. Meanwhile, White will attack the black king from the side with the rook whenever they can. In order for Black's plan to work, Black therefore has to succeed in playing six moves (from a8 to a2) without White getting two consecutive turns (in which case the rook would capture the black king), and then winning the final coin flip when the kings are face to face. The probability of Black winning at least one of two coin flips is $\frac{3}{4}$, and therefore the probability of Black's king getting to a2 without being captured by the rook is $\left(\frac{3}{4}\right)^6$. Black's winning chances are therefore $\left(\frac{1}{2}\right) \cdot \left(\frac{3}{4}\right)^6 = \frac{729}{8192}$. Making the analysis rigorous would require dismissing other moves as inferior, but that is relatively straightforward.

5. Finite *n* thresholds and their limits

We have written a computer program (the code, in the *Processing* language, is available on request) that has calculated the values of all positions with three pieces on the board. The program starts by calculating certain thresholds that we now describe.

For each n and each position P, we can define a threshold $\alpha_n(P) \in [0, 1]$ such that if Black's bankroll is strictly smaller than $\alpha_n(P)$, White can force a capture of the black king in at most n moves, while if Black has strictly more money than $\alpha_n(P)$, the black king can survive for at least n more moves. At the exact threshold, the outcome might depend on the tiebreak rule.

These numbers satisfy the recursion

$$\alpha_{n+1}(P) = \frac{\max_w \alpha_n(P_w) + \min_b \alpha_n(P_b)}{2}, \tag{2}$$

where w and b range over the white and black move options from position P, and P_w and P_b are the positions reached from P by these moves. The "boundary conditions" are given by setting $\alpha_n(P)=0$ or 1 respectively (for all n) in positions where the white or black king has already been captured, and $\alpha_0(P)=0$ otherwise.

Similarly we define thresholds $\beta_n(P)$ as the amount of money that Black needs in order to force a capture of the White king in at most n moves. The β-thresholds satisfy the same recursive equation (2) as α, but start from setting $\beta_0(P)=1$ in positions where both kings remain on the board. Notice that both α and β measure the quality of a position from White's perspective in the sense that a higher value is better for White.

It follows by induction that all these values are dyadic rational numbers, that is, rational numbers with a power of 2 in the denominator. For each position P, the sequence $\alpha_n(P)$ is nondecreasing and bounded above by 1. Therefore there is a limit $\alpha(P)$ which is the smallest number such that if Black's bankroll is below $\alpha(P)$, White can force a win. These limits satisfy the same equations:

$$\alpha(P)=\frac{\max_w \alpha(P_w)+\min_b \alpha(P_b)}{2}. \tag{3}$$

Similarly there is a limit $\beta(P)$ of $\beta_n(P)$ which is the amount of money that Black needs in order to force a capture of the white king. It is clear from the definitions that

$$0\le\alpha_0(P)\le\alpha_1(P)\le\alpha_2(P)\le\cdots\le\alpha(P)\le\beta(P)\le \\ \cdots\le\beta_2(P)\le\beta_1(P)\le\beta_0(P)\le 1. \tag{4}$$

In the examples we have discussed, the values α and β have been equal.

But it may also happen (see Section 13) that $\alpha(P)<\beta(P)$, so that if Black's bankroll is in the interval between $\alpha(P)$ and $\beta(P)$, the game is drawn in the sense that none of the players can force the capture of the opponent's king.

For impartial games, such positions (with so-called *nontrivial Richman intervals*) can occur only in games with infinitely many positions, and an example is demonstrated in [5, Figure 10].

The numbers α and β can also be interpreted as the probabilities that White can win, and avoid losing, respectively, in the random turn game. Notice that the strategy that achieves the maximal probability α of winning may be different from the one that achieves the maximal probability β of not losing.

Our computer program starts by calculating the numbers $\alpha_n(P)$ and $\beta_n(P)$ for all positions with three pieces, and n up to several thousand. This requires working with "big integers" since the values are rational numbers with n-bit numerators (and denominator 2^n), but $\alpha_{1000}(P)$ and $\beta_{1000}(P)$ for instance can be computed in a few minutes without any particular optimization.

6. Rationality of the limits $\alpha(P)$ and $\beta(P)$

In the examples of Section 4, all values were dyadic rational numbers, but this need not always be the case. As we shall see in Section 9, there are positions whose values have non-2-power denominators. However, $\alpha(P)$ and $\beta(P)$ are always rational numbers. This holds in general for games with finitely many positions. Suppose therefore that Black and White play a bidding (or random turn) game defined by a finite set of positions, where each position has prescribed sets of white options and black options (other positions to which White and Black can move respectively). Some positions are designated as winning for one of the players.

Suppose also that $\alpha_n(P)$, $\beta_n(P)$ and their limits $\alpha(P)$ and $\beta(P)$ are defined as in Section 5.

Proposition 1. *For every position P, $\alpha(P)$ and $\beta(P)$ are rational numbers.*

Proof. Consider the system of equations

$$x(P) = \frac{\max_w x(P_w) + \min_b x(P_b)}{2}, \tag{5}$$

with the extra constraints that $0 \le x(P) \le 1$, and that $x(P) = 0$ or 1 respectively for positions defined as winning for one of the players (when the white or black king is already captured). In (5) we have only replaced the symbol α in (3) by x to indicate that these are variables in a system of equations. A solution to (5) (including boundary conditions) will be called a *Richman function*, following [4; 5].

It follows by induction on n that α_n is a lower bound on any Richman function and similarly β_n is an upper bound. Therefore among all Richman functions, α simultaneously minimizes all values, and β simultaneously maximizes them.

Since the right-hand side of (5) involves both a minimization and a maximization, the system is inherently nonlinear. But suppose that for each position P we choose (arbitrarily) move options W and B to positions P_W and P_B respectively. Then in order to check whether there is a Richman function for which $x(P_W) = \max_w x(P_w)$ and $x(P_B) = \min_b x(P_b)$ for every P, we can replace (5) by a linear system of inequalities: For each position P and any white and black options P_w and P_b respectively, we impose the constraints

$$x(P) \ge \frac{x(P_w) + x(P_B)}{2} \quad \text{and} \quad x(P) \le \frac{x(P_W) + x(P_b)}{2}, \tag{6}$$

again together with the boundary conditions that $x(P) = 1$ if White has won and $x(P) = 0$ if Black has won.

Despite the apparent slack in (6), every solution to the system (6) is also a solution to the system (5): Assuming that (6) holds, we have

$$x(P) \le \frac{x(P_W) + \min_b x(P_b)}{2} \le \frac{\max_w x(P_w) + \min_b x(P_b)}{2}$$
$$\le \frac{\max_w x(P_w) + x(P_B)}{2} \le x(P).$$

Conversely, every Richman function will yield, by choosing P_W and P_B as a minimizing and maximizing option respectively, a solution to the system (6).

The system (6) is a set of linear constraints, and whenever the set of solutions is nonempty, it is a polytope with vertices in rational points. Since there are only finitely many ways of choosing P_W and P_B, and each resulting system (if solvable) has rational minimum and maximum values for $x(P)$, it follows that $\alpha(P)$ and $\beta(P)$ are rational for every position P. □

Similar arguments are given in [4; 5], but the situation considered in those papers is slightly simpler since for finite impartial games there is only one Richman function. The example in [5, Figure 10] shows that for a game with infinitely many positions, the minimum and maximum Richman functions are not necessarily rational. Although the authors of [5] seem to have overlooked this, in their example, $r(k) = (\sqrt{5} - 1)^k / 2^{k+1}$.

7. Guessing a rational limit

It is obviously not feasible to solve all the linear systems of the form (6) in order to find the value of a position. On the other hand our computer program will calculate the finite n thresholds $\alpha_n(P)$ and $\beta_n(P)$ for all positions with two or three pieces and all $n \le 1000$ (say) in just a few minutes, and this ought to give a good indication of what the limits $\alpha(P)$ and $\beta(P)$ are. The computation reveals that for all three-piece positions,

$$\beta_{1000}(P) - \alpha_{1000}(P) < 10^{-91}.$$

This clearly suggests that $\alpha(P) = \beta(P)$ for all three-piece positions (and we describe in Section 8 how to verify this). If this is correct, then for sufficiently large n, the common value of $\alpha(P)$ and $\beta(P)$ will be the rational number with the smallest denominator in the interval $[\alpha_n(P), \beta_n(P)]$. We do not know of any simple and useful estimates of how large this n has to be, or of how large the denominators of $\alpha(P)$ and $\beta(P)$ can be (they can be fairly large; see Section 10).

But an obvious thing to do is to let $s_n(P)$ be the simplest rational number (the one with smallest denominator) in the interval $[\alpha_n(P), \beta_n(P)]$, and check whether s_n is a Richman function; in other words, whether $x(P) = s_n(P)$ yields a solution to the system (5). The numbers $s_n(P)$ can be computed from $\alpha_n(P)$

and $\beta_n(P)$ using a standard technique based on comparing the continued fraction expansions of $\alpha_n(P)$ and $\beta_n(P)$.

We let our computer program calculate $s_n(P)$ for reasonably small n using exact rational arithmetic, and then counted for each n the number of equations in the system (5) that were violated when putting $x(P) = s_n(P)$. It turns out that at $n = 2644$, this number drops to zero, and all equations are satisfied. Actually s_n stabilizes for the bishop endgame already at $n = 30$, for the queen endgame at $n = 156$, and for the rook endgame at $n = 331$. The bulk of the computation is then devoted to the knight endgames and a smaller set of pawn endgames potentially leading to knight promotion.

8. Verifying that the guesses are correct

We have now found an explicit Richman function $x(P) = s_{2644}(P)$. This shows that $\alpha(P) \le s_{2644}(P) \le \beta(P)$ for every P. We will show that equality holds, but at this point it is still conceivable that some of these inequalities are strict. Notice that since $s_{2644}(P)$ will be in the interval $[\alpha_n(P), \beta_n(P)]$ for every n, we have $s_n(P) = s_{2644}(P)$ whenever $n \ge 2644$ and we may set $s(P) = s_{2644}(P) = \lim_{n\to\infty} s_n(P)$.

Although the number of violated equations in (5) does not consistently decrease as n increases, once it drops to zero so that the system is satisfied, it must remain zero for all larger n.

Whenever x is a Richman function, it provides a certificate that White cannot win random turn chess from a position P with probability larger than $x(P)$, and that analogously Black cannot win with probability larger than $1 - x(P)$. This is because it provides each player with what we might call an x-*greedy* strategy: Each time it is your turn, you choose to move in such a way that you maximize x if you are White, and minimize x if you are Black.

If White follows an x-greedy strategy from a position $P = P_0$, then no matter what strategy Black adopts, the expectation $\boldsymbol{E}[x(P_n)]$ of the value of x at the position P_n reached after n moves will satisfy

$$\boldsymbol{E}[x(P_n)] \ge x(P_0). \tag{7}$$

Here we use the convention that whenever one of the kings is captured, the resulting terminal position will remain on the board at all subsequent times. We consider the White and Black strategies to be fixed, and the expectation is over the results of the coin flips.

It follows from (7) that the probability of Black having won the game after n moves cannot exceed $1 - x(P)$ for any n. Similarly, if Black follows an x-greedy strategy, White cannot win with probability greater than $x(P)$.

To verify that $\alpha = \beta = x$, we would like to obtain a stronger certificate showing that White can actually win with probability $x(P)$, and that Black can win with probability $1 - x(P)$. When the numbers $x(P)$ have been computed explicitly, this can actually be effectively checked.

We assume that we have computed a table of all positions (with up to three pieces) and the values of a Richman function x (in our case $x = s_{2644}$). We describe how to verify that $\alpha(P) = x(P)$ for every P.

We wish to exhibit a strategy for White in the random turn game which is x-greedy and at the same time has the property that regardless of Black's strategy, the game will terminate with probability 1. If there exists such a strategy, then in view of (7), if we play from a position P, White will win with probability at least $x(P)$. Since this is the best possible probability, it follows that $\alpha = x$.

To find such a strategy we define a sequence of sets of positions as follows: Let T_0 be the set of all terminal positions, and for $n \geq 0$, let T_{n+1} be the set of positions that either belong to T_n, or have an x-greedy white move option to a position T_n, or have *all* their black move options to positions in T_n. These sets are defined by a closure operation and we can therefore effectively compute the sequence of sets until they stabilize, by making a table where each position P is labeled with the smallest n for which $P \in T_n$, if there is such an n. When for some n, $T_{n+1} = T_n$, the process stabilizes and we let $T = T_n$.

The strategy for White now consists in always playing x-greedy moves, and whenever there is an x-greedy move option to T, choosing such a move to a position with minimal label, that is, belonging to T_i for the smallest possible i.

Following this strategy, White will guarantee that the positions in T are *transient* in the sense that they will only be visited a finite number of times. This is because whenever we reach a position in T_i (for $i > 0$), either White has *at least one* move to T_{i-1}, or *all* Black's moves lead to T_{i-1}, and in either case there is (at least) a 50% chance that the next move will lead to a position in T_{i-1}.

Consequently, each time we reach a position in T_n, the probability that the game will terminate in another n moves is at least 2^{-n}. Therefore with probability 1, the game cannot visit such a position infinitely many times.

Proposition 2. *We have $\alpha = x \iff$ all positions P with $x(P) > 0$ belong to T.*

Proof. If all positions where x is nonzero belong to T, then no such position can be visited infinitely many times. Consequently the game will either terminate or end up in an infinite sequence of positions where x takes value zero. Since White plays x-greedily, if the game starts from a position P, in view of (7), White must win with probability at least $x(P)$, showing that $\alpha(P) = x(P)$.

Conversely, if there is a position P with $x(P) > 0$ which is not in T, then every White strategy pretending to win with probability given by x is flawed in one of two ways: Either it consequently plays x-greedy moves, in which case it can't win starting from P (since Black can avoid moving into T). Or it does *not* always play x-greedy moves, in case again it can't always win with probability given by x (provided Black plays x-greedily). □

This shows that once we have verified that $s = s_{2644}$ is a Richman function, we can effectively check whether or not $\alpha = s$, and similarly whether or not $\beta = s$. It turns out that in the three-piece endgames, all positions belong to T, which shows that $\alpha = s$. Notice however that the definition of T is not symmetrical with respect to the two players, so that in order to verify that $\beta = s$, we would need to check a set T' defined similarly but from Black's perspective.

In order to verify that all positions belong to T (and similarly to T'), we actually only need to investigate a small set of positions. Let us say that a position P is *quiescent* (relative to x) if

$$\min_b x(P_b) = \max_w x(P_w).$$

Proposition 3. *If x is a Richman function and all positions that are quiescent with respect to x belong to T, then all positions belong to T.*

Proof. This follows by induction on the number of positions that have x-values larger than $x(P)$ for a given position P. Suppose that all quiescent positions belong to T, and that also all positions Q with $x(Q) > x(P)$ belong to T.

If P is quiescent, then by assumption P belongs to T. If P is not quiescent, then either

$$\min_b x(P_b) < x(P) < \max_w x(P_w) \quad \text{or} \quad \max_w x(P_w) < x(P) < \min_b x(P_b).$$

In the former case there is a white option to a position P_w, which must belong to T since $x(P_w) > x(P)$. In the latter case all Black options lead to positions P_b which similarly must belong to T. In either case, P must belong to T. □

After finding the Richman function $x = s_{2644}$, we let our computer program list all quiescent positions relative to this function. They turn out to fall in four categories, three of which were discussed in Section 4:

- *Bare kings*: Only the two kings on the board.
- *Ghost bishop*: The black king and white bishop on squares of opposite color.
- *Blocked pawn*: The black king on the same file as a white pawn, and in front of it.
- *Cornered king*: There are eight positions similar to the one shown in Figure 5, where the white king is trapped in front of its own pawn near a corner and cannot get out without moving next to the black king. For the king to be trapped on the a-file, the white pawn has to be on a6 or a7. If it is on a6, the

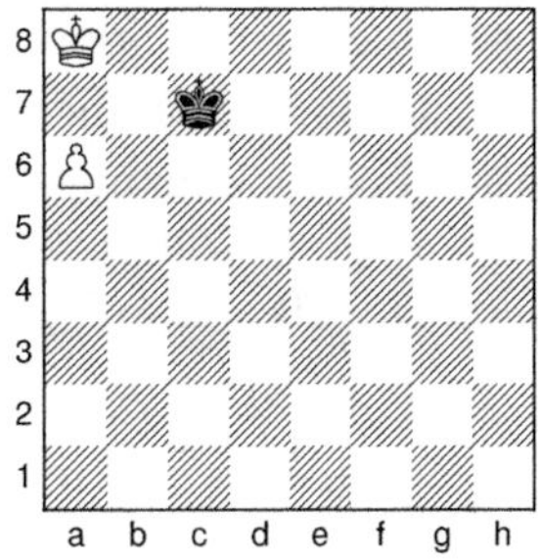

Figure 5. A quiescent position: The white king is cornered and cannot get out without challenging the black king.

black king has to be on c7, and if it is on a7, the black king can be on c7 or c8. And there are four similar positions where the king is trapped on the h-file.

In all these positions, it is easy to see that both White and Black have strategies that win with probability $\frac{1}{2}$ by consistently moving the king towards the opponent's king, except in the *blocked pawn* cases where Black should first capture the white pawn. More precisely, the *bare kings* and *ghost bishop* positions belong to T_7 and T'_7, since a player can capture the opponent's king in at most 7 s-greedy moves with a favorable sequence of coin flips. Similarly the *cornered king* positions belong to T_2 and T'_2, while the *blocked pawn* positions belong to T_7 and T'_{13}, since being restricted to s-greedy moves, it might take Black up to 13 moves to first capture the white pawn and then go after the white king.

We thus conclude that $\alpha(P) = \beta(P) = s_{2644}(P)$ for all three-piece endgames.

9. Nondyadic values

One might have expected from the discussion in Section 4 that all values are dyadic rationals, but this is not the case. The position in Figure 6 with value $\frac{249}{320}$ is the simplest with a non-2-power denominator (on an 8×8 board).

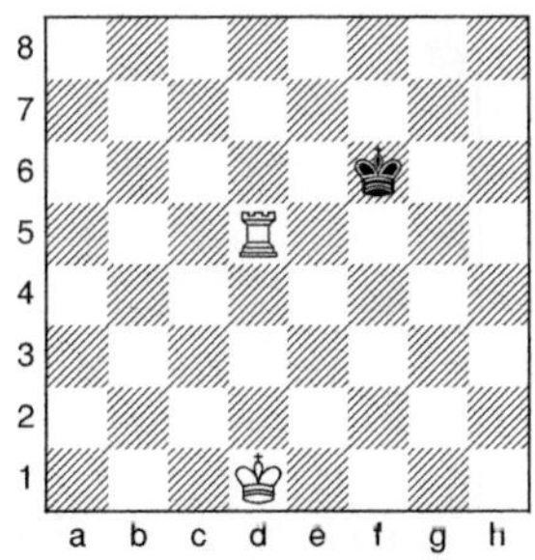

Figure 6. A position with the nondyadic value $\frac{249}{320}$.

Most rook and queen endgames have 2-power denominators. For queen endgames, the largest denominator is 2^{28}, but there are also values with denominators divisible by 3, 5, and 17. For rook endings, the largest denominator is

$$22962750590287872 0 = 2^{40} \cdot 3^3 \cdot 5 \cdot 7 \cdot 13 \cdot 17,$$

but there are also denominators with a factor 251. Bishop endings are relatively simple with only denominators of 1, 2, 4, 8, and 16 occurring.

10. Knight endgames and huge denominators

By far the most complex three-piece endgames are the knight endgames (and some pawn endgames that lead to knight promotion). The largest denominator in a three-piece endgame occurs for the position in Figure 7.

The value of this position is a number with 138-digit numerator and denominator:

$$\frac{\begin{array}{l}118149099210761088839658071450928865980708175943671062283570061370088\\ \qquad\qquad 99029724248731234404879782744818714659268426249519314520276146019737 1\end{array}}{\begin{array}{l}200453006658428905551436939930457127472327950605425153085344343480681\\ \qquad\qquad 727125595119114980629492845444447049929082740309543514434854453248000\end{array}},$$

or approximately 0.5894104617. The denominator factorizes as

$$\begin{aligned}2^{131} \cdot 3^4 \cdot 5^3 \cdot 7^2 \cdot 17 \cdot 211 \cdot 487 & \\ & \cdot 63587 \cdot 68891 \cdot 1894603 \cdot 4248158177642143024 5997\\ & \cdot 240980537473228976453730945188262261414394247399.\end{aligned}$$

It is true that this exact number only governs an idealized version of bidding chess with continuous money, but since the number somehow reflects the pattern of optimal moves, the optimal strategies will likely be very intricate also for a reasonable number of chips (although the optimal moves may vary depending on the number of chips [3]).

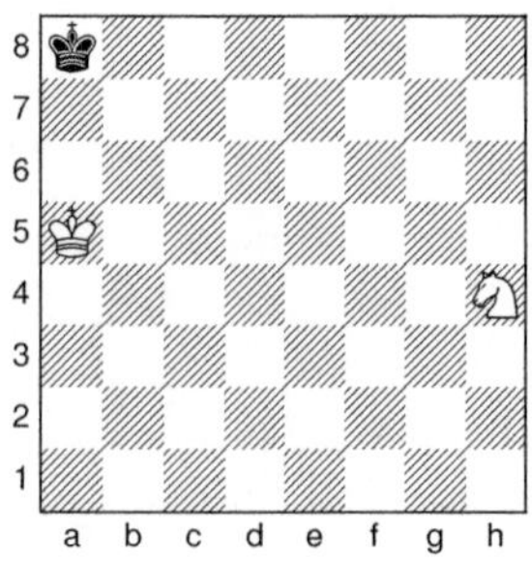

Figure 7. The most complex three-piece ending.

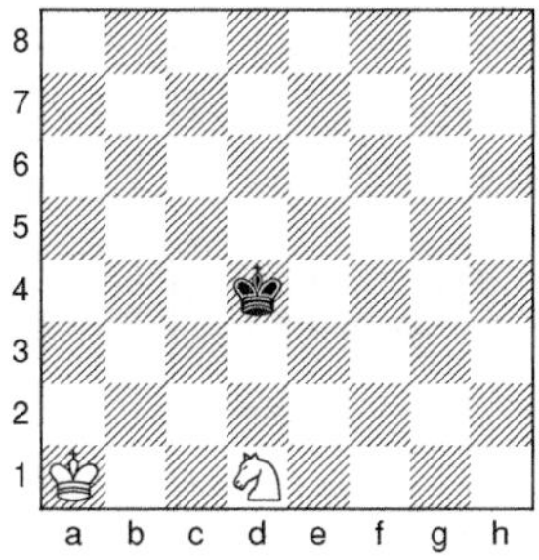

Figure 8. A position of *zugzwang*: neither player wants to move.

11. Zugzwang

It was pointed out in [4] that *impartial* bidding games (so-called Richman games) never require negative bids. This does not hold in general for partizan games [3]. It was speculated in [1] that there might exist positions in bidding chess calling for negative bids, that is, positions where one would prefer the opponent to make the next move. This is indeed the case, and an example is given in Figure 8.

This position has value $\frac{21073}{32256} \approx 0.6533$. White's best move is to "sacrifice" the knight with Nd1–c3, even though this leads to a position of value only $\frac{10489}{16128} \approx 0.6504$. The problem is that a move with the king will bring it closer to the black king, while moving the knight will either put it *en prise* (c3 or e3) or move it further from the black king in the knight's metric (b2 and f2 are four knight moves away from d4). Black's best move is Kd4–c4, bringing the value up to $\frac{21}{32} = 0.65625$.

12. Pawn promotion

Since there is no stalemate in bidding chess, we only need to consider promotion to queen or knight. A rook or bishop can never be better than a queen. In some positions there is only a tiny difference in value between promoting to knight and promoting to a queen. For instance, in the position in Figure 9 (left), White to move should play d8N!, obtaining a position of value $\frac{205}{256} \approx 0.80078$, while a promotion to queen gives a value of only $\frac{3279}{4096} \approx 0.80054$. However, if we move the entire position one step to the right as in Figure 9 (right), the knight-promotion still leads to a position of value $\frac{205}{256}$, while e8Q! gives the slightly higher value of $\frac{3285}{4096} \approx 0.80200$!

13. Positions where $\alpha < \beta$

We have shown that for all three-piece endgames, $\alpha(P) = \beta(P)$, but we have no "theoretical" explanation for why this must be so. There are conditions under

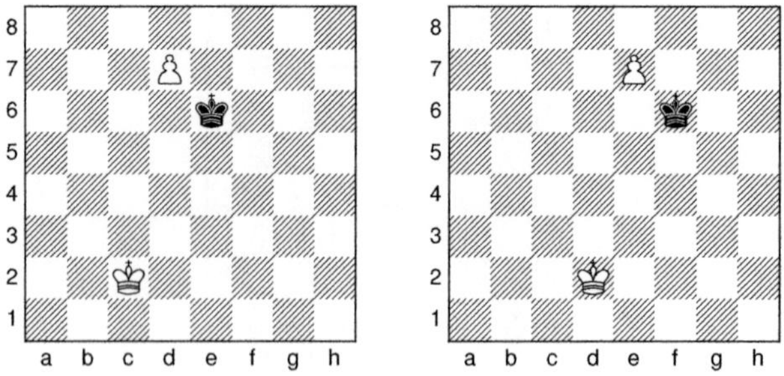

Figure 9. Left: White's best move is to promote to a knight. Right: White should promote to a queen.

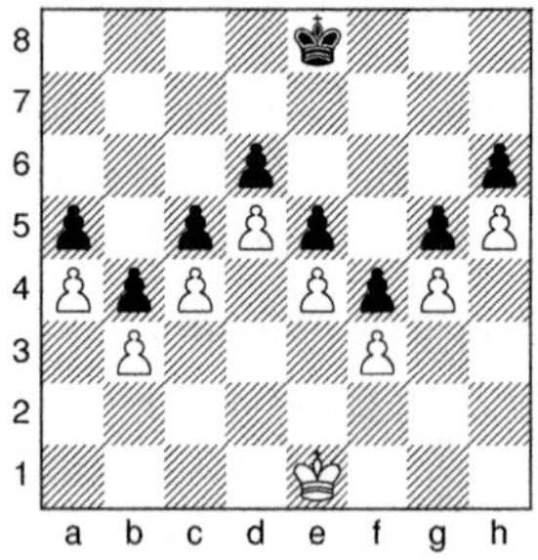

Figure 10. A position where $\alpha < \beta$.

which bidding games must be *sharp* in this sense [4; 5], but such conditions do not seem to be met in chess. And if we allow more pieces on the board, it is easy to construct positions that have so-called *nontrivial Richman intervals*, that is, where $\alpha(P) < \beta(P)$.

An example is given in Figure 10, where we claim that $\alpha \le \frac{1}{4}$ and $\beta \ge \frac{3}{4}$. In other words, a player with more than $\frac{1}{4}$ of the money need not lose. For instance, if White tries to break through the wall of pawns by playing the king to d4 and capturing at e5, Black will go *all in* when the white king has reached d4. Black will then have more money than White after the capture on e5, and will be able to recapture with the d6-pawn.

14. Other board sizes

Mathematically there is of course nothing special about the 8×8 board size, and we have investigated other board sizes as well. The results are similar to those of the 8×8 board. In particular there are no three-piece endgames with $\alpha < \beta$ for any board size smaller than 8×8.

On the 3×4 board, there are quiescent positions of value different than $\frac{1}{2}$. In the position of Figure 11, White cannot improve their position by any move.

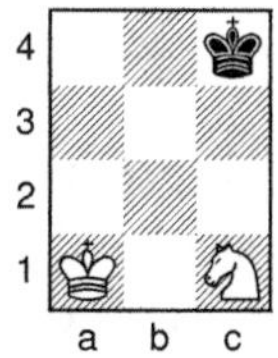

Figure 11. A quiescent position of value $\frac{5}{8}$.

One peculiarity that occurs on a 3×8 board is a value with odd denominator. The following position has the value $\frac{653}{819}$, the denominator factorizing as $3^2 \cdot 7 \cdot 13$.

A curiosity that occurs on a 4×4 board is the following position where White will win the random turn game with probability $\frac{31}{48}$, but where the game (provided it is played optimally) will end in a draw if White wins all the coin flips!

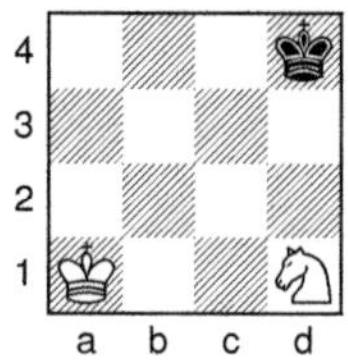

Just like the similar position on the 8×8 board, this is a *zugzwang*, where White would prefer Black to make the next move. As long as the black king stays in the corner, White's problem is that they can't bring their knight to a square where it threatens the black king without first putting it *en prise*. If White has to move, there are three optimal moves, Ka2, Kb1, and Nb2, all three decreasing the value from White's perspective to $\frac{61}{96}$. If White then gets to move again, their best option is to move back to the diagram position (or to the equivalent position with the knight on a4). So as long as White "wins" all the coin flips, they will move back and forth, waiting for Black to have to move their king.

Whenever a position has a nondyadic value, there must be some infinite sequence of coin flips that causes the random turn game to go on forever under optimal play. What is a bit unusual here is that that sequence is one where the same player wins them all.

References

[1] John Beasley, *Bidding chess*, Variant Chess 7 (2008), no. 57, 42–44.

[2] Jay Bhat and Sam Payne, *Bidding chess*, Math. Intelligencer 31 (2009), 37–39.

[3] Mike Develin and Sam Payne, *Discrete bidding games*, Electron. J. Combin. 17 (2010), RP 85.

[4] Andrew J. Lazarus, Daniel E. Loeb, James G. Propp and Daniel Ullman, *Richman games*, in Games of No Chance, MSRI Publications, vol. 29 (1996), 439–449.

[5] Andrew J. Lazarus, Daniel E. Loeb, James G. Propp, Walter R. Stromquist and Daniel H. Ullman, *Combinatorial games under auction play*, Games Econom. Behav. 27 (1999), no. 2, 229–264.

[6] Sam Payne and Elina Robeva, *Artificial intelligence for bidding hex*, in Games of No Chance 4, MSRI Publications, vol. 63 (2015), 207–214.

[7] Yuval Peres, Oded Schramm, Scott Sheffield and David Wilson, *Random-turn hex and other selection games*, Amer. Math. Monthly 114 (2007), 373–387.

urban031@gmail.com *Department of Industrial Engineering and Management, Technion - Israel Institute of Technology, Haifa, Israel*

wastlund@chalmers.se *Department of Mathematical Sciences, Chalmers University of Technology, Gothenburg, Sweden*

Games of No Chance 5
MSRI Publications
Volume **70**, 2017

Phutball draws

SUCHARIT SARKAR

In this short note, we exhibit a draw in the game of Philosopher's Phutball. We construct a position on a 12×10 Phutball board from where either player has a drawing strategy, and then generalize it to an $m \times n$ board with $m-2 \geq n \geq 10$.

Philosophers' Phutball, invented in the towers of Cambridge by Conway and company, and named after a much-beloved Monty Python sketch of a similar name, is a two-player game played on a $m \times n$ board. The official Phutball pitch, as described by Berlekamp–Conway–Guy [BCG03], is 19×15 (that is, there are 19 rows and 15 columns); however, Phutball is usually played on a 19×19 Go board.

The rules of the game are fairly simple. One of the grid points is occupied by *the ball*, usually a black Go stone. Alfred, the first player, wins if the ball crosses the topmost row or ends up in the topmost row at the end of a turn. Betty, going second, wins if the ball crosses the bottommost row or ends up in the bottommost row at the end of a turn. On their turn, the players may either place *a chap*, usually a white Go stone, or move the ball. The ball is moved by jumping over a line of chaps in one of the eight possible directions, and those chaps are immediately removed; and multiple jumps are allowed, although not required. However, in all our diagrams, we will represent the ball by a gray stone and the chaps by black stones (Phutball played with reversed colors is called floodlit Phutball).

Despite the simplicity of the rules, the game is fairly complicated. Phutball on an $n \times n$ board is PSPACE-hard [Der10] and to even check if one has a win in one is NP-complete [DDE02]. Several variants of Phutball have been analyzed in great detail, such as "directional Phutball" [Loo08] when the players are constrained to jump in certain directions, or "one-dimensional Phutball" [GN02] played on an $m \times 1$ board. Even the one-dimensional game is surprisingly complicated: It has only been analyzed fully in [GN02] when the players are forbidden to place "off-parity" chaps. In the general one-dimensional game, sometimes the only winning move for a player is to jump all the way back! This makes the game "loopy" and hard to analyze. It is not even known if there are one-dimensional configurations from where both players have a drawing strategy [Sie09]. Even less is known regarding the usual two-dimensional Phutball. Starting from the

MSC2010: 91A46.
Keywords: philosopher's phutball.

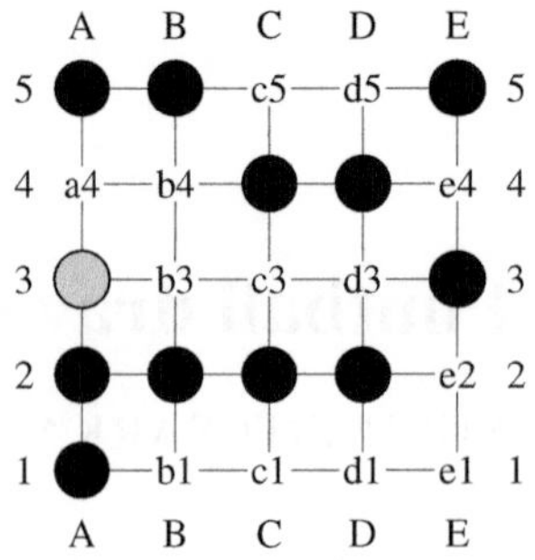

Figure 1. An illustrative Phutball position.

middle on a one-dimensional empty $(2m+1)\times 1$ board, the first player Alfred can at least ensure a draw by the usual strategy-stealing argument, similar to the proof of [Loo08, Observation 3], but even the strategy-stealing argument fails in the two-dimensional case. After filling up his top row with chaps, Alfred runs out of "passes". So it is still unknown if starting from the middle on a two-dimensional empty $(2m+1)\times n$ board the first player Alfred has a drawing strategy.

In this short note, we will study draws in the game of Phutball. The main result, Theorem 1, is that there is a position on a 12×10 board which leads to a draw under optimal play by both the players. The configuration immediately generalizes to an $m\times n$ board with $m-2\geq n\geq 10$; see Corollary 2.

We will use chessboard notation to describe the board. The columns are labeled A, B, C, …, left to right, the rows are numbered 1, 2, 3, …, bottom to top, and the grid points are labeled a1, a2, …, b1, b2, …, accordingly. The k-th move by Alfred will be referred to as $\alpha(k)$, and the k-th move by Betty as $\beta(k)$, with the k in Roman numerals. A chap placement will simply be referred to by the name of the grid point where the chap was placed, while a jump will be described by a (nonempty) sequence of arrows, where the arrows describe the directions of the jump (read left to right).

Allow us to illustrate the rules, the notations, and a few subtleties in the following example. Assume we have the board position from Figure 1 on a 5×5 board. At this point, these are the legal moves: a chap placement on any of the 13 empty points, or any of the following jumps, $\downarrow$, $\searrow$, $\searrow\uparrow$, $\searrow\uparrow\searrow$, $\searrow\uparrow\uparrow$, $\searrow\uparrow\nearrow$. The jumps $\searrow\uparrow\uparrow$ and $\searrow\uparrow\nearrow$ win the game for Alfred, while the jumps $\downarrow$, $\searrow$, and $\searrow\uparrow\searrow$ win it for Betty. Note that during the jumps $\searrow\uparrow$, $\searrow\uparrow\uparrow$, and $\searrow\uparrow\nearrow$, the ball uses the bottom row, but since it does not end up in the bottom row at the end of the turn, it is not a win for Betty. Also note that the jumps $\searrow\nearrow$, $\searrow\uparrow\uparrow\leftarrow$, and $\searrow\uparrow\uparrow\searrow$ are not allowed since they hit the sidelines, but the jump $\searrow\uparrow\nearrow$ is allowed; the corners are considered part of the goal-lines, not the sidelines, and jumping diagonally through the corners is allowed; jumping through the top

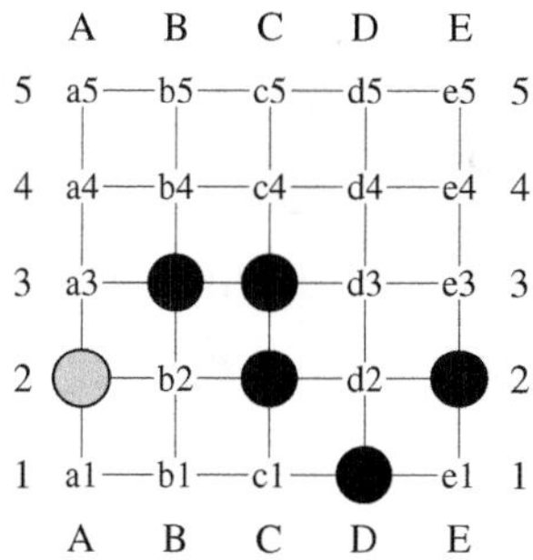

Figure 2. Shots, tackles, and jots.

corners is a win for Alfred while jumping through the bottom corners is a win for Betty. The jump ↘↑↙ is also not allowed, since the chap at b2 is removed immediately after the first jump and cannot be reused during the third jump.

Let us now discuss an elementary Phutball strategy before we proceed. If a player has a winning jump, then it is called *a shot*. There are only two ways to defend against a shot. One may perform a jump of their own, jumping away from the danger; we will call such a jump *a jot*, short for jumping out of trouble. Alternatively, one may place a chap along the route of the winning jump, so that the winning jump no longer exists; such a chap placement is called *a tackle*. If after every jump (respectively, a chap placement), it is still a shot, perhaps by some other route, then the original shot is called an *unjottable* (respectively, *untackleable*) shot. We will annotate shots, unjottable shots, and untackleable shots by !, $*$!, and !!, respectively. If a player has a shot that is both unjottable and untackleable, then he or she has a win in one, and we will indicate this by #. Consider the situation from Figure 2 on the 5×5 board. It is Alfred's turn, and he has to defend against Betty's shot ↗↓. He can jot ↗↓→↑, or he can tackle by placing a chap at c4. In this case, the tackle wins while the jot loses. By tackling at c4, Alfred gets the untackleable unjottable shot ↗. On the other hand, if Alfred had jotted off to e3, then Betty can simply place a chap on e2 to get a untackleable unjottable shot of her own.

Now we are ready for our main result.

Theorem 1. *There are drawn configurations in Phutball played on a* 12×10 *board.*

Proof. Consider the configuration from Figure 3, with Alfred to play. (Notice that the configuration of the chaps has a symmetry, which will of relevance very soon.)

For her next move, Betty is threatening to place a chap at a1 to get an untackleable shot. Alfred must jot away from that threat, so he needs to place a chap next to the ball to create a jump of his own. Placing a chap at a1, b1, or b2 is a

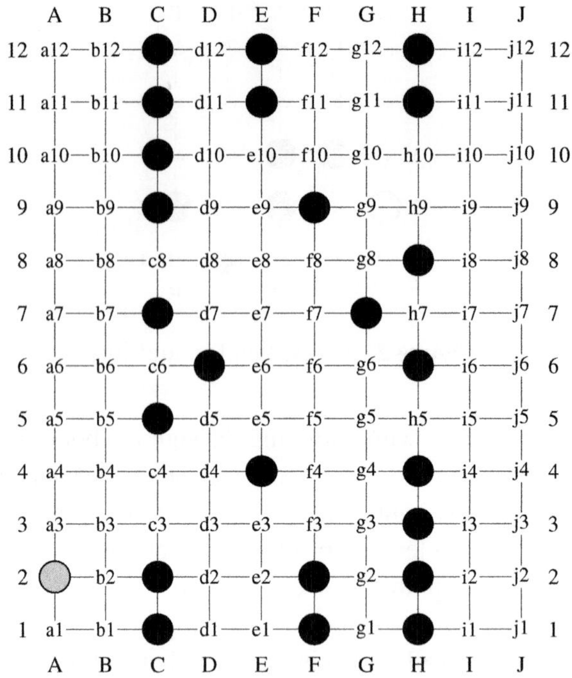

Figure 3. A drawn position in Phutball.

shot for Betty, while placing a chap at a3 is not much better, since the sequence α(i)a3 β(i)a1!! α(ii)↑ β(ii)a3# loses for Alfred. So he must place a chap at b3.

However, this creates a shot for Alfred, so Betty has to defend against it. Jotting off ↗↑→ to e6 is not helpful since Alfred can place a chap back at d6 to get an untackleable unjottable shot. So Betty has to tackle by placing a chap at c4.

The tables turn again. Now Betty has a shot and Alfred has to defend against it. His only available tackle is at d5, and as the following game tree shows, if he does not tackle, he loses.

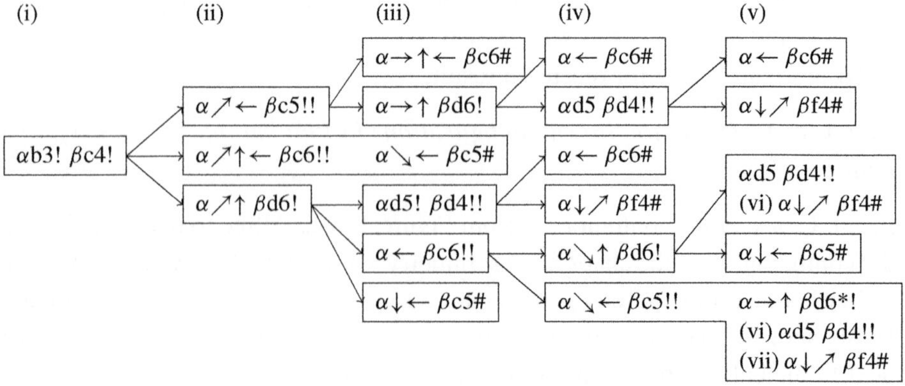

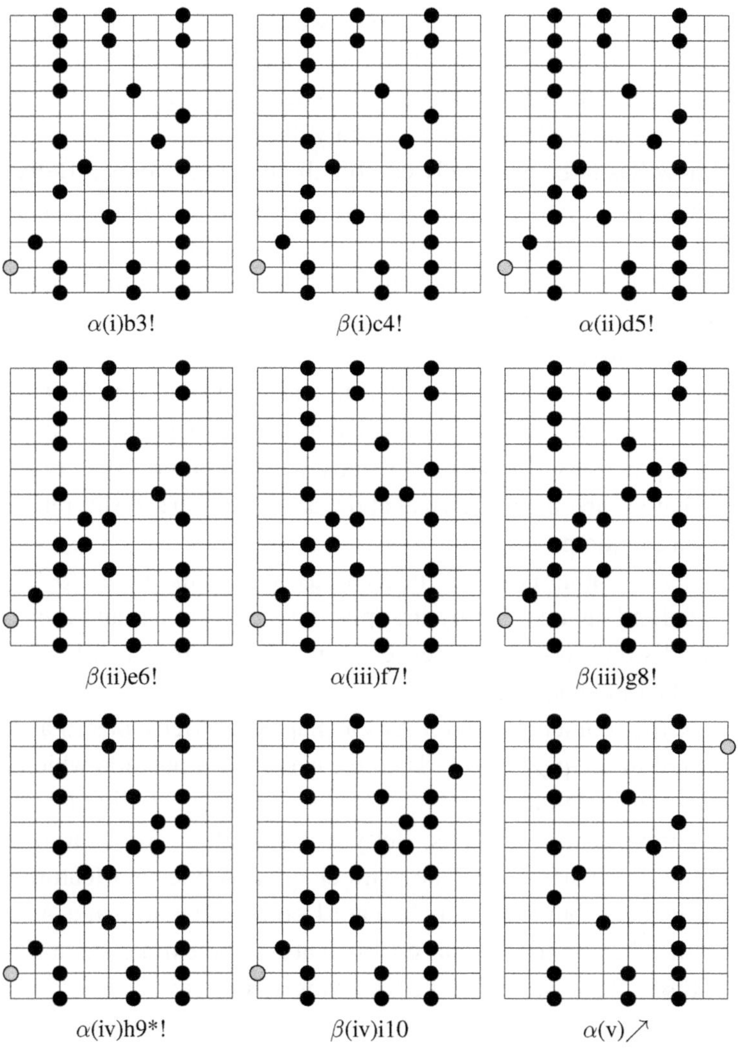

Figure 4. Half of the forced sequence of moves in the draw.

That is, whenever Alfred jumps to b5, b7, d7, or f5, Betty places a chap at c5, c6, d6, or f4, respectively; if the ball is at d7 and there is a chap at d6 and Alfred tackles at d5, then Betty responds by retackling at d4.

Therefore, Alfred is forced to tackle at d5, but once he does, he has a shot of his own. Betty can jot off $\nearrow\leftarrow\downarrow$ to c4, but Alfred can simply respond by placing a chap at c5 to get an untackleable unjottable shot. So Betty is also forced to tackle at e6.

And the tables keep on turning. Now Betty gets a shot and jotting off to h9 does not help, so Alfred tackles at f7. Then he gets a shot, and by the previous analysis and the symmetry of the board, we know that none of Betty's three possible jots to g6, i6, or i8 help. Therefore, Betty tackles at g8 and gets another shot. Alfred's jot to f7 is no good, so he tackles at h9, gaining an unjottable shot. Betty places a chap at i10 which is her only defense.

Alfred now has to be extremely careful about his next move. Betty is threatening a win in two by first placing a chap at j11, which blocks all jumps along the $\nearrow$ direction, and then placing a chap at a1 for a shot. The only way Alfred can prevent Betty from placing a chap at j11 is by jumping there himself. If he does not do the jump, then he has two moves to create a new jump for himself against Betty's threat.

In one of those two moves, Alfred has to place a chap next to the ball. Placing a chap at a1 and b1 are always shots for Betty, so he has to place it at b2 or a3. Placing a chap at b2 is also a shot for Betty with the jump $\rightarrow\swarrow$; if Alfred tries to prevent that by first placing a chap at d2, and then at b2, it still remains a shot for Betty via the jump $\rightarrow\searrow$.

Placing a chap at a3 is a shot for Betty as well using the jump $\uparrow\searrow$; Alfred can try to defend against that by first placing a chap at a4, and then at a3. So this is Alfred's only possible defense against Betty's threat, if he chooses not to jump to j11 immediately.

However, the moment Alfred places a chap at a4, Betty places a chap at a1, forcing Alfred to jump $\nearrow$ to j11. Then we get a board position which is almost symmetric to the original position, except we have two additional chaps at a1 and a4. By the previous analysis, Betty and Alfred are forced to place chaps, one at a time, from i10 to b3. The extra chap at a4 does not feature in any of relevant sequences, but the extra chap at a1 comes back to haunt Alfred at the very end, for when he tries to place a chap at b3, the extra chap at a1 makes it a shot for Betty. That is, we get the following sequence:

$$\alpha(\text{v})\text{a4}\ \beta(\text{v})\text{a1!!}\ \alpha(\text{vi})\nearrow\ \beta(\text{vi})\text{i10!}$$
$$\alpha(\text{vii})\text{h9!}\ \beta(\text{vii})\text{g8!}\ \alpha(\text{viii})\text{f7!}\ \beta(\text{viii})\text{e6!}\ \alpha(\text{ix})\text{d5!}\ \beta(\text{ix})\text{c4\#},$$

which loses for Alfred.

Therefore, Alfred is forced to play $\alpha(\text{v})\nearrow$. It is now Betty's move, and the position is symmetric to the starting position. Consequently, this is a draw. (The optimal sequence of moves is shown in Figure 4.) □

Corollary 2. *There are drawn configurations in Phutball played on an $m \times n$ board with $m - 2 \geq n \geq 10$.*

Proof. Let us just show how to generalize the configuration from Figure 3 to the configuration of Figure 5 on the official 19×15 board, and leave the rest as an exercise to the reader. The reader should note that despite filling up the rightmost

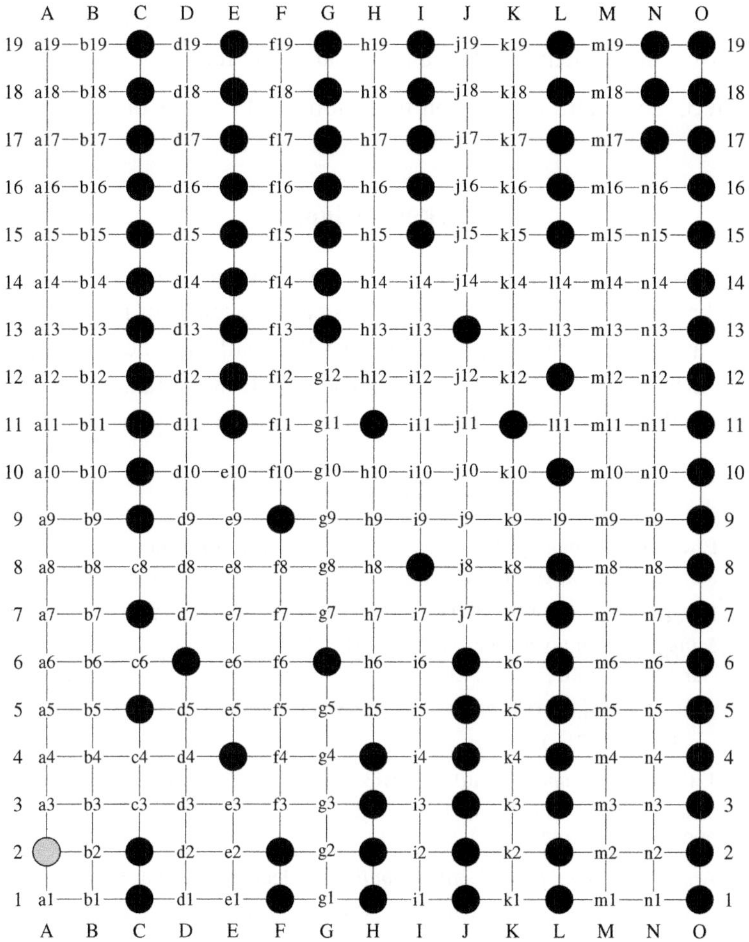

Figure 5. A drawn position in Phutball on the standard pitch.

column with chaps, the sideline subtleties from Figure 1 are not invoked. The reader should additionally note that some of the analysis from the proof of Theorem 1 become more involved. For instance, after α(i)b3! β(i)c4!, if Alfred does not tackle at d5, he still loses, but the following sequence is longer.

α(ii)↗↑ β(ii)d6! α(iii)d5!

β(iii)d4!! α(iv)↓↗ β(iv)f4!! α(v)↗ β(v)h6!! α(vi)↗ β(iv)j8# □

The careful reader will notice that the above configurations do not generalize to the usual 19×19 board since the optimal sequence of chap placements in the draw is done along a diagonal, cf. Figure 4, and the diagonal needs go from one sideline to the other since Alfred is only forced to jump because of the sidelines, and we need to have a row below and a row above to ensure that the game is not already over. So we leave that as a question to the reader.

Question 3. Are there drawn configurations in Phutball played on a 19×19 board?

Acknowledgment

The author is grateful to John Conway for introducing him to Phutball, for teaching him the subtle failure of the strategy-stealing argument in two-dimensional Phutball, and for the many great games of Phutball that they played for hours in John's office (the Fine Hall common room). He would also like to thank the referees for their comments and suggestions. And thank you Simon!

References

[BCG03] Elwyn R. Berlekamp, John H. Conway, and Richard K. Guy, *Winning ways for your mathematical plays*, 2nd ed., vol. 3, A K Peters, Ltd., Natick, MA, 2003.

[DDE02] Erik D. Demaine, Martin L. Demaine, and David Eppstein, *Phutball endgames are hard*, More games of no chance, Math. Sci. Res. Inst. Publ., vol. 42, Cambridge Univ. Press, Cambridge, 2002, pp. 351–360.

[Der10] Dariusz Dereniowski, *Phutball is PSPACE-hard*, Theoret. Comput. Sci. **411** (2010), no. 44–46, 3971–3978.

[GN02] J. P. Grossman and Richard J. Nowakowski, *One-dimensional phutball*, More games of no chance, Math. Sci. Res. Inst. Publ., vol. 42, Cambridge Univ. Press, Cambridge, 2002, pp. 361–367.

[Loo08] Anne M. Loosen, *Directional phutball*, Atl. Electron. J. Math. **3** (2008), no. 1, 30–45.

[Sie09] Aaron N. Sieger, *Coping with cycles*, Games of no chance 3, Math. Sci. Res. Inst. Publ., vol. 56, Cambridge University Press, New York, NY, USA, 1st ed., 2009, pp. 215–232.

sucharit@math.ucla.edu *Department of Mathematics, University of California, Los Angeles, United States*

Games of No Chance 5
MSRI Publications
Volume **70**, 2017

Scoring play combinatorial games

FRASER STEWART

In this paper we will discuss scoring play games. We will give the basic definitions for scoring play games, and show that they form a well-defined set, with clear and distinct outcome classes under these definitions. We will also show that under the disjunctive sum these games form a monoid that is closed and partially ordered. We also show that they form equivalence classes with a canonical form, and even though it is not unique, it is as good as a unique canonical form.

Finally we will define impartial scoring play games. We will then examine the game of nim and all octal games, and define a function that can help us analyse these games. We will finish by looking at the properties this function has and give many conjectures about the behaviour this function exhibits.

1. Introduction

Combinatorial games where the winner is determined by a "score", rather than who moves last, have been largely ignored by combinatorial game theorists. As far as this author is aware, there have been four previous studies of scoring play combinatorial games, all of which focused on the universe of "well-tempered" scoring games.

There are the works of Milnor [9], Hanner [6], Ettinger [3; 4] and most recently, Johnson [7]. The definition of a scoring game that all of them used is the following.

Definition 1. A scoring game is defined as

$$G = \begin{cases} \text{a real number} & \text{if } G^L = G^R = \varnothing, \\ \langle G^L | G^R \rangle, & \text{if } G^L \text{and } G^R \neq \varnothing. \end{cases}$$

The authors would say that a game G, where $G^L = G^R = \varnothing$, is *atomic* and all other games are not atomic. Using this terminology, they were able to define concepts such as the disjunctive sum, and the "left outcome" and "right outcome", which are the score at the end of a game under optimal play, when Left and Right move first respectively. Their mathematical definitions are given here.

Definition 2. The disjunctive sum is defined as follows:

$$G+H=\begin{cases} G+H, & \text{if } G \text{ and } H \text{ are atomic,} \\ \langle G^L+H, G+H^L | G^R+H, G+H^R\rangle, & \text{otherwise.} \end{cases}$$

Definition 3. The left outcome $L(G)$ and right outcome $R(G)$ are defined as follows:

$$L(G)=\begin{cases} G, & \text{if } G \text{ is atomic,} \\ \max_{G^L} R(G^L), & \text{otherwise;} \end{cases}$$

$$R(G)=\begin{cases} G, & \text{if } G \text{ is atomic,} \\ \min_{G^R} L(G^R), & \text{otherwise.} \end{cases}$$

Effectively they showed that under the disjunctive sum, this class of games forms a nontrivial monoid, and that with certain restrictions, it is equivalent to the set of all small normal play combinatorial games.

However, these games all share one thing in common, they are all *dicot* scoring games, or a subset of dicot scoring games. Meaning that if one player has an option, so does the other, and if one player has no options, then neither does his opponent.

For this paper, we will be considering the most general class of scoring games that it is possible to define. The definitions given in this paper, are effectively, equivalent to the definitions given by Milnor, Hanner, Ettinger and Johnson. That is to say, the class of games studied by these four authors is a proper subset of the class of games we will be defining and analysing in this paper.

2. Scoring play games

In this paper, we will be looking at the structure of scoring play games under the disjunctive sum, since it is by far the most commonly used operator in combinatorial game theory. Intuitively, we want all scoring play games to have the following four properties:

(1) The rules of the game clearly define what points are, and how players either gain or lose them.

(2) When the game ends, the player with the most points wins.

(3) For any two games G and H, a points in G are equal to a points in H, where $a \in \mathbb{R}$.

(4) At any stage in a game G, if Left has L points and Right has R points then the score of G is $L - R$, where $L, R \in \mathbb{R}$.

For example, in the game Go you get one point for each of your opponents stones that you capture, and for each piece of area you successfully take. In Mancala you get one point for each bean you place in your Kala. So when comparing these games, we would like one point in Mancala to be worth one point in Go.

Mathematically, scoring games are defined in the following way.

Definition 4. A scoring play game $G = \{G^L \mid G^S \mid G^R\}$, where G^L and G^R are sets of games and $G^S \in \mathbb{R}$, the base case for the recursion is any game G where $G^L = G^R = \varnothing$. $G^L = \{$all games that Left can move to from $G\}$, $G^R = \{$all games that Right can move to from $G\}$, and for all G there is an $S = (P, Q)$ where P and Q are the number of points that Left and Right have on G respectively. Then $G^S = P - Q$, and for all $g^L \in G^L$, $g^R \in G^R$, there is a p^L, $p^R \in \mathbb{R}$ such that $g^{LS} = G^S + p^L$ and $g^{RS} = G^S + p^R$.

A quick note about the notation. One thing the reader will notice, especially after we introduce the disjunctive sum, is that if both G^L and G^R are nonempty, then the value of G^S does not appear to be relevant.

However, it is useful for several reasons. The first is that it tells us how many points a player gains or loses on their turn, i.e., Left gains $G^{LS} - G^S$ points, and Right gains $G^S - G^{RS}$ points. The second is that if we are playing games under the short rule (i.e., the game ends when a player cannot move on any one component), then the value of G^S can change everything. It is also worth keeping it so that we can use "standard" notation for scoring play games.

The reader may also feel that it is perhaps better to write $\{. \mid G^S \mid G^R\}$ as $\{G^{SL} \mid G^R\}$, and likewise if $G^R = \varnothing$. However, we feel that this is simply a matter of personal choice, and from a mathematical perspective, not really relevant.

A concept we will be using throughout this paper is the game tree of a game. While it may be intuitively obvious to the reader, nonetheless, we feel it is important to define it mathematically.

Definition 5. The game tree of a scoring play game $G = \{G^L \mid G^S \mid G^R\}$ is a tree with a root node, and every node has children either on the Left or the Right, which are the Left and Right options of G respectively. All nodes are numbered, and are the scores of the game G and all of its options.

We also need to define a concept that we call the "final score". This is something which hopefully the reader finds relatively intuitive. When the game ends, which it will after a finite amount of time, the score is going to determine whether a player won, lost or tied.

From a combinatorial game theory perspective we want to know "what is the best that a player can do?". Left is trying to maximise the value of the score, while Right is trying to minimise it. Since this is going to be the backbone of our theory it is important to get it right, and so we use the following definition.

Definition 6. We define the following:

- G_F^{SL} is called the Left final score, and is the maximum score — when Left moves first on G — at a terminal position on the game tree of G, if both Left and Right play perfectly.
- G_F^{SR} is called the Right final score, and is the minimum score — when Right moves first on G — at a terminal position on the game tree of G, if both Left and Right play perfectly.

The reason we define it this way is because the terminal position can vary dramatically, depending on the rules of the game and the operator being used. For instance under the long rule the game ends when a player cannot move on all components, but under the short rule the game ends when a player cannot move on any one component. These two rules will clearly give different results when computing the final score of a game.

Since we want our definition to be as general as possible, i.e., cover every possibility, it makes sense to define the final score in this way. For the purposes of this paper we will be using standard combinatorial game theory convention. That is, a game ends when it is a player's turn and he has no options.

It is also important to note that we will only be considering finite games, i.e., for any game G the game tree of G has finite depth and finite width. This means that G_F^{SL} and G_F^{SR} are always computable, and cannot be infinite or unbounded.

There is also the case where a game may have a form of aggregate scoring. For example players may play two games in sequence, and the winner would be the player who gets the most points over both games. This gives scoring play games an additional dynamic, where in the event of a tie after two games, the winner may be determined by the player who managed to accumulate more points in one of the games.

However, as far as this paper is concerned, we will not be considering games of this type. We will only look at games where the winner is determined after one game ends. Games with aggregate scoring would be an interesting area to look at for further research.

There are two conventions that we will be using throughout this paper. The first is that in all examples given we will take the initial score of the game to be 0, unless stated otherwise. The second is that if for a game G, $G^L = G^R = \varnothing$, we will simply write G as G^S, rather than $\{. \mid G^S \mid .\}$. For example the game $G = \{\{. \mid 0 \mid .\} \mid 1 \mid \{. \mid 2 \mid .\}\}$, will be written as $\{0 \mid 1 \mid 2\}$. The game $\{. \mid n \mid .\}$, will be written as n and so on. This is simply for convenience and ease of reading.

2.1. ***An example.*** Before we continue we will give an example of a scoring play game to demonstrate how to use the notation. So consider the game Toad and Frogs from Winning Ways [1], under scoring play. The rules are as follows:

(1) The game is played on a horizontal grid.

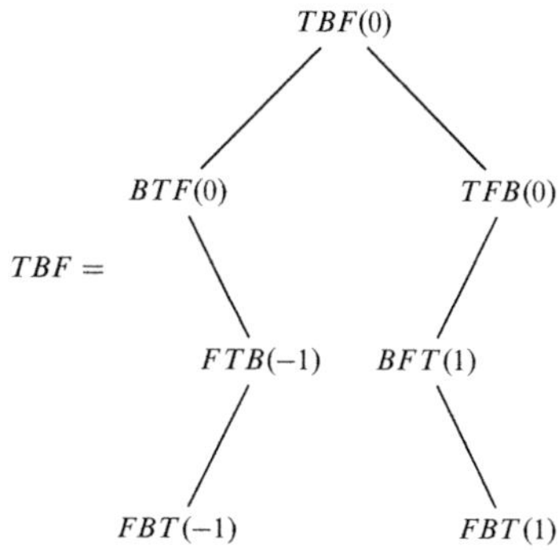

Figure 1. $TBF = \{\{.\mid 0 \mid \{-1 \mid -1 \mid .\}\} \mid 0 \mid \{\{.\mid 1 \mid 1\} \mid 0 \mid .\}\}$.

(2) Left moves Toads and Right moves Frogs.

(3) Toads move from left to right and Frogs move from right to left.

(4) Toads can only jump Frogs and Frogs can only jump Toads.

(5) The player who jumps the most pieces wins.

So consider the game *TBF* as shown in Figure 1, where *B* represents a blank space, *T* represents toads and *F* represents frogs. The numbers in brackets are the current score.

The game in Figure 1 has value $\{\{.\mid 0 \mid \{-1 \mid -1 \mid .\}\} \mid 0 \mid \{\{.\mid 1 \mid 1\} \mid 0 \mid .\}\}$. This game is in "canonical form", that is it neither has a dominated or reversible option. For more details see Section 3.

2.2. ***Outcome classes.*** In combinatorial game theory we would like to know who wins under optimal play, e.g., if $G \in \mathcal{L}$, then that means Left has a winning strategy moving first or second, if he plays his optimal strategy for both normal and misère play. Under scoring play the outcome classes are a little different, since in scoring play we allow ties, i.e., games where neither player wins.

Before we can define what the outcome classes precisely, we first need a new definition. The definition we are about to give is very important for scoring play combinatorial game theory. It, together with the definition of the final score, forms the core of our theory.

Definition 7.

$$L_> = \{G \mid G_F^{SL} > 0\}, \quad L_< = \{G \mid G_F^{SL} < 0\}, \quad L_= = \{G \mid G_F^{SL} = 0\};$$
$$R_> = \{G \mid G_F^{SR} > 0\}, \quad R_< = \{G \mid G_F^{SR} < 0\}, \quad R_= = \{G \mid G_F^{SR} = 0\};$$
$$L_\geq = L_> \cup L_=, \quad L_\leq = L_< \cup L_=, \quad R_\geq = R_> \cup R_=, \quad R_\leq = R_< \cup R_=.$$

Since we would like to classify every game by an outcome class it is also important that every game belongs to exactly one outcome class. So we define the five outcome classes as follows.

Definition 8. The outcome classes of scoring games are defined as:

- $\mathcal{L} = (L_> \cap R_>) \cup (L_> \cap R_=) \cup (L_= \cap R_>)$.
- $\mathcal{R} = (L_< \cap R_<) \cup (L_< \cap R_=) \cup (L_= \cap R_<)$.
- $\mathcal{N} = L_> \cap R_<$.
- $\mathcal{P} = L_< \cap R_>$.
- $\mathcal{T} = L_= \cap R_=$.

The reason that we chose the outcome classes in this way is because if you have a game $G = \{1 \mid 0 \mid 0\}$, then it is more natural to say that it belongs to the outcome $\mathcal{L}$, since Right cannot win, but Left can if he moves first. In this way we also keep the usual convention of calling a game $G \in \mathcal{N}$ a "next player win" and a game $H \in \mathcal{P}$ a "previous player win".

An interesting distinction is that while $\mathcal{L}$ means the set of games where Left can win moving first or second in both normal and misère play, in scoring play, it means that if Left wins moving first he does not lose, and may win, moving second, and vice-versa. Another distinction is the addition of the outcome class $\mathcal{T}$, which of course does not exist in either normal or misère play, and means that the game ends in a tied score regardless of who moves first.

Theorem 9. *Every game G belongs to exactly one outcome class.*

Proof. This is clear since every game belongs to exactly one of $L_>$, $L_<$, $L_=$ and exactly one of $R_>$, $R_<$, $R_=$. Therefore, every game belongs to exactly one of the nine possible intersections of $L_>$, $L_<$, $L_=$ and $R_>$, $R_<$, $R_=$. Since each outcome class is simply the union of one or more of these, then each game can only be in exactly one outcome class. □

2.3. *The disjunctive sum.* As we mentioned earlier, the disjunctive sum is by far the most commonly used operator in combinatorial game theory. This is because many well-known games, such as Go, naturally break up into the disjunctive sum of two or more components. For scoring play the disjunctive sum needs to be defined a little differently; this is because in scoring play games when we combine them together we have to sum the games and the scores separately.

For this reason we will be using two symbols $+_\ell$ and $+$. The ℓ in the subscript stands for "long rule". This comes from [2], and means that the game ends when a player cannot move on any component on his turn. The "short rule" means that the game ends when a player cannot move on at least one component on his turn.

In this paper we will only be considering the disjunctive sum played with the long rule.

Definition 10. The disjunctive sum is defined as follows:

$$G +_\ell H = \{G^L +_\ell H, G +_\ell H^L \mid G^S + H^S \mid G^R +_\ell H, G +_\ell H^R\},$$

where $G^S + H^S$ is the normal addition of two real numbers.

As with the disjunctive sum of normal and misère play games we abuse notation by making the comma mean set union, and $G^L +_\ell H$ means take the disjunctive sum of all $g^L \in G^L$ with H.

We would also like to know when one game is "better" than another one. That is, given several options to play, which one is the best? In normal play and misère play the definitions of "$\geq$" and "$\leq$" are relatively easy to define, since players either win or lose; however, for scoring play we have to take into account tied scores. So for this reason we will redefine "$\geq$" and "$\leq$".

Definition 11. We define the following:

- $-G = \{-G^R \mid -G^S \mid -G^L\}$.
- For any two games G and H, $G = H$ if $G +_\ell X$ has the same outcome as $H +_\ell X$ for all games X.
- For any two games G and H, $G \geq H$ if $H +_\ell X \in O$ implies $G +_\ell X \in O$, where $O = L_\geq$, $R_\geq$, $L_>$ or $R_>$, for all games X.
- For any two games G and H, $G \leq H$ if $H +_\ell X \in O$ implies $G +_\ell X \in O$, where $O = L_\leq$, $R_\leq$, $L_<$ or $R_<$, for all games X.
- $G \cong H$ means G and H have identical game trees.
- $G \approx H$ means G and H have the same outcome.

Theorem 12. *$G \geq H$ if and only if $H \leq G$.*

Proof. First let $G \geq H$, and let $G +_\ell X \in O$ for some game X, where O is one of $L_\leq$, $R_\leq$, $L_<$ or $R_<$. This means that $H +_\ell X \notin O'$, where O' is one of $L_\geq$, $R_\geq$, $L_>$ or $R_>$, since if it was this would mean that $G +_\ell X \in O'$, since $G \geq H$; therefore $H +_\ell X \in O$, and hence $H \leq G$.

A completely identical argument can be used for $H \leq G$, and hence $G \geq H$ if and only if $H \leq G$ and the theorem is proven. □

Theorem 13. *Scoring play games are partially ordered under the disjunctive sum.*

Proof. To show that we have a partially ordered set we need 3 things:

(1) *Transitivity*: If $G \geq H$ and $H \geq J$ then $G \geq J$.

(2) *Reflexivity*: For all games G, $G \geq G$.

(3) *Antisymmetry*: If $G \geq H$ and $H \geq G$ then $G = H$.

(1) Let $G \geq H$ and $H \geq J$. $G \geq H$ means that if $H +_\ell X \in O$ this implies $G +_\ell X \in O$, where $O = L_\geq, R_\geq, L_>$ or $R_>$, for all games X. $H \geq J$ means that if $J +_\ell X \in O$ this implies that $H +_\ell X \in O$. Since $G \geq H$, then this implies that $G +_\ell X \in O$, therefore $J +_\ell X \in O$ implies that $G +_\ell X \in O$ for all games X, and $G \geq J$.

(2) Clearly $G \geq G$, since if $G +_\ell X \in O$ then $G +_\ell X \in O$, where $O = L_\geq, R_\geq$, $L_>$ or $R_>$, for all games X.

(3) First let $G \geq H$ and $H \geq G$. $G = H$ means that $G +_\ell X \approx H +_\ell X$ for all X. So first let $G +_\ell X \in L_=$, then this implies that $H +_\ell X \in L_\geq$, since $H \geq G$. However, $H +_\ell X \in L_=$, since if $H +_\ell X \in L_>$, then this implies that $G +_\ell X \in L_>$, since $G \geq H$, therefore $G +_\ell X \in L_=$ if and only if $H +_\ell X \in L_=$.

An identical argument can be used for all remaining cases, therefore $G +_\ell X \approx H +_\ell X$ for all games X, i.e., $G = H$. □

Theorem 14. *For any three outcome classes $\mathcal{X}$, $\mathcal{Y}$ and $\mathcal{Z}$, there is a game $G \in \mathcal{X}$ and $H \in \mathcal{Y}$ such that $G +_\ell H \in \mathcal{Z}$.*

Proof. Consider the games $G = \{\{\{d \mid c \mid e\} \mid b \mid .\} \mid a \mid .\}$ and $H = \{. \mid f \mid \{. \mid g \mid h\}\}$. The final scores of G are $G_F^{SL} = a$ and $G_F^{SR} = b$, and the final scores of H are $H_F^{SL} = f$ and $H_F^{SR} = g$. Now consider the game $G +_\ell H$ shown in Figure 2.

The final scores of $G +_\ell H$ are $(G +_\ell H)_F^{SL} = e+g$ or $d+h$ and $(G +_\ell H)_F^{SR} = e+h$. Since e, d and h can take any value we can select them so that: $e+g$, $d+h$ and $e+h > 0$ and $G +_\ell H \in \mathcal{L}$; $e+g$, $d+h$ and $e+h < 0$ and $G +_\ell H \in \mathcal{R}$; $e+g$, $d+h > 0$ and $e+h < 0$ and $G +_\ell H \in \mathcal{N}$; $e+g$, $d+h < 0$ and $e+h > 0$ and $G +_\ell H \in \mathcal{P}$ or finally $e+g = d+h = e+h = 0$ and $G +_\ell H \in \mathcal{T}$.

Since the outcomes of G and H depend on the values of a, b, f and g, we can select them so that G and H can be in any outcome class, and thus the theorem is proven. □

From the theory of normal play games, we have the following theorem.

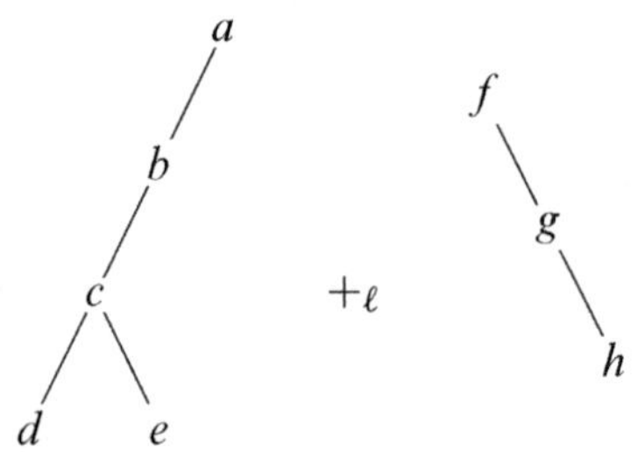

Figure 2. The game $G +_\ell H$, $G = \{\{\{d \mid c \mid e\} \mid b \mid .\} \mid a \mid .\}$ and $H = \{. \mid f \mid \{. \mid g \mid h\}\}$.

Theorem 15 (the greediness principle). *Let* $G = \{G^L \mid G^R\}$ *and* $H = \{H^L \mid H^R\}$ *be two combinatorial games. If* $H^L \subseteq G^L$ *and* $G^R \subseteq H^R$, *then* $G \geq H$.

A direct consequence of Theorem 14 is that this principle will not hold for scoring play games.

Under normal play combinatorial games form an abelian group under the disjunctive sum. The identity that is used is the set $\mathcal{P}$, that is if $I \in \mathcal{P}$ then $G +_\ell I \approx G$ for all games G. In this case the entire set $\mathcal{P}$ has a single unique representative, the game $\{.\mid.\}$. This of course also means that $G = H$ if and only if $G +_\ell (-H) \in \mathcal{P}$.

Under misère play, the identity set contains only one element, which is the same game $\{.\mid.\}$. That is, if $G \not\cong \{.\mid.\}$, then $G \neq \{.\mid.\}$. This was proven by Paul Ottaway (personal communication, 2007). This of course means that there is no easy or equivalent method for determining if two games are equivalent under misère play.

For scoring play games, we have an equivalent theorem. That is our identity set contains only one element, namely the game $\{.\mid 0 \mid .\}$, which we will call 0. It should be clear that $0 +_\ell G \approx G$ for all games G, and so 0 is the identity.

Theorem 16. *For any game* G, *if* $G \not\cong 0$ *then* $G \neq 0$.

Proof. The proof of this is very simple, first let $G^L \neq \varnothing$, since the case $G^R \neq \varnothing$ will follow by symmetry. Next let $P = \{.\mid a \mid b\}$, and note that $P_F^{SL} = a$, since Left has no move on P. So let $a > 0$; if $G = 0$ then this means that $(G +_\ell P)_F^{SL} \approx P$. However, since G is a combinatorial game we know from the definition that G has both finite depth, and finite width. So we can choose $b < 0$ such that $|b|$ is greater than any score on the game tree of G.

Therefore when Left moves first on $G +_\ell P$ he must move to the game $G^L +_\ell P$. Right will respond by moving to $G^L +_\ell b$, since $(G +_\ell P)_F^{SL} < 0$ by choice of b. This implies that $G +_\ell P \not\approx P$, and $G \neq 0$. □

What is interesting is that unlike misère games, some scoring games do have an inverse, namely the set of games $\{.\mid n \mid .\}$, where n is a real number. It should be clear that these are the only games which are invertible under scoring play, and any other nontrivial game cannot be inverted.

Another important consequence of this theorem is that under normal play if we wish to know if $G > H$ for any two games G and H, we simply play $G + (-H)$, where "+" here means the disjunctive sum. However, because no nontrivial scoring games are invertible, we can no longer use this technique to compare them.

3. Canonical forms

Canonical forms are important, because if we can show that these games can be split up into equivalence classes with a unique representative for each class, then it makes these games much easier to analyse and compare. We do not have to consider each game individually, but only the equivalence class to which it belongs.

Theorem 17. *There exist two games G and H such that $G \not\cong H$ and $G = H$.*

Proof. Consider the games G and H, where $a, b, c, d, e, f \in \mathbb{R}$, shown in Figure 3.

This example is a variant of a similar example used to prove the same theorem for misère games in [8].

For any two games G and H, $G = H$ if $G +_\ell X \approx H +_\ell X$ for all games X. The easiest way to prove this is to show that $G \geq H$ and $H \geq G$. Right can do at least as well playing $H +_\ell X$ as he can playing $G +_\ell X$, by simply copying his strategy from $G +_\ell X$ and not playing the left-hand string on H. Right cannot do better on $H +_\ell X$ than he can on $G +_\ell X$, since the string on the left hand side of H can be copied on $G +_\ell X$ by simply not moving to e. So therefore if $H +_\ell X \in O$ then this implies that $G +_\ell X \in O$ where $O = L_{\geq}$, $R_{\geq}$, $L_{>}$ or $R_{>}$, i.e., $G \geq H$.

Left can also do at least as well playing $H +_\ell X$ as he can playing $G +_\ell X$, since if Right can achieve a lower final score playing the left-hand string on $H +_\ell X$, then he can also do so by choosing not to move to e on $G +_\ell X$. Similarly if Right copies his strategy from $G +_\ell X$ onto $H +_\ell X$ then their final scores will be the same. So if $G +_\ell X \in O$ then this implies that $H +_\ell X \in O$ where $O = L_{\geq}$, $R_{\geq}$, $L_{>}$ or $R_{>}$, i.e., $H \geq G$. So therefore, $G = H$ and the proof is finished. □

For both normal and misére play games, the standard way to reduce a game to its canonical form is to use two concepts. These are called domination and reversibility, and are defined as follows.

Definition 18. Let $G = \{A, B, C, \ldots \mid G^S \mid D, E, F, \ldots\}$, if $A \geq B$ or $D \leq E$ we say that A dominates B and D dominates E.

Definition 19. Let $G = \{A, B, C, \ldots \mid G^S \mid D, E, F, \ldots\}$, an option A is reversible if $A^R \leq G$. An option D is also reversible if $D^L \geq G$.

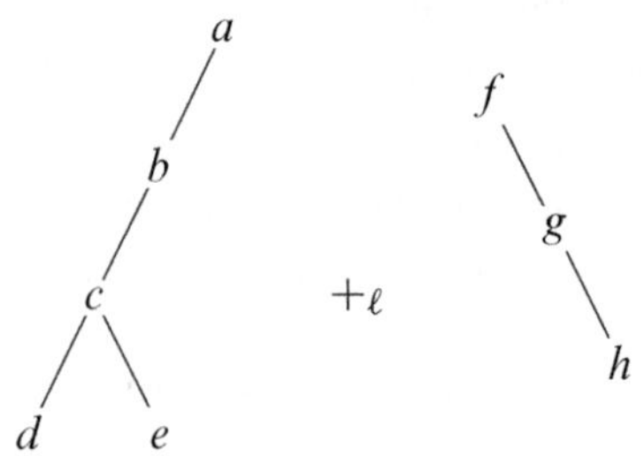

Figure 3. Two games G and H, where $G \not\cong H$, but $G = H$.

Theorem 20. *Let* $G = \{A, B, C, \dots \mid G^S \mid D, E, F, \dots\}$, *and let* $A \geq B$, *then* $G' = \{A, C, \dots \mid G^S \mid D, E, F, \dots\} = G$. *By symmetry if* $D \leq E$ *and* $G'' = \{A, B, C, \dots \mid G^S \mid D, F, \dots\}$ *then* $G'' = G$.

Proof. Let $G = \{A, B, C, \dots \mid G^S \mid D, E, F, \dots\}$ such that $A \geq B$; further let $G' = \{A, C, \dots \mid G^S \mid D, E, F, \dots\}$. First suppose that $G +_\ell X \in O$, where $O = L_\geq, R_\geq, L_>$ or $R_>$ if Left moves to $B +_\ell X$. This implies that $G' +_\ell X \in O$, since $A \geq B$. Hence if $G +_\ell X \in O$ this implies that $G' +_\ell X \in O$, and since the Right options of G and G', this implies that $G' \geq G$.

Next suppose that $G' +_\ell X \in O'$ where $O' = L_\leq, R_\leq, L_<$ or $R_<$. This implies that $G +_\ell X \in O'$, since the only option in G^L that is not in G'^L is B and $B \leq A$, therefore $G' \leq G$, and $G = G'$. So this means that the option B may be disregarded and the proof is finished. □

Theorem 21. *Let* $G = \{A, B, C, \dots \mid G^S \mid D, E, F, \dots\}$, *and let A be reversible with Left options of* $A^R = \{W, X, Y, \dots\}$. *If* $G' = \{W, X, Y, \dots, B, C, \dots \mid G^S \mid D, E, F, \dots\}$, *then* $G = G'$. *By symmetry if D is reversible with Right options of* $D^L = \{T, S, R, \dots\}$. *If* $G'' = \{A, B, C, \dots \mid G^S \mid T, S, R, \dots, D, E, F, \dots\}$, *then* $G = G''$.

Proof. Let $G = \{A, B, C, \dots \mid G^S \mid D, E, F, \dots\}$, where the Left options of $A^R = \{W, X, Y, \dots\}$ and let $G' = \{W, X, Y, \dots, B, C, \dots \mid G^S \mid D, E, F, \dots\}$, further let $A^R \leq G$. If $G +_\ell X \in O$, where $O = L_\geq, R_\geq, L_>$ or $R_>$, when Left does not move to A on G, then clearly $G' +_\ell X$ is also in O, since all other options for Left on G are available for Left on G'.

So consider the case where $G +_\ell X \in O$ if Left moves to $A +_\ell X$, then this implies that $A^R +_\ell X$ must also be in O. This means that $G' +_\ell X \in O$ because $A^{RL} \subset G'^L$, and since all other options on G' are the same as G, then $A^R +_\ell X \in O$ implies that $G' +_\ell X \in O$. Hence if $G +_\ell X \in O$ then this implies that $G' +_\ell X \in O$ for all games X, i.e., $G' \geq G$.

Next assume that $G +_\ell X \in O'$, where $O' = L_\leq, R_\leq, L_<$ or $R_<$, for all games X. However, $A^R \leq G$, i.e. $G +_\ell X \in O'$ implies that $A^R +_\ell X \in O'$, and since $A^{RL} \subset G'^L$, and all other options on G' are identical to options on G, this means that $G +_\ell X \in O'$, implies that $G' +_\ell X \in O'$, for all games X, i.e. $G' \leq G$. Therefore $G = G'$ and the theorem is proven. □

Definition 22. A vertex v on the game tree of a game G is called a termination vertex if there is a game X, such that $G +_\ell X$ ends if the players reach vertex v.

The reason why we called it a termination vertex is because it is a place where a game could potentially end. If both Left and Right have an option at a particular vertex, then under the disjunctive sum a game cannot end at that point.

Definition 23. We say that G is equivalent to H, or $G \equiv H$, if the underlying game trees of G and H are identical, and all termination vertices have the same score.

Theorem 24. *If* $G \equiv H$, *then* $G = H$.

Proof. To prove this, first let $G \equiv H$ and let $G +_\ell X \in O$, where $O = L_>, L_\geq, R_>$ or $R_\geq$. Since the underlying game trees of G and H are identical, and all termination vertices have the same score, then Left can do at least as well on $H +_\ell X$ simply by copying his strategy from $G +_\ell X$. If he does so, then he will arrive at the same termination vertex on $H +_\ell X$ as he did on $G +_\ell X$, and therefore the games will end with identical scores.

Therefore, if $G +_\ell X \in O$ then this implies that $H +_\ell X \in O$, i.e., $H \geq G$. By a totally symmetrical argument we also have that $G \geq H$. So, $G = H$ and the theorem is proven. □

Equivalence is a little stronger than equality, and a little weaker than saying two games are identical. The reason we need it is because it is possible for two games, say $G = \{1 \mid 1 \mid 1\}$ and $H = \{1 \mid 0 \mid 1\}$, to be equal to each other, not identical and neither has a dominated or reversible option.

However, we still want to use domination and reversibility to achieve a "canonical form", so we will say that nontermination vertices are not important in the sense of determining the winner. So while it is not a "true" canonical form in the sense that it is not necessarily unique, it is still useful for studying games.

Definition 25. A game G is in canonical form if it has no dominated or reversible options.

Theorem 26. *For any two games G and H, if $G = H$, and both G and H are in canonical form, then $G \equiv H$.*

Proof. Let G and H be two games such that $G = H$ and neither G nor H has a dominated or reversible option.

So first let $H +_\ell X \in O$, where $O = L_<, R_<, L_\leq$ or $R_\leq$. Since $G = H$, this implies that $G +_\ell X \in O$. However, if Left moves to $G^L +_\ell X$ then $G^{LR} +_\ell X$ cannot be in O. If it were, this would mean that $H +_\ell X \in O$ implies $G^{LR} +_\ell X \in O$, i.e., $G^{LR} \leq H$, and G would have a reversible option, which means that $G^L +_\ell X^R \in O$.

This implies that $H^L +_\ell X^R \notin O'$, where $O' = L_>, R_>, L_\geq$ or $R_\geq$. Since if it were, then H would have a dominated option. Therefore, $G^L +_\ell X^R \in O$ if and only if $H^L +_\ell X^R \in O$, i.e., for all $g^L \in G^L$ there is an $h^L \in H^L$ such that $g^L \leq h^L$, and for all $h^L \in H^L$ there is a $g^{L'} \in G^L$ such that $h^L \leq g^{L'}$.

So that means $g^L \leq h^L \leq g^{L'}$. However, g^L and $g^{L'}$ must be identical, otherwise g^L is a dominated option. So every Left option of G is equivalent to a Left option of H, i.e., $G^L \subseteq H^L$, and by a symmetrical argument $H^L \subseteq G^L$. Therefore, $H^L \equiv G^L$, and similarly $H^R \equiv G^R$.

Since all options of G and H are equivalent, we can conclude that the only differences between the game trees of G and H are on nonterminating vertices. Therefore, $H \equiv G$ and the proof is finished. □

It is also important to note that a game may have more than one canonical form. For example, consider the game $G = \{\{3 \mid 0 \mid 4\}, \{3 \mid 1 \mid 4\} \mid 0 \mid .\}$. This game has two canonical forms, namely $\{\{3 \mid 0 \mid 4\} \mid 0 \mid .\}$ and $\{\{3 \mid 1 \mid 4\} \mid 0 \mid .\}$. However, both of these games are equivalent, so either can be used as the canonical form and it will not affect the analysis of this game.

Theorem 27. *Let $G \to G_1 \to G_2 \to \cdots \to G_n$ represent a series of reductions on a game G to a game G_n, which is in canonical form. Further let $G \to G'_1 \to G'_2 \to \cdots \to G'_m$ represent a different series of reductions on G to a game G'_m which is also in canonical form, then $G_n \equiv G'_m$*

Proof. Since each reduction preserves equality, then $G_n = G'_m$ and they are both in canonical form. By Theorem 26 $G_n \equiv G'_m$, and so the theorem is proven. □

Finally, it is important to note that it is certainly possible to define a unique canonical form, for example we could simply set all nonterminating vertices on a game tree to 0. However, we feel that it is more important to keep the original values as this gives a lot of information about the games.

Consider the games $G = \{3 \mid 0 \mid 4\}$ and $H = \{3 \mid 10 \mid 4\}$. In the game G Left moves and gains 3 points, while Right moves and loses 4, but in the game H Left moves and loses 7 points, while Right moves and gains 6. If we set H^S to zero then this information would be lost. So for this reason, we feel the only ways you should reduce a game is using domination and reversibility.

4. Impartial scoring games

The definition of an impartial scoring play game is less intuitive than for normal and misère play games. The reason for this is because we have to take into account the score; for example, consider the game $G = \{4 \mid 3 \mid 2\}$. On the surface the game does not appear to fall into the category of an impartial game, since Left wins moving first or second, however this game is impartial since both players move and gain a single point, i.e., they both have the same options.

So we will use the following definition for an impartial game.

Definition 28. A scoring game G is impartial if it satisfies the following:

(1) $G^L = \varnothing$ if and only if $G^R = \varnothing$.

(2) If $G^L \neq \varnothing$ then for all $g^L \in G^L$ there is a $g^R \in G^R$ such that $g^L +_\ell - G^S = -(g^R +_\ell - G^S)$.

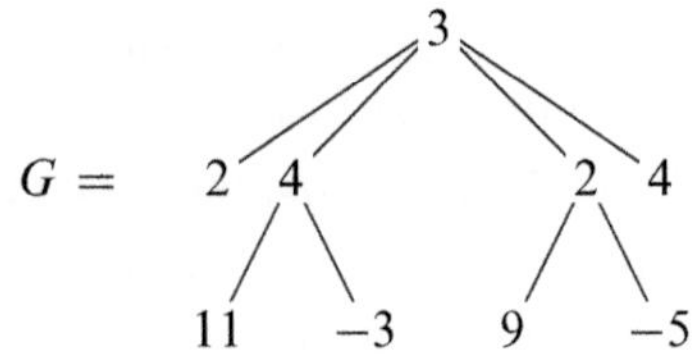

Figure 4. The impartial game $G = \{2, \{11 \mid 4 \mid -3\} \mid 3 \mid, 4, \{9 \mid 2 \mid -5\}\}$.

An example of an impartial game is shown in Figure 4. This game satisfies the definition since

$$2 +_\ell -3 = -(4 +_\ell -3),$$
$$\{11 \mid 4 \mid -3\} +_\ell -3 = -(\{9 \mid 2 \mid -5\} +_\ell -3) = \{8 \mid 1 \mid -6\} \text{ and}$$
$$11 +_\ell -4 = -(-3 +_\ell -4) = 9 +_\ell -2 = -(-5 +_\ell -2) = 7.$$

The reader may be confused about why we choose the name "impartial". The reason for this is because under normal play a game G is impartial if both Left and Right have the same options *at all stages* in G. The phrase "at all stages" is crucial here. If a scoring play game G is impartial, then the options of G must also be impartial. If the reader checks, he will find that our definition is exactly analogous to the definition used for normal play.

As stated in the introduction, all of the work into scoring play games in the past focused exclusively on dicot games. Since impartial games are merely a subset of dicot games, we can deduce much of the structure of these games from Ettinger's work.

In particular [5, Theorem 20, p. 20; Corollary 1, p. 22; Theorem 14, p. 48] give us the structure of impartial games. That is to say, they form a nontrivial monoid. However, we have the following conjecture.

Conjecture 29. Not all impartial scoring play games have an inverse.

To prove this one needs to show that given an impartial game G, for all impartial games Y there is an impartial game P such that $G +_\ell Y +_\ell P \not\approx P$. This is very difficult to show, however it is extremely likely that this conjecture is true because for normal play games the inverse of any game G is $-G$, and as we will now show there are impartial games H where $-H$ is not the inverse.

So consider the game $G = \{2, \{1 \mid 2 \mid 3\} \mid 0 \mid -2, \{-3 \mid -2 \mid -1\}\}$, in this case $-G = G$. If G is the inverse of itself then $G +_\ell G +_\ell 0 \approx 0$; in other words, $G +_\ell G \in \mathcal{T}$. However, $G +_\ell G \in \mathcal{P}$, this is easy to see since if Left moves first and moves to $2 +_\ell G$, then Right can respond by moving to $2 +_\ell \{-3 \mid -2 \mid -1\}$ and Left must move to $2 +_\ell -3$ and loses. If Left moves to $\{1 \mid 2 \mid 3\} +_\ell G$, then Right will move to $\{1 \mid 2 \mid 3\} +_\ell -2$ and Left must move to $1 +_\ell -2$ and again loses. Obviously the opposite will be true if Right moves first on $G +_\ell -G$. So $G +_\ell -G +_\ell 0 \not\approx 0$ and $G +_\ell -G \notin I$.

So, because $-G$ is not the inverse of G in this case then it is very unlikely that any other impartial game could be G's inverse, and while we do not have a proof of that, this simple example shows that it is probably true.

It is also worth noting that impartial scoring games can belong to any of the five outcomes for scoring games, i.e., $\mathcal{L}$, $\mathcal{R}$, $\mathcal{P}$, $\mathcal{N}$ and $\mathcal{T}$. This is in stark contrast to both normal play and misére play games, where impartial games can only belong to either $\mathcal{P}$ or $\mathcal{N}$.

It is easy to see that this is true by considering an impartial game of the form $\{a \mid G^S \mid b\}$. Clearly when $G^S = 0$ then $b = -a$ and the outcome can only be $\mathcal{N}$, $\mathcal{P}$ or $\mathcal{T}$. However, we can set $G^S \neq 0$ and either large enough that both a and b are greater than zero, or less than zero, depending on if we make G^S a very large positive or negative number. In these cases the outcome will either be $\mathcal{L}$ or $\mathcal{R}$.

5. Nim

Nim is a classic combinatorial game. It has been studied under both normal and misère play extensively, and for that reason we wish to study it, or at least variations of it, under scoring play. We will define scoring play nim by the following rules:

(1) The initial score is 0.

(2) The game is played on heaps of beans, and on a players turn he may remove as many beans as he wishes from any one heap.

(3) A player gets 1 point for each bean he removes.

(4) The player with the most points wins.

It should be clear that the best strategy for this game is simply to remove all the beans from the largest possible heap, and keep doing so until the game ends.

Another thing to note is that, under normal play, for every single impartial game G there is a nim heap of size n such that $G = n$. This not the case with scoring play games, but as we will show in the next section, these games are still relatively easy to solve, regardless of the rules and of the scoring method.

5.1. ***Scoring Sprague-Grundy theory.*** Sprague–Grundy theory is a method that is used to solve any variation of a game of nim. The function for normal play $\mathcal{G}(n)$ is defined in a such a way that if for a given heap n, played under some rules, if $\mathcal{G}(n) = m$ then this means that the original heap n is equivalent to a nim heap of size m.

For scoring play games this function is going to be defined slightly differently. Rather than telling us equivalence classes of different games, it will tell us the final scores of games. While this may not be as powerful as normal play Sprague–Grundy theory, it is still a very useful function and can be used to solve many different variations of scoring play nim.

One of the standard variations that have been used widely in books such as Winning Ways [1] are a group of games called octal games. These games cover a very large portion of nim variations, including all subtraction games. For scoring games we will use the following definition.

Definition 30. A scoring play octal game $O = (n_1 n_2 \dots n_k, p_1 p_2 \dots p_k)$, is a set of rules for playing nim where if a player removes i beans from a heap of size n he gets p_i points, $p_i \in \mathbb{R}$, and he must leave $a, b, c \dots$ or j heaps, where $n_i = 2^a + 2^b + 2^c + \cdots + 2^j$.

By convention we will say that a nim heap $n \in O$ means that n is played under the rule set O. We will now define the function that will be the basis of our theory.

Definition 31. Let $n \in O = (t_1 \dots t_f, p_1 \dots p_f)$ and $m \in P = (s_1 \dots s_e, q_1 \dots q_e)$:

- $\mathcal{G}_s(0) = 0$.
- $\mathcal{G}_s(n) = \max_{k,i}\{p_k - \mathcal{G}_s(n_1 +_\ell n_2 +_\ell \cdots +_\ell n_i)\}$, where $n_1 + n_2 + \cdots + n_i = n - k,\ t_k = 2^a + 2^b + \cdots 2^p$ and $i \in \{a, b, \dots, p\}$.
- $\mathcal{G}_s(n +_\ell m) = \max_{k,i,l,j}\{p_k - \mathcal{G}_s(n_1 +_\ell n_2 +_\ell \cdots +_\ell n_i +_\ell m), q_l - \mathcal{G}_s(n +_\ell m_1 +_\ell m_2 +_\ell \cdots +_\ell m_j)\}$, where $n_1 + n_2 + \cdots + n_i = n - k,\ t_k = 2^a + 2^b + \cdots 2^p$ and $i \in \{a, b, \dots, p\},\ m_1 + m_2 + \cdots + m_j = m - l,\ s_l = 2^c + 2^d + \cdots + 2^q$ and $j \in \{c, d, \dots, q\}$.

The first thing to prove is that this function gives us the information we want, namely the final score of a game. So we have the following theorem.

Theorem 32.

$$\mathcal{G}_s(n) = n_F^{SL} = -n_F^{SR} \quad \textit{and} \quad \mathcal{G}_s(n +_\ell m) = (n +_\ell m)_F^{SL} = -(n +_\ell m)_F^{SR}.$$

Proof. The proof of this will be by induction on all heaps $n_1, n_2, \dots, n_i$, $m_1, \dots, m_j$, such that $n_1 + n_2 \cdots + n_i, m_1 + \cdots + m_j \le K$ for some integer K, the base case is trivial since $\mathcal{G}_s(0 +_\ell 0 +_\ell 0 +_\ell \cdots +_\ell 0) = 0$ regardless of how many zeroes there are.

So assume that the theorem holds for all $n_1, n_2, \dots, n_i, m_1 \dots, m_j$, such that $n_1 + n_2 + \cdots + n_i, m_1 + \cdots + m_j \le K$ for some integer K, and consider $\mathcal{G}_s(n +_\ell m)$, where $n + m = K + 1$.

$\mathcal{G}_s(n +_\ell m) = \max_{k,i,l,j}\{p_k - \mathcal{G}_s(n_1 +_\ell n_2 +_\ell \cdots +_\ell n_i +_\ell m), q_l - \mathcal{G}_s(n +_\ell m_1 +_\ell m_2 +_\ell \cdots +_\ell m_j)\}$, but since $n_1 + n_2 + \cdots + n_i + m$ and $n + m_1 + m_2 + \cdots +$

$m_j \leq K$, then by induction

$$\max_{k,i,l,j}\{p_k - \mathcal{G}_s(n_1 +_\ell n_2 +_\ell \cdots +_\ell n_i +_\ell m), q_l - \mathcal{G}_s(n +_\ell m_1 +_\ell m_2 +_\ell \cdots +_\ell m_j)\}$$
$$= \max_{k,i,l,j}\{p_k - (n_1 +_\ell n_2 +_\ell \cdots +_\ell n_i +_\ell m)_F^{SL}, q_l - (n +_\ell m_1 +_\ell \cdots +_\ell m_j)_F^{SL}\}$$
$$= (n +_\ell m)_F^{SL},$$

and the theorem is proven. □

5.1.1. *Subtraction games.* Subtraction games are a very widely studied subset of octal games. A subtraction game is a game of nim where there is a predefined set of integers and a player may only remove those numbers of beans from a heap. This set is called a subtraction set. From our definition of an octal game this means that each n_i is either 0 or 3. In this section we will also say that if a player removes i beans then he gets i points.

Lemma 33. *Let S be a finite subtraction set, then for all $s \in S$, $\mathcal{G}_s(s + 2ik) = k - \mathcal{G}_s(s + (2i - 1)k)$ for all $i \in \mathbb{N}$, where $k = \max\{S\}$.*

Proof. We will split the proof of this into three parts:

Part 1: For all $i \in \mathbb{Z}^+$, $\mathcal{G}_s(r + 2ik) \leq r$.

The first thing to show is that for each $0 \leq r \leq k$, $\mathcal{G}_s(r) \leq r$ and $\mathcal{G}_s(r+2ik) \leq r$ for all $i \in \mathbb{Z}^+$. First let $r \leq k$, $\mathcal{G}_s(r) = \max_j\{j - \mathcal{G}_s(r-j)\}$ and since each j in the set is less than or equal to r, and each $\mathcal{G}_s(r - j) \geq 0$, this implies that $\mathcal{G}_s(r) \leq r$.

Next let $\mathcal{G}_s(r + 2ik) \leq r$ for smaller i, and consider $\mathcal{G}_s(r + 2ik) = \max_j\{j - \mathcal{G}_s(r + 2ik - j)\}$. If $j \leq r$, then since $\mathcal{G}_s(r + 2ik - j) \geq 0$, we have $j - \mathcal{G}_s(r + 2ik - j) \leq j \leq r$. If $j > r$, then $\mathcal{G}_s(r + 2ik - j) = \mathcal{G}_s(r + k - j + (2i - 1)k) \geq k - (r + k - j) = j - r$, by induction, therefore $j - \mathcal{G}_s(r + 2ik - j) \leq j - (j - r) = r$. So therefore $\mathcal{G}_s(r + 2ik) \leq r$ for all i.

Part 2: For all $i \in \mathbb{Z}^+$, $\mathcal{G}_s(r + (2i + 1)k) \geq k - r$.

We also need to show that for each $0 \leq r \leq k$, $\mathcal{G}_s(r + (2i + 1)k) \geq k - r$ for all $i \in \mathbb{N}$. Clearly $\mathcal{G}_s(r + k) \geq k - \mathcal{G}_s(r) \geq k - r$. Again let $\mathcal{G}_s(r + (2i + 1)k) \geq k - r$ for smaller i, then $\mathcal{G}_s(r + (2i + 1)k) \geq k - \mathcal{G}_s(r + 2ik)$ and from above we know that $\mathcal{G}_s(r + 2ik) \leq r$ and hence $\mathcal{G}_s(r + (2i + 1)k) \geq k - \mathcal{G}_s(r + 2ik) \geq k - r$ for all i.

Part 3: For all $s \in S$ and $i \in \mathbb{Z}^+$, $\mathcal{G}_s(s + 2ik) \geq s$ and $\mathcal{G}_s(s + (2i + 1)k) \leq k - s$.

Let $s \in S$, then $\mathcal{G}_s(s) \geq s - \mathcal{G}_s(0) = s$, since we know from Part 1 that $\mathcal{G}_s(s) \leq s$; this means that $\mathcal{G}_s(s) = s$. So consider $\mathcal{G}_s(s + k) = \max_j\{j - \mathcal{G}_s(s + k - j)\}$, if $j \leq s$ then $j - \mathcal{G}_s(s + k - j) \leq j - k + \mathcal{G}(s - j) \leq j - k + s - j \leq s - k \leq k - s$. If $j > s$ then $j - \mathcal{G}_s(s + k - j) \leq j - s + \mathcal{G}_s(k - j) \leq j - s + k - j = k - s$. From Part 2 we know that $\mathcal{G}_s(s + k) \geq k - \mathcal{G}_s(s) = k - s$, so $\mathcal{G}_s(s + k) = k - s$.

So assume that the theorem holds up to $i \geq 1$, and consider $\mathcal{G}_s(s + (2i + 1)k) = \max_j\{j - \mathcal{G}_s(s + (2i + 1)k - j)\}$. If $j \leq s$ then $j - \mathcal{G}_s(s + (2i + 1)k - j) \leq$

n	0	1	2	3	4	5	6	7	8	9	10	11	12	13	14	15
$\mathcal{G}_s(n)$	0	0	0	0	4	5	5	5	5	1	0	0	0	3	4	5

Table 1. A game with subtraction set {4, 5}.

$j-k+\mathcal{G}_s(s+2ik-j)$, and from Part 2 we know that $\mathcal{G}_s(s+2ik-j) \le s-j$; therefore $j-k+\mathcal{G}_s(s+2ik-j) \le j-k+s-j \le s-k \le k-s$.

If $j > s$ then $j-\mathcal{G}_s(s+(2i+1)k-j) = j-\mathcal{G}_s(s+k+2ik-j) \le j-s+\mathcal{G}_s(k-j+2ik) \le j-s+k-j$, by induction, which is equal to $k-s$.

Finally consider $\mathcal{G}_s(s+(2i+2)k) \ge k-\mathcal{G}_s(s+(2i+1)k)$, and from before we know that $\mathcal{G}_s(s+(2i+1)k) \le k-s$; therefore $k-\mathcal{G}_s(s+(2i+1)k) \ge k-(k-s) = s$. So therefore $\mathcal{G}_s(s+(2i+2)k) = s$ and the lemma is proven. □

The obvious question to ask is, does the lemma hold for all n? The answer is no. While it is clear that our function is eventually periodic for subtraction games at least, there are many examples where simply taking the largest number of beans, as in the lemma, is not always the best move. For example consider a game with subtraction set {4, 5}. The table of this game's $G_s(n)$ values are given in Table 1.

In particular consider the value of $\mathcal{G}_s(13)$, this is $\max\{4-\mathcal{G}_s(9), 5-\mathcal{G}_s(8)\} = 4-\mathcal{G}_s(9) = 3$. Therefore, for this game taking 4 beans and gaining 4 points is preferable to taking 5 beans and gaining 5 points. This is a very simple example to illustrate the point that we cannot say playing greedily would always work. In other words, we need to show that if n is large enough then taking the largest number of beans available *is* the best strategy. So we make the following conjecture.

Conjecture 34. Let S be a finite subtraction set, then there exists an N such that $\mathcal{G}_s(n+2k) = \mathcal{G}_s(n)$ for all $n \ge N$, where $k = \max\{S\}$.

It seems plausible that this conjecture is true, given the lemma, however it is also possible that there is an n such that $\mathcal{G}_s(n+2ik) = J$ and $\mathcal{G}_s(n+(2i+1)k) = k-j$, where $J > j$. What we have seen from the data is that often if $n \notin S$ the values of $\mathcal{G}_s(n+2ik)$ and $\mathcal{G}_s(n+(2i+1)k)$ will alternate as in the lemma, but then you will reach an i where the values change, and this switch might happen several times before it settles down.

A proof of the conjecture or a counterexample would be a very big step forward in understanding how the function operates.

5.2. ***Taking-no-breaking games.*** Taking-no-breaking games are a more general version of subtraction games, and cover a fairly wide range of octal games. The rules of these games are fairly basic, when a player removes a certain number of beans from a heap, he will have one of three options:

(1) Leave a heap of size zero, i.e., remove the entire heap.

(2) Leave a heap of size strictly greater than zero.

(3) Leave a heap of size greater than or equal to zero.

From the definition of an octal game this means that each n_i is either 0, 1, 2 or 3, also an octal game $O = (n_1 n_2 \dots n_k, p_1 p_2 \dots p_k)$ is finite if k is finite.

It should be clear that for a fixed $m \in P$ and finite O, where P and O are two taking-no-breaking games, then the function $\mathcal{G}_s(n +_\ell m)$ must always be eventually periodic. The reason is that we always compute each value from a finite number of previous values, and since O is finite this implies that $\mathcal{G}_s(n +_\ell m)$ is bounded, and both of these facts together mean that the function will be eventually periodic.

The real question that one needs to answer however is not "Is it periodic?", but "What is the period?". We believe we can answer that question for a particular class of taking-no-breaking games, and that is the class of games where if you remove i beans you get i points. We make the following conjecture.

Conjecture 35. Let $O = (n_1 n_2 \dots n_t, p_1 p_2 \dots p_t)$ and $P = (m_1 m_2 \dots m_l, q_1 q_2 \dots q_l)$ be two finite taking-no-breaking octal games such that there is at least one $n_s \neq 0$ or 1, and if n_i and $m_j = 1, 2$ or 3 then $p_i = i$ and $q_j = j$, and $p_i = q_j = 0$, otherwise; then

$$\mathcal{G}_s(n + 2k +_\ell m) = \mathcal{G}_s(n +_\ell m),$$

where O is finite and k is the largest entry in O such that $n_k \neq 0, 1$.

There is very strong evidence that this conjecture will hold. Since m is a constant it changes the value of $\mathcal{G}_s(n +_\ell m)$, but not the period. We have checked the theorem for many examples and not yet found a counterexample, which suggests that it is probably true.

Unfortunately, proving it is surprisingly difficult. The conjecture says that if n is large enough, then your best move is to simply remove the maximum available beans from the heap n, so a proof would need to show that for any given m, there are only finitely many places where moving on m or removing fewer than k beans from n is a better move.

There are several problems with this, the first is that the function $\mathcal{G}_s(n +_\ell m)$ only tells us the maximum possible value from the set of possible values. This makes it very difficult to do a proof that first shows $\mathcal{G}_s(n+2k +_\ell m) \geq \mathcal{G}_s(n +_\ell m)$ and vice-versa. The second is to understand *why* removing a lower number of beans would be better than playing greedily, in some instances.

The last problem is that induction is hard, because what may hold for lower values may not hold at higher values, making a proof by induction difficult.

However, since the function is recursively defined an inductive proof seems to be more natural than a deductive proof.

We believe that a proof of this theorem would also help in finding the period, and proving it for the more general case, where i beans are worth k points, $k \in \mathbb{R}$.

Of course, it is natural to ask what happens in the general case. Unfortunately, in the general case the conjecture does not hold. To see why consider the game $O = (3333, 2222)$. The values of $G_s(n)$ are given in the following table:

n	0	1	2	3	4	5	6	7	8	9	10
$\mathcal{G}_s(n)$	0	2	2	2	2	0	2	2	2	2	0

This game has period 5, which does not correspond to a possible value of k, i.e., 1, 2, 3 or 4. While all taking-no-breaking games are periodic as we can see from the example, it is not clear what the period is, since we can take our p_i's to be any real number. So we make the following conjecture.

Conjecture 36. Let $O = (n_1 n_2 \dots n_t, p_1 p_2 \dots p_k)$ and $P = (m_1 m_2 \dots m_l, q_1 q_2 \dots q_l)$ be two finite taking-no-breaking octal games; then there exists a t such that

$$\mathcal{G}_s(n + t +_\ell m) = \mathcal{G}_s(n +_\ell m).$$

5.3. *Taking-and-breaking.* Another type of nim game we can examine are taking-and-breaking games. That is, games where after the player removes some beans from a heap, he must break the remainder into two or more heaps. This is more general than taking-no-breaking games, since taking-no-breaking games are a subset of taking-and-breaking games.

There are several problems with examining taking-and-breaking scoring games. The first is that we cannot even say that the function $\mathcal{G}_s(n +_\ell m)$ is bounded. The reason is that with each iteration you are increasing the number of heaps, which may increase the value of the function as n increases. So we cannot put a bound on the function as we could with subtraction games and taking-no-breaking games.

Another problem is that if we were to say, examine the game 0.26, which means take one bean and leave one nonempty heap, or take two beans and leave either two nonempty heaps, or one nonempty heap, the number of computations required to find $\mathcal{G}_s(n)$ increases exponentially with n. Since a heap of size $n-2$ may be broken into two smaller heaps n_1 and n_2, we must therefore also compute the value of $\mathcal{G}_s(n_1 +_\ell n_2)$.

However, if $n_1 - 2$ or $n_2 - 2$ may also be broken into two smaller heaps, say n_1', n_1'', n_2' and n_2'', then we must compute the value of $\mathcal{G}_s(n_1' +_\ell n_1'' +_\ell n_2)$ and $\mathcal{G}_s(n_1 +_\ell n_2' +_\ell n_2'')$. This process will continue until we have heaps that are too small to be broken up. So this means that computing $\mathcal{G}_s(n)$ for a taking-and-breaking game is a lot harder than for a taking-no-breaking game, simply due to the number of computations involved.

So we have the following conjecture.

Conjecture 37. Let $O = (n_1 n_2 \dots n_k, p_1 p_2 \dots p_k)$ and $P = (m_1 m_2 \dots m_l, q_1 q_2 \dots q_l)$ be two finite octal games; then there exists a t such that

$$\mathcal{G}_s(n + t +_\ell m) = \mathcal{G}_s(n +_\ell m).$$

While we feel that this conjecture may be true, it is certainly not as strong as Conjecture 35, for the reasons previously given. However, studying these games would certainly be interesting, and anything anyone could find out about them would be useful.

Conclusion

We hope that we have given the readers some interesting new ideas about the types of games that can be studied with scoring play theory, as well as opening up a whole new world of impartial games that can be researched. We have simply introduced the ideas, but there is still much to be learned from these fascinating games.

Acknowledgements

I would like to thank my supervisor Keith Edwards for all his thoughtful comments and suggestions, as well as proof reading this paper. He also gave me the idea to use the sets $L_>$, $L_<$ and so on.

References

[1] E. Berlekamp, J. Conway, R. Guy, *Winning Ways for your Mathematical Plays*, Volumes 1-4, A.K. Peters (2002).

[2] J. Conway, *On Numbers and Games*, A.K. Peters (2000).

[3] J. M. Ettinger, *On the Semi-group of Positional Games*. https://tinyurl.com/Ettinger1996, 1996

[4] J. M. Ettinger, *A Metric for Positional Games*, Theoretical Computer Science (Math. Games), 230, pg 207-219, 2000.

[5] J. M. Ettinger, *Topics in Combinatorial Games*, PhD Thesis 1996, University of Wisconsin.

[6] O. Hanner, *Mean Play of Sums of Positional Games*, Pacific Journal of Mathematics, 9(1):81-99, 1959.

[7] W. Johnson, *The Combinatorial Game Theory of Well-Tempered Scoring Games*. Online at http://arxiv.org/abs/1112.3610

[8] G. A. Mesdal, P. Ottaway, *Simplification of Partizan Games in Misère Play*, Integers, Volume 7(2007).

[9] J. W. Milnor, *Sums of Positional Games*, Contributions to the Theory of Games II, 28. Annals of Mathematics Studies, pg 291-301, 1953.

fraseridstewart@gmail.com

Department of Mathematics and Statistics, Xi'An Jiaotong University, Xi'An, China

Games of No Chance 5
MSRI Publications
Volume **70**, 2017

Generalized misère play

MIKE WEIMERSKIRCH

We introduce a new framework by which to view impartial games. Instead of thinking of normal play and misère play as differing in their winning condition, we instead view them as differing in which set of positions are "in the field of play". This leads to a generalization producing an infinite array of *game boards*.

Throughout this paper, the natural numbers $= \{0, 1, 2, \ldots\}$ are denoted $\mathbb{N}$, the set of sequences of natural numbers (integers) are denoted $\mathbb{N}^\infty$ ($\mathbb{Z}^\infty$) and the origin is denoted $\boldsymbol{O}$.

For an introduction to combinatorial game theory, including impartial games, see [WW]. For a detailed description of misère quotient monoids, see [PS]. An algorithm for computing quotient monoids is found in [W].

1. Introduction

Traditionally, normal play impartial games and misère play impartial games are thought to differ by their respective *winning conditions*, that is, the goal of normal play is to make the final move to the empty game, whereas the goal of misère play is to force the opponent to make the final move. Both versions end when the last bean is removed. Here we present a different perspective, that normal play and misère play differ in the set of positions which are legal, so that both end with the same winning condition: having your opponent unable to move to a legal position.

Also, the traditional description of impartial games relies on the notion of the *sum* of two games and the theory is typically presented in terms of arbitrary sums of games. Here we adopt a notation that includes arbitrary sums of games under a particular ruleset as a set of lattice points $\subseteq \mathbb{N}^\infty$. Hence we need not talk about *sums* of positions; all discussion will be localized to legal moves from a particular lattice point, since the notion of sum is inherent in the lattice point definition of a position.

MSC2010: 91A46.
Keywords: combinatorial games, misère, algorithm, Nim, impartial.

Our basic framework is to play on several (but a finite number of) heaps of beans, where the rules allow a heap to be replaced by specified finite multisets of heaps of smaller sizes.

2. Heap games

Our definitions are motivated by the game of Nim, in which the set of all legal positions is generated by the set of heaps of various sizes. We will denote the heap of size i as h_i and call the set of all heap sizes the *heap alphabet*, denoted H. Thus every position in Nim is the sum of (perhaps repeated) elements of H. Therefore, we can represent any position in Nim as a sequence $(x_1, x_2, \ldots)$, where $x_i \in \mathbb{N}$ counts the number of copies of h_i in the sum.

Games not typically played with heaps of beans can still be thought of in this setting (see [GM1, §5] for a general description. Also see the Appendix for a specific example.)

In Nim, any heap can be moved to a heap of a smaller size or removed all together, thus a typical move[1] can be represented as a sequence where all entries are zero except for a single coordinate $= -1$ and perhaps a single coordinate $= 1$ which is to the left of the -1.

Normal play variations of Nim (subtraction games, octal games, etc.) use the same heap alphabet and the same set of legal positions, but have a different set of moves. To ensure our rules obey the termination condition of combinatorial games, we require that a *move* be a sequence with a rightmost, nonzero entry of -1 and all entries to the left of the -1 be elements of $\mathbb{N}$.

Example 1. Let's examine moves in Dawson's Chess, which has octal code 0.137. Imagine a row of bowling pins, and you may remove any bowling pin along with it's neighbors, if any. One pin can be removed only when it is isolated. Two pins can be removed only from the end of a row. Three pins can be removed anywhere, including the possibility of splitting the row into two separate rows. For example, a row of size 5 can be split into two rows of size 1 by removing the middle three pins. Using the notation above, $(2, 0, 0, 0, -1, 0, \ldots)$ is a move.

In normal play, the legal positions are all sequences whose entries are in $\mathbb{N}$. From the position $(1, 1, 0, \ldots)$, the standard way of thinking about possible moves is to list the moves from heaps of size 1 and 2 only, since those are the only heaps present in the position. We wouldn't think that a player could make the move $(0, 0, -1, \ldots)$ since there aren't any heaps of size 3 present in the current position. Such a move would be "off the board". We can get around the difficulty of moves

[1]The notation used here is in the opposite direction from that used in [GM1]. In this paper, when p is a position and m is a move, then the options of p are of the form $p + m$, whereas in [GM1], the options are $p - m$.

to "off the board" positions by declaring that any sequence with a coordinate < 0 is a **D**efeated position, (a "D"-position). We typically think of a normal play contest ending when a player moves to the origin, thereby winning. We now think of the move to the origin as a good strategy, since the opponent has no option but to move to a defeated position, thereby losing. A contest ends when a player moves to a defeated position. By making this adjustment, the moves are now *invariant*.

We call the set of positions which are not defeated positions the *game board*, thus for normal play Nim, the game board is $\mathbb{N}^\infty$. If we wish to use the misère play convention, all we need to do is remove the origin from the game board (thereby making the origin a defeated position). The remainder of the language stays the same. For misère play, the game board is $\mathbb{N}^\infty \setminus \boldsymbol{O}$ and as always, a player loses by moving to a defeated position.

Definition 2. A *heap game* is a triple (H, B, Γ), where H is the *heap alphabet*, B is the *game board*, and $\Gamma \subset \mathbb{Z}^\infty$ is the set of *moves*.

We require the following:

- H is a countable set;
- $B \subseteq \mathbb{N}^\infty$;
- $\forall \gamma \in \Gamma,\ \gamma$ has a rightmost nonzero entry which is -1 and all other entries of $\gamma \in \mathbb{N}$.

2.1. ***Heap games restricted to heaps of a fixed finite size d.*** If we restrict the heap alphabet to heaps of size $\leq d$, we often can employ the quotient monoid approach [PS] to describe the strategy. To each position $p \in B$, the *outcome function*[2] $o : B \to \{P, N\}$ assigns the outcome $o(p) = N$ if there exists an option $p + \gamma$ $(\gamma \in \Gamma)$ with outcome P and $o(p) = P$ if each option $p + \gamma$ has outcome N or D. We then find the *quotient monoid* Q by starting with the free monoid generated by the heap alphabet $H = \{h_1, \ldots, h_d\}$ and modding out by the equivalence $p \equiv r$ if

$$\forall x \in \mathbb{N}^d, o(p + x) = P \iff o(r + x) = P.$$

Finally, we define the *monoid outcome function* $\mathbb{O} : Q \to \{P, N\}$ making the following diagram commute:

$$\begin{array}{ccc} B & \xrightarrow{\ o\ } & \{P, N\} \\ {\scriptstyle \equiv}\big\downarrow & \nearrow_{\mathbb{O}} & \\ Q & & \end{array}$$

[2]In [PS], the normal play outcome function is denoted o^+ and the misère play outcome function is denoted o^-. Here, we have no need to distinguish the two; normal play and misère play differ in their game board, not in their winning condition.

It is commonplace for the game board to be generated by the heaps $\{h_1, h_2, h_3, \ldots\}$, but the quotient monoid Q to be generated by $\{a, b, c, \ldots\}$. As an intermediate step, we may first pass from the game board on heaps of size $\le n$ to the free monoid on n generators via the map

$$h_1 \to a,\ h_2 \to b,\ h_3 \to c \ldots,$$

and then find the quotient produced by the equivalence relation on the generators $\{a, b, c, \ldots\}$.

Aaron Siegel's *MisèreSolver* computer program finds misère quotients when $|Q| < \infty$. An algorithm for finding misère quotients is also given in [W].

If you wish to be proficient at beating your opponent, you will need to know which positions are P-positions, which in the monoid presentation, requires you to memorize $\mathcal{O}^{-1}(P)$, the preimage of P for the monoid outcome function.[3] As a very simple example, in misère Nim[4], if you are to move from the position $(3, 4)$, first reduce the game a^3b^4 by using the relations $a^2 = 1$ and $b^3 = b$ to conclude that a^3b^4 is equivalent to ab^2. You are happy to see that ab^2 is an N-position, since it is in Q, but not in $\mathcal{O}^{-1}(P)$. You now must find an option equivalent to either a or b^2. After a brief search, you realize that $(2, 4) \to a^2b^4$ reduces to b^2, thus you remove a heap of size 1.

Example 3. Normal play Nim restricted to heaps of size ≤ 2 has game board $B = \mathbb{N}^2$ and ruleset

$$\Gamma = \{(-1, 0), (0, -1), (1, -1)\}.$$

Table 1 shows the outcomes of positions $(x_1, x_2) \in B$.

The quotient monoid is $\langle a, b \mid a^2 = 1, b^2 = 1\rangle$ with $\mathcal{O}^{-1}(P) = \{1\}$.

[3] An alternate approach for defining the winning strategy via the generating function of P-positions can be found in [GM2], where in many cases it is called a "rational strategy".

[4] For the presentation of the quotient monoid for misère Nim, see Example 4 below.

$x_1 =$	<0	0	1	2	3	4	5	$\cdots$
$x_2 < 0$	D	D	D	D	D	D	D	$\cdots$
$x_2 = 0$	D	**P**	N	**P**	N	**P**	N	$\cdots$
$x_2 = 1$	D	N	N	N	N	N	N	$\cdots$
$x_2 = 2$	D	**P**	N	**P**	N	**P**	N	$\cdots$
$x_2 = 3$	D	N	N	N	N	N	N	$\cdots$
$x_2 = 4$	D	**P**	N	**P**	N	**P**	N	$\cdots$
$\vdots$	$\vdots$	$\vdots$	$\vdots$	$\vdots$	$\vdots$	$\vdots$	$\vdots$	$\ddots$

Table 1. Normal play Nim.

$x_1 =$	<0	0	1	2	3	4	5	$\cdots$
$x_2 < 0$	D	D	D	D	D	D	D	$\cdots$
$x_2 = 0$	D	D	**P**	N	**P**	N	**P**	$\cdots$
$x_2 = 1$	D	N	N	N	N	N	N	$\cdots$
$x_2 = 2$	D	**P**	N	**P**	N	**P**	N	$\cdots$
$x_2 = 3$	D	N	N	N	N	N	N	$\cdots$
$x_2 = 4$	D	**P**	N	**P**	N	**P**	N	$\cdots$
$x_2 = 5$	D	N	N	N	N	N	N	$\cdots$
$\vdots$	$\vdots$	$\vdots$	$\vdots$	$\vdots$	$\vdots$	$\vdots$	$\vdots$	$\ddots$

Table 2. Misère play Nim.

Example 4. Misère play Nim restricted to heaps of size ≤ 2 has game board $B = \mathbb{N}^2 \setminus \boldsymbol{O}$ and ruleset

$$\Gamma = \{(-1, 0), (0, -1), (1, -1)\}.$$

Table 2 shows the outcomes of positions $(x_1, x_2) \in B$.

The quotient monoid is $\langle a, b \mid a^2 = 1, b^3 = b\rangle$ with $\mathbb{O}^{-1}(P) = \{a, b^2\}$.

3. Generalized misère play

There is no reason why $\mathbb{N}^d$ and $\mathbb{N}^d \setminus \boldsymbol{O}$ are the only choices for B.[5] We could instead play Nim with the restriction that no move can be made which reduces the number of heaps to one.

Example 5. The game board $B = \mathbb{N}^2 \setminus \{(0, 0), (0, 1), (1, 0)\}$ and the ruleset

$$\Gamma = \{(-1, 0), (0, -1), (1, -1)\}.$$

Table 3 shows the outcomes of positions $(x_1, x_2) \in B$.

The quotient monoid is $\langle a, b \mid a^3 = a, ba^2 = b, b^2 = a^2\rangle$ with $\mathbb{O}^{-1}(P) = \{a^2\}$. Note that $|Q| = 5$. Quotients of odd order never occur in either in misère or normal play impartial games (see [PS, Theorem 4.5]).

For another variation, the set of defeated positions in $\mathbb{N}^d$ need not include the origin.

Example 6. The gameboard $B = \mathbb{N}^2 \setminus \{(2, 0)\}$ and the ruleset

$$\Gamma = \{(-1, 0), (0, -1), (1, -1)\}.$$

[5] This concept is also used in [GFL] where the authors restrict the number of heaps instead of restricting the size of the heaps

$x_1 =$	<0	0	1	2	3	4	5	...
$x_2 < 0$	D	D	D	D	D	D	D	...
$x_2 = 0$	D	D	D	**P**	N	**P**	N	...
$x_2 = 1$	D	D	N	N	N	N	N	...
$x_2 = 2$	D	**P**	N	**P**	N	**P**	N	...
$x_2 = 3$	D	N	N	N	N	N	N	...
$x_2 = 4$	D	**P**	N	**P**	N	**P**	N	...
$x_2 = 5$	D	N	N	N	N	N	N	...
⋮	⋮	⋮	⋮	⋮	⋮	⋮	⋮	⋱

Table 3. $B = \mathbb{N}^2 \setminus \{(0,0), (0,1), (1,0)\}$.

$x_1 =$	<0	0	1	2	3	4	5	...
$x_2 < 0$	D	D	D	D	D	D	D	...
$x_2 = 0$	D	**P**	N	D	**P**	N	**P**	...
$x_2 = 1$	D	N	**P**	N	N	N	N	...
$x_2 = 2$	D	N	N	**P**	N	**P**	N	...
$x_2 = 3$	D	**P**	N	N	N	N	N	...
$x_2 = 4$	D	N	**P**	N	**P**	N	**P**	...
$x_2 = 5$	D	N	N	N	N	N	N	...
$x_2 = 6$	D	**P**	N	**P**	N	**P**	N	...
⋮	⋮	⋮	⋮	⋮	⋮	⋮	⋮	⋱

Table 4. $B = \mathbb{N}^2 \setminus \{(2,0)\}$.

Table 4 shows the outcomes of positions $(x_1, x_2) \in B$.

The quotient monoid is $\langle a, b \mid a^4 = a^2, b^3a^3 = b^3a, b^4a^2 = b^4, b^7 = b^5 \rangle$ with $\mathcal{O}^{-1}(P) = \{1, a^3, ba, b^2a^2, b^3, b^4a, b^6\}$.

4. Victorious positions

We achieved misère play by inserting a defeated position inside $\mathbb{N}^d$ at the origin. We achieved other variations by similarly forcing players to avoid certain positions inside $\mathbb{N}^d$. These defeated (D) positions are worse than N positions. In an N position, you are losing, but could still win if your opponent makes a mistake. In a D position, you have lost. We can do the same for P positions. In a P position, you are winning, but might wind up losing if you don't follow optimal play. The counterpart to the defeated position is a position in which you are declared the winner. We will call such a position a *victorious* position. A victorious (V)

position is a position in $\mathbb{N}^d$ such that when a player moves to that position, the contest ends with the player making the move declared the winner.

As a motivation for victorious positions, we can think of this being somewhat analogous to the premature ending of the battle in chess via checkmate compared to the battle in checkers. When two imperfectly programmed computers that never resign play checkers, they battle until one side captures all of the opponent's pieces, similar to whittling a Nim game down to the empty position. The addition of D-positions is akin to "resigning". Once you realize you are going to lose the battle, you may wish to quit the game in a losing manner without fighting the battle to it's conclusion. If the rules of chess treated the king as an ordinary piece whose mate did not end the game, the battle would rage on until one army was vanquished as in checkers. The checkmate provides a position that ends the contest in victory prematurely, despite the winner perhaps having a disadvantage in terms of material and position on the rest of the board.

Formally, we choose a game board $B \subseteq \mathbb{N}^d$, with all positions in $\mathbb{Z}^d \setminus B$ declared to be defeated positions. Then we choose the set of victorious positions $V \subset B$. For all $p \in B \setminus V$, the outcome $\mathbb{O}(p) = N$ if any option of p has outcome P or V and the outcome $\mathbb{O}(p) = P$ if all options of p have outcome N or D. Thus $\mathbb{O}$ is a function $\mathbb{O} : \mathbb{Z}^d \to \{P, N, D, V\}$. We define the equivalence relation

$$p \equiv r \text{ if } \mathbb{O}(p+x) \in \{P, V\} \Longleftrightarrow \mathbb{O}(r+x) \in \{P, V\}, \quad \forall x \in \mathbb{N}^d.$$

Example 7. Let the game board $B = \mathbb{N}^2$ with victorious position $V = \{(2, 1)\}$ and ruleset

$$\Gamma = \{(-1, 0), (0, -1), (1, -1)\}.$$

Table 5 shows the outcomes of positions $(x_1, x_2) \in B$.

The quotient monoid is $\langle a, b \mid a^5 = a^3, b^2a^4 = b^2a^2, b^4a^3 = b^4a, b^6a^2 = b^6, b^9 = b^7\rangle$ with $\mathbb{O}^{-1}(\{P, V\}) = \{1, a^2, a^4ba^2, b^2, b^2a^3, b^3a, b^4a^2, b^5, b^6a, b^8\}$.

5. Open questions

- Example 5 provides a quotient of order 5. Quotients of odd order do not occur in either normal play or misère play. Are there game boards which produce quotients of other odd orders? In particular, is there a game board with $|Q| = 3$? (The only way to produce $|Q| = 1$ is for $B = \varnothing$.) Are there any positive integers k for which it is impossible to find a quotient monoid with $|Q| = k$?
- What happens when the defeated positions in $\mathbb{N}^d$ are not within a neighborhood of the origin? Most misère quotient monoids have the property that once we are far enough away from the origin, the outcomes resemble those in normal play. For example, in misère Nim, once there is a heap of size

$x_1 =$	<0	0	1	2	3	4	5	$\cdots$
$x_2 < 0$	*D*	*D*	*D*	*D*	*D*	*D*	*D*	$\cdots$
$x_2 = 0$	*D*	**P**	*N*	**P**	*N*	**P**	*N*	$\cdots$
$x_2 = 1$	*D*	*N*	*N*	V	*N*	*N*	*N*	$\cdots$
$x_2 = 2$	*D*	**P**	*N*	*N*	**P**	*N*	**P**	$\cdots$
$x_2 = 3$	*D*	*N*	**P**	*N*	*N*	*N*	*N*	$\cdots$
$x_2 = 4$	*D*	*N*	*N*	**P**	*N*	**P**	*N*	$\cdots$
$x_2 = 5$	*D*	**P**	*N*	*N*	*N*	*N*	*N*	$\cdots$
$x_2 = 6$	*D*	*N*	**P**	*N*	**P**	*N*	**P**	$\cdots$
$x_2 = 7$	*D*	*N*	*N*	*N*	*N*	*N*	*N*	$\cdots$
$x_2 = 8$	*D*	**P**	*N*	**P**	*N*	**P**	*N*	$\cdots$
$\vdots$	$\vdots$	$\vdots$	$\vdots$	$\vdots$	$\vdots$	$\vdots$	$\vdots$	$\ddots$

Table 5. $V = \{(2, 1)\}$.

at least two, the outcomes agree with the outcomes in normal play Nim. Will certain patterns of defeated positions extending arbitrarily far from the origin prevent us from being able to find any equivalence relations?

Appendix

We will use the game *Cram* [WW] to illustrate how an arbitrary impartial game can be expressed as a heap game. *Cram* is the impartial version of *Domineering*, which is played by placing 2×1 dominoes on a board (typically starting with an 8×8 board). In *Cram*, either player may place their domino either vertically or horizontally. After two moves for each player, the board may look like this:

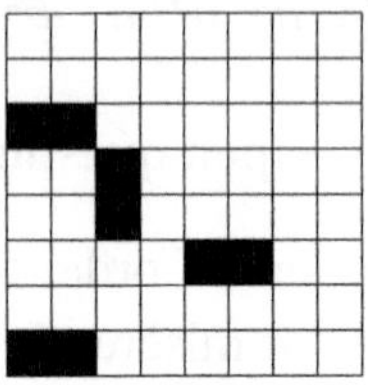

Eventually, the board will be partitioned into connected components, such as this:

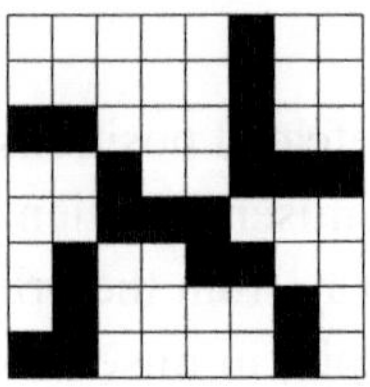

Any *Cram* position can be decomposed into to the sum of its connected components, so the various connected components can serve as a generating set for *Cram* positions. We will have the heap alphabet once we have ordered the connected components.

For example, we could begin by ordering connected components of *Cram* based on the number of cells they contain, and then arbitrarily extending that partial order to a total order. One example being that the first several heaps are

$h_1 =$  $h_2 =$  $h_3 =$ 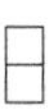 $h_4 =$ 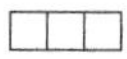 $h_5 =$ 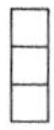 $h_6 =$ 

$h_7 =$  $h_8 =$  $h_9 =$ 

Of course, we could interchange h_2 and h_3. All that is required is that when we play in one connected component, the resulting position is composed of connected components that occur earlier in the heap alphabet, which they will, since the resulting connected components will have fewer cells.

Depending on the setting, we may wish to further simplify the heap alphabet. For example, in normal play, the rotational symmetry allows us to think of h_2 and h_3 as identical, so we only need to include one of them in the heap alphabet. Also, since there is no move from h_1, we don't need to include it in the heap alphabet at all, $h_1 = 0$. Additionally, under normal play, any move from $h_4, \ldots, h_9$ is to h_1, so we don't need to include those in the heap alphabet either, since they act in the same manner as h_2 in that they have a single option to the endgame.

There is, however, no harm in leaving these heaps in from the perspective of the quotient monoid; $h_2, \ldots, h_9$ will be equated in the quotient Q and h_1 and 0 will likewise be equated in Q for normal play.

In misère play, we can declare that arbitrary collections consisting solely of disconnected single cells are also defeated positions and then make the same simplifications so that the heap alphabet begins

$h_1 =$  $=$ 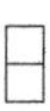 $=$ 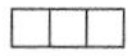 $=$ 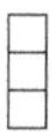 $=$  $=$  $=$  $=$ 

Next come the positions containing four and five cells. Some of them, such as  have moves only to h_1, and as such, are reversible to 0. Others, such as  can be moved only to 0, and are equivalent to h_1. Others have moves to both h_1 and 0 and will be used as the heap h_2:

$h_2 =$ 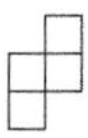 $=$ 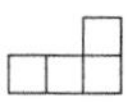 $=$ 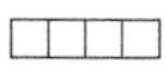 $=$ 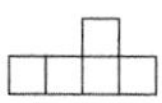 $=$ 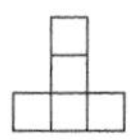 $=$ 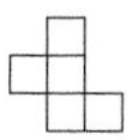

Possible choices for the next two heaps are

$h_3 =$ 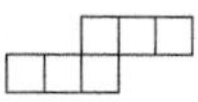 $h_4 =$ 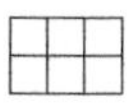

The two positions have the following moves: h_3 can be moved to h_2, $2h_1$, or h_1+0. h_4 can be moved to h_2 or $2h_1$. (Note that $2h_1$ is reversible to 0.)

Using the map from B to Q,

$$h_1 \to a,\ h_2 \to b,\ h_3 \to c,\ h_4 \to d.$$

the quotient monoid is $\langle a, b, c, d \mid a^2 = 1, b^3 = b, c = a, d = ab\rangle$ with $\mathbb{O}^{-1}(P) = \{a, b^2\}$.

References

[GFL] Garrabrant, S., Friedman, E. and Landsberg A. *Cofinite induced subgraphs of impartial combinatorial games: an analysis of CIS-Nim*, arXiv:1201.0405.

[GM1] Guo, A. and Miller, E. *Lattice point methods for combinatorial games*, Adv. Appl. Math. 46: 363-378 (2010). doi:10.1016/j.aam.2010.10.004.

[GM2] Guo, A. and Miller, E. *Algorithms for lattice games*, Int. J. Game Theory (2013) 42: 777. doi:10.1007/s00182-012-0319-9.

[PS] Plambeck, T. and Siegel, A. *Misère quotients for impartial games*, J. Combin. Theory Ser. A, (2008) 115(4), 593–622.

[W] Weimerskirch, M. *An algorithm for computing indistinguishability quotients in misère impartial combinatorial games*, Games of No Chance, vol. 4., MSRI Publications, No. 63 (2015), 267–277.

[WW] Berlekamp, E., Conway J. H. and Guy, R., *Winning ways for your mathematical plays*, Wellesley, Massachusetts: A. K. Peters Ltd., 4 vols., 2001-2004; vol. 1: ISBN 1-56881-130-6; vol. 2: ISBN 1-56881-142-X; vol. 3: ISBN 1-56881-143-8; vol. 4: ISBN 1-56881-144-6.

weim0024@umn.edu

School of Mathematics, University of Minnesota, Minneapolis, MN, United States

Games of No Chance 5
MSRI Publications
Volume **70**, 2017

Retrospective Author Index

The starting page is given in the form **V**:P, where **V** is the volume number in the *Games of No Chance* sequence. Unsurprisingly, **1** and **2** stand for *Games of No Chance* and *More Games of No Chance*.